Operational Mathematics

Operational Mathematics

Third Edition

Ruel V. Churchill
Professor Emeritus of Mathematics
University of Michigan

McGraw-Hill Book Company

New York   St. Louis   San Francisco   Düsseldorf   Johannesburg
Kuala Lumpur   London   Mexico   Montreal   New Delhi   Panama
Rio de Janeiro   Singapore   Sydney   Toronto

Operational Mathematics

Copyright © 1958, 1972 by McGraw-Hill, Inc. All rights reserved.
Copyright 1944 by McGraw-Hill, Inc. All rights reserved.
Printed in the United States of America. No part of this publication
may be reproduced, stored in a retrieval system, or transmitted, in
any form or by any means, electronic, mechanical, photocopying,
recording, or otherwise, without the prior written permission of the
publisher.

Library of Congress Catalog Card Number 70-174611

07-010870-6

    1314151617181920 VBVB   87654

This book was set in Times Roman, and was printed and bound
by Vail-Ballou Press, Inc. The designer was Jo Jones; the
drawings were done by John Cordes, J. & R. Technical Services, Inc.
The editors were Howard S. Aksen and Madelaine Eichberg. Matt
Martino supervised production.

Contents

Preface

This is an extensive revision of the second edition of "Operational Mathematics" published in 1958. Chapters have been added on general integral transforms, finite Fourier transforms, exponential Fourier transforms, Fourier transforms on the half line, Hankel transforms, and on Legendre and other integral transforms. The presentation of theory and applications of the Laplace transformation has been revised. Tables of several of the most useful transforms now appear in the Appendix or in the text. Additional problems illustrate applications of the various integral transformations.

The book is designed as a text and a reference on *integral transforms and their applications to problems in linear differential equations,* to boundary value problems in partial differential equations in particular. It presents the operational properties of the linear integral transformations that are useful in those applications. The selection of a transformation that is adapted to a given problem, by observing the differential forms and boundary conditions that appear in the problem, is emphasized in this edition. The Laplace transformation receives special attention because of its many useful operational properties and the large class of problems to which it applies, including applications outside the field of differential equations.

The applications to problems in physics and engineering are kept on a fairly elementary level. They include problems in vibrations or displacements in elastic bodies, in diffusion or heat conduction, and in static potentials. No previous preparation in the subject of partial differential equations is required of the reader.

This book is a companion volume to "Fourier Series and Boundary Value Problems" and "Complex Variables and Applications." The three books cover, respectively, these principal methods of solving linear boundary value problems in partial differential equations: the operational methods of integral transforms, separation of variables and Fourier series, and conformal mapping. Generalized Fourier series and their applications are presented in Chapter 9 here, with the aid of the theory of the Laplace transformation. A summary of useful theory of functions of a complex variable is given in Chapter 5. All three books are intended to present sound mathematical analysis as well as applications. Conditions of validity of analytical results are kept on a simple and practical level. More elegant conditions may call for training in analysis beyond the level of advanced calculus.

The first four chapters are designed to serve as a text for a short course in real Laplace transforms and their applications.

The impulse symbol or "delta function" is introduced in Sec. 13 in an elementary and careful manner. No theory of distributions, or generalized functions, is included in the book. A satisfactory presentation of the theory would require considerable space and the introduction of concepts not needed elsewhere in the text. Neither can the author justify a presentation of the abstract theory of linear spaces in this book. An intuitive approach from vectors to functions (Secs. 10 and 97) serves as a guide for writing inner products of functions as integrals.

In preparing the three editions of this book the author has taken advantage of improvements suggested by many students and teachers. He is grateful to them for that assistance; also to authors referred to in the Bibliography and footnotes, whose publications have influenced the selection of material.

Ruel V. Churchill

Operational Mathematics

1

The Laplace Transformation

1 INTRODUCTION

The operation of differentiating functions is a transformation from functions $F(t)$ to functions $F'(t)$. If the operator is represented by the letter D, the transformation can be written

$$D\{F(t)\} = F'(t).$$

The function $F'(t)$ is the image, or the transform, of $F(t)$ under the transformation; the function $3t^2$, for example, is the image of the function t^3.

Another transformation of functions that is prominent in calculus is that of integration,

$$I\{F(t)\} = \int_0^x F(t)\, dt.$$

The result of this operation is a functional $f(x)$, the image of $F(t)$ under the transformation. A simpler transformation of functions is the operation of multiplying all functions by the same constant, or by a specified function.

In each of the above examples inverse images exist; that is, when the image is given, a function $F(t)$ exists which has that image.

A transformation $T\{F(t)\}$ is *linear* if for every pair of functions $F_1(t)$ and $F_2(t)$ and for each pair of constants C_1 and C_2, it satisfies the relation

$$(1) \qquad T\{C_1 F_1(t) + C_2 F_2(t)\} = C_1 T\{F_1(t)\} + C_2 T\{F_2(t)\}.$$

Thus the transform of a linear combination of two functions is the same linear combination of the transforms of those functions, if the transformation is linear. Note the special cases of Eq. (1) when $C_2 = 0$ and when $C_1 = C_2 = 1$. The examples cited above represent linear transformations.

The class of functions to which a given transformation applies must generally be limited to some extent. The transformation $D\{F(t)\}$ applies to all differentiable functions, and the transformation $I\{F(t)\}$ to all integrable functions.

Linear integral transformations of functions $F(t)$ defined on a finite or infinite interval $a < t < b$ are particularly useful in solving problems in linear differential equations. Let $K(t,s)$ denote some prescribed function of the variable t and a parameter s. A general *linear integral transformation* of functions $F(t)$ with respect to the kernel $K(t,s)$ is represented by the equation

$$(2) \qquad T\{F(t)\} = \int_a^b K(t,s)F(t)\,dt.$$

It represents a function $f(s)$, the *image*, or *transform*, of the function $F(t)$. The class of functions to which $F(t)$ may belong and the range of the parameter s are to be prescribed in each case. In particular, they must be so prescribed that the integral (2) exists.

We shall see that with certain kernels $K(t,s)$ the transformation (2), when applied to prescribed linear differential forms in $F(t)$, changes those forms into algebraic expressions in $f(s)$ that involve certain boundary values of the function $F(t)$. Consequently, classes of problems in ordinary differential equations transform into algebraic problems in the image of the unknown function. If an inverse transformation is possible, the solution of the original problem may be determined. Boundary value problems in partial differential equations can be simplified in a similar way.

The operational mathematics presented in this book is the theory as well as the application of such linear integral transformations that bears on the treatment of problems in ordinary or partial differential equations. Later on, we shall return to the general transformation (2) and to the question of deciding upon the special cases that may apply to a given problem in differential equations. First we present the special case that is of greatest general importance, the operational mathematics of the Laplace transformation. Other prominent cases include the various Fourier transformations, to be presented later.

When $a = 0$ and $b = \infty$ and $K(t,s) = e^{-st}$, the transformation (2) becomes the Laplace transformation. The direct application of this transformation replaces the earlier symbolic procedure known as Heaviside's operational calculus.[1] The development of the transformation and the accompanying operational calculus was begun before Heaviside's time; Laplace (1749–1827) and Cauchy (1789–1857) were two of the earlier contributors to the subject.[2]

In this chapter we present the basic operational property of the Laplace transformation, the property that gives the image of differentiation of functions as an algebraic operation on the transforms of those functions. In the following chapters further properties of the transformation will be derived and applied to problems in engineering, physics, and other subjects. Applications to boundary value problems in partial differential equations will be emphasized.

Our study of the Laplace transformation leads to the theory of expanding functions in series of the characteristic functions of Sturm-Liouville systems. Such expansions form the basis of the method of solving boundary value problems by separation of variables, a classical method of great importance in partial differential equations.[3] Furthermore, we can use that theory to adapt the integral transformation (2) to certain types of linear boundary value problems.

2 DEFINITION OF THE LAPLACE TRANSFORMATION

If a function $F(t)$, defined for all positive values of the variable t, is multiplied by e^{-st} and integrated with respect to t from zero to infinity, a new function $f(s)$ of the parameter s is obtained; that is,

$$\int_0^\infty e^{-st}F(t)\,dt = f(s).$$

As indicated in the preceding section, this operation on a function $F(t)$ is called the *Laplace transformation* of $F(t)$. It will be abbreviated here by the symbol $L\{F\}$, or by $L\{F(t)\}$; thus

$$L\{F\} = \int_0^\infty e^{-st}F(t)\,dt.$$

The new function $f(s)$ is called the *Laplace transform*, or the *image*, of the *object function* $F(t)$. Wherever it is convenient to do so, we shall denote the

---

[1] Oliver Heaviside, English electrical engineer, 1850–1925.

[2] For historical accounts see J. L. B. Cooper, Heaviside and the Operational Calculus, *Math. Gazette*, vol. 36, pp. 5–19, 1952, and the references given there.

[3] That method is presented in the author's "Fourier Series and Boundary Value Problems," 2d ed., 1963.

object function by a capital letter and its transform by the same letter in lowercase. But other notations that distinguish between functions and their transforms are sometimes preferable; for example,

$$\phi(s) = L\{f(t)\} \quad \text{or} \quad \hat{y}(s) = L\{y(t)\}.$$

For the present, the variable s is assumed to be real. Later on, we shall let it assume complex values. Limitations on the character of the function $F(t)$ and on the range of the variable s will be discussed soon.

Let us note the transforms of a few functions. First, if $F(t) = 1$ when $t > 0$, then

$$L\{F\} = \int_0^\infty e^{-st}\, dt = -\frac{1}{s} e^{-st} \Big]_0^\infty ;$$

hence, when $s > 0$,
$$L\{1\} = \frac{1}{s}.$$

If $F(t) = e^{kt}$ when $t > 0$, where k is a constant, then

$$L\{F\} = \int_0^\infty e^{kt} e^{-st}\, dt = \frac{1}{k-s} e^{-(s-k)t} \Big]_0^\infty ;$$

hence, when $s > k$,
$$L\{e^{kt}\} = \frac{1}{s-k}.$$

With the aid of elementary methods of integration, the transforms of many other functions can be written. For instance,

$$L\{t\} = \frac{1}{s^2}, \quad L\{t^2\} = \frac{2}{s^3},$$

$$L\{\sin kt\} = \frac{k}{s^2 + k^2}, \quad L\{\cos kt\} = \frac{s}{s^2 + k^2},$$

when $s > 0$; but soon we shall have still simpler ways to obtain those transforms.

It follows from elementary properties of integrals that the Laplace transformation is linear in the sense defined by Eq. (1), Sec. 1. We can illustrate the use of this property by writing

$$L\left\{\frac{1}{2} e^{kt} - \frac{1}{2} e^{-kt}\right\} = \frac{1}{2}\frac{1}{s-k} - \frac{1}{2}\frac{1}{s+k};$$

when $s > k$ and $s > -k$; that is,

$$L\{\sinh kt\} = \frac{k}{s^2 - k^2} \qquad\qquad (s > |k|).$$

PROBLEMS

1. Use the linearity property and known transforms to obtain these transformations, where a, b, and c are constants:

(a) $L\{a + bt\} = \dfrac{as + b}{s^2}$ $(s > 0)$;

(b) $L\{a + bt + ct^2\} = L\{(a + bt) + ct^2\}$

$\qquad = \dfrac{as^2 + bs + 2c}{s^3}$ $(s > 0)$;

(c) $L\{a \sin t + b \cos t\} = \dfrac{a + bs}{s^2 + 1}$ $(s > 0)$;

(d) $L\{\cosh ct\} = \dfrac{s}{s^2 - c^2}$ $(s > |c|)$;

(e) $L\{e^{at} - e^{bt}\} = \dfrac{a - b}{(s - a)(s - b)}$ $(s > a$ and $s > b)$.

2. Use trigonometric identities, such as $2 \cos^2 t = 1 + \cos 2t$, $2 \sin at \sin bt = \cos (a - b)t - \cos (a + b)t$, and known transforms to find $f(s)$ when $s > 0$, in case $F(t)$ is

$\qquad$ (a) $\cos^2 t$; (b) $\sin^2 t$; (c) $\sin t \sin 2t$; (d) $\sin t \cos t$; (e) $\sin^3 t = \frac{1}{2}(1 - \cos 2t) \sin t$.

Ans. (a) $\dfrac{s^2 + 2}{s(s^2 + 4)}$; $\quad$ (b) $\dfrac{2}{s(s^2 + 4)}$; $\quad$ (c) $\dfrac{4s}{(s^2 + 1)(s^2 + 9)}$;

$\qquad\qquad\qquad\qquad$ (d) $\dfrac{1}{s^2 + 4}$; $\quad$ (e) $\dfrac{6}{(s^2 + 1)(s^2 + 9)}$.

3. Show that the linearity property (1), Sec. 1, can be extended to linear combinations of three or more functions when all the transforms exist.

4. If for all functions of some class and for every constant C a transformation T satisfies the two conditions

$$T\{F(t) + G(t)\} = T\{F(t)\} + T\{G(t)\},$$

$$T\{CF(t)\} = CT\{F(t)\},$$

prove that the transformation is linear.

3 SECTIONALLY CONTINUOUS FUNCTIONS, EXPONENTIAL ORDER

A function $F(t)$ is *sectionally continuous* on a bounded interval $a < t < b$ if it is such that the interval can be divided into a *finite* number of subintervals interior to each of which F is continuous and has finite limits as t approaches either end point of the subinterval from the interior.

Thus the values of such a function may take at most a finite number of finite jumps on the interval (a,b). The class of sectionally continuous functions includes functions that are continuous on the closed interval $a \leqq t \leqq b$. The integral of every function of this class, over the interval (a,b), exists; it is the sum of the integrals of the continuous functions over the subintervals.

The *unit step function*

$$S_k(t) = 0 \qquad\qquad \text{when } 0 < t < k,$$
$$= 1 \qquad\qquad \text{when } t > k,$$

is an example of a function that is sectionally continuous on the interval $0 < t < T$ for every positive number T (Fig. 1). The Laplace transform of this function is

$$\int_0^\infty S_k(t)e^{-st}\,dt = \int_k^\infty e^{-st}\,dt = -\frac{1}{s}e^{-st}\bigg]_k^\infty ;$$

thus whenever $s > 0$,

$$L\{S_k(t)\} = \frac{e^{-ks}}{s}.$$

A function $F(t)$ is of *exponential order* as t tends to infinity, provided some constant α exists such that the product

$$e^{-\alpha t}|F(t)|$$

is bounded for all t greater than some finite number T. Thus $|F(t)|$ does not grow more rapidly than $Me^{\alpha t}$ as $t \to \infty$, where M is some constant. This is also expressed by saying that $F(t)$ is of the order of $e^{\alpha t}$, or that $F(t)$ is $\mathcal{O}(e^{\alpha t})$.

The function $S_k(t)$ above, as well as the function t^n, is of the order of $e^{\alpha t}$ as $t \to \infty$ for any positive α; in fact, for the first function and, when $n = 0$, for the second, we may write $\alpha = 0$. The function e^{2t} is of exponential order $(\alpha \geqq 2)$; but the function e^{t^2} is not of exponential order.

If a function $F(t)$ is sectionally continuous on each bounded interval $0 < t < T$ and if, for some constant α, F is $\mathcal{O}(e^{\alpha t})$, then the Laplace transform of F exists whenever $s > \alpha$. This follows from a well-known comparison test

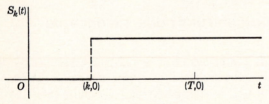

Fig. 1

for the convergence of improper integrals (see Prob. 14, Sec. 5). For in view of the sectional continuity of F, and consequently of the product $e^{-st}F(t)$, that product is integrable over every bounded interval $0 < t < T$. Also, since F is $\mathcal{O}(e^{\alpha t})$, a constant M exists such that for all positive t

$$|e^{-st}F(t)| < Me^{-(s-\alpha)t}.$$

But the integral from 0 to ∞ of $Me^{-(s-\alpha)t}$ exists when $s > \alpha$; consequently, not only the convergence but also the absolute convergence of the Laplace integral

$$\int_0^\infty e^{-st}F(t)\,dt$$

is ensured when $s > \alpha$.

The above conditions for the existence of the transform of a function are adequate for most of our needs; but they are *sufficient* rather than necessary conditions. The function F may have an infinite discontinuity at $t = 0$ for instance, that is, $|F(t)| \to \infty$ as $t \to 0$, provided that positive numbers m, N, and T exist, where $m < 1$, such that $|F(t)| < N/t^m$ when $0 < t < T$. Then if F otherwise satisfies the above conditions, its transform still exists because of the existence of the integral

$$\int_0^T e^{-st}F(t)\,dt.$$

For example, when $F(t) = t^{-\frac{1}{2}}$, its transform can be written, after the substitution of x for $\sqrt{st}$, in the form

(1) $$\int_0^\infty t^{-\frac{1}{2}}e^{-st}\,dt = \frac{2}{\sqrt{s}}\int_0^\infty e^{-x^2}\,dx \qquad (s > 0).$$

The last integral has the value $\sqrt{\pi}/2$ (Prob. 10, Sec. 5); hence

$$L\{t^{-\frac{1}{2}}\} = \left(\frac{\pi}{s}\right)^{\frac{1}{2}} \qquad (s > 0).$$

4   TRANSFORMS OF DERIVATIVES

By a formal integration by parts we have

$$L\{F'(t)\} = \int_0^\infty e^{-st}F'(t)\,dt$$

$$= e^{-st}F(t)\Big]_0^\infty + s\int_0^\infty e^{-st}F(t)\,dt.$$

Let $F(t)$ be of order of $e^{\alpha t}$ as t approaches infinity. Then whenever $s > \alpha$, the bracketed term becomes $-F(0)$, and it follows that

(1)
$$L\{F'(t)\} = sf(s) - F(0),$$

where $f(s) = L\{F(t)\}$.

Therefore in our correspondence between functions *differentiation* of the object function corresponds to the *multiplication* of the result function by its variable s and the addition of the constant $-F(0)$. Formula (1) thus gives the *fundamental operational property* of the Laplace transformation, the property that makes it possible to replace the operation of differentiation by a simple algebraic operation on the transform.

As noted above, formula (1) was obtained only in a *formal*, or manipulative, manner. It is not even correct when $F(t)$ has discontinuities. The following theorem will show to what extent we can rely on our formula.

Theorem 1 *Let the function $F(t)$ be continuous with a sectionally continuous derivative $F'(t)$, over every finite interval $0 \leqq t \leqq T$. Also let $F(t)$ be of order of $e^{\alpha t}$ as $t \to \infty$. Then when $s > \alpha$, the transform of $F'(t)$ exists, and*

(2)
$$L\{F'(t)\} = sL\{F(t)\} - F(0).$$

Since $F(t)$ is continuous at $t = 0$, the number $F(0)$ here is the same as $F(+0)$, the limit of $F(t)$ as t approaches zero through positive values.

To prove this theorem, we note first that

$$L\{F'(t)\} = \lim_{T \to \infty} \int_0^T e^{-st} F'(t)\, dt,$$

if this limit exists. We write the integral here as the sum of integrals in each of which the integrand is continuous. For any given T, let $t_1, t_2, \cdots, t_n$ denote those values of t between $t = 0$ and $t = T$ for which $F'(t)$ is discontinuous (Fig. 2). Then

$$\int_0^T e^{-st} F'(t)\, dt = \int_0^{t_1} e^{-st} F'(t)\, dt + \int_{t_1}^{t_2} e^{-st} F'(t)\, dt + \cdots + \int_{t_n}^T e^{-st} F'(t)\, dt.$$

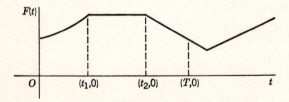

Fig. 2

After integrating each of these integrals by parts, we can write their sum as

$$\left. e^{-st}F(t)\right]_0^{t_1} + \left. e^{-st}F(t)\right]_{t_1}^{t_2} + \cdots + \left. e^{-st}F(t)\right]_{t_n}^{T} + s\int_0^T e^{-st}F(t)\,dt.$$

Now $F(t)$ is continuous so that $F(t_1 - 0) = F(t_1 + 0)$, etc., and hence

(3) $$\int_0^T e^{-st}F'(t)\,dt = -F(0) + e^{-sT}F(T) + s\int_0^T e^{-st}F(t)\,dt.$$

Since $|F(t)| < Me^{\alpha t}$ for large t for some constants α and M, it follows that

$$|e^{-sT}F(T)| < Me^{-(s-\alpha)T},$$

and since $s > \alpha$, this product vanishes as $T \to \infty$. Also the last integral in Eq. (3) approaches $L\{F\}$ as $T \to \infty$ because F is continuous and $\mathcal{O}(e^{\alpha t})$. Hence the limit as $T \to \infty$ of the right-hand member of Eq. (3) exists and equals $-F(0) + sf(s)$; therefore, the same is true of the left-hand member. Thus Theorem 1 is proved.

If F is continuous when $t \geq 0$ except for a finite jump at t_0, where $t_0 > 0$, the other conditions remaining as stated in the theorem, the above proof is easily modified to show that our formula (2) must be replaced by the formula

(4) $$L\{F'(t)\} = sf(s) - F(0) - [F(t_0 + 0) - F(t_0 - 0)]e^{-st_0}.$$

The quantity in brackets is the jump of F at t_0.

We use the symbol F' here and in the sequel to denote the function whose value is the derivative of F wherever the derivative exists. In the case of our step function $S_k(t)$, for instance, $S_k'(t) = 0$ when $0 < t < k$ and when $t > k$, but $S_k'(k)$ has no value.

To obtain the transform of the derivative F'' of the second order, we apply Theorem 1 to the function F'. Let both F and F' be continuous when $t \geq 0$ and $\mathcal{O}(e^{\alpha t})$; also let F'' be sectionally continuous on each bounded interval. Then

$$L\{F''(t)\} = sL\{F'(t)\} - F'(0)$$
$$= s[sL\{F(t)\} - F(0)] - F'(0).$$

Hence we have the transformation

(5) $$L\{F''(t)\} = s^2 f(s) - sF(0) - F'(0).$$

When Theorem 1 is applied to $F^{(n-1)}(t)$ to write

$$L\{F^{(n)}(t)\} = sL\{F^{(n-1)}(t)\} - F^{(n-1)}(0)$$

and again to write $L\{F^{(n-1)}(t)\}$ in terms of $L\{F^{(n-2)}(t)\}$, and so on, the following result is indicated.

Theorem 2 *Let F and each of its derivatives of order up to $n - 1$ be continuous functions when $t \geqq 0$ and $\mathcal{O}(e^{\alpha t})$; also let $F^{(n)}(t)$ be sectionally continuous on each bounded interval $0 < t < T$. Then the transform of the derivative $F^{(n)}(t)$ exists when $s > \alpha$, and it has the following algebraic expression in terms of the transform $f(s)$ of $F(t)$:*

$$(6) \qquad L\{F^{(n)}(t)\} = s^n f(s) - s^{n-1} F(0) - s^{n-2} F'(0)$$
$$- s^{n-3} F''(0) - \cdots - F^{(n-1)}(0) \qquad (s > \alpha).$$

Theorem 2 can be proved by using Theorem 1 and induction. Under the conditions stated, suppose that formula (6) is valid when n is replaced by some integer k where $0 < k < n$. But Theorem 1 expresses $L\{F^{(k+1)}\}$ in terms of $L\{F^{(k)}\}$ to which (6) now applies to show that formula (6) is true when n is replaced by $k + 1$. But the formula is true when $k = 1$, according to Theorem 1; it is therefore true when $k = 2$, hence when $k = 3, \ldots, n$.

The Laplace transformation resolves the differential form $F'(t)$ in terms of $f(s), s,$ and the initial value $F(0)$ when F satisfies the conditions stated in Theorem 1. As a consequence of Theorem 1, Theorem 2 shows how that transformation resolves the iterates $F''(t), F'''(t), \ldots$ of that form in terms of $f(s), s,$ and initial values of F and its derivatives.

5   EXAMPLES.   THE GAMMA FUNCTION

In order to gain familiarity with the above fundamental operational property of the transformation, let us first use it to obtain a few transforms.

Example 1 Find $L\{t\}$.

The functions $F(t) = t$ and $F'(t) = 1$ are continuous, and F is $\mathcal{O}(e^{\alpha t})$ for any positive α. Hence,

$$L\{F'(t)\} = sL\{F(t)\} - F(0) \qquad (s > 0),$$

or
$$L\{1\} = sL\{t\}.$$

Since $L\{1\} = 1/s$, it follows that

$$L\{t\} = \frac{1}{s^2} \qquad (s > 0).$$

Example 2 Find $L\{\sin kt\}$.

The function $F(t) = \sin kt$ and its derivatives are all continuous and bounded, and therefore of exponential order, where $\alpha = 0$. Hence

$$L\{F''(t)\} = s^2 L\{F(t)\} - sF(0) - F'(0) \qquad (s > 0),$$

or
$$-k^2 L\{\sin kt\} = s^2 L\{\sin kt\} - k.$$

Solving for $L\{\sin kt\}$, we see that

$$L\{\sin kt\} = \frac{k}{s^2 + k^2} \qquad (s > 0).$$

Example 3 Find $L\{t^m\}$ where m is any positive integer.

The function $F(t) = t^m$ satisfies all the conditions of Theorem 2 for any positive α. Here

$$F(0) = F'(0) = \cdots = F^{(m-1)}(0) = 0,$$

$$F^{(m)}(t) = m!, \qquad F^{(m+1)}(t) = 0.$$

Applying formula (6) when $n = m + 1$, we find that

$$L\{F^{(m+1)}(t)\} = 0 = s^{m+1} L\{t^m\} - m!,$$

and therefore

(1) $$L\{t^m\} = \frac{m!}{s^{m+1}} \qquad (s > 0).$$

This formula can be generalized to the case in which the exponent is not necessarily an integer. To obtain $L\{t^k\}$ where $k > -1$, we make the substitution $x = st$ in the Laplace integral, giving

$$\int_0^\infty t^k e^{-st}\, dt = \frac{1}{s^{k+1}} \int_0^\infty x^k e^{-x}\, dx \qquad (s > 0).$$

The integral on the right represents the *gamma function*, or factorial function, with the argument $k + 1$. Hence

(2) $$L\{t^k\} = \frac{\Gamma(k + 1)}{s^{k+1}} \qquad (k > -1, s > 0).$$

Formula (1) is a special case of (2) when k is a positive integer (see Prob. 13).

Example 4 Find $L\{\int_0^t F(\tau)\, d\tau\}$ when F is sectionally continuous and of exponential order.

The function

(3) $$G(t) = \int_0^t F(\tau)\, d\tau$$

is continuous (Sec. 14), and $G(0) = 0$. Also $G'(t) = F(t)$, except for those values of t for which $F(t)$ is discontinuous; thus $G'(t)$ is sectionally continuous on each finite interval. If the function $G(t)$ is also $\mathcal{O}(e^{\alpha t})$, then according to Theorem 1,

$$L\{G'(t)\} = sL\{G(t)\} = L\{F(t)\} \qquad (s > \alpha);$$

thus, if $\alpha > 0$ so that $s > 0$,

$$(4) \qquad\qquad L\left\{ \int_0^t F(\tau)\, d\tau \right\} = \frac{1}{s} f(s) \qquad\qquad (s > \alpha > 0).$$

To show that the integral (3) represents a function of exponential order when F is sectionally continuous and of exponential order, we first note that constants α and M exist such that $|F(t)| < Me^{\alpha t}$ whenever $t \geq 0$, and if the number α is not positive, it can be replaced by a positive number. Then

$$|G(t)| \leq \int_0^t |F(\tau)|\, d\tau < M \int_0^t e^{\alpha \tau}\, d\tau = \frac{M}{\alpha}(e^{\alpha t} - 1) \qquad (\alpha > 0),$$

and therefore

$$e^{-\alpha t}|G(t)| < \frac{M}{\alpha}(1 - e^{-\alpha t}) < \frac{M}{\alpha} \qquad\qquad (\alpha > 0).$$

This establishes the exponential order of the function (3).

PROBLEMS

1. State why each of the functions

$$(a)\ F(t) = te^{2t} \text{ and } (b)\ G(t) = \begin{cases} \dfrac{1}{t+1} & \text{when } 0 < t < 2 \\ 1 & \text{when } t > 2, \end{cases}$$

is sectionally continuous on every interval $0 < t < T$ and $\mathscr{O}(e^{\alpha t})$ as $t \to \infty$, where $\alpha > 2$ for F and $\alpha \geq 0$ for G.

2. State why neither of the functions $(a)\ (t - 1)^{-1}$ or $(b)\ \tan t$ is sectionally continuous on the interval $0 < t < 3$.

3. If $S_k(t)$ is the unit step function (Sec. 3), draw graphs of the following step functions and find their transforms when $s > 0$: $(a)\ F(t) = 1 - S_1(t)$; $(b)\ G(t) = S_1(t) - S_2(t)$. Also, note that both F and G vanish except on bounded intervals; thus their Laplace integrals become definite integrals that exist for all $s\,(-\infty < s < \infty)$. Show that $f(0) = 1$ and $g(0) = 1$, and that
   $(a)\ f(s) = (1 - e^{-s})/s \qquad (s \neq 0);$
   $(b)\ g(s) = (e^{-s} - e^{-2s})/s \qquad (s \neq 0).$

4. Obtain these transforms with the aid of Theorem 2:
   $(a)\ L\{\cos kt\} = s/(s^2 + k^2) \qquad (s > 0);$
   $(b)\ L\{\sinh kt\} = k/(s^2 - k^2) \qquad (s > |k|).$

5. Given the transform of e^{kt}, use Theorem 1 to show that

$$L\{te^{kt}\} = \frac{1}{(s - k)^2} \qquad\qquad (s > k).$$

6. If $G(t) = 0$ when $0 \leq t \leq k$ and $G(t) = t - k$ when $t \geq k$, draw graphs of G and G'. Given the transform of $S_k(t)$, (a) apply Theorem 1 to prove that $g(s) = s^{-2}e^{-ks}$ $(s > 0)$; (b) show that $G(t) = \int_0^t S_k(\tau) \, d\tau$ and find $g(s)$ from formula (4), Sec. 5.

7. Prove that the function $F(t) = \sin(e^{t^2})$ is of exponential order $(\alpha \geq 0)$ and that its derivative F' is not of exponential order. Show that Theorem 1 ensures the existence of the Laplace transform of F' when $s > 0$, in this case where F' is *not* of exponential order.

8. (a) Derive Eq. (4), Sec. 4. (b) Illustrate that equation by using it to find the transform of the unit step function $S_k(t)$.

9. (a) If a function F and its derivative F' are both $\mathcal{O}(e^{\alpha t})$ and continuous when $t \geq 0$ except possibly for finite jumps at a point t_0, and if F'' is sectionally continuous on each interval $0 < t < T$, apply Eq. (4), Sec. 4, to derive the formula

$$L\{F''(t)\} = s^2 f(s) - sF(0) - F'(0) - se^{-st_0}[F(t_0 + 0) - F(t_0 - 0)]$$

$$- e^{-st_0}[F'(t_0 + 0) - F'(t_0 - 0)] \qquad (s > \alpha, \, t_0 > 0).$$

(b) Show that the formula applies to the function

$$F(t) = \sin t \text{ when } 0 \leq t \leq \pi, \qquad F(t) = 0 \text{ when } t \geq \pi$$

to give $f(s) = (1 + e^{-\pi s})/(s^2 + 1)$ when $-\infty < s < \infty$.

10. Let J denote the second integral in Eq. (1), Sec. 3; then

$$J^2 = \int_0^\infty e^{-x^2} \, dx \int_0^\infty e^{-y^2} \, dy = \int_0^\infty \int_0^\infty e^{-(x^2 + y^2)} \, dx \, dy.$$

Evaluate the iterated integral here by using polar coordinates and show that $J = \sqrt{\pi}/2$.

11. Prove that each linear combination $AF(t) + BG(t)$ of two functions F and G of exponential order is also of exponential order.

12. Use properties of continuous functions to show that if two functions are sectionally continuous on an interval (a,b) then (a) each linear combination of the two is also sectionally continuous on that interval; (b) the product of the two functions is sectionally continuous on the interval.

13. As noted in Sec. 5, the *gamma function* is defined for positive values of r by the formula

$$\Gamma(r) = \int_0^\infty x^{r-1}e^{-x} \, dx \qquad (r > 0).$$

(a) Integrate by parts to show that the function has the *factorial property* $\Gamma(r + 1) = r\Gamma(r)$.

(b) Show that $\Gamma(1) = 1$, and hence that $\Gamma(n + 1) = n!$ when $n = 1, 2, \ldots$.

(c) From the value of the integral J found in Prob. 10, show that $\Gamma(\frac{1}{2}) = \sqrt{\pi}$. Then use the factorial property to find $\Gamma(\frac{3}{2})$ and formula (2) to show that $L\{\sqrt{t}\} = \frac{1}{2}\sqrt{\pi}/s^{\frac{3}{2}}$.

14. *Improper integrals* Consider only functions that are sectionally continuous on each interval $0 < t < T$.

(a) If $0 \leq g(t) \leq h(t)$ whenever $t > 0$ and if $\int_0^\infty h(t)\, dt$ exists, use the definition of the improper integral

$$\int_0^\infty g(t)\, dt = \lim_{T \to \infty} \int_0^T g(t)\, dt$$

to prove the existence of that integral. [Note that the value of the last integral is non-decreasing as T increases. It never exceeds I where $I = \int_0^\infty h(t)\, dt$ because $\int_0^T g(t)\, dt \leq \int_0^T h(t)\, dt \leq I$, so it has a limit.]

(b) If $\int_0^\infty |p(t)|\, dt$ exists, prove that $\int_0^\infty p(t)\, dt$ exists by writing $p(t) = [p(t) + |p(t)|] - |p(t)|$. Note that $0 \leq p(t) + |p(t)| \leq 2|p(t)|$ and apply the test in part (a) to those two components of $p(t)$.

(c) Prove this *comparison test*: If $|q(t)| \leq r(t)$ whenever $t > 0$ and if $\int_0^\infty r\, dt$ exists, then the improper integral $\int_0^\infty q(t)\, dt$ is absolutely convergent, and the integral itself exists.

6   THE INVERSE TRANSFORM

Let the symbol $L^{-1}\{f(s)\}$ denote a function whose Laplace transform is $f(s)$. Thus if

$$L\{F(t)\} = f(s),$$

then $$F(t) = L^{-1}\{f(s)\}.$$

Using two of the transforms obtained in the foregoing sections, we can write, for instance,

$$L^{-1}\left\{\frac{1}{s-k}\right\} = e^{kt}, \qquad L^{-1}\left\{\frac{k}{s^2+k^2}\right\} = \sin kt.$$

This correspondence between functions $f(s)$ and $F(t)$ is called the *inverse Laplace transformation*, $F(t)$ being the *inverse transform* of $f(s)$.

In the strict sense of the concept of uniqueness of functions, the inverse Laplace transform is not unique. The function $F_1(t) = e^{kt}$ is an inverse transform of $1/(s-k)$; but another, for instance, is the function (Fig. 3)

$$F_2(t) = e^{kt} \qquad \text{when } 0 < t < 2, \text{ or } t > 2,$$

$$= 1 \qquad \text{when } t = 2.$$

For the transform of $F_2(t)$ is

$$\int_0^\infty e^{-st} F_2(t)\, dt = \int_0^2 e^{-st} e^{kt}\, dt + \int_2^\infty e^{-st} e^{kt}\, dt,$$

and this is the same as $L\{e^{kt}\}$. The function $F_2(t)$ could have been chosen equally well as one that differs from $F_1(t)$ at any finite set of values of t, or even at such an infinite set as $t = 1, 2, 3, \ldots$.

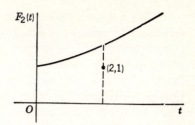

Fig. 3

A theorem on uniqueness of the inverse transform will be proved later (Sec. 69). To state it, we first define a class of functions of exponential order.

Let *the class $\mathscr{E}$* denote the set of all functions $F(t)$ defined on the half line $t > 0$, sectionally continuous on each bounded interval, and defined at each point t_0 where F is discontinuous as the mean value of its limits from the right and left,

$$F(t_0) = \tfrac{1}{2}[F(t_0 + 0) + F(t_0 - 0)] \qquad (t_0 > 0);$$

also, for each individual function F of the set, let constants M and α exist such that $|F(t)| < Me^{\alpha t}$ when $t > 0$. As we have seen, F then has a transform $f(s)$ defined on some half line $s > \alpha$.

The theorem states that *no two functions of class $\mathscr{E}$ can have the same transforms.* Thus if $f(s)$ is the transform of some function $F(t)$ of class $\mathscr{E}$, then $F(t)$ is the unique inverse transform $L^{-1}\{f(s)\}$ in $\mathscr{E}$.

For example, the only function $L^{-1}\{e^{-s}/s\}$ of class $\mathscr{E}$ is the unit step function $S_1(t)$ defined in Sec. 3 if $S_1(1)$ is defined to be $\tfrac{1}{2}$. As another example, the only function $L^{-1}\{1/(s - k)\}$ of class $\mathscr{E}$ is e^{kt}.

It is well to note here that not every function of s is a transform. The kind of functions $f(s)$ that are transforms of functions $F(t)$ of broad classes are limited, as we shall see (Chap. 6), by conditions of regularity that *include* requirements that f be continuous on a half line $s > \alpha$ and that $f(s) \to 0$ as $s \to \infty$.

We have noted that if $L\{F\}$ and $L\{G\}$ exist, then

(1) $$L\{AF(t) + BG(t)\} = Af(s) + Bg(s)$$

whenever A and B are constants. Let us restrict our functions of t to those of class $\mathscr{E}$, and functions of s to transforms of such functions. Then unique inverse transforms exist, and the linearity property (1) can be written

(2) $$L^{-1}\{Af(s) + Bg(s)\} = AF(t) + BG(t)$$

$$= AL^{-1}\{f(s)\} + BL^{-1}\{g(s)\};$$

that is, L^{-1} is also *a linear transformation of functions.*

The most obvious way of finding the inverse transform of a given function of s consists of reading the result from a table of transforms. A fairly extensive table is given in Appendix A. But we shall take up methods of obtaining inverse transforms of certain combinations and modifications of functions of s, as well as methods of resolving such functions into those listed in the tables. With the aid of such procedures, we shall be able to make much use of the transformation. In addition, there are explicit formulas for $L^{-1}\{f(s)\}$. The most useful of these formulas involves an integral in the complex plane. To use this integral, we must let s be a complex variable and we must be prepared to employ some theorems in the theory of functions of a complex variable.

7   A THEOREM ON SUBSTITUTION

Let a function $F(t)$ be such that its Laplace integral converges when $s > \alpha$. Then, replacing the argument of the transform $f(s)$ by $s - a$, where a is a constant, we have

$$f(s - a) = \int_0^\infty e^{-(s-a)t} F(t)\, dt = \int_0^\infty e^{-st} e^{at} F(t)\, dt,$$

when $s - a > \alpha$. Therefore

$$(1) \qquad\qquad f(s - a) = L\{e^{at} F(t)\} \qquad\qquad (s > \alpha + a).$$

Let us state this simple but important property as a theorem.

Theorem 3 *The substitution of $s - a$ for the variable s in the transform corresponds to the multiplication of the object function $F(t)$ by the function e^{at}, as shown in formula (1).*

To illustrate this property, let us recall that

$$\frac{m!}{s^{m+1}} = L\{t^m\} \qquad\qquad (m = 1, 2, \ldots ; s > 0).$$

Hence
$$\frac{m!}{(s - a)^{m+1}} = L\{t^m e^{at}\} \qquad\qquad (s > a).$$

As another illustration,

$$L\{\cos kt\} = \frac{s}{s^2 + k^2} \qquad\qquad (s > 0),$$

and therefore $$L\{e^{-at} \cos kt\} = \frac{s + a}{(s + a)^2 + k^2} \qquad\qquad (s > -a).$$

8 THE USE OF PARTIAL FRACTIONS (TABLE 1)

A few examples will show how the theory of partial fractions can be used in finding inverse transforms of *quotients of polynomials* in *s*. In the next chapter, a more systematic use of this procedure will be introduced.

Example 1 Find $L^{-1}\{(s + 1)/(s^2 + 2s)\}$.

The denominator of the function of *s* here is of higher degree than the numerator and has factors that are linear and distinct. Therefore constants *A* and *B* can be found such that

$$\frac{s + 1}{s(s + 2)} = \frac{A}{s} + \frac{B}{s + 2}$$

for all values of *s* except 0 and -2. Clearing fractions, we have

$$s + 1 = (A + B)s + 2A,$$

and this is an identity if $A + B = 1$ and $2A = 1$. Thus $A = B = \frac{1}{2}$, and hence

$$\frac{s + 1}{s^2 + 2s} = \frac{1}{2}\frac{1}{s} + \frac{1}{2}\frac{1}{s + 2}.$$

Since we know the inverse transforms of the two functions on the right, we have the result

$$L^{-1}\left\{\frac{s + 1}{s^2 + 2s}\right\} = \frac{1}{2} + \frac{1}{2}e^{-2t}.$$

The procedure can be shortened for such a simple fraction by writing

$$s + 1 = \tfrac{1}{2}(s + 2) + \tfrac{1}{2}s,$$

and hence

$$\frac{s + 1}{s(s + 2)} = \frac{1}{2}\frac{1}{s} + \frac{1}{2}\frac{1}{s + 2}.$$

Example 2 Find $L^{-1}\left\{\dfrac{a^2}{s(s + a)^2}\right\}$.

In view of the repeated linear factor, we write

$$\frac{a^2}{s(s + a)^2} = \frac{A}{s} + \frac{B}{s + a} + \frac{C}{(s + a)^2}.$$

Clearing fractions and identifying coefficients of like powers of *s* as before, or else by noting that

$$a^2 = (s + a)^2 - s(s + a) - as,$$

Table 1   A short table of transforms

	$F(t)$	$f(s)$	$\alpha\ (s > \alpha)$
1	1	$\dfrac{1}{s}$	0
2	e^{at}	$\dfrac{1}{s-a}$	a
3	$t^n\ (n = 1, 2, \ldots)$	$\dfrac{n!}{s^{n+1}}$	0
4	$t^n e^{at}\ (n = 1, 2, \ldots)$	$\dfrac{n!}{(s-a)^{n+1}}$	a
5	$\sin kt$	$\dfrac{k}{s^2 + k^2}$	0
6	$\cos kt$	$\dfrac{s}{s^2 + k^2}$	0
7	$\sinh kt$	$\dfrac{k}{s^2 - k^2}$	$|k|$
8	$\cosh kt$	$\dfrac{s}{s^2 - k^2}$	$|k|$
9	$e^{-at}\sin kt$	$\dfrac{k}{(s+a)^2 + k^2}$	$-a$
10	$e^{-at}\cos kt$	$\dfrac{s+a}{(s+a)^2 + k^2}$	$-a$
11	$\sqrt{t}$	$\dfrac{\sqrt{\pi}}{2\sqrt{s^3}}$	0
12	$\dfrac{1}{\sqrt{t}}$	$\sqrt{\dfrac{\pi}{s}}$	0
13	$t^k\ (k > -1)$	$\dfrac{\Gamma(k+1)}{s^{k+1}}$	0
14	$t^k e^{at}\ (k > -1)$	$\dfrac{\Gamma(k+1)}{(s-a)^{k+1}}$	a
15	$S_k(t)$ (Sec. 3)	$\dfrac{e^{-ks}}{s}$	0
16	$e^{at} - e^{bt}\ (a > b)$	$\dfrac{a-b}{(s-a)(s-b)}$	a
17	$\dfrac{1}{a}\sin at - \dfrac{1}{b}\sin bt$	$\dfrac{b^2 - a^2}{(s^2 + a^2)(s^2 + b^2)}$	0
18	$\cos at - \cos bt$	$\dfrac{(b^2 - a^2)s}{(s^2 + a^2)(s^2 + b^2)}$	0

we find that

$$\frac{a^2}{s(s+a)^2} = \frac{1}{s} - \frac{1}{s+a} - \frac{a}{(s+a)^2}.$$

Referring to Table 1, we can now write the result

$$L^{-1}\left\{\frac{a^2}{s(s+a)^2}\right\} = 1 - e^{-at} - ate^{-at}.$$

Example 3 Find

$$L^{-1}\left\{\frac{s}{(s^2+a^2)(s^2+b^2)}\right\} \qquad \text{where } a^2 \neq b^2.$$

Since

$$\frac{s}{(s^2+a^2)(s^2+b^2)} = \frac{s}{a^2-b^2}\frac{(s^2+a^2)-(s^2+b^2)}{(s^2+a^2)(s^2+b^2)}$$

$$= \frac{1}{b^2-a^2}\left(\frac{s}{s^2+a^2} - \frac{s}{s^2+b^2}\right),$$

when $a^2 \neq b^2$, it follows that

$$L^{-1}\left\{\frac{s}{(s^2+a^2)(s^2+b^2)}\right\} = \frac{1}{b^2-a^2}(\cos at - \cos bt).$$

Example 4 Find $F(t)$ if

$$f(s) = \frac{5s+3}{(s-1)(s^2+2s+5)}.$$

In view of the quadratic factor, we write

$$\frac{5s+3}{(s-1)(s^2+2s+5)} = \frac{A}{s-1} + \frac{Bs+C}{s^2+2s+5}.$$

After clearing fractions and identifying coefficients of like powers of s, we find that $A = 1$, $B = -1$, and $C = 2$; thus

$$f(s) = \frac{1}{s-1} - \frac{s-2}{(s+1)^2+4}$$

$$= \frac{1}{s-1} - \frac{s+1}{(s+1)^2+4} + \frac{3}{(s+1)^2+4}.$$

Referring to Table 1, or to Theorem 3, we see that

$$F(t) = e^t - e^{-t}(\cos 2t - \tfrac{3}{2}\sin 2t).$$

9   THE SOLUTION OF SIMPLE DIFFERENTIAL EQUATIONS

The application of the Laplace transformation to the solution of linear ordinary differential equations with *constant coefficients*, or systems of such equations, can now be made clear by means of examples. Such problems can of course be solved also by methods studied in a first course in differential equations. Later on, when we have developed further properties of the transformation, we shall solve problems of this sort with greater efficiency. We shall also be able to solve more difficult problems, especially in partial differential equations.

Example 1   Find the general solution of the differential equation

(1) $$Y''(t) + k^2 Y(t) = 0.$$

Let the value of the unknown function at $t = 0$ be denoted by the constant A and the value of its first derivative at $t = 0$ by the constant B; that is,

$$Y(0) = A, \qquad Y'(0) = B.$$

In view of the differential equation, we can write

$$L\{Y''(t)\} + k^2 L\{Y(t)\} = 0,$$

assuming that Y and Y'' have transforms. If the unknown function Y satisfies the conditions in Theorem 2, then

$$L\{Y''(t)\} = s^2 y(s) - As - B,$$

where $y(s) = L\{Y(t)\}$. Hence $y(s)$ must satisfy the equation

$$s^2 y(s) - As - B + k^2 y(s) = 0,$$

which is a simple *algebraic equation*. Its solution is clearly

$$y(s) = A\frac{s}{s^2 + k^2} + \frac{B}{k}\frac{k}{s^2 + k^2}.$$

Now $Y(t) = L^{-1}\{y(s)\}$, and the inverse transforms of the functions on the right in the last equation are known. Hence

(2) $$Y(t) = A\cos kt + \frac{B}{k}\sin kt,$$

$$= A\cos kt + B'\sin kt,$$

where A and B' are arbitrary constants since the initial values $Y(0)$ and $Y'(0)$ are not prescribed.

To verify our formal result given by formula (2), we need only to find $Y''(t)$ from that formula and substitute into Eq. (1) to see that the differential equation is satisfied regardless of the values of A and B'. Thus it is not necessary to justify the use of Theorem 2. However, our function $A \cos kt + B' \sin kt$ does satisfy all conditions in that theorem, and the order of the steps taken above can be reversed to show in another way that our function satisfies the differential equation. These remarks on verifying the solution apply also to the other examples and problems that follow in this section.

Example 2   Find the solution of the differential equation

(3) $$Y''(t) - Y'(t) - 6Y(t) = 2$$

satisfying the initial conditions

(4) $$Y(0) = 1, \qquad Y'(0) = 0.$$

Applying the transformation to both members of the differential equation, and letting $y(s)$ denote the transform of $Y(t)$, we obtain formally the algebraic equation

$$s^2 y(s) - s - sy(s) + 1 - 6y(s) = \frac{2}{s},$$

where we have used the initial conditions in writing the transforms of $Y''(t)$ and $Y'(t)$. Hence

$$(s^2 - s - 6)y(s) = \frac{s^2 - s + 2}{s},$$

or $$y(s) = \frac{s^2 - s + 2}{s(s-3)(s+2)} = \frac{A}{s} + \frac{B}{s-3} + \frac{C}{s+2}.$$

Evaluating the coefficients A, B, and C as in the preceding section, we find that

$$y(s) = -\frac{1}{3}\frac{1}{s} + \frac{8}{15}\frac{1}{s-3} + \frac{4}{5}\frac{1}{s+2}.$$

Hence $$Y(t) = -\tfrac{1}{3} + \tfrac{8}{15}e^{3t} + \tfrac{4}{5}e^{-2t}.$$

It is easy to verify that this function Y satisfies the differential equation (3) and both conditions (4).

Example 3   Find the functions $Y(t)$ and $Z(t)$ that satisfy the following system of differential equations:

$$Y''(t) - Z''(t) + Z'(t) - Y(t) = e^t - 2,$$
$$2Y''(t) - Z''(t) - 2Y'(t) + Z(t) = -t,$$
$$Y(0) = Y'(0) = Z(0) = Z'(0) = 0.$$

Let $y(s)$ and $z(s)$ denote the transforms of $Y(t)$ and $Z(t)$, respectively. Then in view of the differential equations and the initial conditions, those transforms formally satisfy the following simultaneous algebraic equations:

$$s^2 y(s) - s^2 z(s) + sz(s) - y(s) = \frac{1}{s-1} - \frac{2}{s},$$

$$2s^2 y(s) - s^2 z(s) - 2sy(s) + z(s) = -\frac{1}{s^2}.$$

These equations can be written

$$(s+1)y(s) - sz(s) = -\frac{s-2}{s(s-1)^2},$$

$$2sy(s) - (s+1)z(s) = -\frac{1}{s^2(s-1)}.$$

Eliminating $z(s)$, we find that

$$(s^2 - 2s - 1)y(s) = \frac{s^2 - 2s - 1}{s(s-1)^2}.$$

With the aid of partial fractions, we then find that

$$y(s) = \frac{1}{s(s-1)^2} = \frac{A}{s} + \frac{B}{s-1} + \frac{C}{(s-1)^2}$$

$$= \frac{1}{s} - \frac{1}{s-1} + \frac{1}{(s-1)^2}.$$

Therefore $\qquad Y(t) = 1 - e^t + te^t.$

Likewise we find that

$$z(s) = \frac{2s-1}{s^2(s-1)^2} = -\frac{1}{s^2} + \frac{1}{(s-1)^2},$$

and therefore $\qquad Z(t) = -t + te^t.$

Example 4 Solve the problem

$$Y'''(t) - 2Y''(t) + 5Y'(t) = 0,$$

$$Y(0) = 0, \qquad Y'(0) = 1, \qquad Y\left(\frac{\pi}{8}\right) = 1.$$

Let C denote the unknown initial value $Y''(0)$. Then

$$s^3 y(s) - s - C - 2s^2 y(s) + 2 + 5sy(s) = 0,$$

so that

$$y(s) = \frac{C - 2 + s}{s(s^2 - 2s + 5)}$$

$$= \frac{C - 2}{5}\left[\frac{1}{s} - \frac{s - 1}{(s - 1)^2 + 4}\right] + \frac{C + 3}{10}\frac{2}{(s - 1)^2 + 4}.$$

Thus $$Y(t) = \frac{C - 2}{5} + e^t\left(\frac{C + 3}{10}\sin 2t - \frac{C - 2}{5}\cos 2t\right).$$

Since $Y(\pi/8) = 1$, it follows that

$$1 = \frac{C - 2}{5} + \frac{e^{\pi/8}}{10\sqrt{2}}(C + 3 - 2C + 4),$$

or that $C = 7$. Hence the solution is

$$Y(t) = 1 + e^t(\sin 2t - \cos 2t).$$

PROBLEMS

1.   Use Theorem 3 to (a) obtain entry 9 from entry 5 in Table 1 and (b) show that $L^{-1}\{(s - a)^{-\frac{1}{2}}\} = e^{at}(\pi t)^{-\frac{1}{2}}$, given $L^{-1}\{s^{-\frac{1}{2}}\}$.

2.   Use partial fractions to find the inverse transforms of

(a) $\dfrac{a}{s(s + a)}$;   (b) $\dfrac{a^2}{s(s^2 + a^2)}$;   (c) $\dfrac{a^3}{s(s + a)^3}$.

Ans. (a) $1 - e^{-at}$; (b) $1 - \cos at$; (c) $1 - (1 + at + \frac{1}{2}a^2t^2)e^{-at}$.

3.   Use partial fractions to obtain the inverse transforms shown in (a) entry 16 of Table 1; (b) entry 17 of Table 1.

Solve the following problems and verify that your solutions satisfy the differential equations and any accompanying boundary conditions.

4. $Y''(t) - k^2 Y(t) = 0\ (k \neq 0)$.   Ans. $Y(t) = C_1 e^{kt} + C_2 e^{-kt}$.
5. $Y''(t) - 2kY'(t) + k^2 Y(t) = 0$.   Ans. $Y(t) = e^{kt}(C_1 + C_2 t)$.
6. $Y''(t) + k^2 Y(t) = a$.   Ans. $Y(t) = C_1 \sin kt + C_2 \cos kt + a/k^2$.
7. $Y''(t) + 2Y'(t) + 2Y(t) = 0,\ Y(0) = 0,\ Y'(0) = 1$.   Ans. $Y = e^{-t}\sin t$.
8. $Y''(t) + 4Y(t) = \sin t,\ Y(0) = Y'(0) = 0$.   Ans. $Y(t) = \frac{1}{3}\sin t - \frac{1}{6}\sin 2t$.
9. $Y''(t) + Y'(t) = t^2 + 2t,\ Y(0) = 4,\ Y'(0) = -2$.   Ans. $3Y(t) = t^3 + 6e^{-t} + 6$.
10. $Y'''(t) + Y'(t) = 10e^{2t},\ Y(0) = Y'(0) = Y''(0) = 0$.
Ans. $Y(t) = e^{2t} - 5 - 2\sin t + 4\cos t$.
11. $Y^{(4)}(t) = Y(t),\ Y(0) = Y'(0) = Y''(0) = 0,\ Y'''(0) = 2$.
Ans. $Y(t) = \sinh t - \sin t$.
12. $X'(t) + Y'(t) + X(t) + Y(t) = 1,\ Y'(t) - 2X(t) - Y(t) = 0,\ X(0) = 0,\ Y(0) = 1$.
Ans. $X(t) = e^{-t} - 1,\ Y(t) = 2 - e^{-t}$.

13. $Y''(t) + 2Z'(t) + Y(t) = 0$, $Y'(t) - Z'(t) - 2Y(t) + 2Z(t) = 1 - 2t$, $Y(0) = Y'(0) =$
$Z(0) = 0$.   *Ans.* $Y(t) = 2(1 - e^{-t} - te^{-t})$, $Z(t) = Y(t) - t$.

14. $Y''(t) + Y(t) = t$, $Y'(0) = 1$, $Y(\pi) = 0$.   *Ans.* $Y(t) = t + \pi \cos t$.

15. $y''(x) + y(x) = 1$, $y(0) = 1$, $y(\frac{1}{2}\pi) = 0$.   *Ans.* $y(x) = 1 - \sin x$.

16. $y''(x) + 2y'(x) + y(x) = 0$, $y(0) = 0$, $y(1) = 1$.

10   GENERATION OF THE TRANSFORMATION

We can show in a manipulative way why a linear integral transformation of functions $F(t)$ defined on the half line $t \geq 0$,

$$(1) \qquad\qquad T\{F(t)\} = \int_0^\infty K(t,s)F(t)\, dt = f(s),$$

must be essentially the Laplace transformation if it is to have the basic operational property of that transformation, namely, that *it replaces the simple differential form* $F'(t)$ *by an algebraic form in* $f(s)$, s *and the initial value* $F(0)$ *of the function* F.

  Let primes denote differentiation with respect to t. Then by a formal integration by parts we can write

$$(2) \qquad\qquad T\{F'(t)\} = \int_0^\infty K(t,s)F'(t)\, dt$$

$$= K(t,s)F(t) \Big]_0^\infty - \int_0^\infty K'(t,s)F(t)\, dt$$

$$= -K(0,s)F(0) - \int_0^\infty K'(t,s)F(t)\, dt,$$

provided that $K(t,s)F(t) \to 0$ as $t \to \infty$. The kernel K must involve some parameter s if the transform f is to correspond to a unique function F of some large class; otherwise, f is merely the numerical value of the integral (1) of the product $K(t)F(t)$, a number that is the same for many functions F.

  In order that the final integral in formula (2) will represent $f(s)$ except for a factor $\lambda(s)$, where λ is some function of s, we require that

$$(3) \qquad\qquad K'(t,s) = -\lambda(s)K(t,s).$$

Thus $K(t,s) = ce^{-\lambda t}$. But it is convenient to write $c = 1$ and choose $\lambda(s)$ as the parameter, so we write $\lambda(s) = s$ here. Then

$$(4) \qquad K(t,s) = e^{-st}, \qquad T\{F(t)\} = \int_0^\infty e^{-st}F(t)\, dt = L\{F(t)\},$$

and formula (2) is the basic operational property of the Laplace transformation:

(5) $$T\{F'(t)\} = -F(0) + sT\{F(t)\}.$$

A corresponding procedure will be used later on for the generation of integral transformations that will reduce other differential forms in functions defined on some prescribed interval, in terms of the transform of the function itself and prescribed boundary values of the function or its derivatives at the ends of the interval.

Readers acquainted with the concept of a function $F(t)$ $(a \leqq t \leqq b)$ as a generalized vector[1] can see that a linear integral transformation

(6) $$T\{F(t)\} = \int_a^b K(t,s)F(t)\,dt = f(s)$$

is a *generalization of linear vector transformations.* Our transformation (1) is a special case of (6) in which $a = 0$ and b is infinite.

A generalized vector, or function, F has an infinity of components consisting of the values $F(t)$ for each t; that is, the components are all the ordinates of points on the graph of $F(t)$ over the interval $a \leqq t \leqq b$. The generalized scalar product, called the *inner product,* of two such vectors F and G is the generalized sum of products of corresponding components, namely

$$\int_a^b F(t)G(t)\,dt.$$

Thus for each fixed s, the integral (6) is the inner product of the generalized vectors $K(t,s)$ and $F(t)$.

We begin with ordinary vectors $\mathbf{F}$ in three-dimensional space having rectangular cartesian components F_1, F_2, F_3, which may be written $F(1)$, $F(2)$, $F(3)$. A general linear transformation of $\mathbf{F}$ into a new vector $\mathbf{f}$ can be written in terms of components and constants K_{ij} $(i,j = 1, 2, 3)$. For the first component we write

$$f_1 = K_{11}F_1 + K_{12}F_2 + K_{13}F_3 = \mathbf{K}_1 \cdot \mathbf{F}$$

where $\mathbf{K}_1$ is a fixed vector with components K_{11}, K_{12}, K_{13}, and the dot denotes the scalar product of vectors. Similarly, $\mathbf{K}_2$ and $\mathbf{K}_3$ are fixed vectors and

$$f_2 = \mathbf{K}_2 \cdot \mathbf{F}, \qquad f_3 = \mathbf{K}_3 \cdot \mathbf{F}.$$

In space of n dimensions the general linear vector transformation is determined by a family of n fixed vectors $\mathbf{K}_i$:

(7) $$f_i = \mathbf{K}_i \cdot \mathbf{F} \qquad\qquad (i = 1, 2, \ldots, n).$$

[1] Churchill, R. V., "Fourier Series and Boundary Value Problems," 2d ed., chap. 3, 1963.

The extension to generalized vectors, or functions, is now fairly natural. A fixed function of the variable t and a parameter s, $K(t,s)$, is a family of generalized vectors, one vector for each fixed value of s of some set, say $c \leqq s \leqq d$. The inner product of K and F represents the s component of the transformed vector $f(s)$; that is,

$$(8) \qquad\qquad f(s) = \int_a^b K(t,s)F(t)\, dt.$$

This is the linear integral transformation (6).

Functions of certain classes, as well as vectors, are elements of linear vector spaces. The abstract theory of linear spaces is independent of the kind of elements involved. In that respect the treatment of functions as vectors is not merely an intuitive approach or an analogy. But the abstract theory specifies properties of inner products without suggesting formulas for them. The theory will not be needed here.

2

Further Properties of the Transformation

11 TRANSLATION OF $F(t)$

There are several further operational properties of the Laplace transformation that are important in the applications. Those properties whose derivations and applications do not necessarily involve the use of complex variables will be taken up in this chapter.

We begin with an analog of Theorem 3 of the first chapter. According to that theorem, the multiplication of the object function by an exponential function corresponds to a linear substitution for s in the transform. Now let us note the correspondence arising from the multiplication of the transform by an exponential function.

Let $F(t)$ have a transform,

$$f(s) = \int_0^\infty e^{-st} F(t)\, dt.$$

Then
$$e^{-bs} f(s) = \int_0^\infty e^{-s(t+b)} F(t)\, dt,$$

where b is a constant, assumed to be positive. Substituting $t + b = \tau$, we can write the last integral in the form

$$\int_b^\infty e^{-st} F(\tau - b)\, d\tau = \int_0^b 0 + \int_b^\infty e^{-st} F(\tau - b)\, d\tau.$$

Thus if we define a function $F_b(t)$ as follows,

(1)
$$\begin{aligned} F_b(t) &= 0 && \text{when } 0 < t < b, \\ &= F(t - b) && \text{when } t > b, \end{aligned}$$

we see that
$$f(s)e^{-bs} = \int_0^\infty e^{-st} F_b(\tau)\, d\tau.$$

The following property is therefore established.

Theorem 1 *If $f(s) = L\{F(t)\}$, then for any positive constant b,*

(2)
$$e^{-bs} f(s) = L\{F_b(t)\},$$

where $F_b(t)$ is the function defined by Eq. (1).

The function $F_b(t)$ is illustrated in Fig. 4. Its graph is obtained by translating the graph of $F(t)$ to the right through a distance of b units and making $F_b(t)$ identically zero between $t = 0$ and $t = b$. We can refer to $F_b(t)$ as the translated function.

Our unit step function $S_b(t)$ is the translation of the function $S_0(t) = 1$ ($t > 0$). It serves as a familiar illustration of the above theorem, since its transform is $s^{-1}e^{-bs}$. This step function can be used to describe the translation of any function $F(t)$ by writing

$$F_b(t) = S_b(t)F(t - b) \qquad\qquad (t > 0),$$

provided that $F(t - b)$ is defined where $t > 0$; that is, provided that $F(t)$ has numerical values for those values of t in the range $t > -b$. For example, the function $F(t) = \sin kt$ is defined for all t, so in this case we can write

$$F_b(t) = S_b(t)\sin k(t - b) = L^{-1}\left\{\frac{ke^{-bs}}{s^2 + k^2}\right\} \qquad (t > 0,\ b \geq 0).$$

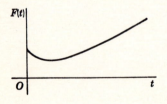
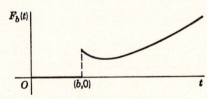

Fig. 4

On some occasions it is convenient to define $F(t)$ as zero for all negative values of t. When that is done, the graph of $F_b(t)$ is simply a translation of the graph of $F(t)$, and

$$F_b(t) = F(t - b) \qquad \text{for all } t.$$

Now consider the linear substitution cs for s in $f(s)$, where c is a positive constant. If the transform $f(s)$ of a function $F(t)$ exists whenever $s > \alpha$, then

$$f(cs) = \int_0^\infty e^{-cs\tau} F(\tau)\, d\tau \qquad \text{whenever } cs > \alpha.$$

The substitution $t = c\tau$ enables us to write that formula as

$$f(cs) = \frac{1}{c} \int_0^\infty e^{-st} F\!\left(\frac{t}{c}\right) dt$$

in terms of the transform of $F(t/c)$, to establish this theorem:

Theorem 2 *If $L\{F(t)\} = f(s)$ whenever $s > \alpha$, and if c is a positive constant, then*

(3) $$f(cs) = \frac{1}{c} L\left\{ F\!\left(\frac{t}{c}\right) \right\} \qquad\qquad \left(s > \frac{\alpha}{c} \right).$$

We can write $a = 1/c$ to obtain an alternate form of (3):

(4) $$L\{F(at)\} = \frac{1}{a} f\!\left(\frac{s}{a}\right) \qquad (a > 0,\ s > a\alpha).$$

Given, for example, that $L\{\cos t\} = s/(s^2 + 1)$ when $s > 0$, it follows that

$$L\{\cos at\} = \frac{1}{a} \frac{s/a}{(s/a)^2 + 1} = \frac{s}{s^2 + a^2} \qquad (s > 0).$$

The effect of a general linear substitution for s can be seen from formula (3) and Theorem 3, Chap. 1, since

(5) $$f(as - b) = f\left[a\!\left(s - \frac{b}{a} \right) \right] = L\left[\frac{1}{a} e^{(b/a)t} F\!\left(\frac{t}{a}\right) \right] \qquad (a > 0).$$

12 STEP FUNCTIONS

When $t \geq 0$, the bracket symbol $[t]$ is used in mathematics to denote the greatest integer, $0, 1, 2, \ldots$, that does not exceed the number t. Thus $[\pi] = 3 = [3]$. The function $[t]$ is therefore a step function of the type that is sometimes called a staircase function with unit rise and run:

$$[t] = 0 \qquad (0 \leq t < 1),$$
$$= 1 \qquad (1 \leq t < 2),$$
$$= 2 \qquad (2 \leq t < 3), \ldots.$$

The staircase function with an arbitrary positive run h, and with a unit rise beginning at the origin $t = 0$ (Fig. 5), is then represented by the symbol, $[1 + t/h]$ or $1 + [t/h]$. The function $c[1 + t/h]$ has the rise c and the run h.

To obtain the transform of the function shown in Fig. 5,

$$(1) \qquad Y(t) = \left[1 + \frac{t}{h} \right] \qquad (h > 0),$$

we may describe the function by means of a *difference equation* of the first order together with an initial condition, as follows.

$$Y(t) = Y(t - h) + 1 \qquad (t \geq 0),$$
$$= 0 \qquad (t < 0).$$

The function $Y(t - h)$ is then the same as the translated function $Y_h(t)$ and, in view of Theorem 1, the transform $y(s)$ of the function $Y(t)$ satisfies the equation

$$y(s) = e^{-hs} y(s) + \frac{1}{s} \qquad (s > 0).$$

Therefore the transform of the staircase function (1) is

$$(2) \qquad y(s) = L\left\{ \left[1 + \frac{t}{h} \right] \right\} = \frac{1}{s} \frac{1}{1 - e^{-hs}} \qquad (s > 0).$$

Another useful step function is the *unit finite impulse function*

$$(3) \qquad I(h, t - t_0) = \frac{1}{h} \text{ when } t_0 < t < t_0 + h,$$

$$= 0 \text{ when } t < t_0 \text{ and when } t > t_0 + h,$$

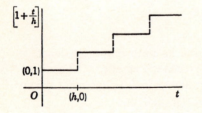

Fig. 5

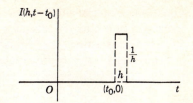

Fig. 6

illustrated in Fig. 6. Let us introduce a *variation of our unit step function* $S_k(t)$, one that is defined to be zero when t is negative; namely,

(4) $$S_0(t - t_0) = 0 \qquad\qquad \text{when } t < t_0,$$
$$= 1 \qquad\qquad \text{when } t > t_0.$$

Then our function I can be written

(5) $$I(h, t - t_0) = \frac{S_0(t - t_0) - S_0(t - t_0 - h)}{h}.$$

It follows at once from the transform of $S_0(t - t_0)$ that

(6) $$L\{I(h, t - t_0)\} = e^{-st_0}\frac{1 - e^{-hs}}{hs} \qquad (t_0 \geqq 0, h > 0, s > 0).$$

It is clear from the definition (3) that

(7) $$\lim_{h\to 0} I(h, t - t_0) = 0 \qquad\qquad \text{when } t \neq t_0.$$

Also, the area under the graph of the function I is unity for every positive h; hence

(8) $$\lim_{h\to 0} \int_{-\infty}^{\infty} I(h, t - t_0)\, dt = 1.$$

The order of the positions of the limit and the integral here is important, for in view of Eq. (7)

$$\int_{-\infty}^{\infty} \lim_{h\to 0} I(h, t - t_0)\, dt = 0.$$

It is interesting to note the behavior of the transform of the finite unit impulse function as h tends to zero. By evaluating the limit of the right-hand member of Eq. (6), we find that

(9) $$\lim_{h\to 0} L\{I(h, t - t_0)\} = e^{-st_0} \qquad (t_0 \geqq 0, h > 0, s > 0).$$

But note that

$$L\{\lim_{h\to 0} I(h, t - t_0)\} = L\{0\} = 0.$$

Equation (9) presents the *exponential function* e^{-st_0}, or $\exp(-st_0)$, not as the transform of a function, but as a limit of transforms of functions of the set $I(h, t - t_0)$ consisting of one function for each positive value of the parameter h. Thus $\exp(-st_0)$ approximates the transform of the impulse function $I(h, t - t_0)$ for small positive values of h, when $t_0 \geq 0$.

Example   A particle of mass m, initially at rest at the origin $X = 0$, is subjected to a force in the form of a unit finite impulse $I(h,t)$ beginning at the instant $t_0 = 0$, in the direction of the X axis. Its displacements $X(h,t)$, which satisfy the conditions

(10) $mX''(h,t) = I(h,t), \qquad X(h,0) = X'(h,0) = 0 \qquad (h > 0),$

where $X' = dX/dt$, could be found here by integration.

But if we use transforms to solve for $X(h,t)$, we write

(11) $ms^2 x(h,s) = L\{I(h,t)\} = \dfrac{1 - e^{-hs}}{hs},$

assuming that X and X' are continuous functions of t of exponential order. Thus

$$mx(h,s) = \frac{1}{2h}\left(\frac{2}{s^3} - \frac{2}{s^3}e^{-hs}\right)$$

and therefore

(12) $mX(h,t) = \dfrac{1}{2h}[t^2 - (t - h)^2 S_0(t - h)];$

that is

(13) $X(h,t) = \dfrac{1}{2mh}t^2$ $\qquad$ when $0 \leq t \leq h,$

$\qquad = \dfrac{t^2 - (t - h)^2}{2mh} = \dfrac{1}{m}\left(t - \dfrac{1}{2}h\right)$ $\quad$ when $t \geq h.$

That function satisfies all conditions (10).

Now if $h \to 0$, the second of formulas (13) applies when $t > 0$ to give the limiting displacements $X(t)$:

(14) $X(t) = \lim_{h \to 0} X(h,t) = \dfrac{t}{m}$ $\qquad (t > 0).$

The displacements $X(t)$ correspond to the idealized unit impulse given to m at the instant $t = 0$. As a verification of the solution (14) of that idealized case, we note that $X''(t) = 0$ whenever $t > 0$, and $X(+0) = 0$.

Furthermore, the particle has the momentum $mX'(t) = 1$ whenever $t > 0$; thus if it is initially at rest, then its momentum jumps suddenly from 0 to 1 at the instant $t = 0$. The instantaneous impulse has the same effect as an initial velocity $1/m$.

13 THE IMPULSE SYMBOL $\delta(t - t_0)$

The above example is a simple illustration of problems in differential equations involving instantaneous impulses. In order to reduce the number of steps in formal solutions of such problems, we shall occasionally use the unit impulse symbol $\delta(t - t_0)$ to a limited extent indicated below.

Let a function $F(t)$, defined for all real t, be continuous on some interval $t_0 \leq t \leq t_0 + k$. Then according to the basic law of the mean for integrals to each $h(0 < h < k)$, there corresponds a number $\theta(0 < \theta < 1)$ such that

$$\int_{-\infty}^{\infty} I(h, t - t_0)F(t)\, dt = \frac{1}{h} \int_{t_0}^{t_0+h} F(t)\, dt = F(t_0 + \theta h).$$

Consequently the impulse function I has the property

(1) $$\lim_{h \to 0} \int_{-\infty}^{\infty} I(h, t - t_0)F(t)\, dt = F(t_0),$$

sometimes called the *sifting property* of selecting that value $F(t_0)$ of $F(t)$.

We abbreviate that property by writing

(2) $$\int_{-\infty}^{\infty} \delta(t - t_0)F(t)\, dt = F(t_0),$$

where the entire left-hand member is a *symbol* that denotes the operation represented by the left-hand member of Eq. (1). That symbol is selected so as to suggest an integral of a function, because if $\delta(t - t_0)$ is replaced by $I(h, t - t_0)$, the symbol does represent an integral whose value is approximately $F(t_0)$ when h is a small positive number.

If $t_0 \geq 0$, then $I(h, t - t_0) = 0$ when $t < 0$ and

(3) $$\lim_{h \to 0} \int_{0}^{\infty} I(h, t - t_0)F(t)\, dt = F(t_0) \qquad (t_0 \geq 0, h > 0)$$

according to Eq. (1); that is, in symbolic form

(4) $$\int_{0}^{\infty} \delta(t - t_0)F(t)\, dt = F(t_0) \qquad (t_0 \geq 0).$$

When $F(t) = 1$, formula (2) becomes

(5) $$\int_{-\infty}^{\infty} \delta(t - t_0)\, dt = 1,$$

the symbolic form of Eq. (8), Sec. 12. When $F(t) = e^{-st}$ and $t_0 \geq 0$, the symbolic integration formula (4) can be written

$$(6) \qquad\qquad L\{\delta(t - t_0)\} = \exp(-st_0),$$

which is the symbolic form of Eq. (9), Sec. 12.

The letter δ used in our symbolic integrals has further significance in differential equations. We use the example treated in Sec. 12 as an illustration. Let us write

$$(7) \qquad\qquad mX''(t) = \delta(t); \qquad X(0) = X'(0) = 0,$$

to signify that X is the limit as $h \to 0$ of the solution $X(h,t)$ of problem (10) there, the problem with $\delta(t)$ replaced by $I(h,t)$. In the transformed problem (11), Sec. 12, we may let h tend to zero and assume that $x(h,s) \to x(s)$ where $x(s) = L\{X(t)\}$; the result is

$$(8) \qquad\qquad ms^2 x(s) = 1.$$

But that result is obtained at once by formally applying the operator L to terms in problem (7) if $L\{\delta(t)\} = 1$, in accordance with formula (6), while the symbol $L\{X''(t)\}$ means the limit as $h \to 0$ of the transform of $X(h,t)$ so that we may be justified in replacing it by $s^2 x(s)$. Thus $mx(s) = 1/s^2$, and our result $mX(t) = t$ follows.

That symbolic method gives limits of solutions, as $h \to 0$, of other problems in differential equations with $I(h, t - t_0)$ as a forcing function. We shall use the method occasionally in the sequel. Results obtained by such manipulations clearly need to be verified if they are to be relied upon. More often we sacrifice the brevity gained by using the δ symbol and use instead the function $I(h, t - t_0)$, then find the limit of the solution as $h \to 0$.

The symbol $\delta(t - t_0)$, often called the *Dirac delta function*, is not a function. In particular, it is not the limit of $I(h, t - t_0)$ as $h \to 0$; that limit is a function with value zero everywhere except at the one point $t = t_0$, a function whose integral is zero (Sec. 12). The symbol $\delta(t - t_0)$ may, however, be replaced by the function $I(h, t - t_0)$, when h is small, to give results that approximate the effect of δ.

Generalizations of functions, called *generalized functions* or *distributions*, have been developed since 1950 which include the symbol $\delta(t - t_0)$ and its generalized derivatives. There are now a variety of theories that give sound developments of distributions. But even the introductory presentations of the theory of distributions seem too lengthy to present here.[1]

---

[1] See, for instance, the article entitled From Delta Functions to Distributions, by A. Erdélyi, in "Modern Mathematics for the Engineer," Second Series, 1961, or A. E. Danese, "Advanced Calculus," vol. 2, pt. 6, 1965, or references listed in those books.

PROBLEMS

1. With the aid of our theorems and known transforms find the inverse transforms $F(t)$ tabulated below and draw a graph of $F(t)$.

$f(s)$	$F(t)$
(a) $\dfrac{e^{-bs}}{s^2}(b > 0)$	$0\,(0 < t < b); t - b\,(t > b)$
(b) $\sqrt{\pi}s^{-\frac{3}{2}}e^{-2s}$	$2\sqrt{t - 2}\,S_2(t)$
(c) $\dfrac{e^{-\pi s}}{s^2 + 1}$	$-S_\pi(t)\sin t$
(d) $\dfrac{1 + e^{-\pi s}}{s^2 + 1}$	$\sin t\,(0 < t < \pi); 0\,(t > \pi)$
(e) $\dfrac{se^{-s/2}}{s^2 + \pi^2}$	$S_0(t - \tfrac{1}{2})\sin \pi t$
(f) $\dfrac{4}{(2s + 1)^2}$	$te^{-t/2}$
(g) $\dfrac{4}{(2s - 1)^2 - 4}$	$e^{t/2}\sinh t = \dfrac{e^{3t/2} - e^{-t/2}}{2}$

2. From the Maclaurin series that represents $(1 - x)^{-1}$ when $|x| < 1$, or from the sum of an infinite geometric series, show that

$$\frac{1}{s}\frac{1}{1 - e^{-hs}} = \frac{1}{s} + \frac{e^{-hs}}{s} + \frac{e^{-2hs}}{s} + \cdots = \sum_{n=0}^{\infty} \frac{e^{-nhs}}{s} \qquad (h > 0, s > 0).$$

Apply the inverse transformation term by term to this infinite series, formally, to obtain the result shown in Eq. (2), Sec. 12; namely,

$$L^{-1}\left\{\frac{1}{s}\frac{1}{1 - e^{-hs}}\right\} = \left[1 + \frac{t}{h}\right] \qquad (h > 0).$$

3. Show that Eq. (2), Sec. 12, can be written in the form

$$L\left\{\left[1 + \frac{t}{h}\right]\right\} = \frac{1}{2s}\left(1 + \coth\frac{hs}{2}\right).$$

4. (a) Use the formal method indicated in Prob. 2 to find $F(t)$ when

$$f(s) = \frac{1}{s}\frac{2}{1 + e^{-s}} \qquad (s > 0).$$

Draw the graph of F to see that F can be written in terms of the bracket symbol $[t]$ in the form

$$F(t) = 1 + (-1)^{[t]}.$$

(b) Show that the function F in part (a) is described by the following difference equation of the first order, when $t > 0$:

$$F(t) + F(t - 1) = 2 \qquad \text{where } F(t) = 0 \qquad \text{when } t < 0.$$

Transform that difference equation to verify the formula given in part (a) for $f(s)$.

5. Apply Theorem 2 to the transformation found in Problem 4 to show that

$$L\left\{1 + (-1)^{\left[\frac{t}{h}\right]}\right\} = \frac{1}{s}\frac{2}{1 + e^{-hs}} \qquad (h > 0, s > 0).$$

6. Find the function $Y(t)$ that satisfies the following difference equation of the first order and the accompanying initial condition.

$$Y(t) - cY(t - h) = F(t)$$

$$Y(t) = 0 \qquad \text{when } t < 0,$$

where c and h are constants and $h > 0$, and where $F(t) = 0$ when $t < 0$. The expansion of $(1 - ce^{-hs})^{-1}$ in powers of ce^{-hs} is helpful here (compare Prob. 2). Show that the solution can be written

$$Y(t) = \sum_{n=0}^{\infty} c^n F(t - nh);$$

but for each fixed value of t this series is a finite series because $F(t - nh) = 0$ when $t - nh < 0$. Thus an alternate form of the solution is

$$Y(t) = F(t) + cF(t - h) + c^2 F(t - 2h) + \cdots + c^m F(t - mh),$$

where $m = 0, 1, 2, \ldots$, when $mh < t < (m + 1)h$. Verify the solution.

7. Find the function $Y(t)$ that satisfies the following difference equation of the second order and the accompanying initial condition.

$$Y(t) - (a + b)Y(t - h) + abY(t - 2h) = F(t)$$

$$Y(t) = 0 \qquad \text{when } t < 0,$$

where $F(t) = 0$ when $t < 0$. The constants a and b are such that $a \neq b$, and the constant h is positive. Note that the solution of the transformed problem can be simplified, with the aid of partial fractions in the variable e^{-hs}, to

$$y(s) = \frac{f(s)}{a - b}\left(\frac{a}{1 - ae^{-hs}} - \frac{b}{1 - be^{-hs}}\right).$$

$$\text{Ans. } Y(t) = \frac{1}{a - b}\sum_{n=0}^{\infty}(a^{n+1} - b^{n+1})F(t - nh).$$

8. Solve Prob. 7 when $b = a$.
$$\text{Ans. } Y(t) = \sum_{n=0}^{\infty}(n + 1)a^n F(t - nh).$$

9. Solve the difference-differential equation

$$Y'(t) - aY(t - 1) = F(t),$$

where $F(t) = b$ when $t > 0$ and $F(t) = 0$ when $t < 0$, under the condition that $Y(t) = 0$ when $t \leq 0$.

$$\text{Ans. } Y(t) = b\left[t + \frac{a}{2!}(t-1)^2 + \cdots + \frac{a^n}{(n+1)!}(t-n)^{n+1}\right],$$

where $n \leq t \leq n + 1$ and $n = 0, 1, 2, \ldots$.

10. Initially a particle of mass m moving along the X axis is at the origin $X = 0$ with a velocity v_0. The only external force that acts on that particle is an instantaneous impulse p_0 at time t_0, in the X direction. Thus the displacement $X(t)$ satisfies the following symbolic equation and initial conditions:

$$mX''(t) = p_0\, \delta(t - t_0), \qquad X(0) = 0,\ X'(0) = v_0.$$

Find the formal solution (writing $L\{X''\} = s^2 x - v_0$)

$$X(t) = v_0 t + \frac{p_0}{m}(t - t_0)S_0(t - t_0) \qquad\qquad (t \geq 0,\ t_0 > 0)$$

and show $X(t)$ graphically. To verify the solution, show that $X''(t) = 0$ when $t \neq t_0$, that $X(0) = 0$ and $X'(0) = v_0$, and that the momentum $mX'(t)$ undergoes a *jump* p_0 at the instant t_0.

11. Replace the instantaneous impulse $p_0\, \delta(t - t_0)$ in Prob. 10 by the finite impulse $p_0 I(h, t - t_0)$ and then find the displacements $X(h,t)$. Show that the limit of $X(h,t)$ as $h \to 0$ is the function $X(t)$ found before.

12. Problem 10 can be stated in terms of functions as follows:

$$mX''(t) = 0 \text{ when } t \neq t_0, \qquad X(0) = 0, \qquad X'(0) = v_0$$

$$mX'(t_0 + 0) - mX'(t_0 - 0) = p_0 \qquad\qquad (t_0 > 0),$$

where X and X' are continuous except for the specified jump in X'. Use our formula (Prob. 9, Sec. 5) for the transform of the second derivative in this case to obtain the formula found before for $X(t)$.

13. Show formally that the solution of the symbolic problem

$$X''(t) + k^2 X(t) = p_0\, \delta(t - t_0), \qquad X(0) = X'(0) = 0, \qquad\qquad (t_0 \geq 0)$$

where k and p_0 are constants, is (if we write $L\{X''\} = s^2 x$ here)

$$X(t) = \frac{p_0}{k}S_0(t - t_0)\sin k(t - t_0).$$

14. Show that the formal solution of the symbolic problem (if we write $L\{X'\} = sx$ here)

$$X'(t) = \delta(t - t_0), \qquad X(0) = 0 \qquad\qquad (t_0 \geq 0)$$

is $X(t) = S_0(t - t_0)$. In this sense, then, $\delta(t - t_0)$ corresponds to the derivative of the unit step function $S_0(t - t_0)$ as we could anticipate from the symbolic formula

$$L\{\delta(t - t_0)\} = e^{-st_0} = sL\{S_0(t - t_0)\}.$$

15. Let ΔI denote the change in $I(h, t - t_0)$ when $\Delta t = -h$. Draw the graph of the function of t represented by the difference quotient

$$\frac{\Delta I}{h} = \frac{I(h, t - t_0) - I(h, t - h - t_0)}{h}.$$

If F and F' are continuous on an interval that includes the points t_0 and $t_0 + 2h$, show that

$$\int_{-\infty}^{\infty} \frac{\Delta I}{h} F(t)\, dt = -\frac{1}{h} \int_{t_0}^{t_0 + h} \frac{F(\tau + h) - F(\tau)}{h}\, d\tau$$

$$= -\frac{1}{h} \int_{t_0}^{t_0 + h} F'(\tau + \theta h)\, d\tau$$

where $0 < \theta < 1$, and hence that

$$\lim_{h \to 0} \int_{-\infty}^{\infty} \frac{\Delta I}{h} F(t)\, dt = -F'(t_0).$$

A symbolic form of that formula is

$$\int_{-\infty}^{\infty} \delta'(t - t_0) F(t)\, dt = -F'(t_0);$$

which is a symbolic integration by parts:

$$\int_{-\infty}^{\infty} \delta'(t - t_0) F(t)\, dt = -\int_{-\infty}^{\infty} \delta(t - t_0) F'(t)\, dt = -F'(t_0).$$

In particular, if $t_0 \geq 0$, note that

$$L\{\delta'(t - t_0)\} = -\frac{d}{dt} e^{-st} \Big]_{t = t_0} = s e^{-s t_0}.$$

16. Let F be continuous $(t \geq 0)$ except for a jump j_0 at t_0 $(t_0 > 0)$. Then $F(t) - j_0 S_0(t - t_0)$ is continuous whenever $t \geq 0$ if $F(t_0)$ is properly defined. If F is exponential order and F' is sectionally continuous, then

$$L\left\{\frac{d}{dt}[F(t) - j_0 S_0(t - t_0)]\right\} = s\left[f(s) - j_0 \frac{e^{-s t_0}}{s}\right] - F(0).$$

Show that this formula agrees with formula (4), Sec. 4, and note that its symbolic form may be written (Prob. 14)

$$L\{F'(t) - j_0\, \delta(t - t_0)\} = sf(s) - F(0) - j_0 e^{-s t_0}.$$

17. Since $I(h, t) = 1/h$ when $0 < t < h$ and vanishes when $t < 0$ and when $t > h$, describe the function $I(h, -t)$ and show that

$$\lim_{h \to 0} \int_{-\infty}^{\infty} F(t) I(h, -t)\, dt = F(0).$$

when F is defined for all t and continuous over some interval $|t| < a$. Symbolically,

$$\int_{-\infty}^{\infty} F(t)\,\delta(-t)\,dt = F(0) = \int_{-\infty}^{\infty} F(t)\,\delta(t)\,dt;$$

thus δ *has the symbolic property*

$$\delta(-t) = \delta(t).$$

14   INTEGRALS CONTAINING A PARAMETER

We now review some properties of integrals that will be useful.

Let x denote a parameter and $f(x,t)$ a continuous function of the two variables x and t together in a region $\alpha(x) \leq t \leq \beta(x)$, $c \leq x \leq d$, where α and β are continuous functions. Then *the integral*

(1)
$$\int_{\alpha(x)}^{\beta(x)} f(x,t)\,dt = g(x) \qquad\qquad (c \leq x \leq d)$$

represents a continuous function of x, which we have called g. If the derivatives $\alpha'(x)$ and $\beta'(x)$ exist, and if $\partial f / \partial x$ as well as f is continuous, then the derivative of the integral (1) is given by the *generalized Leibnitz formula*,

(2)
$$g'(x) = \int_{\alpha(x)}^{\beta(x)} \frac{\partial}{\partial x} f(x,t)\,dt + f[x,\beta(x)]\beta'(x) - f[x,\alpha(x)]\alpha'(x).$$

The foregoing two properties of the integral (1) are established in advanced calculus from fairly elementary considerations. Let us indicate here how the continuity requirements on the integrand can be relaxed in establishing certain properties.

Let a, b, and c denote constants, and let the integrand of the integral

(3)
$$\int_{0}^{ax+b} f(x,t)\,dt = h(x) \qquad\qquad (0 \leq x \leq c)$$

be such that the region $0 \leq t \leq ax + b$, $0 \leq x \leq c$ can be divided into polygonal subregions by a finite number of straight lines (Fig. 7)

$$t = a_i x + b_i \qquad\qquad (i = 1, 2, \ldots, m)$$

and
$$x = x_j \qquad\qquad (j = 1, 2, \ldots, n),$$

where $f(x,t)$ is continuous over each subregion up to its boundary; that is, for each subregion f can be defined on the boundary in such a way that it is continuous in x and t together over the closed subregion. We then say that f is *continuous by subregions*.

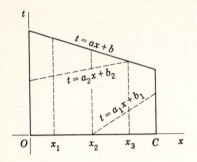

Fig. 7

When $x_1 \leqq x \leqq x_2$ in Fig. 7, for instance, the integral can be written as a sum of integrals of the type (1), such as

$$
(4) \qquad \int_{a_2 x + b_2}^{ax + b} f(x, t)\, dt \qquad\qquad (x_1 \leqq x \leqq x_2),
$$

in which the integrand is continuous so that the integral represents a continuous function of $x(x_1 \leqq x \leqq x_2)$. Similarly when x is between any two adjacent x_j's, $h(x)$ is continuous. Consequently, the integral (3) *represents a sectionally continuous function $h(x)$ on the interval $(0,c)$* when f is continuous by subregions.

As a special case, when the integrand is a sectionally continuous function $f(t)$, independent of x, the integral (3) represents a continuous function of x on the closed interval $[0,c]$.

The integral of the sectionally continuous function (3)

$$
(5) \qquad \int_0^c h(x)\, dx = \int_0^c \int_0^{ax+b} f(x,t)\, dt\, dx
$$

exists. It can be written as the sum of iterated integrals over each of the subregions in which $f(x,t)$ is continuous. When the order of integration is interchanged in each subregion, the sum of the integrals represents the iterated integral, with order interchanged, over the entire region.

Thus our condition of continuity by subregions *permits the interchange of order of integration with respect to t and x.* In particular, when $a = 0$ so that the limits are constants,

$$
(6) \qquad \int_0^c \int_0^b f(x,t)\, dt\, dx = \int_0^b \int_0^c f(x,t)\, dx\, dt.
$$

These conclusions can be reached by the same procedure for regions of more general shapes. The argument used above applied, for instance, when the lower limits of the integrals (5) and (6) are constants other than zero.

15 IMPROPER INTEGRALS

Corresponding properties can be established for convergent improper integrals with the aid of a condition of uniform convergence with respect to the parameter x. The concept of *uniform convergence* of integrals, which applies only to improper integrals that involve a parameter, is illustrated in the following example.

For each value of $x(x \geqq 0)$ the particular Laplace integral

$$(1) \qquad p(x) = \int_0^\infty x e^{-xt}\, dt$$

exists; it has the values

$$(2) \qquad \begin{aligned} p(0) &= 0 \\ p(x) &= 1 \end{aligned} \qquad\qquad \text{when } x > 0.$$

It can be written as a definite integral plus a remainder,

$$p(x) = \int_0^T x e^{-xt}\, dt + \int_T^\infty x e^{-xt}\, dt = 1 - e^{-xT} + R(x,T),$$

where $R(x,T) = \int_T^\infty x e^{-xt}\, dt$; then

$$(3) \qquad \begin{aligned} R(0,T) &= 0 \\ R(x,T) &= e^{-xT} \end{aligned} \qquad\qquad \text{when } x > 0.$$

Uniform convergence of the integral for all x in a prescribed interval means not only that the integral exists for each x in the interval, but also that the remainder can be made arbitrarily small in absolute value by taking T sufficiently large, uniformly for all x in the interval. Thus when $x \geqq 1$ and ϵ is any small positive number, we see that

$$|R(x,t)| = e^{-xT} \leqq e^{-T} < \epsilon$$

when $T > \log(1/\epsilon)$, a value that is independent of x, so that the integral converges uniformly with respect to x when $x \geqq 1$. In the same way we can see that it converges uniformly when $x \geqq x_1$ whenever $x_1 > 0$.

But the convergence is not uniform in the range $x \geqq 0$ or $x > 0$, because the remainder (3) when $x > 0$ is small only if xT is large. Since x can be arbitrarily small, the value of T that makes $R(x,T)$ small must depend on x. A consequence of this lack of uniform convergence in the range $x \geqq 0$ is the discontinuity of the function $p(x)$ at $x = 0$ that is exhibited by Eq. (2).

The *Weierstrass test for uniform convergence* of the integral

$$(4) \qquad \int_0^\infty f(x,t)\, dt = q(x) \qquad\qquad (0 \leqq x \leqq c)$$

can be stated as follows. Let $f(x,t)$ be continuous by subregions, in the sense described in Sec. 14, over the region $0 \leq x \leq c, 0 \leq t \leq T$, for each positive constant T. If a sectionally continuous function $M(t)$, independent of x, exists such that

$$|f(x,t)| \leq M(t) \qquad\qquad (0 \leq x \leq c, t > 0)$$

and such that the integral $\int_0^\infty M(t)\, dt$ exists, then the integral (4) is uniformly convergent with respect to $x(0 \leq x \leq c)$. This test is established by first noting that the integral (4) exists according to the comparison test (Prob. 14, Sec. 5), then that the absolute value of the remainder for the integral, for all x, in the interval, does not exceed the remainder $\int_T^\infty M(t)\, dt$ in the integral of M. The latter remainder is independent of x, and it tends to zero as $T \to \infty$.

The Weierstrass test states useful sufficient, but not necessary, conditions for uniform convergence. Under those conditions the integral (4) is also absolutely convergent.

If the integral (4) is uniformly convergent with respect to x and if its integrand $f(x,t)$ is continuous by subregions over the rectangle $0 \leq x \leq c$, $0 \leq t \leq T$ for each positive T, then *the integral represents a sectionally continuous function* $q(x)$; moreover

$$(5) \qquad \int_0^c q(x)\, dx = \int_0^c \int_0^\infty f(x,t)\, dt\, dx = \int_0^\infty \int_0^c f(x,t)\, dx\, dt;$$

that is, *the order of integration of the definite and improper integrals here can be interchanged.*

In particular, let $f(x,t)$ be continuous in each rectangle $0 \leq x \leq c$, $0 \leq t \leq T$, except possibly for finite jumps across a set of lines $t = t_i(i = 1, 2, \ldots, n)$. Then $q(x)$ is *continuous* if the integral (4) is uniformly convergent. Suppose that $\partial f/\partial x$ as well as f satisfies those continuity requirements and that the *integral $\int_0^\infty (\partial f/\partial x)\, dt$ converges uniformly and the integral (4) exists*. Then the derivative of the latter integral exists and *the order of differentiation with respect to x and integration with respect to t can be interchanged*:

$$(6) \qquad \frac{d}{dx}\int_0^\infty f(x,t)\, dt = \int_0^\infty \frac{\partial}{\partial x} f(x,t)\, dt \qquad\qquad (0 < x < c).$$

The continuity of $q(x)$ can be seen by writing

$$q(x) = \int_0^T f(x,t)\, dt + R(x,T) \qquad\qquad (0 \leq x \leq c).$$

For each fixed T the integral on the right is a sum of integrals of the type

$$\int_{t_{i-1}}^{t_i} f(x,t)\, dt$$

with continuous integrands; hence it is a continuous function of x so that the change in its value is small when the change Δx in x is small. But $|R(x,T)|$ and $|R(x + \Delta x, T)|$ are small for all x and $x + \Delta x$ in the interval $(0,c)$ as long as T is taken sufficiently large, because of the uniform convergence of the integral (4). Thus by taking T large first, then $|\Delta x|$ small, the continuity of $q(x)$ follows; that is, $|q(x + \Delta x) - q(x)|$ is arbitrarily small when $|\Delta x|$ is sufficiently small.

The proof of property (5) follows the plan commonly used in advanced calculus, where $f(x,t)$ is generally assumed to be continuous. The details are left to Prob. 11 at the end of Sec. 17.

Property (6) can be established in the following manner. Because of the uniform convergence of the integral on the right in Eq. (6) and the continuity properties of $\partial f/\partial x$ and f, for each $r(0 < r < c)$ we can write

$$(7) \qquad \int_0^r \int_0^\infty \frac{\partial f}{\partial x}\, dt\, dx = \int_0^\infty \int_0^r \frac{\partial f}{\partial x}\, dx\, dt$$

$$= \int_0^\infty f(r,t)\, dt - \int_0^\infty f(0,t)\, dt,$$

where we have used property (5) to invert the order of integration. Since $\int_0^\infty (\partial f/\partial x)\, dt$ is a continuous function of x, differentiation of the first and last members of Eqs. (7) with respect to r leads to the equation

$$\int_0^\infty \frac{\partial}{\partial x} f(x,t)\bigg]_{x=r} dt = \frac{d}{dr} \int_0^\infty f(r,t)\, dt \qquad (0 < r < c),$$

which is the same as Eq. (6) with x replaced by r.

16 CONVOLUTION

We now determine the operation on two functions of t that corresponds to multiplying their transforms together. This convolution operation, which gives the inverse transform of the product of two transforms directly in terms of the original functions, is one of primary importance in operational mathematics.

Let $F(t)$ and $G(t)$ denote any two functions that are sectionally continuous on each finite interval $0 \leq t \leq T$ and $\mathcal{O}(e^{\alpha t})$, and write

$$f(s) = L\{F(t)\}, \qquad g(s) = L\{G(t)\} \qquad\qquad (s > \alpha).$$

It will be convenient to define $G(t)$ to be zero when $t < 0$; but our final result depends only on the values of $F(t)$ and $G(t)$ when $t > 0$. According to Theorem 1, for each fixed $\tau(\tau \geq 0)$,

$$e^{-s\tau} g(s) = L\{G(t - \tau)\} = \int_0^\infty e^{-st} G(t - \tau)\, dt,$$

where $s > \alpha$. Hence

$$f(s)g(s) = \int_0^\infty F(\tau)e^{-s\tau}g(s)\,d\tau = \int_0^\infty F(\tau)\int_0^\infty e^{-st}G(t-\tau)\,dt\,d\tau;$$

that is,

(1) $$f(s)g(s) = \lim_{T\to\infty}\int_0^T\int_0^\infty F(\tau)e^{-st}G(t-\tau)\,dt\,d\tau.$$

Since $f(s)$ and $g(s)$ exist when $s > \alpha$, the limit here exists.

The integrand of the inner integral in Eq. (1) is continuous by subregions (Sec. 14), over the rectangular region $0 \leq \tau \leq T, 0 \leq t \leq R$, for each pair of positive constants T and R. For if the jumps of $F(\tau)$ and $G(t)$ occur at $\tau = \tau_i (i = 1, 2, \dots, m)$ and $t = t_j (j = 1, 2, \dots, n)$, then the jumps of the integrand occur at the lines $\tau = \tau_i$ and $t = \tau + t_j$. In view of Eq. (5), Sec. 15, the order of integration in Eq. (1) can be interchanged, provided that the improper integral there is uniformly convergent with respect to its parameter $\tau(0 \leq \tau \leq T)$.

The uniform convergence is seen by noting that, owing to the exponential order of F and G, a constant N exists such that

(2) $$|F(\tau)e^{-st}G(t-\tau)| < Ne^{\alpha\tau}e^{-st}e^{\alpha(t-\tau)} = M(t),$$

where $M(t) = N\exp[-(s-\alpha)t]$. The function $M(t)$ satisfies the conditions of the Weierstrass test (Sec. 15); that is, $M(t)$ is independent of τ and integrable from zero to infinity.

Equation (1) can now be written in the form

(3) $$f(s)g(s) = \lim_{T\to\infty}\int_0^\infty e^{-st}\int_0^T F(\tau)G(t-\tau)\,d\tau\,dt$$

$$= \lim_{T\to\infty}[I_1(T) + I_2(T)],$$

where $$I_1(T) = \int_0^T e^{-st}\int_0^T F(\tau)G(t-\tau)\,d\tau\,dt,$$

$$I_2(T) = \int_T^\infty e^{-st}\int_0^T F(\tau)G(t-\tau)\,d\tau\,dt.$$

In view of condition (2)

$$|I_2(T)| < N\int_T^\infty e^{-(s-\alpha)t}\int_0^T d\tau\,dt = \frac{NT}{s-\alpha}e^{-(s-\alpha)T};$$

therefore $$\lim_{T\to\infty}I_2(T) = 0.$$

The region of integration for the integral $I_1(T)$ is the square $0 \leq \tau \leq T$, $0 \leq t \leq T$. But $G(t - \tau) = 0$ when $\tau > t$. Therefore

$$I_1(T) = \int_0^T e^{-st} \int_0^t F(\tau)G(t - \tau)\, d\tau\, dt$$

and Eq. (3) reduces to

(4) $$f(s)g(s) = \lim_{T \to \infty} I_1(T) = \int_0^\infty e^{-st} \int_0^t F(\tau)G(t - \tau)\, d\tau\, dt.$$

The *convolution* $F * G$ of the functions $F(t)$ and $G(t)$ is defined as the function

(5) $$F(t) * G(t) = \int_0^t F(\tau)G(t - \tau)\, d\tau,$$

so that Eq. (4) can be written

(6) $$f(s)g(s) = L\{F(t) * G(t)\}.$$

We summarize our result as follows.

Theorem 3   *If $f(s)$ and $g(s)$ are the transforms of two functions $F(t)$ and $G(t)$ that are sectionally continuous on each interval $0 \leq t \leq T$ and of the order of $e^{\alpha t}$ as t tends to infinity, then the transform of the convolution $F(t) * G(t)$ exists when $s > \alpha$; it is $f(s)g(s)$. Thus the inverse transform of the product $f(s)g(s)$ is given by the formula*

(7) $$L^{-1}\{f(s)g(s)\} = F(t) * G(t).$$

The functions $F(t) = t$ and $G(t) = e^{at}$, for example, satisfy the conditions of Theorem 3. Consequently

$$L^{-1}\left\{\frac{1}{s^2} \frac{1}{s - a}\right\} = t * e^{at} = \int_0^t \tau e^{a(t - \tau)}\, d\tau$$

$$= e^{at} \int_0^t \tau e^{-a\tau}\, d\tau = \frac{1}{a^2}(e^{at} - at - 1).$$

Partial fractions can also be used to obtain that result.

When $G(t) = F(t)$, we have the formula

(8) $$[f(s)]^2 = L\{F * F\}.$$

As an example that will be useful in finding inverse transforms with the aid of partial fractions, we note that

$$(9) \qquad L^{-1}\left\{\frac{1}{(s^2 + k^2)^2}\right\} = \frac{1}{k^2} \sin kt * \sin kt$$

$$= \frac{1}{k^2} \int_0^t \sin k\tau \sin k(t - \tau)\, d\tau$$

$$= \frac{1}{2k^3}(\sin kt - kt \cos kt).$$

The conditions stated in Theorem 3 are narrower than necessary for the validity of formula (7). If $F(t) = t^{-\frac{1}{2}}$ for example, F is not sectionally continuous on an interval $0 < t < T$, but the Laplace integral of $F(t)$ is absolutely convergent and formula (7) is still valid if $G(t)$ satisfies the conditions stated in the theorem.

Thus $\qquad L^{-1}\left\{\frac{1}{\sqrt{s(s-1)}}\right\} = \frac{1}{\sqrt{\pi t}} * e^t = \frac{2e^t}{\sqrt{\pi}}\int_0^t e^{-\tau}\frac{d\tau}{2\sqrt{\tau}}.$

If we make the substitution $r = \sqrt{\tau}$ here, we find that

$$(10) \qquad L^{-1}\left\{\frac{1}{\sqrt{s(s-1)}}\right\} = \frac{2e^t}{\sqrt{\pi}}\int_0^{\sqrt{t}} e^{-r^2}\, dr = e^t \operatorname{erf}(\sqrt{t}),$$

where the *error function* erf(x), also called the *probability integral*, is a tabulated function defined by the equation

$$(11) \qquad \operatorname{erf}(x) = \frac{2}{\sqrt{\pi}}\int_0^x e^{-r^2}\, dr.$$

It was shown in Prob. 10 at the end of Sec. 5 that $\operatorname{erf}(\infty) = 1$.

Since the substitution of $s + 1$ for s in a transform corresponds to multiplication of the object function by e^{-t}, it follows from Eq. (10) that

$$(12) \qquad L^{-1}\left\{\frac{1}{s\sqrt{s+1}}\right\} = \operatorname{erf}(\sqrt{t}).$$

17   PROPERTIES OF CONVOLUTION

By substituting the new variable of integration $\lambda = t - \tau$ in the convolution integral (5), Sec. 16, we find that *the convolution operation is commutative;* that is,

$$(1) \qquad F(t) * G(t) = G(t) * F(t) = \int_0^t F(t - \lambda)G(\lambda)\, d\lambda.$$

The operation is clearly *distributive* with respect to addition:

(2) $$F(t) * [G(t) + H(t)] = F(t) * G(t) + F(t) * H(t).$$

Also, $F * (kG) = k(F * G)$ if k is a constant.

Properties (1) and (2) are valid whenever the functions are sectionally continuous over an interval $(0, T)$ that contains the point t. For such functions *the operation is also associative*:

(3) $$F(t) * [G(t) * H(t)] = [F(t) * G(t)] * H(t).$$

To prove this we first write, in view of Eq. (1),

$$F * (G * H) = \int_0^t F(\tau) \int_0^{t-\tau} G(t - \tau - \lambda)H(\lambda) \, d\lambda \, d\tau$$

and observe that the iterated integral here represents an integration over the triangular region shown in Fig. 8. When the order of integration is reversed, the integral becomes

$$\int_0^t H(\lambda) \int_0^{t-\lambda} F(\tau)G(t - \lambda - \tau) \, d\tau \, d\lambda,$$

which represents $H * (F * G)$ or $(F * G) * H$. Note that for each fixed value of t the integrand of the iterated integral is a function of τ and λ that is continuous by subregions over the region $0 \leq \lambda \leq t - \tau, 0 \leq \tau \leq t$, because of the sectional continuity of the functions F, G, and H. The interchange of order of integration is therefore justified (Sec. 14).

The convolution of a sectionally continuous function $F(t)$ and our unit step function $S_k(t)$ can be written

(4) $$F(t) * S_k(t) = \int_0^t S_k(\tau)F(t - \tau) \, d\tau = 0 \qquad \text{when } 0 \leq t \leq k,$$

$$= \int_k^t F(t - \tau) \, d\tau = \int_0^{t-k} F(\lambda) \, d\lambda \qquad \text{when } t \geq k.$$

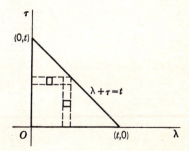

Fig. 8

Since the integral of a sectionally continuous function $F(\lambda)$ is a continuous function of its upper limit when the lower limit is a constant (Sec. 14) and since the function (4) is continuous at $t = k$, it follows that $F * S_k$ is continuous on the interval $0 \leq t \leq T$ if $F(t)$ is sectionally continuous there. A proof, based on the continuity of $F * S_k$, that *the convolution of any pair of sectionally continuous functions is a continuous function*, will be left to the problems.

If two functions $F(t)$ and $G(t)$ are sectionally continuous on each interval $(0,T)$ and of the order of $e^{\alpha t}$ as t tends to infinity, then for some constant M_1

$$|F * G| < M_1 \int_0^t e^{\alpha \tau} e^{\alpha(t - \tau)} \, d\tau = M_1 t e^{\alpha t} \qquad (t > 0).$$

But if ϵ is any positive number and if M_2 represents the maximum value of the function $t e^{-\epsilon t}$ when $t \geq 0$, then

$$M_1 t e^{\alpha t} = M_1 t e^{-\epsilon t} e^{(\alpha + \epsilon)t} \leq M e^{(\alpha + \epsilon)t},$$

where $M = M_1 M_2$. Consequently *$F * G$ is of exponential order*:

(5) $$|F(t) * G(t)| < M e^{(\alpha + \epsilon)t} \qquad (\epsilon > 0, t > 0).$$

Now, if three functions $F(t)$, $G(t)$, and $H(t)$ satisfy the conditions of Theorem 3, then the function $G * H$ satisfies those conditions and its transform is $g(s)h(s)$. Therefore

$$f(s)g(s)h(s) = L\{F(t) * [G(t) * H(t)]\};$$

but since the convolution operation is associative, this equation can be written in the form

(6) $$f(s)g(s)h(s) = L\{F(t) * G(t) * H(t)\}.$$

A similar formula can be written for the product of n transforms.

When $G(t) = 1$ in Theorem 3, we have a result noted earlier (Sec. 5):

Theorem 4   *Division of the transform of a function by s corresponds to integration of the function between the limits 0 and t:*

(7) $$L^{-1}\left\{\frac{1}{s}f(s)\right\} = \int_0^t F(\tau) \, d\tau,$$

(8) $$L^{-1}\left\{\frac{1}{s^2}f(s)\right\} = \int_0^t \int_0^\tau F(\lambda) \, d\lambda \, d\tau,$$

etc., for division by s^n, provided $F(t)$ is sectionally continuous and of the order of $e^{\alpha t}$ $(\alpha > 0)$, where $s > \alpha$.

As examples, we note that

$$L^{-1}\left\{\frac{k}{s(s^2 + k^2)}\right\} = \int_0^t \sin k\tau \, d\tau = \frac{1}{k}(1 - \cos kt),$$

$$L^{-1}\left\{\frac{k}{s^2(s^2 + k^2)}\right\} = \int_0^t \int_0^\tau \sin k\lambda \, d\lambda \, d\tau = \frac{1}{k^2}(kt - \sin kt).$$

PROBLEMS

1. Use transforms to show that

(a) $1 * 1 * 1 = \frac{1}{2}t^2$;   (b) $t * t * t = \dfrac{t^5}{5!}$;

(c) $\displaystyle\int_0^t \tau^2 \cos(t - \tau) \, d\tau = 2t - 2 \sin t.$

2. Show that an alternate form of the transformation (8) Sec. 17, is

$$L^{-1}\left\{\frac{1}{s^2}f(s)\right\} = t * F(t) = t \int_0^t F(\tau) \, d\tau - \int_0^t \tau F(\tau) \, d\tau.$$

3. Show that

$$\frac{1}{\sqrt{s} - 1} = \frac{1}{s - 1} + \frac{1}{\sqrt{s}}\left(1 + \frac{1}{s - 1}\right)$$

and in view of formula (10), Sec. 16, that

$$L^{-1}\left\{\frac{1}{\sqrt{s} - 1}\right\} = e^t + \frac{1}{\sqrt{\pi t}} + e^t \operatorname{erf}(\sqrt{t}).$$

4. Generalize formula (10), Sec. 16, by replacing s by s/a^2 to obtain transform 40, Appendix A, Table A.2.

5. Generalize formula (12), Sec. 16, to obtain transform 44, Appendix A, Table A.2.

6. Find $L^{-1}\{(s + 1)^{-1}s^{-\frac{1}{2}}\}$ and generalize to obtain transform 41, Appendix A, Table A.2.

7. Obtain the inverse transform

$$L^{-1}\left\{\frac{s^{-\frac{3}{2}}}{s - 1}\right\} = e^t \operatorname{erf}(\sqrt{t}) - 2\sqrt{\frac{t}{\pi}}.$$

8. With the aid of transformation (9), Sec. 16, show that

$$L^{-1}\left\{\frac{8}{(s^2 + 1)^3}\right\} = 3(\sin t - t \cos t) - t^2 \sin t.$$

9. Let $t = h$ and $t = k$ be the points of discontinuity of a sectionally continuous function $G(t)$ on an interval $0 < t < T$, and let j_h and j_k denote the jumps in the value of $G(t)$ at those points. With the aid of a graph of $G(t)$, note that the function

$$G_c(t) = G(t) - j_h S_h(t) - j_k S_k(t) \qquad\qquad (0 \leq t \leq T)$$

is continuous if its values at $t = h$ and $t = k$ are properly defined. Write $F * G$ in terms of $F * G_c$ and of convolutions like the one in Eq. (4), Sec. 17, to show that if $F(t)$ is sectionally continuous, then $F * G$ is continuous. Also note how the result can be generalized to permit $G(t)$ to have n jumps in the interval.

10. Show that the integral

$$u(x) = \int_0^\infty \frac{x}{1 + x^2 t^2}\, dt$$

has the values $u(x) = -\pi/2$ when $x < 0$, $u(x) = \pi/2$ when $x > 0$ and $u(0) = 0$. Examine the remainder $R(x,T)$ for this improper integral and show that the integral is not uniformly convergent with respect to x over any interval that contains the point $x = 0$ in its interior or as an end point.

11. Prove property (5), Sec. 15, under the conditions stated there. This can be done by first showing that for each positive number T

$$\int_0^c q\, dx = \int_0^c \int_0^T f\, dt\, dx + \int_0^c R(x,T)\, dx$$

and hence that, corresponding to each positive number ϵ, a number T_ϵ exists such that when $T > T_\epsilon$

$$\left| \int_0^c q\, dx - \int_0^T \int_0^c f\, dx\, dt \right| < c\epsilon.$$

12. Extend property (5), Sec. 15, for the interchange of order of integration of definite and improper integrals, to the following case:

$$\int_b^c \int_x^\infty \phi(x,t)\, dt\, dx = \int_b^c \int_b^t \phi(x,t)\, dx\, dt + \int_c^\infty \int_b^c \phi(x,t)\, dx\, dt$$

where $b \geq 0$ and ϕ is continuous by subregions over the region $b < x < c$, $x < t < T$, whenever $T > c$, and $\int_c^\infty \phi\, dt$ is uniformly convergent with respect to $x (b < x < c)$. *Suggestions:* Show the region graphically. Introduce a function f to which formula (5), Sec. 15, applies by writing

$$f(x,t) = \phi(x,t) \qquad \text{when } b < x < c \text{ and } t > x,$$

$$f(x,t) = 0 \qquad \text{when either } x < b \text{ or } t < x.$$

18   DIFFERENTIAL AND INTEGRAL EQUATIONS

General nonhomogeneous linear differential equations with constant coefficients can be solved with the aid of the convolution property.

Example 1   Find the general solution of the differential equation

(1) $Y''(t) + k^2 Y(t) = F(t)$

in terms of the constant k and the function $F(t)$.

   We assume for the present that $F(t)$ is sectionally continuous and of exponential order, that the derivative $Y''(t)$ of the unknown function satisfies those conditions, and that $Y'(t)$ is continuous and of exponential order. Then the equation in the transforms of $Y(t)$ and $F(t)$ can be written

$$s^2 y(s) - s Y(0) - Y'(0) + k^2 y(s) = f(s),$$

where $Y(0)$ and $Y'(0)$ are arbitrary constants. Hence

$$y(s) = \frac{1}{k} \frac{k}{s^2 + k^2} f(s) + Y(0) \frac{s}{s^2 + k^2} + \frac{Y'(0)}{k} \frac{k}{s^2 + k^2},$$

and therefore

$$Y(t) = \frac{1}{k} (\sin kt) * F(t) + Y(0) \cos kt + \frac{Y'(0)}{k} \sin kt.$$

This general solution of Eq. (1) can be written

$$(2) \quad Y(t) = \frac{1}{k} \int_0^t \sin k(t - \tau) F(\tau) \, d\tau + C_1 \cos kt + C_2 \sin kt,$$

where C_1 and C_2 are arbitrary constants.

The function $C_1 \cos kt + C_2 \sin kt$ is the general solution of the homogeneous equation that arises when $F(t) = 0$ in Eq. (1). To verify the solution (2), it is therefore sufficient to show that the first term on the right is a solution of Eq. (1). If $F(\tau)$ has jumps at $\tau = t_1, t_2, \ldots, t_n$ when $0 < \tau < t$, that term can be written

$$Z(t) = \frac{1}{k} \int_0^{t_1} \sin k(t - \tau) F(\tau) \, d\tau + \cdots + \frac{1}{k} \int_{t_n}^t \sin k(t - \tau) F(\tau) \, d\tau.$$

The Leibnitz formula (2), Sec. 14, applies to each of the integrals in this equation; thus

$$Z'(t) = \int_0^{t_1} \cos k(t - \tau) F(\tau) \, d\tau + \cdots + \int_{t_n}^t \cos k(t - \tau) F(\tau) \, d\tau + 0.$$

By applying that formula again to the integrals here and collecting the resulting integrals, we find that

$$Z''(t) = -k \int_0^t \sin k(t - \tau) F(\tau) \, d\tau + F(t)$$

or $Z''(t) = -k^2 Z(t) + F(t)$. Therefore $Z(t)$ satisfies Eq. (1), and the function (2) is verified as the general solution of that equation for all values of t.

The function $Y(t)$ given by Eq. (2) is continuous, together with $Y'(t)$, when $F(t)$ is sectionally continuous, in view of the above expressions for $Z(t)$ and $Z'(t)$. The condition of exponential order on

$F(t)$ was not used in verifying the solution (2); the condition that $F(t)$ be sectionally continuous, on an interval containing all values of t to be considered, was sufficient.

An equation in which the unknown function occurs inside an integral is called an *integral equation*. In certain applied problems, to be illustrated in the following chapter, the integral in the equation is a convolution integral. Such integral equations of the convolution type transform into algebraic equations.

Example 2   Solve the integral equation

$$Y(t) = at + \int_0^t Y(\tau) \sin(t - \tau)\, d\tau.$$

We can write this equation in the form

$$Y(t) = at + Y(t) * \sin t$$

and transform both members to get the algebraic equation

$$y(s) = \frac{a}{s^2} + y(s)\frac{1}{s^2 + 1},$$

whose solution is $$y(s) = a\left(\frac{1}{s^2} + \frac{1}{s^4}\right).$$

Therefore $$Y(t) = a(t + \tfrac{1}{6}t^3),$$

which can be verified directly as the solution of the above integral equation.

The general *integral equation of convolution type* has the form

(3) $$Y(t) = F(t) + \int_0^t G(t - \tau)Y(\tau)\, d\tau,$$

where the functions $F(t)$ and $G(t)$ are given and $Y(t)$ is to be found. Since the transformed equation is

$$y(s) = f(s) + g(s)y(s),$$

the transform of the unknown function is

(4) $$y(s) = \frac{f(s)}{1 - g(s)}.$$

Even if Eq. (3) is modified by replacing $Y(t)$ by linear combinations of $Y(t)$ and its derivatives, the transform of the modified equation is an algebraic equation in $y(s)$. For instance, the integrodifferential equation

(5)
$$aY(t) + bY'(t) = F(t) + \int_0^t G(t - \tau)Y(\tau)\,d\tau,$$

where a and b are constants, gives rise to the transformed equation

$$(a + bs)y(s) - bY(0) = f(s) + g(s)y(s),$$

which is easily solved for $y(s)$.

The equation

(6)
$$F(t) = \int_0^t (t - \tau)^{-b}Y'(\tau)\,d\tau \qquad (0 < b < 1),$$

is known as *Abel's integral equation*. The unknown function could of course be considered here as $Y'(t)$ instead of $Y(t)$; but no advantage is gained by doing so. The solution of that equation is

(7)
$$Y(t) = Y(0) + \frac{1}{\Gamma(b)\Gamma(1 - b)}t^{b-1} * F(t),$$

valid when $F(t)$ satisfies certain conditions of continuity. The formal derivation of this solution is left to the problems.

PROBLEMS

Solve the following differential equations.

1. $Y''(t) - k^2Y(t) = F(t)$, if $Y(0) = Y'(0) = 0 (k \neq 0)$.

Ans. $2kY(t) = e^{kt}\int_0^t e^{-k\tau}F(\tau)\,d\tau - e^{-kt}\int_0^t e^{k\tau}F(\tau)\,d\tau.$

2. $Y''(t) - 2kY'(t) + k^2Y(t) = F(t)$.

Ans. $e^{-kt}Y(t) = C_1 + C_2t + \int_0^t (t - \tau)e^{-k\tau}F(\tau)\,d\tau.$

3. $2Y''(t) - 3Y'(t) - 2Y(t) = F(t)$, if $Y(0) = Y'(0) = 0$. Verify that your function satisfies the differential equation and the initial conditions.

4. $Y''(t) + 4Y'(t) + 5Y(t) = F(t)$, if $Y(0) = Y(\pi/2) = 0$.

Ans. $e^{2t}Y(t) = \sin t \int_{\pi/2}^t F(\tau)e^{2\tau}\cos \tau\,d\tau - \cos t \int_0^t F(\tau)e^{2\tau}\sin \tau\,d\tau.$

5. $Y'''(t) - Y'(t) = F(t)$. Verify your result.

6. $Y'''(t) - Y''(t) + Y'(t) - Y(t) = F(t)$, if $Y(0) = Y'(0) = Y''(0) = 0$.

Ans. $2Y(t) = \int_0^t F(t - \tau)(e^\tau - \cos \tau - \sin \tau)\,d\tau.$

7. Show that the solution of the system of differential equations

$$X'(t) - 2Y'(t) = F(t), \qquad X''(t) - Y''(t) + Y(t) = 0,$$

under the conditions $X(0) = X'(0) = Y(0) = Y'(0) = 0$, so that $F(0) = 0$, is

$$X(t) = \int_0^t F(\tau)\,d\tau - 2\int_0^t F(\tau)\cos (t - \tau)\,d\tau,$$

$$Y(t) = -\int_0^t F(\tau)\cos (t - \tau)\,d\tau.$$

8. Solve the following system and verify your result:

$$X'(t) + Y(t) = F(t), \qquad Y'(t) + X(t) = 1, \qquad X(0) = 1, \qquad Y(0) = 0.$$

9. Solve for $Y(t)$ and verify your solution:

$$\int_0^t Y(\tau)\,d\tau - Y'(t) = t, \qquad Y(0) = 2.$$

Ans. $Y(t) = 1 + \cosh t$.

10. Derive the solution (7) of Abel's integral equation (6).

11. Solve the integral equation

$$Y(t) = a \sin t - 2 \int_0^t Y(\tau) \cos (t - \tau)\,d\tau. \qquad Ans.\ Y(t) = ate^{-t}.$$

12. Find the solution of the integral equation

$$Y(t) = a \sin bt + c \int_0^t Y(\tau) \sin b(t - \tau)\,d\tau$$

(a) when $b^2 > bc$; (b) when $b = c$.

Ans. (a) $Y(t) = ab(b^2 - bc)^{-\frac{1}{2}} \sin (t\sqrt{b^2 - bc})$;   (b) $Y(t) = abt$.

13. Solve the nonlinear integral equation

$$2Y(t) + \int_0^t Y(\tau)Y(t - \tau)\,d\tau = t + 2. \qquad Ans.\ Y(t) = 1.$$

14. Write $Y''(t) = V(t)$. (a) If $Y(0) = a$ and $Y'(0) = b$ show that

$$Y'(t) = 1 * V(t) + b, \qquad Y(t) = t * V(t) + bt + a;$$

(b) then convert the initial-value problem

$$Y''(t) + \alpha(t)Y'(t) + \beta(t)Y(t) = F(t), \qquad Y(0) = Y'(0) = 0,$$

in $Y(t)$ into the following integral equation in $V(t)$:

$$V(t) + \int_0^t [\alpha(t) + \beta(t)(t - \tau)]V(\tau)\,d\tau = F(t).$$

19 DERIVATIVES OF TRANSFORMS

When the Laplace integral

$$(1) \qquad\qquad f(s) = \int_0^\infty e^{-st}F(t)\,dt$$

is formally differentiated with respect to the parameter s by carrying out the differentiation inside the integral sign, the formula

$$f'(s) = \int_0^\infty e^{-st}(-t)F(t)\,dt = L\{-tF(t)\}$$

is obtained. Another formal manipulation with the integral (1) indicates that $f(s) \to 0$ as $s \to \infty$. We shall establish conditions under which those formulas are valid.

First, we note that if $F(t)$ is of the order of $e^{\alpha t}$ as $t \to \infty$, then the function $t^n F(t)$, where $n = 0, 1, 2, \ldots$, is of exponential order. Let ϵ be a positive number. Then constants N_1 and N_2 exist such that

$$|t^n F(t)| < t^n N_1 e^{\alpha t} = N_1 t^n e^{-\epsilon t} e^{(\alpha + \epsilon)t} \leqq N_1 N_2 e^{(\alpha + \epsilon)t} \qquad (t > 0),$$

where N_2 represents the maximum value of the function $t^n e^{-\epsilon t}$ when $t > 0$. Thus the function $t^n F(t)$ is of the order of $e^{\alpha_0 t}$, where $\alpha_0 = \alpha + \epsilon$.

Also let $F(t)$ be sectionally continuous on each interval $0 < t < T$. Then $t^n F(t)$ has that property. The absolute value of the integrand of the Laplace integral of $t^n F(t)$ satisfies the condition

$$(2) \qquad\qquad |t^n F(t)e^{-st}| < Ne^{-(s-\alpha_0)t} \qquad\qquad (n = 0, 1, 2, \ldots),$$

where N denotes a constant. Consequently when $s > \alpha_0$,

$$(3) \qquad \left| \int_0^\infty t^n F(t)e^{-st}\, dt \right| < N \int_0^\infty e^{-(s-\alpha_0)t}\, dt = \frac{N}{s - \alpha_0};$$

therefore $L\{t^n F(t)\} \to 0$ as $s \to \infty$. Moreover, if $s \geqq \alpha_1$ where $\alpha_1 > \alpha_0$, then according to condition (2),

$$|t^n F(t)e^{-st}| < Ne^{-(\alpha_1 - \alpha_0)t} = M(t),$$

where this exponential function $M(t)$ is independent of s and integrable from zero to infinity. It follows from the Weierstrass test (Sec. 15) that the Laplace integral

$$\int_0^\infty t^n F(t)e^{-st}\, dt \qquad\qquad (n = 0, 1, 2, \ldots)$$

converges uniformly with respect to s when $s \geqq \alpha_1 > \alpha_0$.

Theorem 5 *If $F(t)$ is sectionally continuous and of the order of $e^{\alpha t}$, then each of the Laplace integrals $L\{F(t)\}, L\{tF(t)\}, L\{t^2 F(t)\}, \ldots$, is uniformly convergent when $s \geqq \alpha_1$ where $\alpha_1 > \alpha$; moreover*

$$(4) \qquad \lim_{s \to \infty} f(s) = 0 \quad \text{and} \quad \lim_{s \to \infty} L\{t^n F(t)\} = 0 \quad (n = 1, 2, \ldots).$$

The function $F(t)e^{-st}$ and its partial derivative of each order, with respect to s, satisfy our conditions for the validity of formula (6), Sec. 15. Hence differentiation with respect to s can be performed in Eq. (1) inside the integral sign, and the following theorem is established.

Theorem 6 *Differentiation of the transform of a function corresponds to the multiplication of the function by* $-t$:

$$(5) \qquad\qquad f^{(n)}(s) = L\{(-t)^n F(t)\} \qquad\qquad (n = 1, 2, \ldots);$$

moreover, $f^{(n)}(s) \to 0$ *as* $s \to \infty$. *Those properties hold true whenever* $F(t)$ *is sectionally continuous and of the order of* $e^{\alpha t}$, *if* $s > \alpha$ *in formula* (5).

Since a function is continuous wherever its derivative exists, it is true that $f(s)$ *and each of its derivatives is continuous when* $s > \alpha$.

To illustrate the last theorem, we can note that since

$$\frac{k}{s^2 + k^2} = L\{\sin kt\} \qquad\qquad (s > 0),$$

it follows that

$$\frac{-2ks}{(s^2 + k^2)^2} = L\{-t \sin kt\}.$$

Thus we have a formula that is useful in finding inverse transforms with the aid of partial fractions:

$$(6) \qquad\qquad L\{t \sin kt\} = \frac{2ks}{(s^2 + k^2)^2} \qquad\qquad (s > 0).$$

Transformation (6) can be obtained by another method that is sometimes useful, that of *differentiating a Laplace integral with respect to a parameter* k, independent of s. When $s > 0$, the Laplace integral

$$(7) \qquad\qquad L\{\cos kt\} = \int_0^\infty e^{-st} \cos kt \, dt = \frac{s}{s^2 + k^2}$$

converges, and the integral

$$(8) \qquad\qquad L\left\{\frac{\partial}{\partial k} \cos kt\right\} = \int_0^\infty \frac{\partial}{\partial k}(e^{-st} \cos kt) \, dt$$

$$= -\int_0^\infty e^{-st} t \sin kt \, dt$$

converges uniformly with respect to k for all real k, according to the Weierstrass test. All integrands here are continuous functions of k and t. Hence formula (6), Sec. 15, applies to show that the integral (8) represents the derivative, with respect to k, of the integral (7); that is,

$$-L\{t \sin kt\} = \frac{\partial}{\partial k}\left(\frac{s}{s^2 + k^2}\right) = -\frac{2ks}{(s^2 + k^2)^2},$$

and this is the transformation (6).

20　SERIES OF TRANSFORMS

A useful method of finding inverse transforms $L^{-1}\{f(s)\}$ is that of representing the function f by an infinite series of known transforms, then applying the operator L^{-1} to the terms of the series. For certain types of series in powers of $1/s$, conditions for the validity of the procedure will now be established.

Theorem 7　*Let $f(s)$ denote the sum of an infinite series of positive integral powers of $1/s$ which is absolutely convergent when $s > \alpha$, where $\alpha \geq 0$:*

$$(1) \qquad f(s) = \sum_{n=0}^{\infty} a_n \frac{1}{s^{n+1}} = \frac{a_0}{s} + \frac{a_1}{s^2} + \cdots \qquad (s > \alpha \geq 0).$$

Then the power series in t obtained by applying the operator L^{-1} to that series term by term converges to a function $F(t)$ whose transform is $f(s)$, when $s > \alpha$:

$$(2) \qquad F(t) = \sum_{n=0}^{\infty} a_n \frac{t^n}{n!} = L^{-1}\{f(s)\} \qquad (t \geq 0),$$

where $0! = 1$. Also, F is continuous when $t \geq 0$, and of exponential order $\mathcal{O}(e^{\alpha_1 t})$ whenever $\alpha_1 > \alpha$.

Since series (1) is absolutely convergent when $s \geq \alpha_1$ whenever $\alpha_1 > \alpha \geq 0$, it follows that $|a_n|/s^{n+1} \leq |a_n|/\alpha_1^{n+1} \to 0$ as $n \to \infty$. Therefore a constant M, independent of s, exists such that $|a_n|/s^{n+1} < M$; that is,

$$(3) \qquad |a_n| < Ms^{n+1} \qquad (n = 0, 1, 2, \ldots; s \geq \alpha_1 > \alpha \geq 0).$$

If we write $s = \alpha_1$ here it follows that, when $t \geq 0$,

$$(4) \qquad \left| a_n \frac{t^n}{n!} \right| = |a_n| \frac{t^n}{n!} \leq M\alpha_1 \frac{(\alpha_1 t)^n}{n!}.$$

The series of terms $(\alpha_1 t)^n/n!$ converges to $e^{\alpha_1 t}$, so by the comparison test series (2) is absolutely convergent, and its sum is of exponential order,

$$|F(t)| \leq \sum_{n=0}^{\infty} |a_n| \frac{t^n}{n!} \leq M\alpha_1 e^{\alpha_1 t} \qquad (t \geq 0, \alpha_1 > \alpha).$$

Also, F is continuous because it is the sum of a convergent power series in t.

Let s be fixed $(s > \alpha)$ and let T be a positive number. The series of the terms $a_n e^{-st} t^n/n!$ is uniformly convergent with respect to t over the interval $0 < t < T$ according to the Weierstrass test for infinite series of functions, for in view of condition (3),

$$|a_n| e^{-st} \frac{t^n}{n!} < Ms \frac{(sT)^n}{n!} \qquad (s > \alpha, 0 < t < T);$$

the numbers on the right are independent of t, and they are terms of a convergent series. The series of the continuous functions $a_n e^{-st} t^n / n!$ can therefore be integrated term by term over the bounded interval $(0,T)$ (cf. Prob. 15, Sec. 21); thus

$$\int_0^T e^{-st} F(t)\, dt = \sum_{n=0}^{\infty} \frac{a_n}{n!} \int_0^T e^{-st} t^n \, dt$$

$$= \sum_{n=0}^{\infty} \frac{a_n}{n!} \left[L\{t^n\} - \int_T^{\infty} e^{-st} t^n \, dt \right]$$

$$= \sum_{n=0}^{\infty} \left(\frac{a_n}{s^{n+1}} - \frac{a_n}{n!} \int_T^{\infty} e^{-st} t^n \, dt \right).$$

By subtracting the last series from series (1) which converges to $f(s)$, we find that

$$(5) \qquad f(s) - \int_0^T e^{-st} F(t)\, dt = \sum_{n=0}^{\infty} \frac{a_n}{n!} \int_T^{\infty} e^{-st} t^n \, dt.$$

We are to prove that the difference (5) vanishes as $T \to \infty$; that is, to each positive number ϵ we are to exhibit a corresponding number T_ϵ such that

$$(6) \qquad \left| f(s) - \int_0^T e^{-st} F(t)\, dt \right| < \epsilon \qquad \text{whenever } T > T_\epsilon.$$

It will follow that $f(s) = L\{F(t)\}$; then Theorem 7 will be proved.

The series in Eq. (5) is absolutely convergent because

$$(7) \qquad \frac{|a_n|}{n!} \int_T^{\infty} e^{-st} t^n \, dt \leqq \frac{|a_n|}{n!} L\{t^n\} = \frac{|a_n|}{s^{n+1}}$$

and series (1) is absolutely convergent. Hence if $N = 1, 2, \ldots,$

$$(8) \qquad \left| f(s) - \int_0^T e^{-st} F(t)\, dt \right| \leqq \sum_{n=0}^{N-1} \frac{|a_n|}{n!} \int_T^{\infty} e^{-st} t^n \, dt + R_N(T)$$

where, in view of condition (7),

$$R_N(T) = \sum_{n=N}^{\infty} \frac{|a_n|}{n!} \int_T^{\infty} e^{-st} t^n \, dt \leqq \sum_{n=N}^{\infty} \frac{|a_n|}{s^{n+1}}.$$

Since series (1) is absolutely convergent, it now follows that to each ϵ ($\epsilon > 0$) there corresponds a number N_ϵ, independent of T, such that

$$R_N(T) < \tfrac{1}{2}\epsilon \qquad \text{whenever } N > N_\epsilon.$$

Now consider the sum from $n = 0$ to $N - 1$ in condition (8). Let N be some fixed integer greater than N_ϵ. By successive integrations by parts we find that

$$\frac{1}{n!}\int_T^\infty e^{-st}t^n\,dt = \frac{e^{-sT}}{s}\left[\frac{T^n}{n!} + \frac{T^{n-1}}{(n-1)!s} + \frac{T^{n-2}}{(n-2)!s^2} + \cdots + \frac{1}{s^n}\right].$$

Since $|a_n| < Ms^{n+1}$, it follows that, if $sT > 1$ and $n \le N - 1$,

$$\frac{|a_n|}{n!}\int_T^\infty e^{-st}t^n\,dt < Me^{-sT}\left[\frac{(sT)^n}{n!} + \frac{(sT)^{n-1}}{(n-1)!} + \cdots + 1\right]$$

$$\le Me^{-sT}(sT)^n\left[\frac{1}{n!} + \frac{1}{(n-1)!} + \cdots + 1\right] < Me^{-sT}(sT)^N N.$$

Thus the sum in condition (8) is less than $MN^2e^{-sT}(sT)^N$, which vanishes as $T \to \infty$ for that fixed N. Therefore, there is a number T_ϵ such that $MN^2e^{-sT}(sT)^N < \epsilon/2$ when $T > T_\epsilon$, and condition (6) follows to complete the proof of the theorem.

Theorem 7 can be generalized to the following theorem:

Theorem 8 *Let $g(s)$ be represented by an absolutely convergent series when $s > \alpha$ ($\alpha \ge 0$) of this type:*

$$(9) \qquad\qquad g(s) = \sum_{n=0}^\infty a_{n-1}\frac{1}{s^{n+k}} \qquad (0 < k \le 1, s > \alpha \ge 0).$$

Then $L\{G(t)\} = g(s)$ when $s > \alpha$, where the function G is continuous when $t > 0$, is $\mathcal{O}(e^{\alpha_1 t})$ if $\alpha_1 > \alpha$, and is represented by the convergent series of inverse transforms:

$$(10) \qquad\qquad G(t) = \sum_{n=0}^\infty L^{-1}\left\{\frac{a_{n-1}}{s^{n+k}}\right\} = \sum_{n=0}^\infty a_{n-1}\frac{t^{n+k-1}}{\Gamma(n+k)} \qquad (t > 0).$$

A proof can be based on Theorem 7 by writing

$$g(s) = \frac{a_{-1}}{s^k} + \frac{1}{s^k}\sum_{n=1}^\infty\frac{a_{n-1}}{s^n} = \frac{a_{-1}}{s^k} + \frac{1}{s^k}\sum_{n=0}^\infty\frac{a_n}{s^{n+1}}.$$

The last series here represents the transform of the function

$$(11) \qquad\qquad F(t) = \sum_{n=0}^\infty a_n\frac{t^n}{n!},$$

a function that is continuous ($t \ge 0$), and $\mathcal{O}(e^{\alpha_1 t})$ if $\alpha_1 > \alpha$. Also we know that (Sec. 5)

$$L\{t^{k-1}\} = \frac{\Gamma(k)}{s^k} \qquad\qquad (k > 0, s > 0).$$

Since $k \leq 1$, the function t^{k-1} is bounded when $t \geq T > 0$. Then

$$L^{-1}\{g(s)\} = \frac{a_{-1}}{\Gamma(k)} t^{k-1} + \frac{t^{k-1}}{\Gamma(k)} * F(t) \qquad (0 < k \leq 1).$$

By writing the convolution here as an integral from $\tau = 0$ to $\tau = \epsilon$ plus an integral from ϵ to t, we can modify the proof in Sec. 17 to show that $t^{k-1} * F(t)$, and hence $L^{-1}\{g(s)\}$, is $\mathcal{O}(e^{\alpha_1 t})$ if $\alpha_1 > \alpha$.

In view of Eq. (11) we can write

$$(12) \qquad t^{k-1} * F(t) = \int_0^t (t - \tau)^{k-1} F(\tau) \, d\tau$$

$$= \sum_{n=0}^{\infty} \frac{a_n}{n!} \int_0^t (t - \tau)^{k-1} \tau^n \, d\tau$$

$$= \sum_{n=0}^{\infty} \frac{a_n}{n!} t^{k-1} * t^n,$$

where the term-by-term integration is valid because the power series for $F(\tau)$ converges uniformly with respect to τ over the interval $0 < \tau < t$ (Prob. 15, Sec. 21). By using known transforms, we find easily that

$$t^{k-1} * t^n = \frac{\Gamma(k)n!}{\Gamma(n + 1 + k)} t^{n+k}$$

and therefore

$$\frac{t^{k-1}}{\Gamma(k)} * F(t) = \sum_{n=0}^{\infty} a_n \frac{t^{n+k}}{\Gamma(n + 1 + k)}.$$

It now follows that $L^{-1}\{g\}$ is the function G represented by series (10). That convergent series is the product of t^{k-1}, or $t^{-(1-k)}$, by a power series in t. Hence G is continuous whenever $t > 0$, and Theorem 8 is proved.

Example   To establish transformation 78, Appendix A, Table A.2, when $k = 1$ there, we write the absolutely convergent representation

$$g(s) = s^{-\frac{3}{2}} e^{-1/s} = \sum_{n=0}^{\infty} \frac{(-1)^n}{n!} \frac{1}{s^{n+\frac{3}{2}}} \qquad (s > 0).$$

According to Theorem 8 and the form 6, Appendix A, Table A.2, of the elementary transformation $L^{-1}\{s^{-n-\frac{1}{2}}\}$, then, $L\{G\} = g$ where

$$G(t) = \frac{1}{\sqrt{\pi}} \sum_{n=0}^{\infty} \frac{(-1)^n}{n!} \frac{2^{n+1} t^{n+\frac{1}{2}}}{(1)(3)(5) \cdots (2n + 1)}.$$

The terms of that series reduce to $(-1)^n(2\sqrt{t})^{2n+1}/(2n+1)!$ and therefore

$$G(t) = L^{-1}\left\{\frac{e^{-1/s}}{s\sqrt{s}}\right\} = \frac{1}{\sqrt{\pi}}\sin(2\sqrt{t}).$$

21 DIFFERENTIAL EQUATIONS WITH VARIABLE COEFFICIENTS

We have seen that

$$L\{t^n Y(t)\} = (-1)^n\frac{d^n}{ds^n}L\{Y(t)\} = (-1)^n y^{(n)}(s),$$

and therefore we can write the transform of the product of t^n and any derivative of $Y(t)$ in terms of $y(s)$; for instance,

$$L\{t^2 Y'(t)\} = \frac{d^2}{ds^2}[sy(s) - Y(0)] = sy''(s) + 2y'(s),$$

$$L\{tY''(t)\} = -\frac{d}{ds}[s^2 y(s) - sY(0) - Y'(0)]$$

$$= -s^2 y'(s) - 2sy(s) + Y(0).$$

A linear differential equation in $Y(t)$ whose coefficients are polynomials in t transforms into a linear differential equation in $y(s)$ whose coefficients are polynomials in s. In case the transformed equation is simpler than the original, the transformation may enable us to find the solution of the original equation.

If the coefficients are polynomials of the first degree, the transformed equation is a linear equation of the first order, whose solution can be written in terms of an integral. To find the solution of the original equation, however, the inverse transform of the solution of the new equation must be obtained.

Example 1 Find the solution of the problem

$$Y''(t) + tY'(t) - Y(t) = 0, \qquad Y(0) = 0, \qquad Y'(0) = 1.$$

The transformed equation is

$$s^2 y(s) - 1 - \frac{d}{ds}[sy(s)] - y(s) = 0,$$

or

$$y'(s) + \left(\frac{2}{s} - s\right)y(s) = -\frac{1}{s},$$

which is a linear equation of the first order. An integrating factor is

$$\exp\left[\int\left(\frac{2}{s} - s\right) ds\right] = \exp\left(2\log s - \tfrac{1}{2}s^2\right) = s^2 e^{-\frac{1}{2}s^2},$$

so the equation can be written

$$\frac{d}{ds}[s^2 e^{-\frac{1}{2}s^2} y(s)] = -s e^{-\frac{1}{2}s^2}.$$

Integrating, we have

$$y(s) = \frac{1}{s^2} + \frac{C}{s^2}e^{\frac{1}{2}s^2},$$

where C is a constant of integration. But C must vanish if $y(s)$ is a transform since $y(s)$ must vanish as s tends to infinity. It follows that

$$Y(t) = t,$$

and this is readily verified as the solution.

Example 2   Solve Bessel's equation with index zero,

$$t Y''(t) + Y'(t) + t Y(t) = 0$$

under the conditions that $Y(0) = 1$ and $Y(t)$ and its derivatives have transforms.

The point $t = 0$ is a singular point of this differential equation such that one of the solutions is a function that behaves like $\log t$ near that singular point, and the Laplace transform of the derivative of the function does not exist.

The transformed equation is

$$-\frac{d}{ds}[s^2 y(s) - s - Y'(0)] + s y(s) - 1 - \frac{d}{ds}y(s) = 0,$$

or $$(s^2 + 1)y'(s) + s y(s) = 0.$$

Separating variables, we have

$$\frac{dy}{y} = -\frac{s\, ds}{s^2 + 1},$$

and upon integrating and simplifying, we find that

(2) $$y(s) = \frac{C}{\sqrt{s^2 + 1}},$$

where C is a constant of integration.

Expanding the function for $y(s)$ by the binomial series, we have, when $s > 1$,

$$y(s) = \frac{C}{s}\left(1 + \frac{1}{s^2}\right)^{-\frac{1}{2}} = \frac{C}{s}\left[1 - \frac{1}{2}\frac{1}{s^2} - \frac{1}{2}\left(-\frac{3}{2}\right)\frac{1}{2!s^4} - \cdots\right]$$

$$= \frac{C}{s}\left[1 + \sum_{n=1}^{\infty}(-1)^n\frac{(1)(3)\cdots(2n-1)}{2^n n! s^{2n}}\right] = C\sum_{n=0}^{\infty}\frac{(-1)^n(2n)!}{(2^n n!)^2 s^{2n+1}},$$

where $0! = 1$. The ratio test shows that series in positive powers of $1/s$ are absolutely convergent when $s > 1$. Theorem 7 therefore applies to show that the operator L^{-1} can be applied term by term to that series to represent the continuous function whose transform is $y(s)$ as a convergent series in powers of t:

$$(3) \qquad L^{-1}\{y(s)\} = C\sum_{n=0}^{\infty}\frac{(-1)^n}{(2^n n!)^2}t^{2n}.$$

It is not difficult to verify that for all t the power series (3) is a solution of Bessel's equation (1). If that function is to satisfy the condition $Y(0) = 1$, then $C = 1$, and the solution (3) can be written

$$(4) \qquad Y(t) = J_0(t)$$

where J_0 is *Bessel's function* of the first kind with index zero:

$$(5) \qquad J_0(t) = \sum_{n=0}^{\infty}\frac{(-1)^n}{(n!)^2}\left(\frac{t}{2}\right)^{2n} = 1 - \frac{t^2}{2^2} + \frac{t^4}{(2^2)(4^2)} - \cdots.$$

We have shown above that, when $s > 1$,

$$(6) \qquad L\{J_0(t)\} = \frac{1}{\sqrt{s^2 + 1}},$$

a result that is actually valid whenever $s > 0$, because J_0 can be represented by an integral[1] that shows that $|J_0(t)| \leq 1$.

The differential equation

$$(7) \qquad t^2 Y''(t) + t Y'(t) + (t^2 - n^2)Y(t) = 0$$

is Bessel's equation with index n. It can be verified that the function

$$(8) \qquad J_n(t) = \sum_{k=0}^{\infty}\frac{(-1)^k}{k!(n+k)!}\left(\frac{t}{2}\right)^{n+2k} \qquad (n = 0, 1, 2, \ldots),$$

known as *Bessel's function of the first kind* with index n, is a solution of that equation. In the problems to follow we shall establish the transforms of $J_1(t)$ and $t^n J_n(t)$.

[1] Churchill, R. V., "Fourier Series and Boundary Value Problems," 2d ed., p. 175, 1963.

PROBLEMS

1. Use the transformation (6) and Theorem 2 to find $L\{J_0(at)\}$ listed in Appendix A, Table A.2 (transform 55).

2. If Y, Y', and Y'' are of exponential order and Y, Y' are continuous ($t \geq 0$) while Y'' is sectionally continuous, and if $L\{Y(t)\} = y(s)$, show that

$$L\{t^2 Y''(t)\} = s^2 y''(s) + 4s y'(s) + 2y(s).$$

Solve the following differential equations for Y if Y and its derivatives are to have transforms.

3. $Y''(t) + at Y'(t) - 2a Y(t) = 1$, $Y(0) = Y'(0) = 0$, $a > 0$.   *Ans.* $Y(t) = t^2/2$.

4. $t Y''(t) + (t - 1)Y'(t) + Y(t) = 0$, $Y(0) = 0$.   *Ans.* $Y(t) = Ct^2 e^{-t}$.

5. $t Y''(t) + (2t + 3)Y'(t) + (t + 3)Y(t) = 3e^{-t}$.   *Ans.* $Y(t) = (C + t)e^{-t}$.

6. $t^2 Y''(t) - 2Y(t) = 2t$, $Y(2) = 2$.   *Ans.* $Y(t) = t^2 - t$.

7. When k is a constant, show that the equation

$$t^2 Z''(t) + 2t Z'(t) + k Z(t) = 0$$

leads to the same differential equation in the transform $z(s)$.

8. With the aid of transformation (6) show that

$$L\{J_0'(t)\} = \frac{s}{\sqrt{s^2 + 1}} - 1.$$

Use formula (8) to show that $J_0'(t) = -J_1(t)$, and hence that

$$L\{J_1(t)\} = \frac{\sqrt{s^2 + 1} - s}{\sqrt{s^2 + 1}}.$$

9. Use the transforms of J_0 and J_1 (Prob. 8) to show that

(a) $\displaystyle\int_0^t J_0(\tau)J_0(t - \tau)\, d\tau = \sin t$;

(b) $\displaystyle\int_0^t J_0(\tau)J_1(t - \tau)\, d\tau = J_0(t) - \cos t$.

10. Expand the function $s^{-1}\exp(-s^{-1})$ in powers of s^{-1} and apply Theorem 7 to verify that

$$L^{-1}\left\{\frac{1}{s}e^{-1/s}\right\} = J_0(2\sqrt{t}),$$

and thus obtain the transformation 75, Appendix A, Table A.2.

Apply Theorem 8 to verify the following transformations:

11. $L^{-1}\left\{\dfrac{1}{\sqrt{s}}e^{-1/s}\right\} = \dfrac{1}{\sqrt{\pi t}}\cos(2\sqrt{t}).$

(Cf. transformation 76, Appendix A, Table A.2.)

12. $L^{-1}\left\{\dfrac{1}{\sqrt{s}}e^{1/s}\right\} = \dfrac{1}{\sqrt{\pi t}}\cosh(2\sqrt{t}).$

(Cf. transformation 77, Appendix A, Table A.2.)

13. $L^{-1}\left\{\dfrac{1}{(s^2 + 1)^{n+\frac{1}{2}}}\right\} = \dfrac{t^n J_n(t)}{(1)(3)(5)\cdots(2n-1)}$ $(n = 1, 2, \ldots)$.

(Cf. transformation 57, Appendix A, Table A.2.)

14. $L^{-1}\left\{\dfrac{1}{s^k - 1}\right\} = \dfrac{1}{t} \displaystyle\sum_{n=1}^{\infty} \dfrac{t^{nk}}{\Gamma(nk)}$ $(k > 0, s > 1)$.

15. Given that a power series $\sum_{n=0}^{\infty} A_n \tau^n$ converges uniformly over an interval $0 \le \tau \le t$ to a sum $F(\tau)$, that is, the series converges to $F(\tau)$ there and

$$F(\tau) = \sum_{n=0}^{N-1} A_n \tau^n + R_N(\tau) \qquad (0 \le \tau \le t)$$

where $R_N(\tau) \to 0$ uniformly with respect to τ as $N \to \infty$. If $0 \le c < 1$, prove that

$$t^{-c} * F(t) = \sum_{n=0}^{\infty} A_n t^{-c} * t^n,$$

a result that was used to obtain Eq. (12), Sec. 20. *Suggestion:* Write

$$t^{-c} * F(t) - \sum_{n=0}^{N-1} A_n t^{-c} * t^n = t^{-c} * R_N(t)$$

and prove that the right-hand member vanishes as $N \to \infty$, when t and c are kept fixed.

22 INTEGRATION OF TRANSFORMS

When a function $F(t)$ is sectionally continuous and of the order of $e^{\alpha t}$, then its Laplace integral

$$f(x) = \int_0^{\infty} e^{-xt} F(t)\, dt$$

is uniformly convergent with respect to x in every interval $x \ge \alpha_1$, where $\alpha_1 > \alpha$, according to Theorem 5. It follows from formula (5), Sec. 15, that when $r > s > \alpha$,

$$\int_s^r f(x)\, dx = \int_s^r \int_0^{\infty} e^{-xt} F(t)\, dt\, dx = \int_0^{\infty} F(t) \int_s^r e^{-xt}\, dx\, dt.$$

If the function F is such that $F(t)/t$ has a limit as t tends to zero, then the latter function is also sectionally continuous and of exponential order. Under those conditions the last equation can be written

$$\int_s^r f(x)\, dx = \int_0^{\infty} \frac{F(t)}{t} e^{-st}\, dt - \int_0^{\infty} \frac{F(t)}{t} e^{-rt}\, dt = g(s) - g(r),$$

where $g(s) = L\{F(t)/t\}$. But $g(r) \to 0$ as $r \to \infty$ (Theorem 5); hence

(1) $$\int_s^{\infty} f(x)\, dx = \int_0^{\infty} \frac{F(t)}{t} e^{-st}\, dt \qquad (s > \alpha)$$

and we have established the following theorem:

Theorem 9   *Division of the function* $F(t)$ *by* t *corresponds to integration of the transform* $f(s)$, *in this manner:*

(2)
$$L\left\{\frac{F(t)}{t}\right\} = \int_s^\infty f(x)\, dx.$$

Sufficient conditions for the validity of formula (2) *are that* $F(t)$ *be sectionally continuous and of the order of* $e^{\alpha t}$, *that* $s > \alpha$ *in formula* (2), *and further that the limit of* $F(t)/t$ *exists as* $t \to +0$.

The function $F(t) = \sin kt$, for example, satisfies the above conditions when $\alpha = 0$; in particular, $t^{-1} \sin kt \to k$ as $t \to 0$. Hence when $s > 0$,

(3)
$$L\left\{\frac{\sin kt}{t}\right\} = \int_s^\infty \frac{k\, dx}{x^2 + k^2} = \frac{\pi}{2} - \arctan\frac{s}{k} = \arctan\frac{k}{s}.$$

Recalling how integration with respect to t corresponds to division by s, we can now write the transform of the *sine-integral function*

(4)
$$\text{Si}\, t = \int_0^t \frac{\sin \tau}{\tau}\, d\tau.$$

This function is of some importance in applied mathematics. Its values are tabulated in the more extensive mathematical tables. If $k = 1$ in Eq. (3), it follows from Eq. (4) that

(5)
$$L\{\text{Si}\, t\} = \frac{1}{s}\operatorname{arccot} s = \frac{1}{s}\arctan\frac{1}{s} \qquad (s > 0).$$

As another illustration of Theorem 9 we note that

$$L\left\{\frac{e^{-at} - e^{-bt}}{t}\right\} = \int_s^\infty \left(\frac{1}{x+a} - \frac{1}{x+b}\right) dx = \log\frac{x+a}{x+b}\Bigg]_s^\infty,$$

when $s > -a$ and $s > -b$. Hence

(6)
$$L\left\{\frac{e^{-at} - e^{-bt}}{t}\right\} = \log\frac{s+b}{s+a}.$$

When $a = 0$ and $b = 1$, we have the special case

(7)
$$L\left\{\frac{1 - e^{-t}}{t}\right\} = \log\left(1 + \frac{1}{s}\right) \qquad (s > 0).$$

23   PERIODIC FUNCTIONS

Let a function F be periodic with period a over the half line $t > 0$ and sectionally continuous over a period $0 < t < a$. That periodicity can be described by either of the two statements

$$F(t + a) = F(t) \quad \text{when } t > 0, \quad \text{or} \quad F(t) = F(t - a) \quad \text{when } t > a.$$

The function is bounded over the half line and sectionally continuous over each bounded subinterval, so it has a transform $f(s)$ when $s > 0$.

For convenience in examining the transform, we write

$$F(t) = 0 \quad \text{when } t < 0$$

and introduce a function F_0 that is the same as F when $0 < t < a$ and zero elsewhere; thus

$$F_0(t) = [1 - S_0(t - a)]F(t), \quad f_0(s) = L\{F_0\} = \int_0^a e^{-st} F(t) \, dt.$$

Then F is described for all t by the difference equation

$$(1) \qquad\qquad F(t) - F(t - a) = F_0(t),$$

as we can see either directly from graphs of $F(t)$ and $F(t - a)$ or else by noting that when $t > a$, the equation becomes $F(t) = F(t - a)$, and when $t < a$, it becomes $F(t) = F_0(t)$. Therefore

$$f(s) - e^{-as}f(s) = f_0(s) \quad \text{and} \quad f(s) = \frac{f_0(s)}{1 - e^{-as}}.$$

Theorem 10   *If $F(t)$ is periodic with period a when $t > 0$, and sectionally continuous, then*

$$(2) \qquad\qquad f(s) = \frac{\int_0^a e^{-st} F(t) \, dt}{1 - e^{-as}} \qquad\qquad (s > 0).$$

Let us apply that formula to the function

$$M(c,t) = 1 \qquad\qquad \text{when } 0 < t < c,$$

$$= -1 \qquad\qquad \text{when } c < t < 2c,$$

$$M(c, t + 2c) = M(c,t).$$

This is sometimes called the *square-wave function* (Fig. 9).

Since

$$\int_0^{2c} e^{-st} M(c,t) \, dt = \int_0^c e^{-st} \, dt - \int_c^{2c} e^{-st} \, dt$$

$$= \frac{1}{s}(1 - e^{-cs})^2,$$

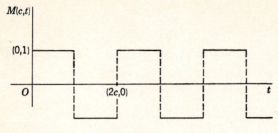

Fig. 9

the transform of $M(c,t)$ is

$$\frac{(1 - e^{-cs})^2}{s(1 - e^{-2cs})} = \frac{1 - e^{-cs}}{s(1 + e^{-cs})}.$$

Hence

(3) $$L\{M(c,t)\} = \frac{1}{s}\tanh\frac{cs}{2}$$ $(s > 0)$.

The integral of the function M from 0 to t is the function $H(c,t)$ defined as follows:

$$H(c,t) = t \qquad\qquad \text{when } 0 < t < c,$$
$$= 2c - t \qquad \text{when } c < t < 2c,$$
$$H(c, t + 2c) = H(c,t).$$

This function, whose graph is the *triangular wave* shown in Fig. 10, has the transform

(4) $$L\{H(c,t)\} = \frac{1}{s^2}\tanh\frac{cs}{2}$$ $(s > 0)$.

Let G denote an *antiperiodic function* when $t > 0$:

$G(t + c) = -G(t)$ when $t > 0$,   or $G(t) = -G(t - c)$ when $t > c$,

which is sectionally continuous over the interval $0 < t < c$.   Then

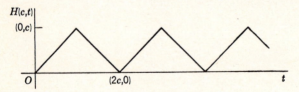

Fig. 10

$G(t + 2c) = -G(t + c) = G(t)$; that is, G is periodic with period $2c$. $M(c,t)$ is an example of such a function, as is $\sin t$ when $c = \pi, 3\pi, 5\pi, \ldots$,

If $G_0(t) = G(t)$ when $0 < t < c$ and G_0 vanishes for all t outside that interval, and if we write $G(t) = 0$ when $t < 0$, we can see that G is described by the difference equation

(5) $$G(t) + G(t - c) = G_0(t).$$

The transform $g(s)$ therefore satisfies the equation

$$g(s) + e^{-cs}g(s) = g_0(s) = \int_0^c e^{-st}G(t)\, dt,$$

and we have established the following theorem:

Theorem 11　*If G is sectionally continuous over an interval $(0,c)$ and $G(t + c) = -G(t)$ when $t > 0$, then*

(6) $$g(s) = \frac{\int_0^c e^{-st}G(t)\, dt}{1 + e^{-cs}} \qquad (s > 0).$$

For the periodic function G_1 with period $2c$ such that

$$G_1(t) = G(t) \qquad \text{when } 0 < t < c,\ G_1(t) = 0 \qquad \text{when } c < t < 2c,$$

where G is the above antiperiodic function, formula (2) becomes

(7) $$L\{G_1(t)\} = \frac{\int_0^c e^{-st}G(t)\, dt}{1 - e^{-2cs}} = \frac{g(s)}{1 - e^{-cs}} \qquad (s > 0).$$

For example, when $G(t) = \sin t$ and $c = \pi$, $g(s) = (s^2 + 1)^{-1}$ and the function G_1 shown in Fig. 11 has the transform

(8) $$L\{G_1(t)\} = L\left\{\frac{\sin t + |\sin t|}{2}\right\} = \frac{1}{(s^2 + 1)(1 - e^{-\pi s})} \qquad (s > 0).$$

Similarly, the periodic function G_2 such that

$$G_2(t) = G(t) \qquad \text{when } 0 < t < c,\ G_2(t + c) = G_2(t)$$

$$\text{where } G(t + c) = -G(t),$$

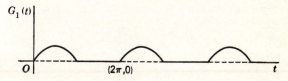

Fig. 11

has the transform

$$g_2(s) = \frac{\int_0^c e^{-st}G(t)\,dt}{1 - e^{-cs}} = \frac{1 + e^{-cs}}{1 - e^{-cs}}g(s);$$

therefore

(9) $$L\{G_2(t)\} = g(s)\coth\frac{cs}{2} \qquad (s > 0).$$

When $G(t) = \sin t$, for example, then $G_2(t) = |\sin t|$ if $c = \pi$ and

(10) $$L\{|\sin t|\} = \frac{1}{s^2 + 1}\coth\frac{\pi s}{2} \qquad (s > 0).$$

In case the antiperiodic function G has nonnegative values over the interval $(0,c)$, then $G(t) \leq 0$ when $c < t < 2c$ and the function G_1 described above is the *half-wave rectification* of G. It replaces the negative values of G by zero. Also, in that case G_2 is the *full-wave rectification* of G, and we can write

$$G_1(t) = \tfrac{1}{2}[G(t) + |G(t)|], \qquad G_2(t) = |G(t)|.$$

24 PARTIAL FRACTIONS

We shall now systematize the procedure of finding inverse transforms of *quotients of polynomials* in s. Let $p(s)$ and $q(s)$ denote polynomials in the variable s with no factor in common, and let the degree of $p(s)$ be lower than that of $q(s)$. We shall see that the inverse transform of the function $f(s) = p(s)/q(s)$ exists and that it can be found when the elementary factors of $q(s)$ can be determined.

Consider first the case in which $q(s)$ has a linear factor $s - a$, *not repeated*. Let $\phi(s)$ denote the function that is left after removing that factor from the denominator of $f(s)$; that is,

(1) $$f(s) = \frac{p(s)}{q(s)} = \frac{\phi(s)}{s - a}.$$

Note that $\phi(s)$ may be a quotient of polynomials. According to the theory of partial fractions, a constant C exists such that

(2) $$\frac{\phi(s)}{s - a} = \frac{C}{s - a} + h(s),$$

where $h(s)$ represents the sum of the partial fractions that correspond to the other linear and quadratic factors of $q(s)$, any of which may be repeated.

In order to determine the value of C, we multiply both members of Eq. (2) by $s - a$, when $s \neq a$, to obtain the equation

$$\phi(s) = C + (s - a)h(s),$$

which is satisfied identically for all values of s in a neighborhood of the point $s = a$, except possibly at that point. But both members of the equation are continuous functions of s at that point; thus their limits as $s \to a$ are the same as their values when $s = a$. Therefore $C = \phi(a)$. The inverse transform of the partial fraction corresponding to the factor $s - a$, or the term in $F(t)$ corresponding to that factor, is $\phi(a)e^{at}$.

In view of Eq. (1), we can also write

$$\lim_{s \to a} \phi(s) = \lim_{s \to a} \left[p(s) \frac{s - a}{q(s)} \right] = p(a) \frac{1}{q'(a)},$$

where we have evaluated the limit of $(s - a)/q(s)$ as the limit of the quotient of the derivatives of $s - a$ and $q(s)$, since $q(a) = 0$ and $q'(a) \neq 0$ because $s = a$ is a simple zero of $q(s)$. Consequently $\phi(a) = p(a)/q'(a)$.

When all the factors of $q(s)$ are linear and not repeated and when $q(s)$ is written in the form

(3) $$q(s) = (s - a_1)(s - a_2) \cdots (s - a_m),$$

where all the constants a_n are distinct, we can write the inverse transform of $f(s)$ in full. Let $q_n(s)$ denote the product of all the factors on the right of Eq. (3) except the factor $s - a_n$, so that the function $\phi(s)$ corresponding to that factor is $p(s)/q_n(s)$. Then

(4) $$L^{-1}\left\{ \frac{p(s)}{q(s)} \right\} = \sum_{n=1}^{m} \frac{p(a_n)}{q_n(a_n)} e^{a_n t} = \sum_{n=1}^{m} \frac{p(a_n)}{q'(a_n)} e^{a_n t}.$$

The second sum here is sometimes called Heaviside's expansion.

The principal results are stated in the following theorem.

Theorem 12 *If f is the quotient $p(s)/q(s)$ of two polynomials such that $q(s)$ has the higher degree and contains the factor $s - a$ which is not repeated, then the term in $F(t)$ corresponding to that factor can be written in either of these two forms:*

(5) $$\phi(a)e^{at} \qquad or \qquad \frac{p(a)}{q'(a)} e^{at},$$

where $\phi(s)$ is the quotient of $p(s)$ divided by the product of all factors of $q(s)$ except $s - a$.

Theorem 12 is valid when the constant a is any complex number.

For complex arguments the exponential function is defined by the equation (Sec. 55)

$$e^{x+iy} = e^x(\cos y + i \sin y) \qquad\qquad (x \text{ and } y \text{ real}).$$

Example   Find $F(t)$ when

$$f(s) = \frac{2s^2 - 4s}{(2s + 1)(s^2 + 1)}.$$

We display the factors of the type $s - a$ by writing

$$f(s) = \frac{s^2 - 2s}{(s + \frac{1}{2})(s - i)(s + i)}.$$

Using the first of the two forms (5), or the first of the expansions (4), we find that

$$F(t) = \frac{\frac{5}{4}}{\frac{5}{4}}e^{-t/2} - \frac{1 + 2i}{(i + \frac{1}{2})2i}e^{it} + \frac{-1 + 2i}{(-i + \frac{1}{2})(-2i)}e^{-it}$$

$$= e^{-t/2} - 2\frac{e^{it} - e^{-it}}{2i} = e^{-t/2} - 2\sin t.$$

PROBLEMS

1. Show that Theorem 9 applies and find the transform of: (a) $(1 - \cos at)/t$; (b) $(1 - \cosh at)/t$; (c) $(e^t - \cos t)/t$.
> Ans. (a) (b) See transforms 105 and 106, Appendix A, Table A.2;
> (c) $\frac{1}{2}\log[(s^2 + 1)/(s - 1)^2](s > 1)$.

2. The condition that $F(t)/t$ has a limit as $t \to +0$ was used in proving Theorem 9. If it is replaced by the condition that for some *positive* constants k and M, $|F(t)| < Mt^k$ over an interval $0 < t < T$, then $g(s)$, the transform of $F(t)/t$, exists and vanishes as $s \to \infty$. Thus formula (2) in the theorem is still valid. Use this fact to establish transform 36, Appendix A, Table A.2:

$$L\left\{\frac{e^{bt} - e^{at}}{2\sqrt{\pi t^3}}\right\} = \sqrt{s - a} - \sqrt{s - b} \qquad (s > a \text{ and } s > b),$$

by showing that $|e^{bt} - e^{at}|/\sqrt{t} < M\sqrt{t}$ on an interval $(0,T)$.

3. (a) Use Theorem 11 to derive the transform (3), Sec. 23, of $M(c,t)$.

(b) Apply formula (7), Sec. 23, to obtain the transform 68, Appendix A, Table A.2, of the half-wave rectification of $M(k,t)$.

(c) Use formula (9), Sec. 23, to verify that the transform of the full-wave rectification of $M(c,t)$ is $1/s$.

4. Sketch the graph of the periodic function F for which $F(t) = t$ when $-1 < t < 1$ and $F(t + 2) = F(t)$ for all t, and show that the transform of F is

$$f(s) = \frac{1}{s^2} - \frac{1}{s \sinh s} \qquad\qquad (s > 0).$$

5. Let F be the periodic function such that $F(t) = t$ when $0 < t < 1$ and $F(t + 1) = F(t)$, and let G be this antiperiodic function: $G(t) = t$ when $0 < t < 1$ and $G(t + 1) = -G(t)$. Show graphically the functions F, G, and the half-wave rectification G_1 of G, and obtain the transforms

(a) $f(s) = \dfrac{1}{s^2} - \dfrac{e^{-s}}{s(1 - e^{-s})}$ $\qquad\qquad\qquad$ $(s > 0)$,

(b) $g(s) = \dfrac{1}{s^2} - \left(\dfrac{2}{s^2} + \dfrac{1}{s}\right)\dfrac{e^{-s}}{1 + e^{-s}}$ $\qquad\qquad$ $(s > 0)$,

(c) $g_1(s) = \dfrac{f(s)}{1 + e^{-s}}$ $\qquad\qquad\qquad\qquad\qquad$ $(s > 0)$.

6. Use Theorem 12 to find the inverse transforms tabulated below, where a, b, and c are distinct constants.

$\qquad\qquad\qquad$ $f(s)$ $\qquad\qquad\qquad\qquad\qquad\qquad$ $F(t)$

(a) $\dfrac{1}{(s - a)(s - b)}$ $\qquad\qquad\qquad\qquad$ $\dfrac{e^{at} - e^{bt}}{a - b}$

(b) $\dfrac{s}{(s - a)(s - b)}$ $\qquad\qquad\qquad\qquad$ $\dfrac{ae^{at} - be^{bt}}{a - b}$

(c) $\dfrac{-1}{(s - a)(s - b)(s - c)}$ $\qquad\qquad$ $\dfrac{(b - c)e^{at} + (c - a)e^{bt} + (a - b)e^{ct}}{(a - b)(b - c)(c - a)}$

7. Use Theorem 12 to find the inverse transforms of the following functions.

(a) $f(s) = \dfrac{4s + 1}{(s^2 + s)(4s^2 - 1)}.$ $\qquad$ *Ans.* $F(t) = e^{t/2} - e^{-t/2} + e^{-t} - 1.$

(b) $f(s) = \dfrac{s}{s^2 - k^2}.$ $\qquad\qquad\qquad\qquad$ (c) $f(s) = \dfrac{s}{s^2 + k^2}.$

(d) $f(s) = \dfrac{3s^2}{s^3 + s^2 - 4(s + 1)}.$ $\qquad$ *Ans.* $F(t) = e^{2t} + 3e^{-2t} - e^{-t}.$

Use Theorem 12 in solving the following differential equations.

8. $Y''(t) - Y(t) = 1 + e^{3t}.$ $\qquad\qquad$ *Ans.* $Y(t) = C_1 e^t + C_2 e^{-t} - 1 + \frac{1}{8}e^{3t}.$

9. $Y'''(t) + Y''(t) - 4Y'(t) - 4Y(t) = F(t)$, if $Y(0) = Y''(0) = 0$ and $Y'(0) = 2.$
$\qquad\qquad$ *Ans.* $Y(t) = \sinh 2t + \frac{1}{12}F(t) * (e^{2t} + 3e^{-2t} - 4e^{-t}).$

10. $Y^{(4)}(t) - 2Y'''(t) - Y''(t) + 2Y'(t) = 6F(t)$, if $Y(t)$ and its first three derivatives are zero when $t = 0.$

25 REPEATED LINEAR FACTORS

We now consider partial fractions for the case in which the polynomial $q(s)$ contains a repeated linear factor $(s - a)^{n+1}$. We write

(1) $$f(s) = \frac{p(s)}{q(s)} = \frac{\phi(s)}{(s - a)^{n+1}},$$

where $\phi(s)$ is the quotient of polynomials obtained by removing the factor $(s - a)^{n+1}$ from the denominator of the fraction $p(s)/q(s)$. As before, the degree of the polynomial $p(s)$ is assumed to be lower than the degree of $q(s)$. Note that $\phi(s)$ and its derivatives are continuous functions at the point $s = a$.

The representation of $f(s)$ in partial fractions now has the form

$$(2) \quad \frac{\phi(s)}{(s-a)^{n+1}} = \frac{A_0}{s-a} + \frac{A_1}{(s-a)^2} + \cdots + \frac{A_r}{(s-a)^{r+1}} + \cdots + \frac{A_n}{(s-a)^{n+1}} + h(s),$$

where the numbers A_r are independent of s, and $h(s)$ is the sum of the partial fractions corresponding to the remaining factors of $q(s)$. It follows from Eq. (2) that

$$(3) \quad \phi(s) = A_0(s-a)^n + \cdots + A_r(s-a)^{n-r} + \cdots + A_n + (s-a)^{n+1}h(s)$$

in a deleted neighborhood of the point $s = a$. When $s \to a$, we see that $A_n = \phi(a)$.

To find the remaining coefficients A_r, we differentiate both members of Eq. (3) with respect to s, $n - r$ times, in order to isolate the number A_r. When $s \to a$ in the resulting equation, we find that

$$\phi^{(n-r)}(a) = (n - r)!A_r.$$

Equation (2) can now be written in the form

$$(4) \quad f(s) = \sum_{r=0}^{n} \frac{\phi^{(n-r)}(a)}{(n-r)!} \frac{1}{(s-a)^{r+1}} + h(s),$$

where $0! = 1$ and $\phi^{(0)}(a) = \phi(a)$.

If $H(t) = L^{-1}\{h(s)\}$, it follows that the inverse transform of $f(s)$ is

$$(5) \quad F(t) = \sum_{r=0}^{n} \frac{\phi^{(n-r)}(a)}{(n-r)!r!} t^r e^{at} + H(t).$$

This equation can be simplified by recalling the formula for the derivative of order n of the product of two functions $u(s)$ and $v(s)$, namely,

$$\frac{d^n}{ds^n}(uv) = \sum_{r=0}^{n} \frac{n!}{(n-r)!r!} u^{(n-r)}(s)v^{(r)}(s).$$

When $u = \phi(s)$ and $v = e^{st}$, then $\partial^r v/\partial s^r = t^r e^{st}$ and Eq. (5) reduces to the form

$$(6) \quad F(t) = \frac{1}{n!}\left\{\frac{\partial^n}{\partial s^n}[\phi(s)e^{st}]\right\}_{s=a} + H(t).$$

This result can be stated as follows.

Theorem 13 *If $f(s)$ is the quotient $p(s)/q(s)$ of two polynomials such that $q(s)$ has the higher degree and contains the factor $(s - a)^{n+1}$, then the term in $F(t)$ corresponding to that factor is $\Phi_n(a,t)$, where*

(7) $$\Phi_n(s,t) = \frac{1}{n!}\frac{\partial^n}{\partial s^n}[\phi(s)e^{st}]$$

and $\phi(s)$ is the function indicated by Eq. (1).

The term in $F(t)$ corresponding to a factor $(s - a)^2$ in $q(s)$, for instance, is

(8) $$\Phi_1(a,t) = [\phi'(a) + \phi(a)t]e^{at},$$

and the term corresponding to a factor $(s - a)^3$ is

(9) $$\Phi_2(a,t) = \tfrac{1}{2}[\phi''(a) + 2\phi'(a)t + \phi(a)t^2]e^{at}.$$

As an example, if

$$f(s) = \frac{1}{(s - 1)(s - 2)^2},$$

then the term in $F(t)$ corresponding to the factor $s - 1$ is e^t. To correspond with the factor $(s - 2)^2$, we have $\phi(s) = (s - 1)^{-1}$ and $\phi'(s) = -(s - 1)^{-2}$ so that $\phi(2) = 1$ and $\phi'(2) = -1$. In view of formula (8) the term in $F(t)$ is $(-1 + t)e^{2t}$. Hence

$$F(t) = e^t + (t - 1)e^{2t}.$$

Since the number a may be imaginary and since a factorization of every polynomial into linear factors, real or imaginary, exists, Theorems 12 and 13 provide a systematic way of finding inverse transforms of quotients of polynomials in all cases where the factors of the denominator can be determined. If, however, imaginary factors are present, the results are given in terms of imaginary functions. The reduction of the results to real forms is sometimes tedious. To obtain the real form of the inverse transform directly, and to observe the character of that function in general, we may proceed as follows.

26 QUADRATIC FACTORS

In this section we assume that the polynomials $p(s)$ and $q(s)$ have real coefficients. The imaginary zeros of $q(s)$ then occur in pairs, each pair consisting of some complex number $a + ib$ and its conjugate. The corresponding linear factors of q are

(1) $$q_1 = s - a - ib, \qquad q_2 = s - a + ib \qquad (b \neq 0),$$

where a and b are real numbers. The product of those factors is the real quadratic factor $(s - a)^2 + b^2$.

Let q have the factors (1), *not repeated*, and let ϕ_1 and ϕ_2 be the quotients of polynomials obtained by removing the factors q_1 and q_2 in turn from the fraction $p(s)/q(s)$; thus

$$(2) \qquad f(s) = \frac{p(s)}{q(s)} = \frac{\phi_1(s)}{s - a - ib} = \frac{\phi_2(s)}{s - a + ib}.$$

According to Theorem 12 the sum of terms in $F(t)$ corresponding to those two linear factors is

$$(3) \qquad G(t) = e^{at}[\phi_1(a + ib)e^{ibt} + \phi_2(a - ib)e^{-ibt}].$$

This is the component of F corresponding to the quadratic factor $(s - a)^2 + b^2$. We shall represent it in terms of real-valued functions.

The rational function $\phi_1(s)$ is continuous at the point $s = a + ib$. In a neighborhood of that point excluding the point itself

$$(4) \qquad \phi_1(s) = (s - a - ib)f(s).$$

Similarly, ϕ_2 is continuous at the point $a - ib$ and

$$(5) \qquad \phi_2(s) = (s - a + ib)f(s)$$

throughout a neighborhood of that point with that point deleted. From elementary properties of complex conjugates we can see that the conjugate of $f(s)$, a rational function with real coefficients, is $f(\bar{s})$. Consequently

$$\phi_2(\bar{s}) = (\bar{s} - a + ib)f(\bar{s}) = \overline{(s - a - ib)}\,\overline{f(s)} = \overline{\phi_1(s)}.$$

In view of the continuity of ϕ_1 and ϕ_2 it follows that

$$(6) \qquad \phi_2(a - ib) = \overline{\phi_1(a + ib)}.$$

Since $e^{-ibt} = \cos bt - i \sin bt$, it is the conjugate of e^{ibt}. Thus formula (3) can be written

$$G(t) = e^{at}[\phi_1(a + ib)e^{ibt} + \overline{\phi_1(a + ib)e^{ibt}}].$$

The factor in brackets is twice the real part of the first term, $2\,\text{Re}[\phi_1(a+ib)e^{ibt}]$. So if we let r_1 and θ_1 represent polar coordinates of the point representing the complex number $\phi_1(a + ib)$, which is never zero,

$$(7) \qquad \phi_1(a + ib) = r_1 e^{i\theta_1} = r_1(\cos \theta_1 + i \sin \theta_1),$$

we can write the component G in the real form

$$(8) \qquad G(t) = 2e^{at}r_1 \cos (bt + \theta_1) \qquad\qquad (b \neq 0, r_1 > 0).$$

If $a = 0$, the component $G(t)$ is the simple periodic function $2r_1 \cos (bt + \theta_1)$, and if $a < 0$, it is a damped periodic function. Those components, corresponding to cases $a \leq 0$, represent *stable oscillations* in the theory of control of mechanical and electrical systems where t denotes time. Stable oscillations are bounded as $t \to \infty$. When $a > 0$, the component $G(t)$ represents an unstable oscillation.

To illustrate the use of formula (8), we find $F(t)$ when

$$f(s) = \frac{2s - 2}{(s + 1)(s^2 + 2s + 5)} = \frac{2s - 2}{(s + 1)[(s + 1)^2 + 4]} = \frac{\phi_1(s)}{s + 1 - 2i}$$

where

$$\phi_1(s) = \frac{2s - 2}{(s + 1)(s + 1 + 2i)}.$$

Then

$$\phi_1(-1 + 2i) = \frac{-4 + 4i}{2i(4i)} = \frac{1 - i}{2} = \frac{\sqrt{2}}{2}e^{-i\pi/4}$$

and $G(t) = \sqrt{2}e^{-t}\cos(2t - \pi/4) = e^{-t}(\cos 2t + \sin 2t)$. After adding the term corresponding to the linear factor $s + 1$, we find that

$$F(t) = e^{-t}(\cos 2t + \sin 2t - 1).$$

In case q contains the *square* of the quadratic factor $(s - a)^2 + b^2$, we write

(9) $$f(s) = \frac{p(s)}{q(s)} = \frac{\phi_1(s)}{(s - a - ib)^2} = \frac{\phi_2(s)}{(s - a + ib)^2}$$

and apply formula (8), Sec. 25, to get the corresponding component of $F(t)$:

$$G(t) = e^{at}[\phi_1'(a + ib) + t\phi_1(a + ib)]e^{ibt}$$
$$+ e^{at}[\phi_2'(a - ib) + t\phi_2(a - ib)]e^{-ibt}.$$

But $\phi_2(s) = (s - a + ib)^2 f(s)$, and as before we find that $\phi_2(\bar{s}) = \overline{\phi_1(s)}$, also that $\phi_2'(\bar{s}) = \overline{\phi_1'(s)}$, so that

(10) $$G(t) = 2e^{at}\,\mathrm{Re}\,[\phi_1'(a + ib)e^{ibt} + t\phi_1(a + ib)e^{ibt}].$$

We use absolute values and arguments of the complex numbers $\phi_1(a + ib)$ and $\phi_1'(a + ib)$ displayed by the polar forms

(11) $$\phi_1(a + ib) = r_1 e^{i\theta_1}, \quad \phi_1'(a + ib) = \rho_1 e^{i\psi_1},$$

where $r_1 \neq 0$ because $\phi_1(a + ib) \neq 0$. Then formula (10) becomes

(12) $$G(t) = 2e^{at}[\rho_1 \cos(bt + \psi_1) + r_1 t \cos(bt + \theta_1)] \qquad (r_1 > 0).$$

Our results can be summarized as follows.

Theorem 14   *When $f(s) = p(s)/q(s)$ where $p(s)$ and $q(s)$ are polynomials with real coefficients and q has the higher degree and contains a nonrepeated real quadratic factor $(s - a)^2 + b^2$, where $b \neq 0$, the component of $F(t)$ corresponding to that factor is the function $G(t)$ given by formula (8).*

If q contains the square of that quadratic factor, the component of F(t) is given by formula (12).

We note that the component (12) represents a *stable oscillation only if the real part of the imaginary zeros* $a \pm ib$ *of q is negative:* $a < 0$. If $a = 0$, the component contains the term $2r_1 t \cos(bt + \theta_1)$ representing an unstable oscillation.

When the quadratic factor appears to a degree $n + 1$ ($n = 2, 3, \ldots$), we can see from the procedure we used when $n = 1$ and from formula (5), Sec. 25, that the component in F will contain a term of the type $Ce^{at}t^n \cos(bt + \theta_1)$, where C is a constant.

When p and q are not polynomials in s, partial fractions in s cannot be used. Such cases will be treated in Chap. 6 with the aid of residues and contour integrals.

27 TABLES OF OPERATIONS AND TRANSFORMS

Appendix A, Table A.1, contains a list of operations on $F(t)$ with corresponding operations on $f(s)$. That table of operations summarizes several of our results on the theory of the Laplace transformation.

The table of Laplace transforms (Appendix A, Table A.2) gives a fairly extensive list of transforms of particular functions. Derivations of a number of them have been presented above. References to some more extensive tables can be found in the Bibliography.

Transforms 82 to 84, Appendix A, Table A.2, of functions that are prominent in problems of diffusion and conduction of heat, will now be derived. A more direct derivation will be made later on in Chap. 6 with the aid of contour integrals and functions of a complex variable; but we wish to use those transforms earlier, in Chap. 4.

First we perform formal manipulations that will suggest the inverse transforms of the two functions

$$(1) \qquad\qquad y(s) = \frac{1}{\sqrt{s}} e^{-k\sqrt{s}} \qquad\qquad (k \geq 0, s > 0),$$

$$(2) \qquad\qquad z(s) = e^{-k\sqrt{s}} \qquad\qquad (k > 0, s > 0).$$

We see that $(\sqrt{s}\,y)' = -kz/(2\sqrt{s})$ and $2z' = -ky$, so that y and z satisfy this system of differential equations with coefficients that are linear in s:

$$(3) \qquad 2sy'(s) + y(s) + kz(s) = 2z'(s) + ky(s) = 0.$$

The corresponding system for the inverse transforms $Y(t)$ and $Z(t)$ is therefore

$$2(-tY)' + Y + kZ = 0, \qquad -2tZ + kY = 0.$$

Thus $-2tY' - Y + kZ = 0$ and

(4) $$2Y'(t) = \left(\frac{k^2}{2t^2} - \frac{1}{t}\right) Y(t), \qquad Z(t) = \frac{k}{2t} Y(t).$$

The solution of system (4) is

$$Y(t) = \frac{C}{\sqrt{t}} \exp\left(-\frac{k^2}{4t}\right), \qquad Z(t) = \frac{Ck}{2\sqrt{t^3}} \exp\left(-\frac{k^2}{4t}\right).$$

But when $k = 0$, then $y(s) = 1/\sqrt{s}$, and we know that $Y(t)$ is then $1/\sqrt{\pi t}$, so if C is independent of k, it follows that $C = 1/\sqrt{\pi}$, and our formal results can be written

(5) $$L\left\{\frac{1}{\sqrt{\pi t}} \exp\left(-\frac{k^2}{4t}\right)\right\} = \frac{1}{\sqrt{s}} e^{-k\sqrt{s}} \qquad (k \geqq 0),$$

(6) $$L\left\{\frac{k}{2\sqrt{\pi t^3}} \exp\left(-\frac{k^2}{4t}\right)\right\} = e^{-k\sqrt{s}} \qquad (k > 0).$$

Let us now prove that transformations (5) and (6) are correct when $k = 1$ and $s > 0$. The function

(7) $$Y_1(t) = \frac{1}{\sqrt{\pi t}} \exp\left(-\frac{1}{4t}\right) \qquad \text{when } t > 0,\ Y(0) = 0,$$

is continuous when $t \geqq 0$, and bounded. Hence when $s > 0$,

$$\sqrt{\pi} y_1(s) = \int_0^\infty e^{-st} \exp\left(-\frac{1}{4t}\right) \frac{dt}{\sqrt{t}} = e^{-\sqrt{s}} \int_0^\infty \exp\left[-\left(\sqrt{st} - \frac{1}{2\sqrt{t}}\right)^2\right] \frac{dt}{\sqrt{t}}.$$

We substitute τ for $2\sqrt{t}$ to write

(8) $$\sqrt{\pi} e^{\sqrt{s}} y_1(s) = \int_0^\infty \exp\left[-\left(\frac{\sqrt{s}}{2}\tau - \frac{1}{\tau}\right)^2\right] d\tau,$$

and make a further substitution $\frac{1}{2}\sqrt{s}\tau = 1/\lambda$ to see that

(9) $$\sqrt{\pi} e^{\sqrt{s}} y_1(s) = \frac{2}{\sqrt{s}} \int_0^\infty \exp\left[-\left(\frac{\sqrt{s}}{2}\lambda - \frac{1}{\lambda}\right)^2\right] \frac{d\lambda}{\lambda^2}.$$

By first adding corresponding members of equations (8) and (9), then substituting x for $\frac{1}{2}\sqrt{s}\lambda - 1/\lambda$, we can write

$$2\sqrt{\pi} e^{\sqrt{s}} y_1(s) = \frac{2}{\sqrt{s}} \int_0^\infty \exp\left[-\left(\frac{\sqrt{s}}{2}\lambda - \frac{1}{\lambda}\right)^2\right]\left(\frac{\sqrt{s}}{2} + \frac{1}{\lambda^2}\right) d\lambda$$

$$= \frac{2}{\sqrt{s}} \int_{-\infty}^\infty \exp(-x^2)\, dx = 2\sqrt{\frac{\pi}{s}}.$$

Therefore

$$y_1(s) = L\left\{\frac{1}{\sqrt{\pi t}}\exp\left(-\frac{1}{4t}\right)\right\} = \frac{1}{\sqrt{s}}e^{-\sqrt{s}} \qquad (s > 0),$$

and when $k > 0$, transformation (5) follows by writing $y_1(k^2 s)$ and applying Theorem 2. Formula (5) was established earlier when $k = 0$.

If $k > 0$, the function Z, defined as follows,

$$Z(t) = \frac{k}{2\sqrt{\pi t^3}}\exp\left(-\frac{k^2}{4t}\right) \qquad \text{when } t > 0, Z(0) = 0,$$

is continuous and bounded when $t \geq 0$; also $Z = \frac{1}{2}kt^{-1}Y$ where

$$Y(t) = \frac{1}{\sqrt{\pi t}}\exp\left(-\frac{k^2}{4t}\right) \qquad \text{and} \qquad y(s) = \frac{1}{\sqrt{s}}e^{-k\sqrt{s}} \qquad (s > 0).$$

Theorem 9 therefore applies to give the transformation

$$z(s) = \frac{k}{2}\int_s^\infty e^{-k\sqrt{x}}\frac{dx}{\sqrt{x}} = e^{-k\sqrt{s}} \qquad (s > 0),$$

which is transformation (6).

Finally, in view of transformation (6), we note that

$$L^{-1}\left\{\frac{1}{s}e^{-k\sqrt{s}}\right\} = \frac{k}{2\sqrt{\pi}}\int_0^t e^{-k^2/(4\tau)}\tau^{-\frac{3}{2}}\,d\tau = \frac{2}{\sqrt{\pi}}\int_{k/(2\sqrt{t})}^\infty e^{-\lambda^2}\,d\lambda$$

$$= \frac{2}{\sqrt{\pi}}\int_0^\infty e^{-\lambda^2}\,d\lambda - \frac{2}{\sqrt{\pi}}\int_0^{k/(2\sqrt{t})} e^{-\lambda^2}\,d\lambda.$$

Therefore

(10)
$$L^{-1}\left\{\frac{1}{s}e^{-k\sqrt{s}}\right\} = 1 - \operatorname{erf}\left(\frac{k}{2\sqrt{t}}\right) \qquad (k \geq 0, s > 0),$$

where $\operatorname{erf}(x)$ is the error function defined in Sec. 16. Equation (10) can be written

(11)
$$L\left\{\operatorname{erfc}\left(\frac{k}{2\sqrt{t}}\right)\right\} = \frac{1}{s}e^{-k\sqrt{s}} \qquad (k \geq 0, s > 0),$$

where the *complementary error function* $\operatorname{erfc}(x)$ is defined as

(12)
$$\operatorname{erfc}(x) = 1 - \operatorname{erf}(x) = \frac{2}{\sqrt{\pi}}\int_x^\infty e^{-\lambda^2}\,d\lambda.$$

PROBLEMS

1. Find the inverse transforms tabulated below:

$f(s)$ — $F(t)$

(a) $\dfrac{s + a}{(s + b)(s + c)^2}$ — $\dfrac{a - b}{(b - c)^2} e^{-bt} + \left[\dfrac{a - c}{b - c} t - \dfrac{a - b}{(b - c)^2} \right] e^{-ct}$

(b) $\dfrac{s + 3}{(2s + 1)(s^2 + 2s + 2)}$ — $e^{-\frac{1}{2}t} - e^{-t} \cos t$

(c) $\dfrac{2b^4}{(s^2 + b^2)s^3}$ — $2 \cos bt + b^2 t^2 - 2$

(d) $\dfrac{s^2 - b^2}{(s^2 + b^2)^2}$ — $t \cos bt$

(e) $\dfrac{25s^2}{(s + 1)^2(s^2 + 4)}$ — $(5t - 8)e^{-t} + 6 \sin 2t + 8 \cos 2t$

2. Without finding $F(t)$, determine whether the oscillations of that function are stable or unstable when its transform is

(a) $f(s) = \dfrac{2s^2 + 1}{2s^3 + 2s^2 + s}$. — *Ans.* Stable.

(b) $f(s) = \dfrac{s}{(2s^2 + 4s + 3)(s^2 + 2)^2}$. — *Ans.* Unstable.

3. Solve the following differential equations:
   (a) $Y''(t) - 2Y'(t) + Y(t) = 1$. — *Ans.* $Y(t) = (C_1 + C_2 t)e^t + 1$.
   (b) $Y''(t) + Y(t) = 2 \sin t$, if $Y(0) = 0$ and $Y'(0) = -1$. — *Ans.* $Y(t) = -t \cos t$.
   (c) $4Y'''(t) + 4Y''(t) + Y'(t) = F(t)$.
   (d) $Y^{(4)}(t) + 2Y''(t) + Y(t) = 0$, if $Y(0) = 0$, $Y'(0) = 1$, $Y''(0) = 2$, and $Y'''(0) = -3$. — *Ans.* $Y(t) = t(\sin t + \cos t)$.

4. In Sec. 26, if we write $\phi_1(a + ib) = \alpha_1 + i\beta_1$, show that formula (8) for the component G can be written

$$G(t) = 2e^{at}(\alpha_1 \cos bt - \beta_1 \sin bt).$$

5. With the aid of transformation (5), Sec. 27, find $L^{-1}\{s^{-\frac{3}{2}} \exp(-k\sqrt{s})\}$ listed as transformation 85 in Appendix A, Table A.2.

6. Solve for $Y(t)$:
   (a) $(t - t^2)Y''(t) + 2Y'(t) + 2Y(t) = 6t$, $Y(0) = Y(2) = 0$.
      Ans. $Y(t) = t^2 - 8t^{-1}(t - 1)^3 S_0(t - 1)$.
   (b) $Y(t) + 2\int_0^t Y(x) \cos(t - x)\, dx = 9e^{2t}$.
      Ans. $Y(t) = 5e^{2t} + 4e^{-t} - 6te^{-t}$.

7. Draw the graph of the ramp function Y such that

$$Y(t) = t - Y(t - 1) \quad \text{and} \quad Y(t) = 0 \quad \text{when } t \leq 0,$$

and show that $Y(t) = t - n$ when $2n \leq t \leq 2n + 1$, $Y(t) = n + 1$ when $2n + 1 \leq$

$2n + 2, n = 0, 1, 2, \ldots$, and that

$$L\{Y(t)\} = \frac{1}{s^2(1 + e^{-s})} \qquad (s > 0).$$

8. When F, G, and H are sectionally continuous, show that

$$F(t) * G(t) * H(t) = \int_0^t F(t - x) \int_0^x G(x - y)H(y)\, dy\, dx.$$

9. When F is sectionally continuous and G and its derivative G' are continuous over an interval $0 \le t \le T$, show that, at each point $t(0 < t < T)$ where F is continuous,

$$\frac{d}{dt}[F(t) * G(t)] = F(t) * G'(t) + G(0)F(t).$$

10. *Under the following set of conditions the limit of the integral*

$$q(x) = \int_0^\infty f(x,t)\, dt \qquad (x > c),$$

as $x \to \infty$, is the same as the integral of the limit. (a) Let $f(x,t)$ be continuous over each rectangle $c \le x \le C, 0 \le t \le T$, except possibly for finite jumps across a finite number of lines $t = t_i$ in each rectangle, and (b) let the above integral converge uniformly with respect to x when $x > c$. Also, (c) let f have a limit

$$F(t) = \lim_{x \to \infty} f(x,t)$$

uniformly with respect to $t(t \ne t_i)$ on each interval $0 < t < T$; that is, for each positive ϵ there is a number N_ϵ independent of t such that

$$|F(t) - f(x,t)| < \epsilon \qquad \text{whenever} \qquad x > N_\epsilon \qquad (0 < t < T),$$

and (d) let $\int_0^\infty F(t)\, dt$ exist. Then prove that

$$\lim_{x \to \infty} q(x) = \int_0^\infty F(t)\, dt.$$

Note: The difference of the last integral and $q(x)$ can be written

$$\int_0^T [F(t) - f(x,t)]\, dt + \int_T^\infty F(t)\, dt - \int_T^\infty f(x,t)\, dt = \Delta q.$$

Take T large first, independent of x, then x large, to make $|\Delta q|$ small.

11. Use a remainder to prove that the integral

$$q(x) = \int_0^\infty \frac{x}{t^2 + x^2}\, dt$$

is not uniformly convergent with respect to x on the half line $x > 1$ (cf. Prob. 10). Show that $\lim_{x \to \infty} q(x) = \pi/2$ while the integral of the limit of the integrand, as $x \to \infty$, is zero.

12. *If $F(t) \ge G(t)$ when $t > 0$, then $f(s) \ge g(s)$ for every real value of s for which the transforms f and g exist.* Prove that property, and illustrate it with some particular functions. Note that the case $F(t) = 2\sin t$, $G(t) = \sin 2t$ shows that the converse of the property is not always valid.

13: *Initial-value theorem*   If F is continuous over the half line $t \geq 0$ except possibly for a finite number of finite jumps, while F' is sectionally continuous over each bounded interval there, and if F and F' are of exponential order, then the limit of the product $sf(s)$, as $s \to \infty$, exists and equals the initial value of F:

$$\lim_{s \to \infty} sf(s) = F(0).$$

Use a formula for $L\{F'\}$ and Theorem 5 to prove that theorem.[1]

14.   Illustrate the initial-value theorem (Prob. 13) with these particular functions: (a) $F(t) = A$; (b) $F(t) = S_0(t - t_0)$; (c) $F(t) = t$; (d) $F(t) = \cos kt$; (e) $F(t) = e^{kt}[1 - S_0(t - t_0)]$.

15.   If $f(s) = p(s)/q(s)$ where p and q are polynomials whose terms of highest degree are $a_n s^n$ and $b_n s^{n+1}$, respectively, use the initial-value theorem (Prob. 13) to prove that $F(0) = a_n/b_n$.

16.   Let F and F' be continuous while F'' is sectionally continuous over each interval $0 \leq t \leq T$, and let all three functions be of exponential order. Then if $s^2 f(s)$ has a limit as $s \to \infty$, prove that

$$\lim_{s \to \infty} s^2 f(s) = F'(0) \qquad \text{and} \qquad F(0) = 0.$$

17.   Illustrate the modified initial-value theorem presented in Prob. 16 in case $F(t)$ is (a) $\sin kt$; (b) $at + bt^2$; (c) $1 - \cosh t$.

18.   *A final-value theorem*   Let both F and F' be bounded over the half line $t \geq 0$ and continuous there except possibly for a finite number of finite jumps; also let $|F'(t)|$ be integrable from $t = 0$ to $t = \infty$. Then

$$\lim_{s \to +0} sf(s) = \lim_{t \to \infty} F(t).$$

Prove that theorem. In the formula for $L\{F'\}$ note that our conditions ensure the continuity of that Laplace integral with respect to s when $s \geq 0$ (Sec. 15).

19.   Illustrate the final value theorem (Prob. 18) in case (a) $F(t) = A + Be^{-t}$; (b) $F(t) = S_0(t - t_0)$; (c) $F(t) = t - (t - 1)S_0(t - 1)$.

20.   *Generalized convolution*   For a function $F(t,\tau)$ of two variables, defined over the quadrant $t > 0$, $\tau > 0$, let $f(s,\tau)$ denote its transform with respect to t. Using the same parameter s, let $\hat{f}(s,s)$ denote the *iterated transform*:

$$\hat{f}(s,s) = \int_0^\infty e^{-s\tau} f(s,\tau) \, d\tau = \int_0^\infty e^{-s\tau} \int_0^\infty e^{-st} F(t,\tau) \, dt \, d\tau.$$

Indicate formally why $\hat{f}(s,s)$ is the transform of the generalized convolution $F * (t)$ of $F(t,\tau)$ defined below:

$$\hat{f}(s,s) = L\{F * (t)\} \qquad \text{where} \qquad F * (t) = \int_0^t F(t - \tau, \tau) \, d\tau.$$

It is convenient to write $F(t,\tau) = 0$ when either $t < 0$ or $\tau < 0$. (See Prob. 21 for conditions of validity of the generalized-convolution property.) When $F(t,\tau) = F(t)G(t)$, note that the property reduces to the convolution property (6), Sec. 16.

21.   Prove that the generalized-convolution property written in Prob. 20 is valid when $s > \alpha$ under these conditions on F: $|F(t,\tau)| < M \exp[\alpha(t + \tau)]$ for all positive

---

[1] Similar theorems with conditions on $f(s)$ rather than $F(t)$ will be found in Chap. 6.

t and τ, and F is continuous by subregions (Sec. 14) over each rectangle, $0 \le t \le T$, $0 \le \tau \le R$ (cf. Sec. 16).

22.   When $F(t,\tau) = S_0(t - \tau)$ in Prob. 20, show that $\hat{f}(s,s) = \frac{1}{2}s^{-2}$ when $s > 0$, also that $F * (t) = \frac{1}{2}t$, and thus verify the generalized-convolution property in this case.

23.   When $F(t,\tau) = J_0(2\sqrt{t\tau})$, we found (Prob. 10, Sec. 21) that $f(s,t) = s^{-1}\exp(-\tau/s)$. Apply the generalized convolution property (Prob. 20) to derive the integration formula

$$\int_0^t J_0[2\sqrt{\tau(t - \tau)}]\, d\tau = \sin t \qquad\qquad (t \ge 0).$$

24.   When τ is fixed and positive, let $Z(t,\tau)$ denote the solution of the differential equation with constant coefficients (independent of t and τ)

$$Z(t,\tau) + \sum_{n=1}^m a_n \frac{d^n}{dt^n} Z(t,\tau) = F(\tau)$$

such that Z and d^nZ/dt^n $(n = 1, 2, \ldots, m - 1)$ all vanish when $t = 0$. Write the corresponding equation for the iterated transform $\hat{z}(s,s)$ of $Z(t,\tau)$ (Prob. 20) and compare it with the equation for the transform $y(s)$ of $Y(t)$, where

$$Y(t) + \sum_{n=1}^m a_n Y^{(n)}(t) = F(t), \qquad Y(0) = Y'(0) = \cdots = Y^{(m-1)}(0) = 0,$$

to show that $y(s) = s\hat{z}(s,s)$. Thus deduce the following formula for the solution of the problem in Y in terms of the solution of the simpler problem in Z:

$$Y(t) = \int_0^t \frac{\partial}{\partial t} Z(t - \tau, \tau)\, d\tau.$$

25.   Use the formula in Prob. 24 to find the solution of the problem

$$Y''(t) - Y(t) = F(t), \qquad Y(0) = Y'(0) = 0,$$

in the form $Y(t) = \int_0^t F(\tau) \sinh(t - \tau)\, d\tau$. Note that $-F(\tau)$ is a particular solution of the differential equation in $Z(t,\tau)$.

3

Elementary Applications

The properties of the Laplace transformation that we have derived up to this point enable us to solve many problems in engineering and physics involving ordinary linear and partial differential equations. In this chapter we shall solve a number of problems in elastic vibrations involving ordinary differential equations. They are problems in which our method is very convenient, although not essential. We shall also treat some simple applications of integral equations.

The next chapter contains applications that involve partial differential equations. The solution of problems of this type is a primary objective of this book. In later chapters we shall extend our treatment of such problems.

28. FREE VIBRATIONS OF A MASS ON A SPRING

Let a body of mass m attached to the end of a coil spring (Fig. 12) be given an initial displacement and an initial velocity and allowed to vibrate. The other end of the spring is assumed to be kept fixed, and the spring is assumed to

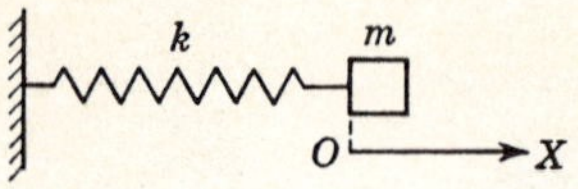

Fig. 12

obey Hooke's law, so that the force exerted by the free end is proportional to the displacement of that end. The factor k of proportionality is called the *spring constant*. We also assume that the mass of the spring can be neglected in comparison with the mass m and that no frictional forces or external forces act on m.

Let X denote the displacement of m from the position of equilibrium; that is, let the origin O denote the position of m when the spring is not deformed. Then according to Newton's second law of motion,

$$(1) \qquad m\frac{d^2 X}{dt^2} = -kX \qquad (k > 0).$$

Let x_0 denote the initial displacement and v_0 the initial velocity, so that the function $X(t)$ satisfies the conditions

$$(2) \qquad X(0) = x_0, \qquad X'(0) = v_0.$$

We can determine the function $X(t)$ by applying the Laplace transformation to both members of Eq. (1) and using the conditions (2). Thus if $x(s)$ denotes the transform of $X(t)$, it follows that

$$m[s^2 x(s) - sx_0 - v_0] = -kx(s),$$

and therefore
$$x(s) = x_0 \frac{s}{s^2 + (k/m)} + v_0 \frac{1}{s^2 + (k/m)}.$$

Hence

$$(3) \qquad X(t) = x_0 \cos \omega_0 t + \frac{v_0}{\omega_0} \sin \omega_0 t$$

$$= \sqrt{x_0{}^2 + \left(\frac{v_0}{\omega_0}\right)^2} \sin (\omega_0 t + \alpha),$$

where $\omega_0 = \sqrt{\dfrac{k}{m}}$, $\tan \alpha = \dfrac{x_0 \omega_0}{v_0}$, $\sin \alpha = \dfrac{\omega_0 x_0}{\sqrt{\omega_0{}^2 x_0{}^2 + v_0{}^2}}$.

The motion described by formula (3) is a simple vibration with angular frequency ω_0, called the *natural frequency* of this system, and phase angle α, and with the amplitude $[x_0{}^2 + (v_0/\omega_0)^2]^{\frac{1}{2}}$.

If a *viscous damping force* proportional to the velocity also acts upon the mass m, as indicated by the presence of a dashpot c in Fig. 13, the equation of motion becomes

$$(4) \qquad mX''(t) = -kX(t) - cX'(t),$$

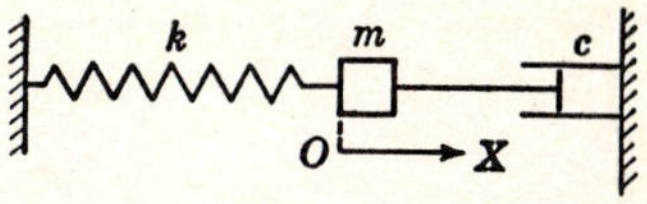

Fig. 13

where the *coefficient of damping* c is a positive constant. Let the mass start from the origin with initial velocity v_0:

$$X(0) = 0, \qquad X'(0) = v_0.$$

The equation in the transform $x(s)$ becomes

$$ms^2 x(s) - mv_0 = -kx(s) - csx(s)$$

or, if we write $2b = c/m$, and again write $\omega_0{}^2 = k/m$,

$$(5) \qquad x(s) = \frac{v_0}{s^2 + 2bs + \omega_0{}^2} = \frac{v_0}{(s+b)^2 + \omega_0{}^2 - b^2}.$$

If $b^2 < \omega_0{}^2$, that is, if the coefficient of damping is small enough that

$$c^2 < 4km,$$

then the formula for the displacement is

$$(6) \qquad X(t) = v_0(\omega_0{}^2 - b^2)^{-\frac{1}{2}} e^{-bt} \sin(t\sqrt{\omega_0{}^2 - b^2}).$$

In the case of critical damping, that is, when $\omega_0 = b$, or

$$c^2 = 4km,$$

it follows from Eq. (5) that

$$(7) \qquad X(t) = v_0 t e^{-bt}.$$

We can see from this formula that the mass m moves in the direction of v_0 until the time $t = 1/b$, then reverses its direction and approaches O as t tends to infinity.

When $c^2 > 4km$, a similar motion of the mass takes place. The discussion of this case and the case of other initial conditions is left to the problems.

The mathematical problem treated in this section can be interpreted also as a problem in electric circuits. This well-known analogy between problems in vibrations of mechanical systems and electric-circuit theory will be observed for other problems in this chapter. Naturally, the notation and terminology differ in the two types of problems.

In the electric circuit shown in Fig. 14, let Q be the charge accumulated in the capacitor C at time t, and I the current in the circuit, so that

$$(8) \qquad I(t) = Q'(t).$$

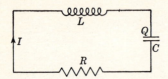

Fig. 14

This equation requires that $I(t)$ be positive when $Q(t)$ is increasing. If the positive sense of flow $I(t)$ of positive charges is taken in the clockwise direction in Fig. 14, then $Q(t)$ measures the charge on the upper plate of the capacitor.

The circuit has a resistance R and a coil of inductance L. Since the sum of the three voltage drops in the circuit is zero in this case of no impressed voltage, then

$$(9) \qquad\qquad LI'(t) + \frac{1}{C}Q(t) + RI(t) = 0.$$

The system of first-order differential equations (8) and (9) can be used directly to determine $I(t)$ and $Q(t)$ in terms of initial values of those functions. By eliminating $I(t)$ from the two equations, however, we see that

$$(10) \qquad\qquad LQ''(t) + RQ'(t) + \frac{1}{C}Q(t) = 0.$$

Except for the notation used, this equation is the same as Eq. (4). When the resistance is negligible, $R = 0$, the equation reduces to our Eq. (1).

The initial conditions in the electrical problem can be made the same as those in the mechanical problem. For example, if the capacitor has an initial charge Q_0 and if the initial current is I_0, then

$$Q(0) = Q_0, \qquad Q'(0) = I_0,$$

which are the same as the initial conditions (2) in our first mechanical problem.

29. FORCED VIBRATIONS WITHOUT DAMPING

Let an external force $F(t)$ act upon the mass in the mechanical system of the last section, assuming there is no damping (Fig. 15). The displacement $X(t)$ of the mass m then satisfies the differential equation

$$(1) \qquad\qquad mX''(t) = -kX(t) + F(t).$$

If the initial conditions are

$$(2) \qquad\qquad X(0) = x_0, \qquad X'(0) = v_0,$$

Fig. 15

the equation in the transform $x(s)$ becomes

$$m[s^2 x(s) - s x_0 - v_0] = -kx(s) + f(s),$$

where $f(s)$ is the transform of the force function $F(t)$. Let ω_0 again denote the natural frequency of the system,

$$\omega_0 = \sqrt{\frac{k}{m}}.$$

Then we can write

(3)
$$x(s) = \frac{x_0 s + v_0}{s^2 + \omega_0^{\,2}} + \frac{1}{m} f(s) \frac{1}{s^2 + \omega_0^{\,2}}.$$

Hence the displacement for any $F(t)$ can be written, with the aid of the convolution, as

(4)
$$X(t) = x_0 \cos \omega_0 t + \frac{v_0}{\omega_0} \sin \omega_0 t + \frac{1}{m\omega_0} \int_0^t \sin \omega_0 (t - \tau) F(\tau)\, d\tau,$$

a result that satisfies conditions (1) and (2) above.

But the motion of the mass under particular external forces $F(t)$ is more interesting than the general formula (4). In these special cases it is often easier to refer to the transform (3) than to formula (4).

When $F(t)$ is a constant F_0, as in the case when the X axis is vertical and the force of gravity acts on m, Eq. (1) can be written

$$mX''(t) = -k\left[X(t) - \frac{F_0}{k} \right].$$

If $Y = X - F_0/k$, this becomes $mY'' = -kY$; so the motion is the same as free vibrations if displacements are measured from a new origin at a distance of F_0/k units from O. Note that F_0/k is the static displacement X_0 of the free end of the spring due to the applied force F_0, since $F_0 = kX_0$.

In case

(5)
$$F(t) = F_0 \qquad\qquad \text{when } 0 < t < t_0,$$
$$= 0 \qquad\qquad \text{when } t > t_0,$$

then
$$f(s) = F_0\left[\frac{1}{s} - \frac{\exp(-t_0 s)}{s} \right].$$

If $x_0 = v_0 = 0$, it follows from Eq. (3) that

(6)
$$x(s) = \frac{F_0}{m}\left[\frac{1}{s(s^2 + \omega_0{}^2)} - \frac{\exp(-t_0 s)}{s(s^2 + \omega_0{}^2)}\right].$$

Now

$$L^{-1}\left\{\frac{1}{s(s^2 + \omega_0{}^2)}\right\} = \frac{1}{\omega_0{}^2}(1 - \cos \omega_0 t) = \frac{2}{\omega_0{}^2}\sin^2\frac{1}{2}\omega_0 t,$$

and if we write $\qquad\qquad \psi(t) = \sin^2 \tfrac{1}{2}\omega_0 t \qquad\qquad$ when $t > 0$,

$$= 0 \qquad\qquad \text{when } t < 0,$$

it follows from Eq. (6) that

(7)
$$X(t) = \frac{2F_0}{k}[\psi(t) - \psi(t - t_0)].$$

The graph of this function can be drawn easily by composition of ordinates. When t_0 is approximately $\tfrac{1}{2}\pi/\omega_0$, the graph is the full-drawn curve in Fig. 16.

When $t_0 = 2\pi/\omega_0$, it follows from Fig. 16 that the mass m performs one oscillation and then remains at the origin (Fig. 17).

The step function (5) is proportional to the unit finite impulse function (3), Sec. 12, with impulse starting at $t = 0$. The area under the graph of our function is $F_0 t_0$. If this is kept fixed, $F_0 t_0 = M_0$, while $t_0 \to 0$ and $F_0 \to \infty$, the force $F(t)$ becomes the impulse represented formally by the equation

(8)
$$F(t) = M_0\delta(t),$$

where $\delta(t)$ is the unit impulse symbol (Sec. 13). The constant M_0 represents the increase in momentum of the mass m at the instant $t = 0$.

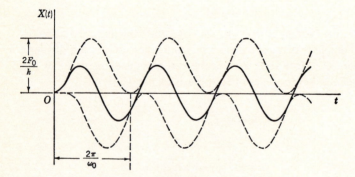

Fig. 16

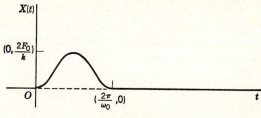

Fig. 17

The formal transform of $M_0\,\delta(t)$ was found to be M_0. This suggests that, in view of Eq. (3), when $x_0 = v_0 = 0$,

$$x(s) = \frac{M_0}{m}\frac{1}{s^2 + \omega_0^2},$$

and hence that

(9)
$$X(t) = \frac{M_0}{m\omega_0}\sin \omega_0 t.$$

It will be left as an exercise to show that formula (9) follows rigorously as a limiting case of formula (7). But the formal solution can be verified as well by noting that, according to formula (9), the momentum of the mass m,

$$M(t) = mX'(t) = M_0 \cos \omega_0 t,$$

satisfies the condition $M(+0) = M_0$. If the mass started from rest, then its momentum jumped to the value M_0 at the instant $t = 0$.

30. RESONANCE

Let the external force in the problem of the last section be

$$F(t) = F_0 \sin \omega t,$$

where F_0 and ω are positive constants. Then according to Eq. (3), Sec. 29,

(1)
$$x(s) = \frac{x_0 s + v_0}{s^2 + \omega_0^2} + \frac{F_0}{m}\frac{\omega}{(s^2 + \omega_0^2)(s^2 + \omega^2)}$$

and, if $\omega \neq \omega_0$,

(2)
$$X(t) = x_0 \cos \omega_0 t + \frac{1}{\omega_0}\left[v_0 + \frac{F_0\omega}{m(\omega^2 - \omega_0^2)}\right]\sin \omega_0 t$$

$$- \frac{F_0}{m(\omega^2 - \omega_0^2)}\sin \omega t.$$

That is, the motion is the superposition of two simple harmonic motions, one with frequency ω_0 and known as the *natural component* of vibration, and the other with frequency ω which is called the *forced component* of the vibration. Note that the natural vibrations are not present in case

$$x_0 = 0, \qquad v_0 = \frac{F_0 \omega}{m(\omega_0^2 - \omega^2)}.$$

However, if the frequency of the periodic force F is the same as the natural frequency of the oscillator, $\omega = \omega_0$, then

$$(3) \qquad x(s) = \frac{x_0 s + v_0}{s^2 + \omega_0^2} + \frac{F_0}{m} \frac{\omega_0}{(s^2 + \omega_0^2)^2}.$$

The presence of the repeated quadratic factor in the denominator here shows that $X(t)$ will contain an unstable component (Sec. 26) having the form of the product of t by a cosine function. In fact,

$$(4) \quad X(t) = x_0 \cos \omega_0 t + \frac{1}{\omega_0^2}\left(v_0 \omega_0 + \frac{F_0}{2m} \right) \sin \omega_0 t - \frac{F_0}{2m\omega_0} t \cos \omega_0 t.$$

In view of the last term here, the amplitude of the oscillations of m increases indefinitely.

In this case the force $F(t)$ is said to be in resonance with the system. We note in particular that if $x_0 = 0$ and $v_0 = -F_0/(2m\omega_0)$ the resonance type of motion reduces to

$$X(t) = -\frac{F_0}{2m\omega_0} t \cos \omega_0 t,$$

shown in Fig. 18.

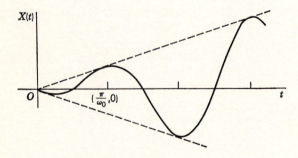

Fig. 18

PROBLEMS

1. A good mechanical oscillator for demonstrating vibrations can be constructed by selecting a length of coil spring such that when a weight of $\frac{1}{2}$ lb ($\frac{1}{2} = mg$, $g = 32$ ft/sec^2) hangs at rest from the free end, the spring is extended by 0.2 ft. Show that the spring constant then has the value $k = \frac{5}{2}$ lb/ft, and that the natural frequency of the oscillator is $\omega_0 = 4\sqrt{10} = 12.65$ rad/sec, or about 2 cycles/sec, and hence the period T of oscillation, $T = 2\pi/\omega_0$, is approximately $\frac{1}{2}$ sec.

2. Let the supported upper end of the spring in Prob. 1 be moved vertically with displacement $Z(t)$, where $Z(0) = 0$. Let $X(t)$ denote the vertical displacement of the mass m from its position on the lower end of the spring at time $t = 0$ when the spring has its natural length. Also let Z and X be positive for displacements downward. If damping forces are neglected, note why $X(t)$ then satisfies the equation

$$mX''(t) = -k[X(t) - Z(t)] + mg.$$

Let $Y(t)$ be the displacement of the mass m from its static position when supported by the spring while $Z(t) = 0$. Show that

$$mY''(t) = -k[Y(t) - Z(t)] = -kY(t) + kZ(t),$$

and hence that Y also represents the displacements in a horizontal oscillator with an external force $kZ(t)$ applied to the mass.

3. The upper end of the spring in the oscillator described in Probs. 1 and 2 is suddenly lowered 1 in. at the instant $t = 0$ and kept in that position during twice the natural period $2T = 4\pi/\omega_0$ sec (about 1 sec), then suddenly returned to its initial position, so that

$$Z(t) = \frac{1}{12}[1 - S_0(t - 2T)] \qquad (t > 0)$$

If the mass m (of weight $\frac{1}{2}$ lb) initially hangs at rest, show that it then makes two oscillations 2 in. downward from and back to its initial position, then remains there after the instant $t = 2T$, since

$$Y(t) = \frac{1}{6}\sin^2\frac{\pi t}{T} \quad \text{when } 0 \leq t \leq 2T, \qquad Y(t) = 0 \quad \text{when } t \geq 2T.$$

4. In Prob. 2 if $Z(t) = v_0 t$ and $Y(0) = Y'(0) = 0$, show that

$$Y(t) = v_0 t - \frac{v_0}{\omega_0}\sin \omega_0 t$$

and describe the motion of the upper end of the spring and the mass.

5. Let the upper end of the spring in the oscillator described in Probs. 1 and 2 be moved vertically by means of a rotating eccentric or cam with the periodic displacement $Z = \frac{1}{12}\sin \omega_0 t$ ft, with the natural frequency $\omega_0 = 4\sqrt{10}$ rad/sec of that oscillator. If $Y(0) = Y'(0) = 0$, show that

$$Y(t) = \frac{1}{24}\sin \omega_0 t - \frac{\omega_0}{24}t \cos \omega_0 t.$$

Note that the vibration of the mass is unstable under that periodic 2-in. oscillation of the support, and that within only four cycles of the oscillations ($t = 8\pi/\omega_0 < 2$ sec), displacements of the mass mount to approximately 1 ft.

6. If the coefficient of damping for the oscillator shown in Fig. 13 is so large that $c^2 > 4km$ and if $X(0) = 0$ and $X'(0) = v_0$, (a) show that

$$X(t) = \frac{v_0}{a} \sinh(at) \exp\left(-\frac{ct}{2m}\right), \quad \text{where } a = \frac{\sqrt{c^2 - 4km}}{2m}.$$

(b) Find $X'(t)$ and show that the mass m moves in the direction of v_0 until the instant

$$t = \frac{1}{2a} \log \frac{c + 2am}{c - 2am},$$

when it turns and approaches the origin.

7. Let the mass m in the damped oscillator shown in Fig. 13 be released from rest with an initial displacement $X(0) = x_0$, and write $2b = c/m$. (a) If $c^2 < 4km$ and $\omega_1 = (\omega_0^2 - b^2)^{\frac{1}{2}}$, show that

$$X(t) = x_0 \frac{\omega_0}{\omega_1} e^{-bt} \cos(\omega_1 t - \alpha) \quad \text{where } \sin \alpha = \frac{b}{\omega_0}, \qquad 0 \le \alpha < \frac{\pi}{2};$$

(b) if $c^2 = 4km$ show that

$$X(t) = x_0 e^{-bt}(1 + bt),$$

and hence that the mass never moves across the origin.

8. When the force on the undamped oscillator shown in Fig. 15 is the periodic function $F(t) = F_0 \sin(\omega t + \theta)$, where θ is a constant, show that resonance occurs when $\omega = \omega_0$.

9. When $t = 0$, let the current in the circuit shown in Fig. 14 be zero and let the capacitor have a positive charge Q_0. If $R > 2\sqrt{L/C}$, derive the formula

$$I(t) = -\frac{2Q_0}{\alpha C} \exp\left(-\frac{Rt}{2L}\right) \sinh \frac{\alpha t}{2L} \qquad \left[\alpha = \left(R^2 - \frac{4L}{C}\right)^{\frac{1}{2}}\right]$$

for the current. Show that $I(t) \le 0$ and $I(\infty) = 0$.

10. The current I, and the charge Q on the capacitor in the circuit shown in Fig. 19, are functions of t that satisfy the conditions

$$L\frac{dI}{dt} + \frac{Q}{C} + RI = E_0, \qquad Q = \int_0^t I(\tau)\, d\tau, \qquad I(0) = 0,$$

when Q and I are initially zero, where t is the time after closing the switch K, and the electromotive force E_0 is constant. If $b = R(2L)^{-1}$ and $\omega_1^2 = (LC)^{-1} - b^2 > 0$, derive the formula

$$I = \frac{E_0}{\omega_1 L} e^{-bt} \sin \omega_1 t.$$

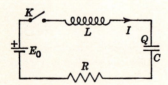

Fig. 19

31. FORCED VIBRATIONS WITH DAMPING

When an external force $F(t)$ acts parallel to the X axis on the mass in the damped oscillator shown in Fig. 13, the equation of motion becomes

$$(1) \qquad mX''(t) = -kX(t) - cX'(t) + F(t).$$

If $X(0) = X'(0) = 0$, the solution of the transformed equation is

$$(2) \qquad x(s) = \frac{1}{m} f(s) \frac{1}{(s + b)^2 + \omega_0{}^2 - b^2} \qquad \left(b = \frac{c}{2m}, \omega_0 = \sqrt{\frac{k}{m}}\right).$$

Again, consider the periodic force

$$F(t) = F_0 \sin \omega t$$

and the case $c^2 < 4km$. Then $b < \omega_0$, and Eq. (2) becomes

$$(3) \qquad x(s) = \frac{F_0}{m} \frac{\omega}{(s^2 + \omega^2)[(s + b)^2 + \omega_1{}^2]} \qquad (\omega_1{}^2 = \omega_0{}^2 - b^2).$$

It follows from Theorem 14, Sec. 26, that $X(t)$ consists of terms of the types $A \cos(\omega t + \theta)$ and $Be^{-bt} \cos(\omega_1 + \theta_1)$, where A, B, θ, and θ_1 are constants. Consequently the component of oscillation with frequency ω_1 is nearly damped out after a sufficiently long time, and the periodic forced component of vibration

$$(4) \qquad X_p(t) = A \cos(\omega t + \theta) \qquad (A > 0)$$

remains. Its amplitude A is, according to Sec. 26, $2|\phi_1(i\omega)|$ where $\phi_1(s) = (s - i\omega)x(s)$. Thus

$$(5) \qquad A = \frac{|F_0|}{m} \frac{1}{|(i\omega + b)^2 + \omega_1{}^2|}.$$

The square of the absolute value in the denominator can be written as $|\omega_0{}^2 - \omega^2 + 2ib\omega|^2$, which is $(\omega^2 - \omega_0{}^2)^2 + 4b^2\omega^2$, and by expanding and completing the square, we can write that polynomial in ω as $(\omega^2 + 2b^2 - \omega_0{}^2)^2 + 4b^2(\omega_0{}^2 - b^2)$. Thus

$$(6) \qquad A = \frac{|F_0|}{m[(\omega^2 + 2b^2 - \omega_0{}^2)^2 + 4b\omega_1{}^2]^{\frac{1}{2}}}.$$

The amplitude A of the periodic vibration depends on the frequency ω of the force $F_0 \sin \omega t$, and it varies directly with the amplitude $|F_0|$ of that force. When F_0 is fixed, the value of ω for which A is greatest is the resonance frequency ω_r of our damped oscillator under that periodic force.

Let the coefficient of damping be small enough that $c^2 < 2km$. Then $2b^2 - \omega_0{}^2 < 0$, and we can see from formula (6) that A is greatest when

$\omega^2 = \omega_0{}^2 - 2b^2$. Thus the resonance frequency in that case is

(7) $$\omega_r = \sqrt{\omega_0{}^2 - 2b^2} \qquad \left(2b = \frac{c}{m}, c^2 < 2km\right),$$

a frequency only slightly less than ω_0 when c is small. When $\omega = \omega_r$, the amplitude A of the undamped component of vibration has the value

(8) $$A_r = \frac{|F_0|}{2bm\omega_1} = \frac{|F_0|}{c\omega_1}.$$

We can see that the expression (3) for $x(s)$ cannot contain a repeated factor of the type $(s^2 + \omega^2)^2$ in its denominator unless $c = 0$ so that $b = 0$. Hence no value of the frequency ω of our periodic force will induce a component of $X(t)$ of type $t \cos(\omega t + \theta)$ as it did for the undamped oscillator.

The amplitude A of the forced component of vibration is still given by formula (5) when the force $F_0 \sin \omega t$ is replaced by a force $F_0 \sin(\omega t + \alpha)$ (Prob. 3, Sec. 32).

In case $c^2 \geqq 2km$, it turns out that the amplitude A of the forced vibration tends toward an upper bound $|F_0|/k$ as ω tends to zero. When ω is small and t is large enough that ωt is near an odd multiple of $\pi/2$, the magnitude of the force $F_0 \sin \omega t$ is then near $|F_0|$. Under static conditions the force $|F_0|$ would displace the end of the spring a distance $|F_0|/k$. The formula for $X(t)$ when $c^2 = 2km$, found in Prob. 4, Sec. 32, shows that when ω is small the mass may oscillate slowly with an amplitude near $|F_0|/k$.

32. A VIBRATION ABSORBER

We have seen that for the simple damped oscillator with an exciting force $F_0 \sin \omega t$ the forced component of vibration $A \cos(\omega t + \theta)$ remains undamped. Let another spring and mass be connected in series with the original mass (Fig. 20), where that second oscillator is undamped. We shall see that if the spring constant and mass of the auxiliary oscillator are chosen so that the natural frequency of that oscillator coincides with the fixed frequency ω of the exciting force, then the forced vibrations of the first mass will be eliminated.

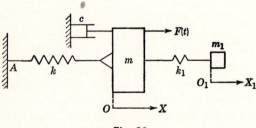

Fig. 20

Let X and X_1 denote the displacements of the masses m and m_1, respectively, from the positions they have when both springs have their natural lengths. If the exciting force is

$$F(t) = F_0 \sin \omega t,$$

and if m and m_1 are initially at rest at their respective origins, then the displacements $X(t)$ and $X_1(t)$ satisfy the following system of differential equations:

$$m\frac{d^2X}{dt^2} = -kX + k_1(X_1 - X) - c\frac{dX}{dt} + F_0 \sin \omega t$$

$$m_1\frac{d^2X_1}{dt^2} = -k_1(X_1 - X),$$

$$X(0) = X'(0) = X_1(0) = X_1'(0) = 0.$$

The transforms $x(s)$ and $x_1(s)$ of $X(t)$ and $X_1(t)$ therefore satisfy the simultaneous algebraic equations

$$(ms^2 + cs + k + k_1)x(s) - k_1x_1(s) = \frac{F_0\omega}{s^2 + \omega^2},$$

$$k_1x(s) - (m_1s^2 + k_1)x_1(s) = 0.$$

Eliminating $x_1(s)$, we find that

(1) $$x(s) = F_0\omega \frac{m_1s^2 + k_1}{(s^2 + \omega^2)p(s)},$$

where

(2) $$p(s) = (m_1s^2 + k_1)(ms^2 + cs + k + k_1) - k_1{}^2.$$

In view of the presence of the quadratic factor $s^2 + \omega^2$ in the denominator of the right-hand member of Eq. (1), it follows that $X(t)$ will contain a term of the type

(3) $$C \cos(\omega t + \theta)$$

unless $k_1/m_1 = \omega^2$, in which case the numerator of the above fraction cancels with the factor in the denominator, leaving

(4) $$x(s) = F_0\omega m_1 \frac{1}{p(s)}.$$

The inverse transform $X(t)$ of the function (4) represents a damped oscillation of the mass m if each of the four roots, real or imaginary, of the equation $p(s) = 0$ has a negative real part (Secs. 24 to 26). Now the polynomial $p(s)$ is the determinant of this system of homogeneous equations in x and x_1:

(5)
$$(ms^2 + cs + k + k_1)x - k_1 x_1 = 0,$$
$$-k_1 x + (m_1 s^2 + k_1)x_1 = 0.$$

When the determinant vanishes, that system is satisfied by numbers x and x_1, both different from zero.

Let s be any one of the four roots of the equation $p(s) = 0$ and let z, z_1 denote a pair of nonvanishing roots of the system (5) that corresponds to that value of s, where z and z_1 may be either real or imaginary numbers. We now write $x = z$ and $x_1 = z_1$ in Eqs. (5), then multiply the members of those equations by the complex conjugates $\bar{z}$ and $\bar{z}_1$, respectively, and add to obtain the equation

(6)
$$(mz\bar{z} + m_1 z_1 \bar{z}_1)s^2 + cz\bar{z}s + B = 0,$$

where
$$B = kz\bar{z} + k_1(z\bar{z} - z\bar{z}_1 - z_1\bar{z} + z_1\bar{z}_1)$$
$$= k|z|^2 + k_1|z - z_1|^2 > 0.$$

If we write $A = m|z|^2 + m_1|z_1|^2$, then Eq. (6) becomes

(7)
$$As^2 + c|z|^2 s + B = 0,$$

where the coefficients A, $c|z|^2$, and B are positive numbers that depend on the value of s. However, the number s is given in terms of those positive coefficients by the quadratic formula

(8)
$$s = \frac{1}{2A}(-c|z|^2 \pm \sqrt{c^2|z|^4 - 4AB}).$$

Whether the radical here is real or imaginary, it follows from formula (8) that the real part of s is negative. This completes the proof that the oscillation $X(t)$ is damped.

Thus the forced component of the vibration of the main mass m is eliminated by the system m_1, k_1, if the natural frequency of that system coincides with the frequency of the exciting force:

(9)
$$\sqrt{\frac{k_1}{m_1}} = \omega.$$

Since all components of the vibration of m are then damped, that mass approaches a fixed position as t increases.

This is the principle of the Frahm vibration absorber, which has been used in such practical appliances as electric hairclippers.[1] Note that, in

[1] Den Hartog, J. P., "Mechanical Vibrations," 4th ed., pp. 87 ff., 1956.

view of Eq. (9), the absorber is designed for a fixed frequency ω of the exciting force $F_0 \sin \omega t$.

By solving the above equations for $x_1(s)$, we can see that the mass m_1 has an undamped component of vibration of the type (3).

PROBLEMS

1. If the initial conditions $X(0) = X'(0) = 0$ used in Sec. 31 for the damped oscillator with exciting force $F(t)$ are replaced by the conditions $X(0) = x_0$, $X'(0) = v_0$, show that the additional terms in the formula for $X(t)$ represent damped oscillations.

2. When $F(t) = F_0\delta(t)$ and $X(0) = X'(0) = 0$ for the damped oscillator considered in Sec. 31, where $c^2 < 4km$, show formally that

$$X(t) = \frac{F_0}{m\omega_1}e^{-bt}\sin\omega_1 t \qquad \left(b = \frac{c}{2m}, \ \omega_1{}^2 = \frac{k}{m} - b^2\right),$$

and note that the jump in the momentum of the mass m at the instant $t = 0$ is F_0.

3. Show that formula (6), Sec. 31, for the amplitude A of the forced component of vibration for the damped oscillator is valid (a) when the exciting force $F(t)$ is $F_0 \cos \omega t$; (b) when $F(t) = F_0 \sin(\omega t + \alpha)$.

4. If $c^2 = 2km$ for the damped oscillator considered in Sec. 31, and if $F(t) = F_0 \sin \omega t$ and $X(0) = X'(0) = 0$, show that

$$X(t) = \frac{F_0}{k}[\cos\theta_2 \cos(\omega t + \theta_1) - e^{-bt}\cos\theta_1 \cos(bt + \theta_2)]$$

where $2b = c/m$ and θ_1 and θ_2 are these arguments of complex numbers:

$$\theta_1 = \arg[-2b\omega + i(\omega^2 - 2b^2)], \qquad \theta_2 = \arg(2b^2 - i\omega^2).$$

Hence when ω is small and ωt is large, show that $X(t)$ is approximately $(F_0/k)\sin\omega t$.

5. In Sec. 31, if $c^2 = 4km$, $F(t) = F_0 \sin \omega t$, and $X(0) = X'(0) = 0$, show that $X(t)$ has the form $A \cos(\omega t + \alpha) + (B + Ct)e^{-bt}$, and when ω is small that $|A|$ is near its upper bound $|F_0|/k$.

6. Let the oscillator described in Probs. 1 and 2, Sec. 30, for which $k = \frac{5}{2}$ lb/ft and $mg = \frac{1}{2}$ lb, be subject to a damping force $-cY'(t)$ where $c = \frac{1}{64}$ lb/ft/sec. (a) If the upper end of the spring is fixed so that $Z(t) = 0$, and if $Y(0) = y_0$ and $Y'(0) = 0$, show that after about nine cycles of oscillation of the mass the amplitude of the displacements $Y(t)$ is reduced to approximately $0.1|y_0|$. (b) When the upper end is oscillated so that $Z(t) = z_0 \sin \omega t$, show that resonance occurs if $\omega = \omega_r$ where $\omega_r = \sqrt{159.5}$ rad/sec, a frequency slightly less than ω_0 (cf. Sec. 31). (c) When $\omega = \omega_r$ in part (b), show that the amplitude A of the undamped forced component of vibration exceeds $12|z_0|$.

7. Let the two masses in the system treated in Sec. 32 have arbitrary initial displacements and velocities. When condition (9) is satisfied by the elements of the absorber, show that the vibration of the main mass again contains no undamped component.

8. (a) If the exciting force $F_0 \sin \omega t$ in the system of Sec. 32 is replaced by the force $F_0 \sin(\omega t + \alpha)$, where α is any constant, show that when condition (9) is satisfied the vibration of m is again entirely damped.

(b) If the exciting force is replaced by $A_1 \sin \omega_1 t + A_2 \sin \omega_2 t$, where the A's and ω's are constants and $\omega_1 \neq \omega_2$, show that the values of m_1 and k_1 cannot be adjusted so that all undamped vibrations of m are absorbed.

9.   Let the force $F(t)$ be removed from the mass m in Fig. 20 and let the end A of the spring k be moved horizontally so that its distance from the wall is $F(t)/k$. Show that the differential equations of motion of m and m_1 are then the same as in the original problem.

10.   A system of two masses and two springs in series, with no damping, is shown in Fig. 21. The weights of the two masses are $m_1 g = 8$ lb and $m_2 g = 2$ lb ($g = 32$ ft/sec^2). The spring constants have the values $k_1 = 24$ lb/ft and $k_2 = 8$ lb/ft. If a periodic force $F = F_0 \cos 2\pi c t$ acts on m_2, find the frequencies c (cycles per second) of the force for which resonance occurs in the vibration of m_1. Assume zero initial displacements and velocities. Also show that resonance occurs in the vibration of m_2 for those same values of c.   *Ans.* $c = 4\pi^{-1}, 4\sqrt{3}\,\pi^{-1}$ cycles/sec.

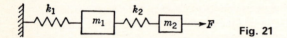

Fig. 21

11.   For the system shown in Fig. 21, if $m_1 = m_2 = \frac{1}{15}$, $k_1 = \frac{3}{15}$, $k_2 = \frac{2}{15}$, and if both masses start from rest in their equilibrium positions, find the formula for the displacement $X_1(t)$ of m_1 when the force $F(t)$ on m_2 is arbitrary.

Ans. $X_1(t) = (6 \sin t - \sqrt{6} \sin t \sqrt{6}) * F(t)$.

12.   Initially the two masses shown in Fig. 22 are at rest and the spring has its natural length. Then a constant force $F = F_0(t > 0)$ acts on m_1. If the system is free from damping, find the formula for the displacement of m_2 and note that the displacement consists of a simple harmonic motion superimposed upon a uniformly accelerated motion. Also show that if $F_0 > 0$, the spring is always compressed by an amount proportional to $1 - \cos \omega t$, where $\omega^2 = k/m_1 + k/m_2$.

Ans. $2\omega^2(m_1 + m_2)X_2(t) = F_0(\omega^2 t^2 - 2 + 2 \cos \omega t)$.

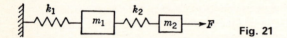

Fig. 22

13.   If the force $F = 0$ in the system shown in Fig. 22 and if the masses are initially released at rest from positions $X_1 = a_1$ and $X_2 = a_2$, show that the frequency of vibration of the masses has the value $\omega = (k/m_1 + k/m_2)^{\frac{1}{2}}$. When $a_2 = -a_1 m_1/m_2$, show that $X_1 = a_1 \cos \omega t$ and $X_2 = a_2 \cos \omega t$ and describe the vibration.

14.   Let the force in the system shown in Fig. 22 be the periodic force $F = F_0 \sin \omega t$, where $\omega^2 = k/m_2$, and let a viscous damping force $-cX_1'(t)$ act on m_1. Find the steady-state vibration of m_2 and note that it does not depend on c.   *Ans.* $-F_0 k^{-1} \sin \omega t$.

15.   In Fig. 23 the end E of the first spring has a periodic displacement $Y = A \sin \omega t$. The two springs are identical. Find the value of ω for which this damped system is in resonance, when $c^2 < 4km$.   *Ans.* $2m^2\omega^2 = 4km - c^2$.

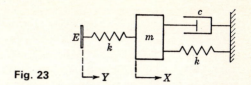

Fig. 23

16: In Fig. 24 the two masses are unit masses ($m = 1$) and the two springs are identical. Initially, the springs have their natural length, the first mass has velocity v_0, and the second mass is at rest. Find the undamped component of vibration of those masses.

Ans. $\frac{1}{2}(v_0/\sqrt{k}) \sin t \sqrt{k}$.

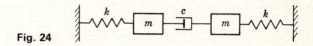

Fig. 24

17. For the undamped system of two equal masses and three identical springs shown in Fig. 25, the initial conditions are $X_1(0) = a_1$, $X_2(0) = a_2$, $X'_1(0) = X'_2(0) = 0$. (a) If $|a_1| \neq |a_2|$, show that the components of vibration of each mass have frequencies $\sqrt{k/m}$ and $\sqrt{3k/m}$. (b) If $a_2 = a_1$, show that both masses vibrate in unison with frequency $\sqrt{k/m}$. (c) If $a_2 = -a_1$, show that the masses vibrate in opposite directions with frequency $\sqrt{3k/m}$.

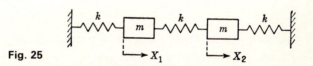

Fig. 25

18. *A damped absorber* Figure 26 shows an absorber with elements m_1, k_1, and c in which the damping is located in the absorber itself. When a force $F_0 \sin \omega t$ acts on the main mass m, the undamped vibrations of m cannot be completely absorbed here, but the coefficient of damping c can be adjusted to give an optimum range of the amplitudes of those vibrations. Let m and m_1 be initially at rest in their positions of equilibrium. Derive the formula

$$x(s) = \frac{F_0 \omega}{s^2 + \omega^2} \frac{m_1 s^2 + cs + k_1}{(m_1 s^2 + cs + k_1)(ms^2 + cs + k + k_1) - (cs + k_1)^2}$$

for the transform of the displacement $X(t)$ of m, then show that the amplitude A of the forced component $A \cos(\omega t + \theta)$ of $X(t)$ is given by the formula

$$\frac{A^2}{F_0^2} = \frac{(m_1 \omega^2 - k_1)^2 + c^2 \omega^2}{[(m\omega^2 - k)(m_1\omega^2 - k_1) - m_1 k_1 \omega^2]^2 + c^2\omega^2[(m + m_1)\omega^2 - k]^2}.$$

(A graphical examination of A^2 as a function of ω^2 would indicate a value of c for which the range of values of A^2 will be as small as possible for all ω.[1])

[1] See Den Hartog, *op. cit.*, pp. 93 ff., for a detailed discussion.

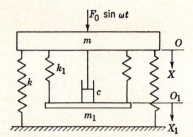

Fig. 26

33. ELECTRIC CIRCUITS

In Sec. 28 we used *Kirchhoff's voltage law*: for each closed circuit in an electrical network the impressed voltage equals the sum of the voltage drops across the elements in that circuit. His *law of currents* states that at each junction point of branches of a network the total current into the point equals the total current away from the point. Let us now present additional applications of these laws for instantaneous behavior of voltage and current, in setting up differential equations for networks.

An electrical analog of the Frahm vibration absorber discussed in Sec. 32 is indicated in Fig. 27, where the impressed voltage is

$$V = V_0 \sin \omega t.$$

According to the law of currents, applied at the junction P,

$$(1) \qquad\qquad I(t) = I_1(t) + I_2(t).$$

We apply the voltage law to the circuit on the left to see that

$$(2) \qquad V = LI'(t) + RI(t) + \frac{1}{C}Q(t) + \frac{1}{C_1}Q_1(t),$$

where $Q'(t) = I(t)$ and $Q'_1(t) = I_1(t)$; then to the circuit on the right to see that

$$(3) \qquad\qquad \frac{1}{C_1}Q_1(t) = L_1 I'_2(t).$$

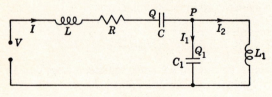

Fig. 27

Let $Q_2(t)$ denote the difference between quantities of charge on the two capacitors, so that

(4) $$Q_2(t) = Q(t) - Q_1(t), \qquad Q_2'(t) = I(t) - I_1(t) = I_2(t).$$

In terms of Q and Q_2 Eqs. (2) and (3) take the form

(5)
$$LQ''(t) = -\frac{1}{C}Q(t) + \frac{1}{C_1}[Q_2(t) - Q(t)] - RQ'(t) + V_0 \sin \omega t,$$
$$L_1 Q_2''(t) = -\frac{1}{C_1}[Q_2(t) - Q(t)].$$

Except for differences in notation the system (5) of differential equations in $Q(t)$ and $Q_2(t)$ is precisely the system written in Sec. 32 for the displacements $X(t)$ and $X_1(t)$ of the masses m and m_1 in the absorber (Fig. 20). Since all the electrical coefficients here, R, L, etc., are positive constants, the signs of the coefficients in Eqs. (5) match those in the equations for X and X_1. The initial conditions for the network match those chosen for the absorber, and the analogy is complete if

(6) $$Q(0) = Q_1(0) = I(0) = I_2(0) = 0.$$

Under conditions (6) the transforms of $I(t)$ and $I_2(t)$ are $sq(s)$ and $sq_2(s)$, which correspond to $sx(s)$ and $sx_1(s)$ for the absorber. According to the results found in Sec. 32 therefore, the current $I(t)$ has an undamped component of type $B \cos(\omega t + \alpha)$ unless

(7) $$\frac{1}{L_1 C_1} = \omega^2.$$

But when the elements of the $L_1 C_1$ circuit satisfy condition (7) for a prescribed frequency ω of the impressed voltage V, the current $I(t)$ contains only damped components; thus $I(t) \to 0$ as $t \to \infty$. The currents $I_1(t)$ and $I_2(t)$ in the $L_1 C_1$ circuit, however, do have undamped components of type $B \cos(\omega t + \alpha)$.

Note that Eqs. (1) to (3) with conditions (6) can be solved directly for the currents either by replacing $Q(t)$ by $\int_0^t I(\tau)\, d\tau$ or $q(s)$ by $i(s)/s$, etc. Thus the transformations of Eqs. (2) and (3) gives the equations

(8)
$$\left(Ls + R + \frac{1}{Cs}\right)i(s) + \frac{1}{C_1 s}i_1(s) = v(s),$$
$$L_1 s\, i(s) - \left(L_1 s + \frac{1}{C_1 s}\right)i_1(s) = 0.$$

Upon eliminating $i_1(s)$, we find that

(9) $$i(s) = v(s)y(s) \qquad \text{if} \qquad y(s) = \frac{s(L_1 s^2 + C_1^{-1})}{p(s)},$$

where $p(s)$ is the polynomial introduced in Sec. 32:

$$p(s) = (L_1 s^2 + C_1^{-1})(Ls^2 + Rs + C^{-1} + C_1^{-1}) - C_1^{-2}.$$

The function $y(s)$ depends only upon the characteristics of the network, not upon the impressed voltage. Since the formal transform of the unit impulse symbol $\delta(t)$ is unity, it follows from Eq. (9) that $y(s) = i(s)$ when $V(t) = \delta(t)$ formally; that is, $Y(t)$ represents the current I produced by the voltage $\delta(t)$. When the network is considered as a system with input $V(t)$ and output $I(t)$, then, in the language of systems analysis, $y(s)$ is the *transfer function* for the system; in view of Eq. (9) the output corresponding to any input $V(t)$ is given by the formula

(10) $$I(t) = V(t) * Y(t).$$

PROBLEMS

1. If $I_1(0) = I_2(0) = 0$ in the network shown in Fig. 28, while the capacitors have the same initial charge Q_0, show that

$$I_1(t) = I_2(t) = -Q_0 \omega \sin \omega t, \qquad \text{where } \omega = (CL)^{-\frac{1}{2}}.$$

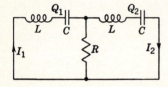

Fig. 28

2. If $I_1(0) = I_2(0) = 0$ in the network shown in Fig. 28 and if $Q_1(0) = a_1$ and $Q_2(0) = a_2$, find the undamped component $\bar{I}_1(t)$ of the current $I_1(t)$, the steady-state current through the first inductance coil.   *Ans.* $\bar{I}_1 = -\frac{1}{2}(a_1 + a_2)\omega \sin \omega t,\ \omega = (CL)^{-\frac{1}{2}}.$

3. Show that the network indicated in Fig. 28 represents an electrical analog of the mechanical system shown in Fig. 24 where Q_1 and Q_2 correspond to the displacements of the two masses.

4. In Fig. 29 the currents $I_1(t)$ and $I_2(t)$ and the charge $Q(t)$ are initially zero.

(a) Show that the current $I_1(t)$ is unstable under the impressed voltage $V = V_0 \sin(\omega t + \alpha)$ if $\omega^2 = (L_1 C)^{-1} + (L_2 C)^{-1}$.

(b) Show that the system indicated in Fig. 22 is a mechanical analog to the network here with $X_1(t)$ and $X_2(t)$ corresponding to $Q_1(t)$ and $Q_2(t)$, where $Q_1'(t) = I_1(t)$, $Q_2'(t) = I_2(t)$, and $Q(t) = Q_1(t) - Q_2(t)$.

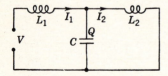

Fig. 29

5. A prescribed current $I(t)$ is supplied to the network shown in Fig. 30. The initial values of $I_1(t)$ and $Q(t)$ are zero. Use the following notation: $V(t)$ is the resulting potential drop from the junction A to the junction B, $\omega_0^2 = (LC)^{-1}$, $2b = (RC)^{-1}$, and $\omega_1^2 = \omega_0^2 - b^2$, given that $b^2 < \omega_0^2$.

(a) Show that the transfer function from input $I(t)$ to output $I_1(t)$ is

$$y(s) = \frac{\omega_0^2}{s^2 + 2bs + \omega_0^2},$$

thus that $I_1(t) = I(t) * Y(t)$, where $Y(t) = \omega_0^2 \omega_1^{-1} e^{-bt} \sin \omega_1 t$.

(b) Find $V(t)$ when $I(t) = I_0$, a constant.   *Ans.* $V(t) = I_0(C\omega_1)^{-1} e^{-bt} \sin \omega_1 t$.

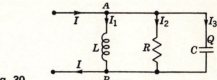

Fig. 30

6. In a simple LC circuit with inductance coil and capacitor in series, let the impressed voltage be a periodic function, with period T, of the type

$$V(t) = a_0 + \sum_{n=1}^{N} (a_n \cos n\omega t + b_n \sin n\omega t),$$

where $\omega = 2\pi/T$. Show that the current $I(t)$ in the circuit is unstable if the frequency ω of $V(t)$ has any one of the values

$$\omega = \frac{1}{m} \frac{1}{\sqrt{LC}} \qquad\qquad (m = 1, 2, \dots, N),$$

provided that a_m and b_m are not both zero.

7. In the network shown in Fig. 31 all currents and charges are zero at the instant $t = 0$ when the switch K is closed. If the impressed voltage is given by the equation $V = V_0 \sin \omega t$, where $\omega^2 = 2(LC)^{-1}$, find the steady-state value $\bar{I}_2(t)$ of the current $I_2(t)$ and note that it is independent of the value of the resistance R.

Ans. $\bar{I}_2(t) = -V_0 C\omega \cos \omega t$.

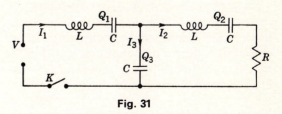

Fig. 31

8. In the system shown in Fig. 25 apply an exciting force to the first mass and viscous damping to the second mass. Show that the resulting system is a mechanical analog of the network shown in Fig. 31.

9. Initially the currents and the charge on the capacitor are zero in the network shown in Fig. 32. The voltage drop from A_1 to B_1 caused by the mutual inductance of the two coils is $MI_2'(t)$, and the drop from A_2 to B_2 caused by that mutual inductance is $MI_1'(t)$, where $M^2 < L_1 L_2$. The impressed voltage is given by the equation $V = V_0 \cos \omega t$. When the values of C and L_2 are adjusted so that $L_2 C = \omega^{-2}$, show that the current $I_1(t)$ in this idealized resistance-free circuit has a simple periodic variation at all times with a frequency greater than ω.

$$\text{Ans. } I_1(t) = V_0 C \frac{L_2}{L_1} \omega_1 \sin \omega_1 t, \qquad \omega_1 = \omega \left(\frac{L_1 L_2}{L_1 L_2 - M^2} \right)^{\frac{1}{2}}.$$

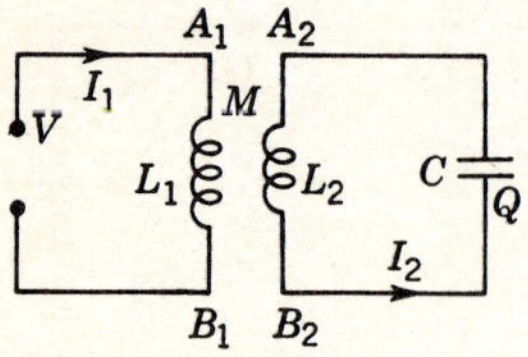

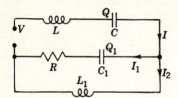

Fig. 32

10. Show that the network in Fig. 33 represents an analog of the damped absorber shown in Fig. 26 if initial currents and charges are zero, where charges Q and $Q - Q_1$ correspond to displacements X and X_1 and the voltage V corresponds to the force $F_0 \sin \omega t$.

Fig. 33

34.   EVALUATION OF INTEGRALS

Laplace transforms can be used to evaluate some types of integrals containing a parameter. The manipulative procedure is often direct and simple, but, as the examples and exercises here will indicate, proofs of the reliability of those formal results can be challenging. In our proofs we use this special case of the theorem on uniqueness stated in Sec. 6: A function $f(s)$ cannot have more than one inverse transform $F(t)$ that is continuous over the half line $t \geqq 0$ and of exponential order.

Example 1   Evaluate the integral

(1)
$$F(t) = \int_0^\infty \frac{\cos tx}{x^2 + a^2} \, dx.$$

Here $F(t)$ is the Fourier cosine transform (Chap. 10) of $(x^2 + a^2)^{-1}$. Note that F is bounded: $|F(t)| \leq \int_0^\infty (x^2 + a^2)^{-1}\, dx = \pi/|2a|$; also, $F(-t) = F(t)$.

Let us first proceed formally to transform with respect to t and interchange the order of integration with respect to x and t:

$$(2) \qquad f(s) = \int_0^\infty L\{\cos tx\} \frac{dx}{x^2 + a^2} = \int_0^\infty \frac{s\, dx}{(x^2 + a^2)(x^2 + s^2)}$$

$$= \frac{s}{s^2 - a^2} \int_0^\infty \left(\frac{1}{x^2 + a^2} - \frac{1}{x^2 + s^2} \right) dx = \frac{\pi}{2a} \frac{1}{s + a}$$

if $a > 0$ and $s > 0$. Thus the integral has the value

$$(3) \qquad\qquad\qquad F(t) = \frac{\pi}{2a} e^{-at} \qquad\qquad (a > 0, t \geq 0).$$

To see that the formal step (2) is sound, we first note that, whenever T and s are positive constants,

$$(4) \qquad \int_0^T e^{-st} \int_0^\infty \frac{\cos tx}{x^2 + a^2} \, dx \, dt = \int_0^\infty \frac{1}{x^2 + a^2} \int_0^T e^{-st} \cos tx \, dt \, dx,$$

because the absolute value of the entire integrand does not exceed $(x^2 + a^2)^{-1}$, a function independent of t whose integral from $x = 0$ to $x = \infty$ exists. The integral $F(t)e^{-st}$ therefore converges uniformly with respect to t when $t \geq 0$ by the Weierstrass test and (Sec. 15) the interchange of order of integration with respect to x and t made in step (4) is valid. Also, F is continuous ($t \geq 0$) and bounded so, according to the theorem on uniqueness, if $L\{F\}$ is given by formula (2), then F must be the exponential function (3).

The final integral in Eq. (4) can be evaluated by integration by parts, and the equation can be written

$$(5) \qquad\qquad \int_0^T e^{-st} F(t) \, dt = \int_0^\infty g(x,T) \, dx,$$

where
$$g(x,T) = \frac{e^{-st}(x \sin Tx - s \cos Tx) + s}{(x^2 + a^2)(x^2 + s^2)}.$$

This continuous function of x and T has a limit $g(x,\infty)$,

$$g(x,\infty) = \lim_{T \to \infty} g(x,T) = s(x^2 + a^2)^{-1}(x^2 + s^2)^{-1},$$

uniformly with respect to x when $x \geq 0$ since

$$|g(x,T) - g(x,\infty)| < e^{-sT} \frac{x + s}{(x^2 + a^2)(x^2 + s^2)} \leq M e^{-sT},$$

where M is the maximum value of the quotient of polynomials in x.

To each positive number ϵ there corresponds a number T_ϵ, independent of x, such that $Me^{-sT} < \epsilon$ when $T > T_\epsilon$. Moreover, the second integral in Eq. (5) converges uniformly with respect to T, according to the Weierstrass test. The limit of the integral (5) as $T \to \infty$ is therefore $\int_0^\infty g(x,\infty)\,dx$ (Prob. 10, Sec. 27), since the latter integral exists; thus Eq. (2) follows from Eq. (4).

Example 2   Evaluate the integral

(6) $$F(t) = \int_0^\infty \frac{\sin tx}{x}\,dx.$$

The formal procedure is simple. When $t > 0$ and $s > 0$

(7) $$f(s) = \int_0^\infty L\{\sin tx\}\frac{dx}{x} = \int_0^\infty \frac{dx}{x^2 + s^2} = \frac{\pi}{2s};$$

hence $F(t) = \pi/2$. In view of Eq. (6), $F(t)$ is an *odd function*: $F(-t) = -F(t)$, and $F(0) = 0$. Therefore

(8) $$F(t) = \frac{\pi}{2} \quad (t > 0); \qquad F(t) = -\frac{\pi}{2} \quad (t < 0); \qquad F(0) = 0.$$

The Weierstrass test cannot be used here to justify step (7). In fact, if our formal result (8) is correct, then F is discontinuous at the origin $t = 0$, and the integral (6) cannot converge uniformly over an interval that includes the origin.

Now the function $S(r) = \sin r/r$, $r \neq 0$, $S(0) = 1$, that is,

(9) $$S(r) = 1 - \frac{r^2}{3!} + \frac{r^4}{5!} - \cdots \qquad (-\infty < r < \infty)$$

is continuous, and $|S(r)| \leq 1$, for all real r. Note that the sum of the alternating series (9) is positive and not greater than unity when $|r| \leq 1$, and if $|r| > 1$, then $|\sin r/r| < 1$. To see that the integral (6) exists, we note that its integrand $tS(tx)$ is continuous in t and x and, if t_0, x_0, and x_1 are positive constants, an integration by parts shows that

$$\int_{x_0}^{x_1} \frac{\sin tx}{x}\,dx = \frac{\cos tx_0}{tx_0} - \frac{\cos tx_1}{tx_1} - \int_{x_0}^{x_1} \frac{\cos tx}{tx^2}\,dx \qquad (t \geq t_0),$$

and this has a limit as $x_1 \to \infty$. Thus $F(t)$ exists because

$$F(t) = \int_0^{x_0} \frac{\sin tx}{x}\,dx + \frac{\cos tx_0}{tx_0} - \int_{x_0}^\infty \frac{\cos tx}{tx^2}\,dx \qquad (t > 0),$$

and because $F(0) = 0$ and $F(-t) = -F(t)$.

When $t > 0$, the substitution $r = tx$ shows that our integral has a constant value, $F(t) = \int_0^\infty r^{-1} \sin r \, dr$. Hence F is bounded, continuous when $t > 0$, and $F(+0)$ exists. If we establish formula (7) for $f(s)$, then F must be the step function (8).

The remainder for integral (6) can be written

$$R(t,X) = \int_X^\infty \frac{\sin tx}{x} \, dx = \frac{\cos tX}{tX} - \int_X^\infty \frac{\cos tx}{tx^2} \, dx$$

and we can see that, when $t \geq t_0 > 0$, for each positive number ϵ there is a number X_ϵ independent of t such that

$$|R(t,X)| \leq \frac{1}{tX} + \frac{1}{tX} \leq \frac{2}{t_0 X} < \epsilon \qquad \text{when } X > X_\epsilon.$$

Thus integral (6) converges uniformly with respect to t $(t \geq t_0 > 0)$, and we can write, with the aid of an elementary integration,

$$\int_{t_0}^T e^{-st} F(t) \, dt = \int_0^\infty \frac{1}{x} \int_{t_0}^T e^{-st} \sin tx \, dt \, dx$$

$$= \int_0^\infty h(T,x) \, dx - \int_0^\infty h(t_0,x) \, dx$$

where h is this continuous function:

$$h(t,x) = -\frac{e^{-st}}{x^2 + s^2}\left(st\frac{\sin tx}{tx} + \cos tx\right) \qquad (s > 0).$$

The integral $\int_0^\infty h(t_0,x) \, dx$ converges uniformly with respect to $t_0 (0 \leq t_0 < 1)$ according to the Weierstrass test, so it is continuous in t_0 at the point $t_0 = 0$; thus

(10) $$\int_0^T e^{-st} F(t) \, dt = \int_0^\infty h(T,x) \, dx + \int_0^\infty \frac{dx}{x^2 + s^2} \qquad (s > 0).$$

Let $N(s)$ denote the maximum value of $(sT + 1)e^{-sT}$. Then

(11) $$|h(T,x)| \leq \frac{N(s)}{x^2 + s^2} \qquad \text{and} \qquad |h(T,x)| \leq \frac{sT + 1}{s^2} e^{-sT}.$$

The uniform convergence of the integral $\int_0^\infty h(T,x) \, dx$ and the fact that $h(T,x) \to 0$ as $T \to \infty$, uniformly with respect to x, follow from conditions (11). Thus as $T \to \infty$, that integral vanishes, and Eq. (10) leads to step (7), so the bounded continuous function F has the transform $\frac{1}{2}\pi/s$, and therefore $F(t) = \frac{1}{2}\pi$ when $t > 0$. This completes the proof.

Incidentally, we have shown that $\mathrm{Si}(\infty) = \frac{1}{2}\pi$, for when $t > 0$, we found that $F(t) = \frac{1}{2}\pi$, and Eq. (6) can be written

(12) $$F(t) = \int_0^\infty \frac{\sin r}{r} \, dr = \mathrm{Si}\,(\infty) = \frac{\pi}{2}.$$

35.   EXPONENTIAL- AND COSINE-INTEGRAL FUNCTIONS

Convenient forms of the *exponential-integral function* are

$$(1) \qquad E_1(t) = \int_t^\infty \frac{e^{-y}}{y}\, dy = \int_1^\infty \frac{e^{-tx}}{x}\, dx \qquad (t > 0).$$

Formally, the transform of the second integral is

$$(2) \qquad \int_1^\infty \frac{dx}{x(x+s)} = \frac{1}{s}\int_1^\infty \left(\frac{1}{x} - \frac{1}{x+s}\right) dx = \frac{1}{s}\log \frac{x}{x+s}\Big]_1^\infty \qquad (s > 0).$$

The proof that expression (2) represents $L\{E_1(t)\}$ is left to the problems. The basic form of transform 100, Appendix A, Table A.2, is, therefore,

$$(3) \qquad L\{E_1(t)\} = \frac{1}{s}\log(s + 1) \qquad (s > 0).$$

In Sec. 22 we found that the transform of the sine-integral function, Si t, is s^{-1} arccot s. Now let us write the transform of the *cosine-integral function*

$$(4) \qquad \text{Ci}\, t = -\int_t^\infty \frac{\cos r}{r}\, dr = -\int_1^\infty \frac{\cos tx}{x}\, dx \qquad (t > 0).$$

An integration by parts, first over a bounded interval $(1,X)$, shows that

$$(5) \qquad \text{Ci}\, t = \frac{\sin t}{t} - \frac{1}{t}\int_1^\infty \frac{\sin tx}{x^2}\, dx \qquad (t > 0);$$

thus Ci t is continuous when $t > 0$, and Ci $(\infty) = 0$.

The procedure used in Example 2, Sec. 34, can be applied to the final integral in formula (4) to prove that

$$(6) \qquad L\{\text{Ci}\, t\} = -s\int_1^\infty \frac{dx}{x(x^2 + s^2)} = -\frac{1}{2s}\log(s^2 + 1) \qquad (s > 0).$$

Next let us derive the transform (No. 98, Appendix A, Table A.2) of the function $\cos t$ Si $t - \sin t$ Ci t which can be written

$$\cos t \int_0^t \frac{\sin r}{r}\, dr + \sin t \int_t^\infty \frac{\cos r}{r}\, dr$$

$$= \cos t \int_0^\infty \frac{\sin r}{r}\, dr - \int_t^\infty (\sin r \cos t - \cos r \sin t)\frac{dr}{r}$$

when $t > 0$. By replacing Si (∞) here by $\frac{1}{2}\pi$ (Sec. 34) and substituting tx for $r - t$, we can write the last expression in the form

$$(7) \qquad \frac{\pi}{2}\cos t - \int_t^\infty \frac{\sin(r-t)}{r}\, dr = \frac{\pi}{2}\cos t - \int_0^\infty \frac{\sin tx}{x+1}\, dx.$$

When we transform that final form (7), we find that

$$(8) \qquad L\{\cos t \operatorname{Si} t - \sin t \operatorname{Ci} t\} = \frac{\log s}{s^2 + 1} \qquad (s > 0).$$

In like manner we can derive the transformation (No. 99, Appendix A, Table A.2)

$$(9) \qquad L\{\cos t \operatorname{Ci} t + \sin t \operatorname{Si} t\} = -\frac{s \log s}{s^2 + 1} \qquad (s > 0).$$

Finally, let us evaluate the Laplace integral

$$(10) \qquad g(x) = \int_0^\infty \frac{e^{-tx}}{t^2 + 1} \, dt \qquad (x > 0).$$

Its transform with respect to x is found to be

$$\hat{g}(s) = \frac{\pi}{2} \frac{s}{s^2 + 1} - \frac{\log s}{s^2 + 1} \qquad (s > 0).$$

In view of formula (8) we have the integration formula

$$(11) \qquad g(x) = (\tfrac{1}{2}\pi - \operatorname{Si} x) \cos x + \operatorname{Ci} x \sin x,$$

which is transformation 120, Appendix A, table A.2:

$$(12) \qquad L\left\{\frac{1}{t^2 + 1}\right\} = \left(\frac{1}{2}\pi - \operatorname{Si} s\right) \cos s + \operatorname{Ci} s \sin s \qquad (s > 0).$$

PROBLEMS

Establish the integration formulas in Probs. 1 to 3, where $t \geq 0$.

1. $\displaystyle\int_0^\infty \frac{\sin tx}{x(x^2 + 1)} \, dx = \frac{\pi}{2}(1 - e^{-t})$,

the Fourier sine transform of $x^{-1}(x^2 + 1)^{-1}$ (Chap. 13).

2. $\displaystyle\int_{-\infty}^\infty \frac{\sin^2 tx}{x^2} \, dx = \int_0^\infty \frac{1 - \cos 2tx}{x^2} \, dx = \pi t$.

3. $\displaystyle\int_0^\infty \frac{1 - \exp(-tx^2)}{x^2} \, dx = \sqrt{\pi t}$.

Obtain formally the integration formulas in Probs. 4 to 7.

4. $\displaystyle\int_{-\infty}^\infty \frac{x \sin tx}{x^2 + a^2} \, dx = \pi e^{-at}$ $\qquad (a > 0, t > 0).$

5. $\displaystyle\int_0^\infty \frac{e^{-x\tau}}{(\tau + 1)\sqrt{\tau}} \, d\tau = \pi e^x \operatorname{erfc} \sqrt{x}$ $\qquad (x \geq 0).$

(cf. transformation 111, Appendix A, Table A.2.)

6. $\displaystyle\int_0^\infty \frac{\sin tx}{\sqrt{x}}\,dx = 2\int_0^\infty \sin ty^2\,dy = \left(\frac{\pi}{2t}\right)^{\frac{1}{2}}$ $(t > 0)$.

[Note that $y^4 + s^2 = (y^2 + s)^2 - 2sy^2 = (y^2 - y\sqrt{2s} + s)(y^2 + y\sqrt{2s} + s)$.]

7. $\displaystyle\int_0^\infty \exp(-tx^2)\cos 2tx\,dx = \frac{1}{2}\sqrt{\frac{\pi}{t}}\,e^{-t}$ $(t > 0)$.

8.   When $T > t_0 > 0$ and $s > 0$, prove that

$$\int_{t_0}^T e^{-st}\int_1^\infty \frac{e^{-tx}}{x}\,dx\,dt = \int_1^\infty \frac{\exp[-(x+s)t_0] - \exp[-(x+s)T]}{x(x+s)}\,dx;$$

then prove that the limit of that integral as $t_0 \to 0$ and $T \to \infty$ is the integral (2), Sec. 35, which represents $L\{E_1(t)\}$.

9.   Use the transformation (3), Sec. 35, to show that, when $a > 0$,

$$L\{e^{at}E_1(at)\} = \frac{\log s - \log a}{s - a}\qquad (s > a;\ \text{cf. transformation 97, Appendix A, Table A.2}).$$

Obtain formally the transforms given in Probs. 10 to 12, if $a > 0$ and $b > 0$ (see Prob. 9).

10. $\displaystyle\int_0^\infty \frac{e^{-s\tau}}{\tau + a}\,d\tau = L\left\{\frac{1}{t + a}\right\} = e^{as}E_1(as)$

(cf. transformation 118, Appendix A, Table A.2.)

11. $L\left\{\dfrac{a - b}{(t + a)(t + b)}\right\} = e^{bs}E_1(bs) - e^{as}E_1(as)$ $(a \neq b)$.

12. $L\left\{\dfrac{a}{(t + a)^2}\right\} = 1 - ase^{as}E_1(as)$ (cf. transformation 119, Appendix A, Table A.2.)

13.   Prove that the remainder for the integral $\operatorname{Ci} t = -\int_1^\infty x^{-1}\cos tx\,dx$ satisfies the condition $|R(t,X)| < 2/(t_0 X)$ when $t \geq t_0 > 0$ and hence that the integral converges uniformly when $t \geq t_0$. Then complete the proof of transformation (6), Sec. 35.

When $a > 0$ and $t > 0$, obtain formally the Fourier transformations of $(x + a)^{-1}$ stated in Probs. 14 and 15.

14. $\displaystyle\int_0^\infty \frac{\sin tx}{x + a}\,dx = \left(\frac{\pi}{2} - \operatorname{Si} at\right)\cos at + \operatorname{Ci} at \sin at$.

15. $\displaystyle\int_0^\infty \frac{\cos tx}{x + a}\,dx = \left(\frac{\pi}{2} - \operatorname{Si} at\right)\sin at - \operatorname{Ci} at \cos at$.

16.   Complete the derivation of formula (11), Sec. 35.

17.   Obtain formally the transformation (9), Sec. 35.

18.   The *beta function* is defined as follows:

$$B(x,y) = \int_0^1 r^{x-1}(1 - r)^{y-1}\,dr\qquad (x > 0, y > 0).$$

Since $L\{t^{x-1}\} = \Gamma(x)s^{-x}$, note that

$$L\{t^{x-1} * t^{y-1}\} = \Gamma(x)\Gamma(y)s^{-x-y} = \frac{\Gamma(x)\Gamma(y)}{\Gamma(x+y)}L\{t^{x+y-1}\}.$$

Write the convolution here in terms of the integral representing $B(x,y)$ to show that the beta function can be written in terms of gamma functions (Sec. 5) as

$$B(x,y) = \frac{\Gamma(x)\Gamma(y)}{\Gamma(x+y)} \qquad\qquad (x > 0, y > 0).$$

36.   STATIC DEFLECTION OF BEAMS

Let $Y(x)$ denote the static transverse displacement of a point at distance x from one end of a uniform beam, caused by a load distributed in any manner along the beam (Fig. 34). Under certain idealizing assumptions, primarily that displacements $Y(x)$ and slopes $Y'(x)$ are small, it is shown in mechanics that the internal bending moment $M(x)$ exerted by any span of the beam upon an adjacent span, through their common cross section A_x, is proportional to the curvature of the beam at position x. In fact

(1) $$M(x) = EI\,Y''(x),$$

where E is Young's modulus, the modulus of elasticity in tension and compression for the material, and I is the moment of inertia of the area A_x with respect to the neutral axis of the cross section, the line in A_x about which A_x turns when the beam bends. Note that $Y''(x)$ is approximately the curvature since $Y'(x)$ is small.

If $F(x)$ denotes the internal shearing force at A_x, it can be seen that $F(x)\,dx = dM(x)$; that is (Prob. 14, Sec. 39), $F(x) = M'(x)$, or

(2) $$F(x) = EI\,Y'''(x).$$

Let $W(x)$ represent the transverse load per unit length along the beam. Then $W(x)\,dx = dF(x)$, and it follows from Eq. (2) that

(3) $$Y^{(4)}(x) = aW(x) \qquad\qquad (a^{-1} = EI).$$

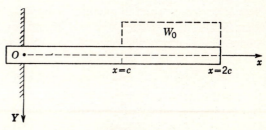

Fig. 34

The shear $F(x)$ is a continuous function except at points where concentrated loads or supports act. When $W(x)$ is sectionally continuous then, in view of Eqs. (2) and (3), $Y'''(x)$ is continuous and $Y^{(4)}(x)$ is sectionally continuous. These are just the continuity conditions that are implied by our formula for the transform, with respect to x, of $Y^{(4)}(x)$.

Although the simple differential Eq. (3) can be solved by successive integrations, the requirement that the solution $Y(x)$ and its derivatives of the first three orders be continuous at all points often involves tedious labor when $W(x)$ is sectionally continuous. The simplicity of Eq. (3) enables us to use the Laplace transformation even though that transformation is not adapted to the two-point boundary conditions which will accompany that equation. Certain Fourier transformations specified by the boundary conditions and the differential form d^4Y/dx^4 are properly adapted to such problems (Chap. 11).

Let us determine the displacements $Y(x)$ in a beam whose end $x = 0$ is built into a rigid support while the end $x = 2c$ is free, or unsupported (Fig. 34). The load $W(x)$ per unit length is zero over the span $0 < x < c$ and a constant w_0 over the span $c < x < 2c$. The total load W_0 is w_0c. Equation (3) can now be written

$$(4) \qquad\qquad Y^{(4)}(x) = aw_0S_c(x) \qquad\qquad (0 < x < 2c),$$

where $S_c(x)$ is our unit step function. The end conditions are

$$(5) \qquad\qquad Y(0) = Y'(0) = 0, \qquad Y''(2c) = Y'''(2c) = 0,$$

since no bending or shear acts on the end $x = 2c$.

Let $y(s)$ denote the Laplace transform of $Y(x)$ when $Y(x)$ satisfies Eq. (4) on the semi-infinite range $x > 0$, together with the first pair of boundary conditions (5). In view of the continuity conditions to be satisfied by $Y(x)$ and its derivatives, we may expect that

$$s^4y(s) - s^3(0) - s^2(0) - sY''(0) - Y'''(0) = aw_0\frac{1}{s}e^{-cs}.$$

We have used a convenient extension, $aw_0S_c(x)$ when $x > 2c$, of the function $aW(x)$ in the right-hand member of Eq. (4) where $0 < x < 2c$. The particular extension used is immaterial because we first seek a function $Y(x)$ that satisfies Eq. (4) when $0 < x < 2c$ and conditions $Y(0) = Y'(0) = 0$, and contains two arbitrary constants $A = Y''(0)$ and $B = Y'''(0)$. Since

$$y(s) = \frac{A}{s^3} + \frac{B}{s^4} + aw_0\frac{1}{s^5}e^{-cs},$$

a function that satisfies those requirements is

$$(6) \qquad\qquad Y(x) = \tfrac{1}{2}Ax^2 + \tfrac{1}{6}Bx^3 + \tfrac{1}{24}aw_0(x - c)^4S_c(x).$$

When $x > c$, it follows from Eq. (6) that

$$Y''(x) = A + Bx + \tfrac{1}{2}aw_0(x - c)^2, \qquad Y'''(x) = B + aw_0(x - c).$$

The last two of conditions (5) are then satisfied if

(7) $$A = \tfrac{3}{2}aw_0c^2, \qquad B = -aw_0c,$$

and the displacements in the beam are given by the equation

(8) $$Y(x) = aw_0(\tfrac{3}{4}c^2x^2 - \tfrac{1}{6}cx^3 + \tfrac{1}{24}(x - c)^4 S_c(x)] \quad (0 \leqq x \leqq 2c).$$

This function satisfies Eqs. (4) and (5) and the continuity conditions.

Since $Y'''(0) = B = -aw_0c$, it follows from Eq. (2) that the shear at $x = 0$ has the value $-w_0c$, or $F(0) = -W_0$ as we should expect, since the magnitude of the vertical force exerted by the support must be the same as the total load on the beam. Actually, the displacements $Y(x)$ for this cantilever beam can be found by first noting that the shear $-Y'''(x)/a$ at each section is the total load on the span to the right of that section.

The bending moment at the end $x = 0$ is $Y''(0)/a$ or A/a:

$$M(0) = \tfrac{3}{2}cW_0 \qquad\qquad\qquad (W_0 = cw_0).$$

If the end $x = 2c$ were pin-supported (hinged or simply supported) so that it can rotate freely about a fixed axis, then the conditions at that end become

$$Y(2c) = Y''(2c) = 0.$$

37. THE TAUTOCHRONE

We shall now discuss a problem in mechanics that leads to a simple integral equation of the convolution type.

The problem is that of determining a curve through the origin in a vertical xy plane such that the time required for a particle to slide down the curve to the origin is independent of the starting position. The particle slides freely from rest under the action of its weight and the reaction of the curve on which it is constrained to move. The required curve is called the tautochrone.

Let σ denote the length of arc of the curve, measured from the origin O, and let (x,y) be the starting point and (ξ,η) any intermediate point (Fig. 35). Equating the gain in kinetic energy to the loss of potential energy, we have

$$\frac{1}{2}m\left(\frac{d\sigma}{dt}\right)^2 = mg(y - \eta),$$

where m is the mass of the particle and t is time. Thus

$$d\sigma = -\sqrt{2g}\sqrt{y - \eta}\,dt,$$

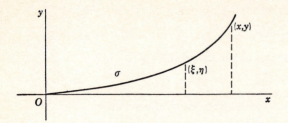

Fig. 35

and upon separating variables and integrating from $\eta = y$ to $\eta = 0$, we have

$$T \sqrt{2g} = \int_{\eta=0}^{\eta=y} \frac{d\sigma}{\sqrt{y-\eta}},$$

where T is the fixed time of descent. Now

$$\sigma = H(\eta),$$

where the function $H(\eta)$ depends upon the curve, and therefore

(1) $$T \sqrt{2g} = \int_0^y (y-\eta)^{-\frac{1}{2}} H'(\eta) \, d\eta.$$

This is an integral equation of convolution type in the unknown function $H'(y)$. We may write it in the form

$$T \sqrt{2g} = y^{-\frac{1}{2}} * H'(y).$$

Let $h(s)$ be the Laplace transform of $H(y)$ with respect to the variable y. Since $H(0) = 0$, it follows formally from Eq. (1) that

$$T \sqrt{2g}\frac{1}{s} = s\, h(s) L\left\{\frac{1}{\sqrt{y}}\right\} = s\, h(s) \sqrt{\frac{\pi}{s}}.$$

That is, $$s\, h(s) = T \sqrt{\frac{2g}{\pi}} \frac{1}{\sqrt{s}};$$

hence

(2) $$H'(y) = \frac{T}{\pi} \sqrt{2g}\, \frac{1}{\sqrt{y}}.$$

We can see that this function does satisfy our integral equation (1) by substituting it into that equation and performing the integration.

Since $$H'(y) = \frac{d\sigma}{dy} = \sqrt{1 + \left(\frac{dx}{dy}\right)^2},$$

the differential equation of the curve in terms of the variables x and y is, according to Eq. (2),

$$1 + \left(\frac{dx}{dy}\right)^2 = \frac{2gT^2}{\pi^2 y} = \frac{a}{y},$$

where $a = 2gT^2/\pi^2$. Separating variables here, we have

$$dx = \sqrt{\frac{a - y}{y}} \, dy,$$

and the necessary integration can be performed easily by substituting $y = a \sin^2 \frac{1}{2}\theta$, for we then find that

$$dx = a \cos^2 \frac{1}{2}\theta \, d\theta = \frac{a}{2}(1 + \cos\theta) \, d\theta.$$

Noting that $x = 0$ when $y = 0$, we see that the parametric equations of the tautochrone are therefore

$$(3) \qquad\qquad x = \frac{a}{2}(\theta + \sin\theta), \qquad y = \frac{a}{2}(1 - \cos\theta).$$

These equations represent the cycloid generated by a point P on a circle of radius $\frac{1}{2}a$ as the circle rolls along the lower side of the line $y = a$. The parameter θ is the angle through which the radius drawn to the point P has turned, where the initial position of P is at the origin. Our tautochrone is of course just one arch of this cycloid. Since $a = 2gT^2/\pi^2$, the diameter of the generating circle is determined by the time T of descent.

The above problem can be generalized in various ways so as to lead to other interesting questions; in fact, it was a generalization of the problem of the tautochrone that led the great Norwegian mathematician Niels Abel (1802–1829) to introduce the subject of integral equations.[1]

If the time T of descent is a function $F(y)$, for example, our integral equation (1) becomes

$$(4) \qquad\qquad \sqrt{2g}\,F(y) = \int_0^y (y - \eta)^{-\frac{1}{2}} H'(\eta) \, d\eta.$$

38. SERVOMECHANISMS

Simple integral equations as well as differential equations arise in the theory of *automatic control*. As a special case we consider servomechanisms that force the angle of turn $\Theta_0(t)$ of a rotating shaft to follow closely the angle of turn $\Theta_i(t)$ of a pointer or indicator, where t denotes time. The shaft and

[1] See Bôcher, M., "Integral Equations," p. 6, 1909.

material rigidity attached to it have a total moment of inertia I that is much greater than that of the pointer. Mechanisms of that general type are used, for instance, in directing antiaircraft guns and in aircraft control.

Let $\Phi(t)$ be the angle of deviation between shaft and pointer or the difference between output angle and input angle:

$$(1) \qquad \Phi(t) = \Theta_0(t) - \Theta_i(t).$$

An auxiliary system or servomechanism can be designed to measure $\Phi(t)$ and feed back to the shaft a component of torque that is proportional to the deviation $\Phi(t)$. The servomechanism may contain its own source of power, motors, generators, and other electrical equipment. In order to provide damping in the system, let the servo also supply a component of torque proportional to the rate of deviation $\Phi'(t)$. Then, since the product of I by the angular acceleration of the shaft equals the torque applied to the shaft,

$$(2) \qquad I\Theta_0''(t) = -k\Phi(t) - c\Phi'(t),$$

where k and c are positive constants.

If the shaft is initially at rest and its angle of turn is measured from the initial position, then $\Theta_0(0) = \Theta_0'(0) = 0$. Also, in view of Eq. (1), $\Phi(0) = -\Theta_i(0)$ and in terms of transforms Eq. (2) becomes

$$Is^2\theta_0(s) = -(k + cs)\phi(s) - c\Theta_i(0),$$

where we have assumed that $\Theta_0(t)$, $\Theta_0'(t)$, and $\Phi(t)$, and therefore $\Theta_i(t)$, are continuous when $t \geqq 0$. Since $\theta_0(s) = \phi(s) + \theta_i(s)$, it follows that

$$(3) \qquad \phi(s) = -\frac{Is^2\theta_i(s) + c\Theta_i(0)}{Is^2 + cs + k}.$$

Under an input $\Theta_i(t) = At$, it follows from Eq. (3) that

$$\Phi(t) = -\frac{A}{\omega}e^{-bt}\sin \omega t \qquad \left(b = \frac{c}{2I}, \omega^2 = \frac{k}{I} - \frac{c^2}{4I^2}\right).$$

If $k > c^2/(4I)$, the angle of deviation $\Phi(t)$ has a damped oscillation with initial value zero. Since $|\sin \omega t/(\omega t)| < 1$ and $(be)^{-1}$ is the maximum value of te^{-bt}, that oscillation is small at all times when b is large, because

$$|\Phi(t)| < |A|te^{-bt} \leqq \frac{|A|}{be}.$$

If in addition to the two components of torque shown in Eq. (2), a component proportional to the accumulated angle of deviation is produced by the servo, then

$$(4) \qquad I\Theta_0''(t) = -k\Phi(t) - c\Phi'(t) - b\int_0^t \Phi(\tau)\,d\tau,$$

where b is a positive constant. When $\Theta_0(0) = \Theta_0'(0) = 0$ therefore,

$$Is^2[\phi(s) + \theta_i(s)] = -\left(k + cs + \frac{b}{s}\right)\phi(s) - c\Theta_i(0).$$

In terms of positive numbers B, C, and K, where

$$IC = c, \qquad IK^2 = k, \qquad IB^3 = b,$$

the last equation can be written

(5)
$$\phi(s) = -\frac{s^3\theta_i(s) + Cs\Theta_i(0)}{s^3 + Cs^2 + K^2s + B^3}.$$

In the special case $\Theta_i(t) = At$, $K^2 = BC$, Eq. (5) becomes

(6)
$$\phi(s) = -\frac{As}{(s + B)[s^2 + (C - B)s + B^2]}$$

and since the polynomials in the numerator and denominator here are of degrees one and three in s, respectively, it follows from our earlier observations (see the problems, Sec. 27) that $\Phi(0) = 0$; that is, the initial value of the angle of deviation is zero. When $C > B$, all values of s that make the polynomial in the denominator vanish have negative real parts, and consequently (Sec. 26) $\Phi(t)$ contains only damped components. But when $C < B$, $\Phi(t)$ has an unstable oscillation.

39. MORTALITY OF EQUIPMENT

Let the function $F(t)$ denote the number of pieces of equipment on hand at time t, where the number is large enough that we can consider it as a continuous variable instead of a variable that takes on only integral values. The equipment wears out in time, or is lost from service for other reasons, so that, out of n_0 pieces of new equipment introduced at time $t = 0$, the number $N(t)$ in service at time t is given by the formula

(1)
$$N(t) = n_0H(t),$$

where $H(t)$ is a function that determines the surviving equipment after t units of time. Note that $H(0) = 1$, necessarily.

If $R(\tau)$ is the total number of replacements up to time τ, then $R'(\tau)\,d\tau$ is the number of replacements during the time interval from $t = \tau$ to $t = \tau + d\tau$ and the number of survivals at any future time t, out of these replacements, is

$$R'(\tau)H(t - \tau)\,d\tau.$$

The total amount of equipment in service at time t is the sum of these survivals from the replacements during every time interval $d\tau$ between

$\tau = 0$ and $\tau = t$, increased of course by the survivals from the new equipment on hand at time $t = 0$. Therefore

$$(2) \qquad F(t) = F(0)H(t) + \int_0^t R'(\tau)H(t - \tau)\,d\tau.$$

We have assumed here that the equipment $F(0)$ on hand at time $t = 0$ is all new; then $R(0) = 0$.

If the amount $F(t)$ that must be in service at each instant is known and if the survival factor $H(t)$ is known, then Eq. (2) is an integral equation of convolution type in $R'(t)$. Its solution gives the formula by which replacements must be made.

The equation is an integral equation in the survival factor $H(t)$ when $F(t)$ and $R(t)$ are known.

In either case, the transformed equation is

$$(3) \qquad f(s) = F(0)h(s) + s\,r(s)h(s).$$

Then

$$(4) \qquad r(s) = \frac{f(s) - F(0)h(s)}{s\,h(s)},$$

and $R(t)$ is the inverse transform of that function.

Suppose the mortality is exponential in character so that

$$H(t) = e^{-kt}$$

and that the amount of equipment on hand is to be a constant,

$$F(t) = b.$$

Then $h(s) = 1/(s + k)$ and $f(s) = b/s$, and it follows from Eq. (4) that

$$r(s) = bk\frac{1}{s^2}.$$

Therefore replacements must be made at such a rate that the total equipment replaced up to time t is

$$R(t) = bkt,$$

a result that is easily verified as the solution of Eq. (2). Thus replacements must be made at the rate of bk pieces per unit time.

PROBLEMS

1. In the example solved in Sec. 36 let the end $x = 2c$ of the beam be pin-supported, rather than free, with no other changes in conditions. Find the vertical force exerted by the beam on the pin, and the vertical force and bending moment exerted on the support at $x = 0$. *Ans.* $\frac{41}{64}W_0$; $\frac{23}{64}W_0$; $\frac{7}{32}cW_0$.

2. Both ends $x = 0$ and $x = 2c$ of a beam are pin-supported. Find the vertical force on each support when a transverse load W_0 is distributed uniformly over the span $c < x < 2c$.   Ans. $W_0/4$; $3W_0/4$.

3. Solve Prob. 2 if both ends are built in rather than pin-supported.

Ans. $\frac{3}{16}W_0$; $\frac{13}{16}W_0$.

4. The end $x = 0$ of a beam is built in and the end $x = 2c$ is pin-supported. The load per unit length is bx on the span $0 < x < c$, and $b(2c - x)$ on the span $c < x < 2c$. Find the vertical force on each support in terms of the total load W_0 on the beam.

Ans. $\frac{21}{32}W_0$; $\frac{11}{32}W_0$.

5. In addition to a distributed load $W(x)$ a single concentrated load W_1 acts between the ends of a beam, at position $x = b$. Then the shear $F(x)$ is continuous except for a jump W_1 at $x = b$. Apply formula (4), Sec. 4, to the function $Y'''(x)$ to show that the transform of Eq. (3), Sec. 36, is

$$s^4 y(s) - s^3 Y(0) - s^2 Y'(0) - s Y''(0) - Y'''(0) - aW_1 e^{-bs} = aw(s).$$

Compare this equation in $y(s)$ with the one obtained formally by replacing $W(x)$ in Eq. (3), Sec. 36, by $W(x) + W_1\delta(x - b)$ and transforming as if $Y'''(x)$ were continuous, where $\delta(x)$ is the unit impulse symbol.

6. In Sec. 37, let the time T of descent be proportional to $\sqrt{y}$, $T\sqrt{2g} = 2B\sqrt{y}$, where $B > 1$. Show that the curve of descent is the line $x = y\sqrt{B^2 - 1}$.

7. If $c = k = 2I$ in Eq. (2), Sec. 38, find the output angle $\Theta_0(t)$ corresponding to the constant input angle $\Theta_i(t) = 1(t > 0)$, and compare them graphically. Assume that $\Theta_0(0) = \Theta_0'(0) = 0$. Also, note that the value of the output lags behind that of the input until $t = 3\pi/4$.   Ans. $\Theta_0(t) = 1 - \sqrt{2}e^{-t} \sin(t + \pi/4)$.

8. For the servomechanism corresponding to Eqs. (4) and (5), Sec. 38, consider this special case: $\Theta_i(t) = At$, $B^3 = CK^2$. If α represents the argument of the complex number $K + Ci$, derive the formula

$$\Phi(t) = \frac{A}{C^2 + K^2}[Ce^{-Ct} - \sqrt{C^2 + K^2} \sin(Kt + \alpha)]$$

for the angle of deviation, and note the undamped component of $\Phi(t)$.

9. Let the servomechanism discussed in Sec. 38 supply only a corrective torque proportional to the deviation angle $\Phi(t)$ while the shaft itself is subject to a damping torque proportional to its angular velocity. If $\Theta_0(0) = \Theta_0'(0) = 0$, show that

$$\theta_0(s) = \frac{k}{Is^2 + cs + k}\theta_i(s),$$

where c, k, and I are positive constants. When $\Theta_i(t) = A$, show that $\Theta_0(t)$ approaches A as t increases.

10. When $H(t) = e^{-kt}$, where $H(t)$ is the survival factor in Sec. 39 and k is a positive constant, (a) find the replacement function $R(t)$ corresponding to an arbitrary amount $F(t)$ of equipment on hand; (b) find $R(t)$ when $F(t) = At + B$.

Ans. (b) $R(t) = (A + Bk)t + \frac{1}{2}Akt^2$.

11. When $H(t)$ in Sec. 39 is the step function $1 - S_k(t)$ so that every piece of new equipment survives for k units of time, (a) show that the number of replacements $R(t)$

up to time t required to maintain $F(t)$ pieces in service at that time is

$$R(t) = F(t) - F(+0) + F(t - k) + F(t - 2k) + F(t - 3k) + \cdots$$

if we define $F(t)$ to be zero when $t < 0$. (b) When $F(t) = A(t > 0)$, show that $R(t) = A[t/k]$, where $[t]$ is the bracket symbol, and draw the graph of $R(t)$.

12.   A particle mass m moves on a vertical X axis under two forces: the force of gravity and a resistance proportional to the velocity. If the axis is taken positive downward, the equation of motion is

$$mX''(t) = mg - kX'(t).$$

Show that its solution, under the conditions $X(0) = 0$, $X'(0) = v_0$, is

$$X(t) = \frac{1}{b^2}[(bv_0 - g)(1 - e^{-bt}) + bgt],$$

where $b = k/m$, and discuss the motion.

13.   Each one of a set of radioactive elements E_1, E_2, E_3, and E_4 disintegrates into the succeeding one at a rate proportional to the number of atoms present, except for the end product E_4 which is a stable element. If $N_i(t)$, where $i = 1, 2, 3, 4$, denotes the number of atoms of element E_i present at time t and if the distinct positive constants c_1, c_2, and c_3 are the respective coefficients of decay of the first three elements, then

$$N_1'(t) = -c_1 N_1(t), \qquad N_2'(t) = -c_2 N_2(t) + c_1 N_1(t),$$

$$N_3'(t) = -c_3 N_3(t) + c_2 N_2(t), \qquad N_4'(t) = c_3 N_3(t).$$

When only M atoms of E_1 are present initially, derive the formula

$$\frac{N_4(t)}{M} = 1 - \frac{c_2 c_3 e^{-c_1 t}}{(c_2 - c_1)(c_3 - c_1)} - \frac{c_1 c_3 e^{-c_2 t}}{(c_1 - c_2)(c_3 - c_2)} - \frac{c_1 c_2 e^{-c_3 t}}{(c_1 - c_3)(c_2 - c_3)}.$$

14.   In Sec. 36 the shear $F(x)$ is the internal force in the direction of the Y axis exerted at a cross section A_x of the beam upon the span to the right of that section, and $M(x)$ is the bending moment exerted on that span there. If $F(0) = F_0$, $M(0) = M_0$, and $W(x)$ is the load per unit length, apply equilibrium conditions to a span extending x units to the right of the section A_0, at $x = 0$, to show that

$$F_0 + \int_0^x W(\xi)\,d\xi - F(x) = 0,$$

$$M_0 + F_0 x + \int_0^x (x - \xi)W(\xi)\,d\xi - M(x) = 0.$$

Thus derive these relations between shear, load, and bending moment:

$$F'(x) = W(x), \qquad M'(x) = F(x).$$

4
Problems in Partial Differential Equations

40 THE WAVE EQUATION

Several functions in physics and engineering satisfy the partial differential equation

(1)
$$\frac{\partial^2 Y}{\partial t^2} = a^2 \frac{\partial^2 Y}{\partial x^2},$$

known as the wave equation in two independent variables x and t. We shall use literal subscripts to indicate partial derivatives; then Eq. (1) can be written

$$Y_{tt}(x,t) = a^2 Y_{xx}(x,t).$$

Brief derivations of this equation for some elementary physical functions will now be given. The derivations are helpful in writing modifications of the equation and in setting up boundary conditions for specific problems.

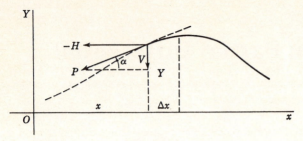

Fig. 36

First, let $Y(x,t)$ denote the displacement at time t away from the x axis of a point (x,Y) of a string in the xY plane under tension P (Fig. 36). The string is assumed to be flexible enough that all bending moments transmitted between its elements can be neglected; thus each element pulls tangentially on an adjacent element with a force of magnitude P. Suppose further than conditions are such that the magnitude H of the x component of the tension remains essentially constant at all times for all points (x,Y); in particular all displacements $Y(x,t)$ are assumed small compared with the length of the string. Finally let the slope angle α remain small in order that each small element of length of the string may be approximated by the length of its projection Δx on the x axis. All these idealizing assumptions are satisfied, for instance, by strings of musical instruments under ordinary conditions of operation.

The vertical component of tension is the vertical force V exerted by the part of the string to the left of point (x,Y) on the part to the right of the point. It is proportional to the slope of the string, $-V/H = \tan \alpha$ (Fig. 36); that is,

$$(2) \qquad\qquad V(x,t) = -HY_x(x,t).$$

This is the basic formula for deriving the equation of motion.

Now consider an element of length of the string whose projection on the x axis is Δx. If ρ denotes the mass per unit length, the mass of the element is approximately $\rho \, \Delta x$. If no external forces act on the string, the application of Newton's second law to the element gives, in view of formula (2),

$$(3) \qquad \rho \, \Delta x \, Y_{tt}(x,t) = -HY_x(x,t) + HY_x(x + \Delta x, t)$$

approximately, when Δx is small; that is,

$$Y_{tt}(x,t) = \frac{H}{\rho} \frac{Y_x(x + \Delta x, t) - Y_x(x,t)}{\Delta x}.$$

When we let Δx approach zero, this becomes Eq. (1), where

$$(4) \qquad\qquad a^2 = \frac{H}{\rho}.$$

If in addition to the internal force a force $F(x,t)$ *per unit of mass* acts in the Y direction along the string, then the additional term $\rho \, \Delta x \, F(x,t)$ appears on the right in Eq. (3). The resulting modification of Eq. (1) is

(5)
$$Y_{tt}(x,t) = a^2 Y_{xx}(x,t) + F(x,t).$$

When the Y axis points vertically upward and the weight of the string is to be taken into account, for instance, $F(x,t) = -g$, where g is the acceleration of gravity.

As another physical example consider the longitudinal displacements in a cylindrical or prismatic elastic bar. The values of the variable x are marked on the bar so as to designate the cross section that is x units from one end when the bar is neither stretched nor compressed (Fig. 37). For each value of x the longitudinal displacement $Y(x,t)$ is measured from a fixed origin outside the bar, an origin in the plane occupied by the cross section at x when the bar is unstrained and in some position of reference. Thus, if the bar is moved lengthwise as a rigid body, $Y(x,t)$ is a constant at each time t.

Since $Y(x + \Delta x, t)$ is the displacement of the cross section at $x + \Delta x$, an element of the bar whose natural length is Δx is stretched by the amount $Y(x + \Delta x, t) - Y(x,t)$ at time t. According to Hooke's law the force exerted by the bar upon the left-hand end of the element to produce that extension is

$$-AE\frac{Y(x + \Delta x, t) - Y(x,t)}{\Delta x}.$$

where A is the area of the cross section and E is Young's modulus of elasticity of the material, the modulus in tension and compression. When Δx tends to zero, it follows that the internal force from left to right at the cross section is

(6)
$$F(x,t) = -AEY_x(x,t).$$

This basic formula corresponds to Eq. (2) for the string.

Let ρ denote the mass of the material per unit volume. When we apply Newton's second law to an element of the bar,

(7)
$$\rho A \, \Delta x \, Y_{tt}(x,t) = -AEY_x(x,t) + AEY_x(x + \Delta x, t),$$

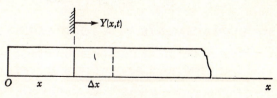

Fig. 37

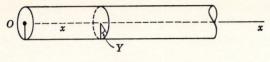

Fig. 38

we find as before that $Y(x,t)$ satisfies Eq. (1), where

$$(8) \qquad\qquad a^2 = \frac{E}{\rho}.$$

When the elastic bar is replaced by a column of air, Eq. (1) has further applications in acoustics.[1]

Again, the function $Y(x,t)$ may represent the angle of turn of a cross section x units from one end of an elastic cylindrical shaft under torsion (Fig. 38). Let I denote the moment of inertia of the cross section with respect to its axis, E_s the modulus of elasticity of the material in shear, and ρ the density of the material. Then, by steps analogous to those used above, we find that the internal torque τ acting through a cross section at position x is

$$(9) \qquad\qquad \tau(x,t) = -IE_s Y_x(x,t)$$

and that Eq. (1) is satisfied by the angle $Y(x,t)$, where

$$(10) \qquad\qquad a^2 = \frac{E_s}{\rho}.$$

Finally, it should be remarked that Eq. (1) is a special case of the *telegraph equation*

$$(11) \qquad Y_{xx}(x,t) = KLY_{tt}(x,t) + (RK + SL)Y_t(x,t) + RSY(x,t),$$

where $Y(x,t)$ represents either the electric potential or the current at time t at a point x units from one end of a transmission line or cable.[2] Here the elements of resistance, inductance, etc., are *distributed* along the cable or through the circuit, in contrast to the *lumped* elements in elementary circuits (Sec. 33) that lead to ordinary differential equations in the currents. The cable has resistance R, electrostatic capacity K, leakage conductance S, and self-inductance L, all per unit length. When R and S are so small that their effect can be neglected, Eq. (11) reduces to Eq. (1), where $a^2 = (KL)^{-1}$.

41   DISPLACEMENTS IN A LONG STRING

Let $Y(x,t)$ represent the transverse displacements of the points of a semi-infinite stretched string, a string having one end fixed so far out on the x axis

[1] Lord Rayleigh, "Theory of Sound," vols. 1 and 2, Dover, 1945.
[2] A derivation of Eq. (11) is outlined in Prob. 6, Sec. 93.

that the end may be considered infinitely far from the origin, and having its other end looped around the Y axis. The loop, initially at the origin, is moved in some prescribed manner along the Y axis (Fig. 39) so that $Y = F(t)$ when $x = 0$ and $t \geq 0$, where $F(t)$ is a prescribed continuous function and $F(0) = 0$. If the string is initially at rest on the x axis, let us find the formula for $Y(x,t)$.

The above conditions on $Y(x,t)$ can be written

(1) $$Y_{tt}(x,t) = a^2 Y_{xx}(x,t) \qquad\qquad (x > 0, t > 0),$$

(2) $$Y(x,0) = Y_t(x,0) = 0 \qquad\qquad (x > 0),$$

(3) $$Y(0,t) = F(t), \qquad \lim_{x \to \infty} Y(x,t) = 0 \qquad\qquad (t \geq 0),$$

where $a^2 = H/\rho$ (Sec. 40). The equation of motion (1) implies that no external forces act along the string.

A problem composed of such conditions is called a *boundary value problem* in partial differential equations. We shall use a formal procedure to solve the problem and then show how our result can be verified as a solution.

If $y(x,s)$ is the Laplace transform of $Y(x,t)$, then, in view of the initial conditions (2), $L\{Y_{tt}\} = s^2 y$. Also,

$$L\{Y_{xx}(x,t)\} = \int_0^\infty \frac{\partial^2}{\partial x^2}[e^{-st}Y(x,t)]\,dt$$

$$= \frac{\partial^2}{\partial x^2} \int_0^\infty e^{-st}Y(x,t)\,dt = y_{xx}(x,s),$$

if the function $e^{-st}Y(x,t)$ satisfies conditions under which the indicated interchange of order of integration with respect to t and differentiation with respect to x is valid (Sec. 15). When both members of the partial differential equation (1) are transformed and conditions (2) are used, we therefore obtain the equation $s^2 y = a^2 y_{xx}$ in the transform of our unknown function.

From conditions (3) we find that $y(0,s) = f(s)$, where $f(s)$ is the transform of $F(t)$, and $\lim_{x \to \infty} y(x,s) = 0$, provided that the order of integrating with respect to t and taking the limit as $x \to \infty$ can be interchanged. The

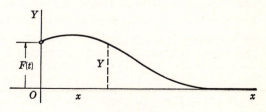

Fig. 39

transformed boundary value problem in $y(x,s)$ is therefore

(4)
$$\frac{d^2 y}{dx^2} - \frac{s^2}{a^2} y = 0 \qquad\qquad (x > 0),$$

(5)
$$y(0,s) = f(s), \qquad \lim_{x \to \infty} y(x,s) = 0.$$

Here we have used the symbol for ordinary rather than partial differentiation because s is only a parameter in the problem; no differentiation with respect to s is involved.

A convenient form of the general solution of Eq. (4) is

$$y(x,s) = C_1 e^{-sx/a} + C_2 e^{sx/a},$$

where C_1 and C_2 may be functions of s. This solution could of course be obtained by transforming the members of Eq. (4) with respect to x. We consider s as positive, since Laplace transforms generally exist for all s greater than some fixed number. Then $C_2 = 0$, if $y(x,s)$ is to approach zero as x tends to infinity. The first of conditions (5) is also satisfied if $C_1 = f(s)$, and the solution of the transformed problem is

(6)
$$y(x,s) = e^{-(x/a)s} f(s).$$

The translation property (Sec. 11) enables us to write the inverse transform of $y(x,s)$ at once:

(7)
$$Y(x,t) = F\left(t - \frac{x}{a}\right) \qquad\qquad \text{when } t \geqq \frac{x}{a},$$

$$= 0 \qquad\qquad \text{when } t \leqq \frac{x}{a}.$$

Since $F(t)$ is continuous and $F(0) = 0$, we can see that the function $Y(x,t)$ described by formula (7) is continuous when $x \geqq 0$ and $t \geqq 0$, including points on the line $x = at$ in the xt plane. The function clearly satisfies the boundary conditions (2) and (3).

Any function of $t - x/a$ is easily seen to be a solution of the wave equation (1) when its derivative of the second order exists. Let $F(t)$ satisfy these additional conditions: $F'(t)$ and $F''(t)$ are continuous when $t \geqq 0$ except possibly for finite jumps at $t = t_i (i = 1, 2, \ldots)$. The function $Y(x,t)$ described by formula (7) then satisfies Eq. (1) except possibly at points (x,t) on the lines $t - x/a = t_i$ and $t - x/a = 0$ in the quadrant $x > 0, t > 0$. In this sense, then, the function (7) is verified as a solution of our boundary value problem *regardless of the validity of formal steps* taken to obtain that function. In case $F''(t)$ is continuous whenever $t \geqq 0$, and $F(0) = F'(0) = F''(0) = 0$ so that $Y(x,t)$ and its partial derivatives of first and second order are also continuous on the line $t = x/a$, then the function (7) satisfies Eq. (1)

with no exceptions when $x > 0$ and $t > 0$ and represents a solution of the boundary value problem in the ordinary sense. Note that conditions of exponential order are not involved in the verification of our solution.

According to formula (7), a point of the string x units from the origin remains at rest until the time $t = x/a$. Starting at that time, it executes the same motion as the loop at the left-hand end. The retarding time x/a is the time taken by a disturbance to travel the distance x with velocity a. Since $a = (H/\rho)^{\frac{1}{2}}$, note that it has the physical dimensions of velocity. The vertical force from left to right at a point (Sec. 40) is $-HY_x(x,t)$; thus the vertical force exerted on the loop to make it move in the prescribed manner, as found from formula (7), is

$$V(0,t) = \frac{H}{a}F'(t) = \sqrt{\rho H}\,F'(t).$$

Finally, let us observe instantaneous positions of the string corresponding to the loop movement

(8) $$F(t) = \sin \pi t \qquad \text{when } 0 \leq t \leq 1,$$
$$= 0 \qquad \text{when } t \geq 1;$$

thus the loop is lifted to the position $Y = 1$ and returned to $Y = 0$, where it remains after time $t = 1$. In this case $F(t - x/a) = 0$ when $t - x/a \geq 1$, that is, when $x \leq a(t - 1)$. Also $F(t - x/a) = \sin \pi(t - x/a)$ when $0 \leq t - x/a \leq 1$, that is, when $a(t - 1) \leq x \leq at$. Then from formula (7) we find that

(9) $$Y(x,t) = 0 \qquad \text{when } x \leq a(t - 1) \text{ or when } x \geq at,$$
$$= \sin \pi(t - x/a) \qquad \text{when } a(t - 1) \leq x \leq at.$$

Thus the string coincides with the x axis except on an interval of length a, if $t > 1$, where it forms one arch of a sine curve ending at $x = at$. As t increases, the arch moves to the right with velocity a (Fig. 40).

It is interesting to note that for each fixed x in formula (9) the function $Y_t(x,t)$ has a jump π at $t = x/a$ and at $t = 1 + x/a$ and hence $s^2 y(x,s)$ is not the transform of $Y_{tt}(x,t)$. On the other hand it can be shown that d^2y/dx^2 is not the transform of $Y_{xx}(x,t)$, because of the discontinuities of $Y_x(x,t)$; but the transform of $Y_{tt} - a^2 Y_{xx}$ is zero.

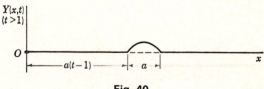

Fig. 40

The function $Y(x,t)$ above can also represent the longitudinal displacements in a semi-infinite elastic bar (Sec. 40), initially at rest and unstrained. The distant end of the bar is held fixed and the end $x = 0$ is displaced in a prescribed manner, $Y(0,t) = F(t)$. When $F(t)$ is given by formula (8), the solution (9) or Fig. 40 shows that at time t the section of the bar from $x = a(t - 1)$ to $x = at$ is strained while the remainder is unstrained and at rest.

42 A LONG STRING UNDER ITS WEIGHT

Let the semi-infinite string be stretched along the positive half of a horizontal x axis with its end $x = 0$ fastened at the origin and with its distant end looped around a vertical support that exerts no vertical force on the loop (Fig. 41). In view of formula (2), Sec. 40, for the vertical force at points of the stretched string, $Y_x(x,t)$ vanishes at the distant end. The string is initially supported at rest along the x axis. At the instant $t = 0$ the support is removed and the string moves downward under the action of gravity. Let us find the displacements $Y(x,t)$.

As noted in Sec. 40, Eq. (5), the equation of motion is

$$(1) \qquad\qquad Y_{tt}(x,t) = a^2 Y_{xx}(x,t) - g \qquad\qquad (x > 0, t > 0).$$

The boundary conditions are

$$(2) \qquad\qquad Y(x,0) = Y_t(x,0) = 0 \qquad\qquad (x \geqq 0),$$

$$(3) \qquad\qquad Y(0,t) = 0, \qquad \lim_{x \to \infty} Y_x(x,t) = 0 \qquad\qquad (t \geqq 0).$$

The transformed problem is found formally to be

$$(4) \qquad\qquad a^2 y''(x,s) - s^2 y(x,s) = \frac{g}{s},$$

$$(5) \qquad\qquad y(0,s) = 0, \qquad \lim_{x \to \infty} y'(x,s) = 0,$$

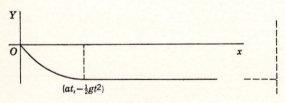

Fig. 41

where the primes denote derivatives with respect to x. Since the constant $-g/s^3$ is a particular solution of Eq. (4), the general solution of that equation is

$$y(x,s) = C_1 e^{-sx/a} + C_2 e^{sx/a} - \frac{g}{s^3}.$$

When conditions (5) are applied, we find that

(6)
$$y(x,s) = -g\left(\frac{1}{s^3} - \frac{1}{s^3} e^{-(x/a)s}\right).$$

The displacements $Y(x,t)$ can be written in the form

(7)
$$Y(x,t) = -\frac{g}{2a^2}(2axt - x^2) \qquad \text{when } x \leq at,$$

$$= -\tfrac{1}{2}gt^2 \qquad \text{when } x \geq at.$$

The details of this step and the verification of the solution (7) are left to the problems. An instantaneous position of the string is shown in Fig. 41. We note that up to time t the segment of the string to the right of the point $x = at$ has moved like a freely falling body.

PROBLEMS

Solve these boundary value problems and verify your results:

1. $Y_x(x,t) + xY_t(x,t) = 0$; $Y(x,0) = 0$, $Y(0,t) = t$.

Ans. $Y = 0$ if $t \leq \tfrac{1}{2}x^2$, $Y = t - \tfrac{1}{2}x^2$ if $t \geq \tfrac{1}{2}x^2$.

2. $Y_x(x,t) + 2xY_t(x,t) = 2x$; $Y(x,0) = Y(0,t) = 1$.

Ans. $Y = 1 + t$ if $t \leq x^2$, $Y = 1 + x^2$ if $t \geq x^2$.

3. $xY_x(x,t) + Y_t(x,t) + Y(x,t) = xF(t)$; $Y(x,0) = Y(0,t) = 0$.

Ans. $Y = xe^{-2t}\displaystyle\int_0^t e^{2\tau}F(\tau)\, d\tau$.

4. $W_{xx}(x,t) + W_{tx}(x,t) - 2W_{tt}(x,t) = 0$ $(x > 0, t > 0)$;
$W(x,0) = W_t(x,0) = \lim_{x \to \infty} W(x,t) = 0$, $W(0,t) = F(t)$.

Ans. $W(x,t) = F(t - 2x)$, where $F(\tau) = 0$ if $\tau < 0$.

5. $W_{xx}(x,t) - 2W_{tx}(x,t) + W_{tt}(x,t) = 0$ $(0 < x < 1, t > 0)$;
$W(x,0) = W_t(x,0) = W(0,t) = 0$; $W(1,t) = F(t)$ $(t > 0)$.

Ans. $W(x,t) = xF(x + t - 1)$, where $F(\tau) = 0$ if $\tau < 0$.

6. In the problem on the semi-infinite string falling under its own weight (Sec. 42), (a) give the details in deriving formula (7) for displacements $Y(x,t)$ and verify the result fully; (b) use formula (7) to verify that the vertical force at time t exerted on the string by the support at the origin equals the weight $g\rho at$ of the curved segment of the string.

7. In Sec. 42 let the weight of the string be replaced by a general vertical force of $F(t)$ units per unit mass of string, so that another special case of Eq. (5), Sec. 40, is involved.

Show that the displacements are

$$Y(x,t) = G(t) - G\left(t - \frac{x}{a}\right)$$

where $G(t) = \int_0^t \int_0^r F(\tau)\, d\tau\, dr$ if $t \geq 0$, $G(t) = 0$ if $t \leq 0$.

8. The end $x = 0$ of a semi-infinite stretched string is looped around the Y axis which exerts no vertical force on the loop, and the distant end is fixed on the x axis. The string is displaced into the position $Y = e^{-x}(x \geq 0)$ and released from rest in that position at the instant $t = 0$. If no external forces act on the string, set up the boundary value problem for the displacements $Y(x,t)$ and find its solution in the form $Y = e^{-at}\cosh x$ if $x \leq at$, $Y = e^{-x}\cosh at$ if $x \geq at$. *Suggestion*: Find $A(s)$ so that $A(s)e^{-x}$ is a particular solution of the nonhomogeneous equation $a^2 y'' - s^2 y = -se^{-x}$ in $y(x,s)$.

9. The force per unit area on the end $x = 0$ of a uniform semi-infinite elastic bar (Fig. 42) is $F(t)$: $-EY_x(0,t) = F(t)$. If the infinite end is fixed and the initial displacement and velocity of each cross section are zero, set up the boundary value problem for longitudinal displacements $Y(x,t)$ and derive the solution

$$Y(x,t) = \frac{a}{E} G\left(t - \frac{x}{a}\right)$$

where $G(r) = \int_0^r F(\tau)\, d\tau$ if $r \geq 0$, $G(r) = 0$ if $r \leq 0$; thus $Y(x,t) = 0$ when $t \leq x/a$. Note that the displacement of the end $x = 0$ is $Y(0,t) = (a/E)G(t)$.

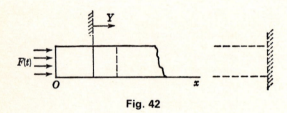

Fig. 42

10. When the pressure $F(t)$ in Prob. 9 is a finite impulse: $F(t) = EF_0$ when $t < t_0$, $F(t) = 0$ when $t > t_0$, show that $Y = 0$ when $t \leq x/a$, $Y = (at - x)F_0$ when $x/a \leq t \leq t_0 + x/a$, and $Y = F_0 a t_0$ when $t \geq t_0 + x/a$. Study this function $Y(x,t)$ graphically.

11. Under the instantaneous impulse of pressure $F(t) = I\delta(t)$ in Prob. 9 show formally that

$$Y(x,t) = \frac{aI}{E} \quad \text{if } x < at, \qquad Y(x,t) = 0 \quad \text{if } x > at.$$

Note that in this hypothetical case a part of the bar is displaced into a region already occupied by another part.

12. Let the pressure F in Prob. 9 be constant, $F(t) = F_0$.

(a) Show formally that $Ey(x,s) = aF_0 s^{-2}\exp(-sx/a)$ and hence

$$Y(x,t) = F_0 E^{-1}(at - x)S_0(at - x).$$

Verify that result as a solution of the boundary value problem.

(b) Draw graphs of $Y(x,t)$ here as a function of x with t fixed, and as a function of t with x fixed. Note the jumps in Y_x and Y_t where $x = at$ and the removable discontinuities in the functions Y_{xx} and Y_{tt} there.

13. For the function $Y(x,t)$ and its transform $y(x,s)$ found in Prob. 12, show that $L\{Y_{tt}\}$ is zero, not $s^2 y$ as assumed in the formal solution; also that $L\{Y_{xx}\}$ is zero, not $d^2 y/dx^2$. Note, however, that $L\{Y_{tt} - a^2 Y_{xx}\} = 0 = s^2 y - a^2 \, d^2 y/dx^2$.

14. If a rigid mass m is attached to the end $x = 0$ of the semi-infinite bar in Prob. 9 (Fig. 42), when the bar has unit cross-sectional area, and the force $F(t)$ is a constant F_0, show why the boundary condition at that end becomes

$$m Y_{tt}(0,t) = E Y_x(0,t) + F_0.$$

Write $b = E/am$ and derive the formula

$$Y(x,t) = \frac{aF_0}{E}\left[t - \frac{x}{a} - \frac{1}{b} + \frac{1}{b}e^{-b(t - x/a)} \right] S_0\!\left(t - \frac{x}{a} \right).$$

15. An unstrained semi-infinite elastic bar is moving lengthwise with velocity $-v_0$ when, at the instant $t = 0$, the end $x = 0$ is suddenly brought to rest, the other end remaining free (Fig. 43). Set up and solve the boundary value problem for the longitudinal displacements $Y(x,t)$. Also show that the force per unit area exerted by the support at $x = 0$ upon the end of the bar is Ev_0/a.

Ans. $Y = -v_0 t$ when $t \leq x/a$, $Y = -v_0 x/a$ when $t \geq x/a$.

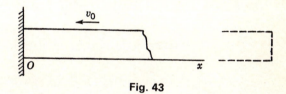

Fig. 43

16. Let the end $x = 0$ of the bar in Prob. 15 meet an *elastic support* at the instant $t = 0$, such that the pressure exerted by that support upon the end is proportional to the displacement of that end: $-EY_x(0,t) = -BY(0,t)$, or

$$Y_x(0,t) = bY(0,t) \qquad\qquad (b = B/E).$$

Show that the velocity of that end is $-v_0 e^{-abt}$ since $abY(0,t) = -v_0(1 - e^{-abt})$.

43   THE LONG STRING INITIALLY DISPLACED

Let the ends of a semi-infinite string stretched along the positive x axis be kept fixed, and let the string be given some prescribed displacement $Y = \Phi(x)$ initially and released from that position with initial velocity zero. Here $\Phi(0) = \Phi(\infty) = 0$. Then the boundary value problem in the transverse displacements $Y(x,t)$ is

$$Y_{tt}(x,t) = a^2 Y_{xx}(x,t) \qquad\qquad (x > 0, t > 0),$$

(1) $$Y(x,0) = \Phi(x), \qquad Y_t(x,0) = 0,$$
$$Y(0,t) = 0, \qquad \lim_{x \to \infty} Y(x,t) = 0.$$

The problem in the transform $y(x,s)$ is therefore

(2) $$s^2 y(x,s) - s\Phi(x) = a^2 y_{xx}(x,s) \qquad\qquad (x > 0),$$

(3) $$y(0,s) = 0, \qquad \lim_{x \to \infty} y(x,s) = 0,$$

where $y_{xx}(x,s) = d^2y/dx^2$. We shall solve the ordinary differential equation (2) by using the Laplace transformation with respect to x. Let $u(z,s)$ denote that transform of $y(x,s)$; that is,

$$u(z,s) = \int_0^\infty e^{-zx} y(x,s)\, dx.$$

Since $y(0,s) = 0$, when we transform both members of Eq. (2), we obtain the equation

$$s^2 u(z,s) - s\varphi(z) = a^2[z^2 u(z,s) - y_x(0,s)],$$

where $\varphi(z)$ is the transform of $\Phi(x)$. Let the unknown function of $s, y_x(0,s)$ be denoted by C. Then the solution of the last equation can be written

$$u(z,s) = \frac{C}{z^2 - (s^2/a^2)} - \frac{s}{a^2}\varphi(z)\frac{1}{z^2 - (s^2/a^2)},$$

and performing the inverse transformation with respect to z, with the aid of the convolution, we have

(4) $$y(x,s) = \frac{aC}{s}\sinh\frac{sx}{a} - \frac{1}{a}\int_0^x \Phi(\xi)\sinh\frac{s}{a}(x - \xi)\, d\xi.$$

In view of the condition requiring $y(x,s)$ to vanish as x tends to infinity, it is necessary that the coefficient of $e^{sx/a}$ on the right of Eq. (4) should vanish as x becomes infinite. Writing the hyperbolic sines in terms of exponential functions, we find that coefficient to be

$$\frac{aC}{2s} - \frac{1}{2a}\int_0^x \Phi(\xi)e^{-(s\xi/a)}\, d\xi.$$

Since the limit of this function is to be zero as $x \to \infty$, we have

(5) $$\frac{aC}{s} = \frac{1}{a}\int_0^\infty \Phi(\xi)e^{-(s\xi/a)}\, d\xi.$$

Substituting this into Eq. (4), we can write the result in the form

$$2ay(x,s) = \int_x^\infty \Phi(\xi)e^{-[s(\xi-x)]/a}\,d\xi - \int_0^\infty \Phi(\xi)e^{-[s(x+\xi)]/a}\,d\xi$$

$$+ \int_0^x \Phi(\xi)e^{-[s(x-\xi)]/a}\,d\xi.$$

The integrals here can be reduced to Laplace integrals. In the first one we substitute $\tau = (\xi - x)/a$, in the second $\tau = (\xi + x)/a$, and in the third $\tau = (x - \xi)/a$ to get

$$(6) \qquad 2y(x,s) = \int_0^\infty \Phi(x + a\tau)e^{-s\tau}\,d\tau$$

$$- \int_{x/a}^\infty \Phi(-x + a\tau)e^{-s\tau}\,d\tau + \int_0^{x/a} \Phi(x - a\tau)e^{-s\tau}\,d\tau.$$

In order to combine the last two integrals, let $\Phi_1(x)$ represent the *odd extension* of the function $\Phi(x)$:

$$(7) \qquad\qquad\qquad \phi_1(x) = \Phi(x) \qquad\qquad\qquad \text{if } x \geq 0,$$

$$= -\Phi(-x) \qquad\qquad \text{if } x \leq 0;$$

thus $\Phi_1(-x) = -\Phi_1(x)$ for all x. Then Eq. (6) can be written

$$2y(x,s) = L\{\Phi(x + at)\} + L\{\Phi_1(x - at)\}$$

and therefore

$$(8) \qquad\qquad Y(x,t) = \tfrac{1}{2}[\Phi(x + at) + \Phi_1(x - at)].$$

It is easily seen that a twice-differentiable function of the variable $x + at$, or of the variable $x - at$, as well as each linear combination of such functions, satisfies the linear homogeneous wave equation $Y_{tt} = a^2 Y_{xx}$. Our continuous function (8) therefore satisfies the wave equation for all x and t where $\Phi''(x + at)$ and $\Phi_1''(x - at)$ exist. It satisfies the condition $Y_t(x,0) = 0$ when $x > 0$ wherever $\Phi'(x)$ exists, and it satisfies the remaining conditions of problem (1) because $\Phi_1(x) = \Phi(x)$ when $x > 0$, $\Phi_1(-at) = -\Phi(at)$ and $\Phi(\infty,t) = 0$.

Instantaneous positions of the string can be sketched by adding ordinates as suggested by formula (8). When t is fixed, the graph of $\tfrac{1}{2}\Phi_1(x - at)$, for instance, is obtained by translating the graph of the function $\tfrac{1}{2}\Phi_1(x)$, defined for all real x, to the right through the distance at.

44   A BAR WITH A PRESCRIBED FORCE ON ONE END

Let the end $x = 0$ of an elastic bar of length c be kept fixed, and let $F(t)$ denote a prescribed force per unit area acting parallel to the bar at the end $x = c$ (Fig. 44). If the bar is initially unstrained and at rest, the boundary

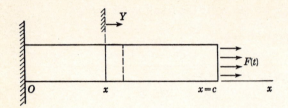

Fig. 44

value problem in the longitudinal displacements $Y(x,t)$ is the following,

$$Y_{tt}(x,t) = a^2 Y_{xx}(x,t) \qquad (0 < x < c, t > 0),$$

$$Y(x,0) = Y_t(x,0) = 0,$$

$$Y(0,t) = 0, \qquad EY_x(c,t) = F(t), \qquad \left(a^2 = \frac{E}{\rho}\right).$$

The transform of $Y(x,t)$ therefore satisfies the conditions

$$s^2 y(x,s) = a^2 y_{xx}(x,s),$$

$$y(0,s) = 0, \qquad Ey_x(c,s) = f(s),$$

and the solution of this transformed problem is readily found to be

(1) $$y(x,s) = \frac{a}{E} f(s) \frac{\sinh (sx/a)}{s \cosh (sc/a)}.$$

CONSTANT FORCE

Consider first the case

(2) $$F(t) = F_0.$$

Then

(3) $$y(x,s) = \frac{aF_0}{E} \frac{\sinh (sx/a)}{s^2 \cosh (sc/a)}$$

and, when $x = c$, $$y(c,s) = \frac{aF_0}{E} \frac{1}{s^2} \tanh \frac{sc}{a}.$$

In Sec. 23 we found that $s^{-2} \tanh (bs/2)$ is the transform of the triangular wave function $H(b,t)$ of period $2b$ (Fig. 10). Therefore the displacement of the end $x = c$ is

(4) $$Y(c,t) = \frac{aF_0}{E} H\left(\frac{2c}{a}, t\right);$$

that is, the end moves by jerks as indicated in Fig. 45.

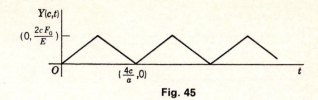

Fig. 45

To find the displacement of an arbitrary point using formula (3) we may write, when $s > 0$,

$$\frac{\sinh(sx/a)}{\cosh(sc/a)} = \frac{e^{-[(c-x)s]/a} - e^{-[(c+x)s]/a}}{1 + e^{-(2cs/a)}}$$

$$= [e^{-[(c-x)s]/a} - e^{-[(c+x)s]/a}] \sum_{n=0}^{\infty} (-1)^n e^{-(2ncs/a)},$$

since $(1 + z)^{-1} = \sum_0^{\infty} (-1)^n z^n$ when $0 < z < 1$. Therefore

$$y(x,s) = \frac{aF_0}{E} \sum_0^{\infty} (-1)^n \left\{ \frac{1}{s^2} \exp\left[-s\frac{(2n+1)c - x}{a} \right] \right.$$

$$\left. - \frac{1}{s^2} \exp\left[-s\frac{(2n+1)c + x}{a} \right] \right\}.$$

Formally applying the inverse transformation to the terms of the infinite series, and using braces to denote this translation of the linear function τ:

(5) $$\{\tau - k\} = (\tau - k)S_0(\tau - k),$$

we obtain the formula

(6) $$Y(x,t) = \frac{F_0}{E}(\{at - c + x\} - \{at - c - x\} - \{at - 3c + x\}$$

$$+ \{at - 3c - x\} + \{at - 5c + x\} - \{at - 5c - x\} - \cdots).$$

For any fixed t the series (6) is finite since each of the braces is to be replaced by zero when the quantity inside is negative. The number of nonvanishing terms in the series increases as t increases.

The function (6) can be verified directly as the solution of our problem. Its graph, for a fixed value of x, is shown in Fig. 46.

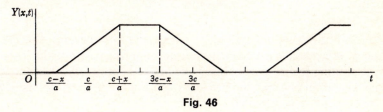

Fig. 46

A GENERAL FORCING FUNCTION

When the pressure on the end $x = c$ is $F(t)$, let us write

$$G(t) = \int_0^t F(\tau) \, d\tau \text{ if } t \geq 0, \qquad G(t) = 0 \text{ if } t \leq 0.$$

Then $L\{G\} = f(s)/s$, and formula (1) can be written

$$y(x,s) = \frac{a}{E} g(s) \frac{\sinh (sx/a)}{\cosh (sc/a)}.$$

Thus if we write $q_n(x) = [(2n + 1)c - x]/a$ then, as in the above case,

$$y(x,s) = \frac{a}{E} \sum_{n=0}^{\infty} (-1)^n \{\exp [-sq_n(x)] - \exp [-sq_n(-x)]\} g(s)$$

and formally,

(7) $$Y(x,t) = \frac{a}{E} \sum_{n=0}^{\infty} (-1)^n \{G[t - q_n(x)] - G[t - q_n(-x)]\}.$$

AN IMPULSE OF PRESSURE

In the hypothetical case where

(8) $$F(t) = I \, \delta(t),$$

$f(s) = I$ in formula (1) and

(9) $$y(x,s) = \frac{aI}{E} \frac{\sinh(sx/a)}{s \cosh (sc/a)}.$$

The displacements $Y(x,t)$ can be represented by a series of the step functions $S_0\{t - [(2n + 1)c \pm x]/a\}$. Those displacements are not continuous.

To see the behavior of the end $x = c$, we note that

$$y(c,s) = \frac{aI}{E} \frac{1}{s} \tanh \frac{sc}{a}.$$

The displacement of that end is therefore represented by the square-wave function (Sec. 23, Fig. 9),

(10) $$Y(c,t) = \frac{aI}{E} M\left(\frac{2c}{a}, t\right).$$

Thus the end jumps suddenly back and forth between two fixed positions. It is possible to demonstrate a close approximation to that behavior by substituting for the bar a loosely wound coil spring. If, when the spring is hanging from one end, the free lower end is given a sharp tap, the lower end tends to move as indicated.

PROBLEMS

1. A string is stretched between two fixed points $(0,0)$ and $(c,0)$. If it is displaced into the curve $Y = b \sin (\pi x/c)$ and released from rest in that position at time $t = 0$, set up and solve the boundary value problem for the displacements $Y(x,t)$. Verify the result fully and describe the motion of the string.   *Ans.* $Y(x,t) = b \cos (\pi at/c) \sin (\pi x/c)$.

2. If the initial displacement of the string in Prob. 1 is changed to

$$Y(x,0) = b \sin \frac{n\pi x}{c} \qquad\qquad (0 \leq x \leq c),$$

where n is any integer, derive the formula

$$Y(x,t) = b \cos \frac{n\pi at}{c} \sin \frac{n\pi x}{c}.$$

Note that the sum of two or more of these functions with different values of n and b is a solution of the equation of motion that satisfies all the boundary conditions if the initial displacement $Y(x,0)$ corresponding to such a *superposition* of solutions is a certain linear combination of the functions $\sin (n\pi x/c)$.

3. As a special case of the initially displaced semi-infinite string considered in Sec. 43, where $Y(0,t) = Y(\infty,t) = 0$, let the end segment $0 \leq x \leq 1$ be plucked so that the initial displacement has this form:

$$Y(x,0) = \sin \pi x \quad \text{if } 0 \leq x \leq 1, \qquad Y(x,0) = 0 \quad \text{if } x \geq 1,$$

where again $Y_t(x,0) = 0$ when $x > 0$. Show the displacement $Y(x,t)$ graphically as a function of x when t is fixed and $at > 1$, and note that the displacement is a wave formed by letting one cycle of the curve $Y = \frac{1}{2} \sin \pi x$, $-1 < x < 1$, move to the right with velocity a. (Boys have used such plucking of a kite string to dislodge parachutes hooked onto the string and blown toward the kite.)

4. A long string stretched along the entire x axis is released at rest from a prescribed initial displacement. Thus

$$Y_{tt}(x,t) = a^2 Y_{xx}(x,t) \qquad\qquad (-\infty < x < \infty, t > 0),$$

$$Y(x,0) = \Phi(x), \qquad Y_t(x,0) = 0 \qquad\qquad (-\infty < x < \infty),$$

$$\lim_{x \to -\infty} Y(x,t) = 0, \qquad \lim_{x \to \infty} Y(x,t) = 0 \qquad\qquad (t \geq 0).$$

Assuming $\Phi''(x)$ exists for all x, derive the formula

$$Y(x,t) = \tfrac{1}{2}[\Phi(x + at) + \Phi(x - at)]$$

and verify that solution.

5. A semi-infinite string stretched along the positive x axis with ends fixed, $Y(0,t) = \lim_{x \to \infty} Y(x,t) = 0$, is given an initial velocity $Y_t(x,0) = g(x)$, where g is sectionally continuous on each bounded interval, $\lim_{x \to \infty} g(x) = 0$, and g is integrable from zero to infinity. (*a*) If g_1 is the odd extension of g, so that $g_1(r) = g(r)$ and $g_1(-r) = -g(r)$ if $r > 0$, show that the transverse displacements can be written

$$Y(x,t) = \frac{1}{2a} \int_{x-at}^{x+at} g_1(r)\, dr = \frac{1}{2a}[G(x + at) - G(x - at)],$$

where $G(r) = \int_0^r g_1(\rho)\,d\rho$ [cf. Eqs. (2), (3), and (6), Sec. 43]. Verify that result. (b) Show $Y(x,t)$ graphically as a function of x when $at > 1$ in case g is the step function $1 - S_0(x - 1)$.

6. In Sec. 44, when the force per unit area at the end $x = c$ of the bar (Fig. 44) is constant, $F(t) = F_0$, and $Y(x,0) = Y_t(x,0) = Y(0,t) = 0$, show that the force per unit area exerted by the bar on the support at the end $x = 0$ is the function shown in Fig. 47. Note that the force becomes twice the applied force during regularly spaced intervals of time.

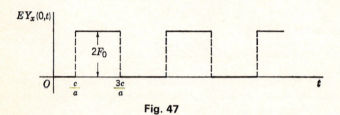

Fig. 47

7. Let the constant force F_0 in Prob. 6 be replaced by a finite impulse of duration $4c/a$:

$$F(t) = F_0 \quad \text{if } 0 < t < \frac{4c}{a}, \qquad F(t) = 0 \quad \text{if } t > \frac{4c}{a}.$$

(a) Show that the force per unit area exerted by the bar on the support at the end $x = 0$ is that shown in Fig. 47 up to the time $t = 3c/a$ after which time the force is zero.

(b) Show that the end $x = c$ moves with constant velocity $v_0 = aF_0/E$ during the time interval $0 < t < 2c/a$, then with velocity $-v_0$ during an equal time interval and that $Y(c,t) = 0$ when $t \geq 4c/a$.

8. A constant longitudinal force F_0 per unit area is applied at the end $x = c$ of an elastic bar (Fig. 48) whose end $x = 0$ is free and which is initially at rest and unstrained. Set up the boundary value problem for the longitudinal displacements $Y(x,t)$ and show that $Ey(x,s) = aF_0 s^{-2} \cosh(sx/a)/\sinh(sc/a)$.

(a) Derive the formula

$$Y_t(0,t) = v_0 \sum_{n=0}^{\infty} S_0\left[t - (2n + 1)\frac{c}{a}\right] \qquad \text{where } v_0 = \frac{2aF_0}{E};$$

hence note that the free end moves with velocity v_0 from time $t = c/a$ until $t = 3c/a$, then with velocity $2v_0$ until $t = 5c/a$, etc.

(b) In terms of the function $\{\tau - k\} = (\tau - k)S_0(\tau - k)$, show that

$$Y(x,t) = \frac{F_0}{E} \sum_{n=0}^{\infty} [\{at + x - c(2n + 1)\} + \{at - x - c(2n + 1)\}].$$

Fig. 48

9.   In Prob. 8, let the force F_0 be replaced by the force

$$F(t) = F_0 \cos \omega t,$$

where $\omega = \pi a/(2c)$. Show that [formula (10), Sec. 23]

$$\pi E Y(c,t) = 2c F_0 |\sin \omega t|.$$

10.   The end $x = 0$ of a bar or heavy coil spring (Fig. 49) is free. The end $x = c$ is displaced in a prescribed manner, $Y(c,t) = G(t)$. If the bar is initially unstrained and at rest, find the transform $y(x,s)$ of longitudinal displacements.

Fig. 49

(a) When $G(t) = F_0 t$, show that the displacement $Y(0,t)$ of the free end is represented by the function obtained by integrating the function $E Y_x(0,\tau)$ in Fig. 47 from 0 to t and show the displacement graphically.

(b) If $G(t) = bt$ when $t \le 4c/a$ and $G(t) = 4bc/a$ when $t \ge 4c/a$, show that the free end moves with a uniform velocity $2b$ to a new position and remains there, as shown in Fig. 50.

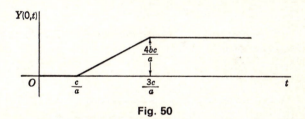

Fig. 50

11.   An unstrained elastic bar is moving lengthwise with velocity v_0 when at the instant $t = 0$ its end $x = c$ meets and adheres to a rigid support (Fig. 51).

(a) If A is the area of the cross section of the bar and M denotes the square-wave function (Fig. 9), show that the force on the support can be written

$$-AE Y_x(c,t) = \frac{AE v_0}{a} M\left(\frac{2c}{a},t\right) \qquad \left(a^2 = \frac{E}{\rho}\right).$$

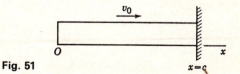

Fig. 51

(b) In terms of the function $\{t - k\} = (t - k)S_0(t - k)$, show that

$$\frac{1}{v_0}Y(x,t) = t - \left\{t - \frac{c - x}{a}\right\} - \left\{t - \frac{c + x}{a}\right\} + \left\{t - \frac{3c - x}{a}\right\} + \left\{t - \frac{3c + x}{a}\right\} - \cdots,$$

and hence that $Y(x,2c/a) = 0$ and $Y_t(x,2c/a) = -v_0$ for all x $(0 < x < c$. Thus if the end $x = c$ is free to leave the support, the bar will move after time $t = 2c/a$ as a rigid unstrained body with velocity $-v_0$.

12. A steel bar 10 in. long is moving lengthwise with velocity v_0 when one end strikes a rigid support squarely (Fig. 51). Find the length of time of contact of the end of the bar with the support and note that the time is independent of v_0. For steel take E to be 30×10^6 lb/in.^{2}, and mass per unit volume ρ such that $g\rho = 0.28$ lb/in.^{3}, where $g = 384$ in./sec^2. (See Prob. 11.)   *Ans.* 0.0001 sec.

13. An unstrained cylindrical shaft is rotating with angular velocity ω when it ends $x = \pm c$ are suddenly clamped (Fig. 52). Derive the formula

$$\theta(x,s) = \frac{\omega}{s^2}\left[1 - \frac{\cosh(sx/a)}{\cosh(sc/a)}\right]$$

for the transform of the angular displacements $\Theta(x,t)$ of the cross sections, where $a^2 = E_s/\rho$ (Sec. 40).

(a) If H denotes the triangular-wave function (Fig. 10), show that the displacement of the middle cross section is

$$\Theta(0,t) = \omega t \qquad\qquad \text{if } t \leq \frac{c}{a},$$

$$\Theta(0,t) = \frac{\omega c}{a} - \omega H\left(\frac{2c}{a}, t - \frac{c}{a}\right) \qquad\qquad \text{if } t \geq \frac{c}{a}.$$

(b) Describe a stretched-string analog for $\Theta(x,t)$ here, as seen from the boundary value problem.

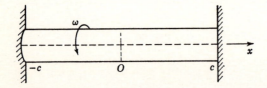

Fig. 52

14. Let M denote the square-wave function (Fig. 9). Show that the torque acting on the support at the end $x = c$ of the shaft in Prob. 13 is $a^{-1}E_s I\omega M(2c/a, t)$.

15. The end $x = 0$ of a cylindrical shaft is kept fixed. The end $x = c$ is rotated through an angle θ_0 and, when all parts have come to rest, that end is released; thus $\Theta(x,0) = \theta_0 x/c$ (Fig. 53). Show that the angular displacement of the free end at each instant can be written

$$\Theta(c,t) = \theta_0 - \theta_0\frac{a}{c}H\left(\frac{2c}{a}, t\right),$$

where H is the triangular-wave function (Fig. 10).

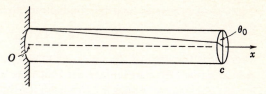

Fig. 53

45 EQUATIONS OF DIFFUSION

Let $U(x,y,z,t)$ denote the temperature at a point (x,y,z) within a solid at time t. Let S be a plane or curved surface passed through that point and let the coordinate n represent directed distance along the line normal to S at the point. Then the flux of heat, the quantity of heat per unit area per unit time transferred across S at the point by conduction, is given by the formula

$$(1) \qquad \Phi(x,y,z,t) = -K\frac{d}{dn}U(x,y,z,t),$$

where the coefficient K is the *thermal conductivity* of the material. The negative sign that appears with this directional derivative of U gives the sense of flow, where positive flux signifies flow in the positive direction of the normal.

Formula (1) is an empirical law or *basic postulate* for the conduction of heat in solids. It is likewise a basic postulate for the flux of a substance undergoing simple diffusion into a porous solid if U represents the concentration of the diffusing substance and K the *coefficient of diffusion*. If the temperature or concentration is a function $U(x,t)$ of x and t only, then the flux in the x direction is

$$(2) \qquad \Phi(x,t) = -KU_x(x,t).$$

When the temperature function has the simple form $U(x,t)$, consider a portion of the solid in the shape of a prism parallel to the x axis (Fig. 54). Since there is no temperature variation with respect to distance normal to the lateral surface of the prism, no heat is conducted across that surface. According to formula (2), an element of the prism extending from position

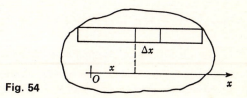

Fig. 54

x to $x + \Delta x$ receives heat by conduction through its two bases at the rate

$$- AKU_x(x,t) + AKU_x(x + \Delta x, t),$$

where A is the area of the cross section of the prism. This is the quantity of heat gained by the element per unit time if no heat is generated or lost inside the element. Another expression for that rate can be written in terms of the coefficient c of *heat capacity per unit volume*, the quantity of heat required to raise the temperature of a unit volume of the material by one degree. Then when Δx is small, $cA \, \Delta x \, U(x,t)$ is a measure of the instantaneous heat content of the element, and

(3) $$cA \, \Delta x \, U_t(x,t) = A[KU_x(x + \Delta x, t) - KU_x(x,t)].$$

If K depends on x, then the first factor K in Eq. (3) is actually $K(x + \Delta x, t)$ and the second, $K(x,t)$. When we divide by Δx and let Δx tend to zero, we obtain the partial differential equation

(4) $$cU_t(x,t) = [K(x,t)U_x(x,t)]_x.$$

When K is either constant or independent of x, the equation becomes

(5) $$U_t(x,t) = kU_{xx}(x,t),$$

where $k = K/c$ and k is called the *thermal diffusivity* of the material. Equation (5) is the simple form of the *heat equation* or the *equation of diffusion*.

When $U(x,t)$ represents the concentration of a diffusing substance in weight per unit volume, Eq. (3) follows as before if $c = 1$ and K is the coefficient of diffusion. Thus, when K is constant, $U(x,t)$ satisfies Eq. (5) where $k = K$.

The heat equation or equation of diffusion for $U(x,y,z,t)$

(6) $$U_t = k(U_{xx} + U_{yy} + U_{zz}),$$

when coefficients are constant, can be derived in a similar manner by considering a three-dimensional element of volume $\Delta x \, \Delta x \, \Delta z$.[1]

Modifications of the heat equation are easily written. For example, let $U(x,t)$ represent temperatures in a slender wire along the x axis. Let heat be generated at a time rate $R(x,t)$ per unit volume of the wire and let heat loss from the lateral surface take place at a rate proportional to the difference between the temperature of the wire and the temperature U_0 of the surroundings. The modification of Eq. (3) then shows that

(7) $$U_t(x,t) = kU_{xx}(x,t) + \frac{1}{c}R(x,t) - h[U(x,t) - U_0],$$

where h is a positive *coefficient of surface heat transfer*.

[1] See chap. 1 of the author's "Fourier Series and Boundary Value Problems," 2d ed., listed in the Bibliography, for a full derivation of the heat equation.

In the problems to follow we assume that the coefficients K, k, c, and h are constants. Also, we often consider semi-infinite solids, where results are relatively simple. Such bodies are not necessarily large in any absolute sense. In the diffusion of hardening materials into steel, for instance, test bars of length 1 in. or even less may be represented quite accurately as semi-infinite solids.

46 TEMPERATURES IN A SEMI-INFINITE SOLID

Let us now derive a formula for the temperatures $U(x,t)$ in a semi-infinite solid $x \geq 0$, initially at temperature zero, when a constant flux of heat is maintained at the boundary $x = 0$ (Fig. 55). In this idealized case of a thick slab of material, we shall substitute for the thermal condition at the right-hand boundary the condition that U tends to zero as x tends to infinity. The boundary value problem is then

(1) $$U_t(x,t) = kU_{xx}(x,t) \qquad (x > 0, t > 0),$$

(2) $$U(x,+0) = 0 \qquad (x > 0),$$

(3) $$-KU_x(0,t) = \phi_0, \qquad \lim_{x \to \infty} U(x,t) = 0 \qquad (t > 0).$$

Let $u(x,s)$ be the transform, with respect to t, of the temperature function $U(x,t)$. Transforming the members of Eqs. (1) and (3), we have the following problem in ordinary differential equations which $u(x,s)$ must satisfy:

$$su(x,s) = ku_{xx}(x,s) \qquad (x > 0),$$

$$-Ku_x(0,s) = \frac{\phi_0}{s}, \qquad \lim_{x \to \infty} u(x,s) = 0.$$

The solution of that problem is

$$u(x,s) = \frac{\phi_0\sqrt{k}}{Ks\sqrt{s}}e^{-x\sqrt{s/k}}.$$

According to formula (5), Sec. 27, we can write

$$L^{-1}\left\{\frac{1}{\sqrt{s}}e^{-x\sqrt{s/k}}\right\} = \frac{1}{\sqrt{\pi t}}\exp\left(-\frac{x^2}{4kt}\right),$$

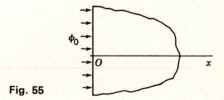

Fig. 55

and in view of the factor $1/s$ in our formula for $u(x,s)$ it follows that

$$U(x,t) = \frac{\phi_0}{K}\sqrt{\frac{k}{\pi}}\int_0^t \exp\left(-\frac{x^2}{4k\tau}\right)\frac{d\tau}{\sqrt{\tau}}$$

$$= \frac{\phi_0 x}{K\sqrt{\pi}}\int_{x/(2\sqrt{kt})}^\infty \frac{1}{\lambda^2}\exp(-\lambda^2)d\lambda,$$

where the second integral is obtained from the first by the substitution $\lambda = x/(2\sqrt{k\tau})$. Upon integrating the last integral by parts, we find that

$$U(x,t) = \frac{\phi_0}{K\sqrt{\pi}}\left[2\sqrt{kt}\exp\left(-\frac{x^2}{4kt}\right) - 2x\int_{x/(2\sqrt{kt})}^\infty \exp(-\lambda^2)\,d\lambda\right].$$

We can therefore write our formula in terms of the complementary error function (Sec. 27) in the form

(4) $$U(x,t) = \frac{\phi_0}{K}\left[2\sqrt{\frac{kt}{\pi}}\exp\left(-\frac{x^2}{4kt}\right) - x\,\mathrm{erfc}\left(\frac{x}{2\sqrt{kt}}\right)\right].$$

We can show that the function (4) satisfies all our conditions (1), (2), and (3). We observe that

$$U(0,t) = \frac{2\phi_0\sqrt{k}}{K\sqrt{\pi}}\sqrt{t}.$$

Thus the temperature of the face of the solid must vary as $\sqrt{t}$ in order that the flux of heat through the face shall be constant.

47   PRESCRIBED SURFACE TEMPERATURE

Let the temperature of the face of a semi-infinite solid $x \geq 0$ be a prescribed function $F(t)$ of time. If the initial temperature is zero, the temperature function $U(x,t)$ is the solution of the boundary value problem

$$U_t(x,t) = kU_{xx}(x,t) \qquad\qquad (x > 0, t > 0),$$

$$U(x,0) = 0 \qquad\qquad (x > 0),$$

$$U(0,t) = F(t), \qquad \lim_{x\to\infty} U(x,t) = 0 \qquad\qquad (t > 0).$$

The transform $u(x,s)$ of $U(x,t)$, therefore, satisfies the conditions

$$su(x,s) = ku_{xx}(x,s) \qquad\qquad (x > 0),$$

$$u(0,s) = f(s), \qquad \lim_{x\to\infty} u(x,s) = 0,$$

where $f(s)$ is the transform of $F(t)$. It follows that

(1) $$u(x,s) = f(s)e^{-x\sqrt{s/k}}.$$

Let us first study the flux of heat through the face of the solid,

$$\Phi(t) = -KU_x(0,t).$$

The transform of this function, $-Ku_x(0,s)$, according to formula (1), is

(2) $$\varphi(s) = \frac{K}{\sqrt{k}}\sqrt{s}f(s) = \frac{K}{\sqrt{k}}sf(s)\frac{1}{\sqrt{s}}.$$

Since $s^{-\frac{1}{2}} = L\{(\pi t)^{-\frac{1}{2}}\}$ and $sf(s) = L\{F'(t)\} + F(+0)$, assuming that F is a continuous function, then

$$\varphi(s) = \frac{K}{\sqrt{k}}\left[\frac{F(+0)}{\sqrt{s}} + L\{F'(t)\}\frac{1}{\sqrt{s}}\right]$$

and with the aid of the convolution it follows that

(3) $$\Phi(t) = \frac{K}{\sqrt{\pi k}}\left[\frac{F(+0)}{\sqrt{t}} + \int_0^t \frac{F'(t-\tau)}{\sqrt{\tau}}\,d\tau\right],$$

when F is continuous. If $F(+0) \neq 0$, the flux is infinite initially, of the order of $t^{-\frac{1}{2}}$ as $t \to 0$. In particular if $F(t) = F_0$, a constant, then

$$\Phi(t) = KF_0(\pi kt)^{-\frac{1}{2}}.$$

The total amount of heat that has been absorbed by the solid through a unit area of the face at time t is

$$Q(t) = \int_0^t \Phi(\tau)\,d\tau.$$

Hence $$q(s) = \frac{1}{s}\varphi(s) = \frac{K}{\sqrt{k}}\frac{1}{\sqrt{s}}f(s),$$

and therefore

(4) $$Q(t) = \frac{K}{\sqrt{\pi k}}\int_0^t \frac{F(\tau)}{\sqrt{t-\tau}}\,d\tau = \frac{K}{\sqrt{\pi k}}\int_0^t \frac{F(t-\tau)}{\sqrt{\tau}}\,d\tau.$$

For example, when

(5) $$F(t) = F_0 \text{ if } t < t_0 \quad \text{and} \quad F(t) = 0 \text{ if } t > t_0,$$

it follows from formula (4) that

$$Q(t) = \frac{2KF_0}{\sqrt{\pi k}}\sqrt{t} \qquad\qquad \text{when } t \leq t_0,$$

$$= \frac{2KF_0}{\sqrt{\pi k}}(\sqrt{t} - \sqrt{t-t_0}) \qquad\qquad \text{when } t \geq t_0.$$

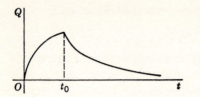

Fig. 56

This function is shown in Fig. 56. Its greatest value is

$$2KF_0\sqrt{\frac{t_0}{\pi k}},$$

which is assumed at the instant $t = t_0$.

Returning to formula (1) for $u(x,s)$ and noting that, according to formula (6), Sec. 27,

$$e^{-x\sqrt{s/k}} = L\left\{\frac{x}{2\sqrt{\pi k t^3}}\exp\left(-\frac{x^2}{4kt}\right)\right\},$$

we can write, with the aid of the convolution,

$$U(x,t) = \frac{x}{2\sqrt{\pi k}}\int_0^t \frac{F(t-\tau)}{\tau^{\frac{3}{2}}}\exp\left(-\frac{x^2}{4k\tau}\right)d\tau.$$

Substituting a new variable of integration, we have for the general temperature formula

$$(7) \qquad U(x,t) = \frac{2}{\sqrt{\pi}}\int_{\frac{1}{2}x/\sqrt{kt}}^{\infty} F\left(t - \frac{x^2}{4k\lambda^2}\right)e^{-\lambda^2}\,d\lambda.$$

When the temperature of the surface is constant,

$$(8) \qquad\qquad\qquad F(t) = F_0,$$

the temperature within the solid is therefore

$$(9) \qquad\qquad U(x,t) = F_0\,\mathrm{erfc}\left(\frac{x}{2\sqrt{kt}}\right).$$

Since this is a function of F_0 and $x/\sqrt{kt}$ only, it follows that the rapidity of heating is proportional to k; for if k is increased and t decreased so that kt is unchanged, the temperature at any given distance x from the face is the same. It is also interesting to note that for a fixed k two points x_1 and x_2 will have equal temperatures at times t_1 and t_2 provided $x_1/\sqrt{t_1} = x_2/\sqrt{t_2}$, that is, if

$$(10) \qquad\qquad\qquad \frac{x_1}{x_2} = \sqrt{\frac{t_1}{t_2}}.$$

This is sometimes called the law of times in the conduction of heat in semi-infinite solids.

If F_0 is positive, formula (9) assigns a positive value to $U(x,t)$ for each x and t because the complementary error function has only positive values. At each interior point of the solid $U(x,t)$ is small when t is sufficiently small, but the temperature does not retain its initial value $U(x,0) = 0$ during an interval of time after $t = 0$. That instantaneous though highly attenuated transfer of the thermal disturbance at the surface can be noted in other cases. In Sec. 46, for example, the flux

$$-KU_x(x,t) = \phi_0 \operatorname{erfc}\left(\frac{1}{2}\frac{x}{\sqrt{kt}}\right)$$

is clearly different from zero immediately after the thermal disturbance took place at the surface $x = 0$. The instantaneous transfer is a consequence of the basic postulate on the flux: $\Phi = -K\,dU/dn$ (Sec. 45).

PROBLEMS

1. The initial temperature of a semi-infinite solid $x \geq 0$ is zero. The surface temperature is this step function:

$$U(0,t) = F_0 \text{ if } 0 < t < t_0, \qquad U(0,t) = 0 \qquad \text{if } t > t_0.$$

Obtain the temperature distribution formula

$$U(x,t) = F_0 \operatorname{erfc}\frac{x}{2\sqrt{kt}} \qquad (t \leq t_0)$$

$$= F_0\left[\operatorname{erf}\frac{x}{2\sqrt{k(t-t_0)}} - \operatorname{erf}\frac{x}{2\sqrt{kt}}\right] \qquad (t \geq t_0).$$

2. A thick slab of iron with thermal diffusivity $k = 0.15$ cgs (centimeter-gram-second) unit is initially at $0°C$ throughout. Its surface is suddenly heated to a temperature of $500°C$ and maintained at that temperature for 5 min, after which the surface is kept chilled to $0°C$. (See Prob. 1.) Find the temperature to the nearest degree at a depth of 10 cm below the surface (a) at the end of 5 min; (b) at the end of 10 min.

Ans. $146°C$; (b) $82°C$.

3. Solve Prob. 2 if the slab is made of firebrick with diffusivity $k = 0.007$ cgs unit.

Ans. (a) $0°C$; (b) $0°C$.

4. The surface of a thick slab of concrete for which $k = 0.005$ cgs unit, initially at $0°C$, undergoes temperature changes described in Prob. 2. Show that at each instant the temperature at any depth x_1 in the concrete slab is the same as the temperature at the depth $x_2 = \sqrt{30}x_1$ in the iron slab. Generalize this result for materials with diffusivities k_1 and k_2 and any common time interval t_0 of heating the surfaces of the slabs.

5. At time $t = 0$, the brakes of an automobile are applied, bringing the automobile to a stop at time t_0. Assuming that the rate of generating heat at the surface of the brake bands varies linearly with the time, then

$$U_x(0,t) = A(t - t_0) \qquad (0 \leqq t \leqq t_0),$$

where A is a positive constant and x is the distance from the face of the band. If t_0 is not large, the band can be assumed to be a semi-infinite solid $x \geqq 0$. If the initial temperature of the band is taken as zero show that the temperature at the face is

$$U(0,t) = \frac{2A}{3}\sqrt{\frac{k}{\pi}}\sqrt{t}(3t_0 - 2t) \qquad (0 \leqq t \leqq t_0).$$

Hence show that this temperature is greatest at the instant $t = \frac{1}{2}t_0$ and that this maximum temperature is $\sqrt{2}U(0,t_0)$.

6. The initial temperature of a semi-infinite solid $x \geqq 0$ is zero. The inward flux of heat through the face $x = 0$ is a prescribed function $\Phi(t)$ of time and the distant face is kept at temperature zero. Derive the temperature formula

$$U(x,t) = \frac{x}{K\sqrt{\pi}}\int_{\frac{1}{2}x/\sqrt{kt}}^{\infty} \Phi\left(t - \frac{x^2}{4k\lambda^2}\right)\frac{e^{-\lambda^2}}{\lambda^2}\,d\lambda \qquad (x > 0).$$

Also show that the temperature of the face can be written

$$U(0,t) = \frac{1}{K}\sqrt{\frac{k}{\pi}}\int_0^t \Phi(\tau)(t - \tau)^{-\frac{1}{2}}\,d\tau.$$

7. An impulse of heat of quantity Q_0 per unit area is introduced through the face $x = 0$ of a semi-infinite solid $x \geqq 0$ at the instant $t = 0$ and the face is insulated when $t > 0$, so that $-KU_x(0,t) = Q_0\,\delta(t)$. (That instantaneous source of heat may be realized approximately by burning a layer of highly combustible fuel at the face.) If $U(x,0) = 0$ when $x > 0$, show formally that the subsequent temperature is

$$U(x,t) = \frac{Q_0}{K}\sqrt{\frac{k}{\pi t}}\exp\left(-\frac{x^2}{4kt}\right) \qquad (t > 0).$$

(See Prob. 8 for a verification.) Observe that $U(x,t) > 0$ whenever $t > 0$.

8. The function $\psi(x,t) = t^{-\frac{1}{2}}\exp[-x^2/(4kt)]$ and its derivative ψ_x are known as *fundamental solutions of the heat equation* $\psi_t = k\psi_{xx}$.

(a) Verify that ψ satisfies that equation when $t > 0$, that $\psi_x(0,t) = 0$ and $\psi(\infty,t) = 0$ when $t > 0$, and that $\psi(x,+0) = 0$ when $x > 0$; then complete a verification of the formal solution of Prob. 7 by showing that the total heat content of the solid per unit area of face, $\int_0^\infty cU(x,t)\,dx$, relative to the initial heat content, is Q_0.

(b) Show that the functions ψ and ψ_x are unbounded in a neighborhood of the point $x = t = 0$ when $x > 0$ and $t > 0$. Note that functions $A\psi(x,t)$, where A is an arbitrary constant, could be added to the solution (4), Sec. 46, to produce a family of solutions of the problem in that section if unbounded solutions were permitted.

9. For a solid whose temperature is a function $U(x,t)$ show that the total quantity of heat transmitted across a unit area of a plane $x = x_0$ can be written, formally,

$$Q(x_0) = -K\int_0^\infty U_x(x_0,t)\,dt = -K\lim_{s\to 0} u_x(x_0,s).$$

For the solid in Sec. 47 show that the formal result is

$$Q(x_0) = \sqrt{cK} \lim_{s \to 0} \sqrt{s f(s)} \qquad\qquad (K = ck).$$

For the solid in Prob. 7, show that $Q(x_0) = Q_0$.

10. The initial temperature distribution of a solid $x \geq 0$ is prescribed:

$$U(x,0) = g(x) \qquad\qquad (x > 0).$$

If $U(0,t) = 0$ when $t > 0$ and $U_x(\infty,t) = 0$, show formally that, when $t > 0$,

$$U(x,t) = \frac{1}{2\sqrt{\pi kt}} \int_0^\infty g(r) \left\{ \exp\left[-\frac{(r-x)^2}{4kt} \right] - \exp\left[-\frac{(r+x)^2}{4kt} \right] \right\} dr.$$

11. If the initial temperature is uniform in Prob. 10, $g(x) = U_0$, show that

$$U(x,t) = U_0 \operatorname{erf}\left(\frac{x}{2\sqrt{kt}} \right).$$

12. Show that the sum of the temperature function found in Prob. 10 and the function (7), Sec. 47, represents the temperature in the solid $x \geq 0$ with initial temperature $g(x)$ and face temperature $F(t)$.

48 TEMPERATURES IN A SLAB

The initial temperature of a slab of homogeneous material bounded by the planes $x = 0$ and $x = l$ is u_0. Let us find a formula for the temperatures in this solid after the face $x = 0$ is insulated and the temperature of the face $x = l$ is reduced to zero (Fig. 57).

The temperature function $U(x,t)$ satisfies the following conditions

$$U_t(x,t) = kU_{xx}(x,t) \qquad\qquad (0 < x < l, t > 0),$$
$$U(x,0) = u_0 \qquad\qquad (0 < x < l),$$
$$U_x(0,t) = 0, \qquad U(l,t) = 0 \qquad\qquad (t > 0).$$

The transform therefore satisfies the conditions

(1) $$su(x,s) - u_0 = ku_{xx}(x,s),$$

(2) $$u_x(0,s) = 0, \qquad u(l,s) = 0.$$

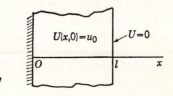

$$U(x,0) = u_0 \qquad U = 0$$

O $\qquad$ l $\qquad$ x

Fig. 57

The solution of the ordinary differential equation (1) that satisfies the first of the conditions (2) is

$$u(x,s) = \frac{u_0}{s} + C \cosh x \sqrt{\frac{s}{k}},$$

where C is determined by the second of conditions (2). Thus

(3)
$$u(x,s) = u_0 \left[\frac{1}{s} - \frac{1}{s} \frac{\cosh x\sqrt{s/k}}{\cosh l\sqrt{s/k}} \right].$$

Let us write
$$q = \sqrt{\frac{s}{k}},$$

and note that

$$\frac{\cosh xq}{\cosh lq} = e^{-lq}(e^{xq} + e^{-xq})\frac{1}{1 + e^{-2lq}}$$

$$= [e^{-(l-x)q} + e^{-(l+x)q}] \sum_{0}^{\infty} (-1)^n e^{-2nlq}$$

$$= \sum_{n=0}^{\infty} (-1)^n \{\exp[-q(ml - x)] + \exp[-q(ml + x)]\},$$

where $m = 2n + 1$. We have seen (Sec. 27) that

$$\frac{1}{s} e^{-\alpha\sqrt{s}} = L\left\{ \text{erfc}\left(\frac{\alpha}{2\sqrt{t}} \right) \right\} \qquad (\alpha \geq 0).$$

Therefore it follows formally from Eq. (3) that

(4)
$$U(x,t) = u_0 - u_0 \sum_{n=0}^{\infty} (-1)^n \left\{ \text{erfc}\left[\frac{(2n + 1)l - x}{2\sqrt{kt}} \right] \right.$$

$$\left. + \text{erfc}\left[\frac{(2n + 1)l + x}{2\sqrt{kt}} \right] \right\}.$$

We shall not take up the verification of this formula since a complete discussion would be lengthy. However, it is not difficult to show with the aid of the ratio test that the series converges uniformly with respect to x and t and that the series can be differentiated term by term. Since the value of the complementary error function here decreases rapidly as n increases, the convergence of the series is rapid, especially when t is small. Moreover the error function is one that is tabulated so that the series is well adapted to computation.

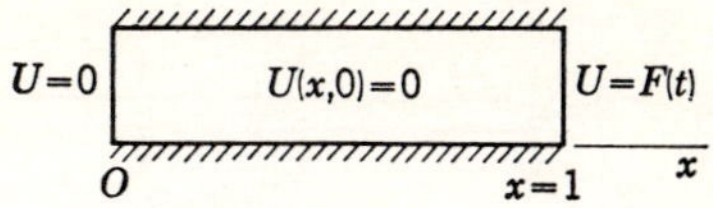

$U=0$ ┌─────────────────────┐ $U=F(t)$
         $U(x,0)=0$
Fig. 58 O $x=1$ x

49   A BAR WITH VARIABLE END TEMPERATURE

Let us determine a formula for the temperature $U(x,t)$ in a bar with its lateral surface insulated against the flow of heat when the initial temperature is zero and one end is kept at temperature zero while the temperature of the other end is a prescribed function of t.

If we take the unit of length as the length of the bar (Fig. 58) and select the unit of time such that $(1/k)\,\partial U/\partial t'$ becomes $\partial U/\partial t$, that is, so that $t = kt'$ where t' is the original and t the new variable, our boundary value problem can be written as follows:

$$U_t(x,t) = U_{xx}(x,t) \qquad\qquad (0 < x < 1,\ t > 0),$$

$$U(x,0) = 0 \qquad\qquad (0 < x < 1),$$

$$U(0,t) = 0, \qquad U(1,t) = F(t) \qquad\qquad (t > 0).$$

The solution of the transformed problem is found to be

$$u(x,s) = f(s)\frac{\sinh x\sqrt{s}}{\sinh \sqrt{s}}.$$

Proceeding as in the last section, and using the convolution, we find that

$$(1) \qquad U(x,t) = \frac{2}{\sqrt{\pi}} \sum_{n=0}^{\infty} \left\{ \int_{\frac{1}{2}(m-x)/\sqrt{t}}^{\infty} F\left[t - \frac{(m-x)^2}{4\lambda^2}\right] e^{-\lambda^2}\, d\lambda \right.$$

$$\left. - \int_{\frac{1}{2}(m+x)/\sqrt{t}}^{\infty} F\left[t - \frac{(m+x)^2}{4\lambda^2}\right] e^{-\lambda^2}\, d\lambda \right\} \qquad (m = 2n + 1).$$

When the temperature of the face $x = 1$ is constant,

$$(2) \qquad\qquad\qquad F(t) = F_0,$$

our formula can be written

$$(3) \qquad U(x,t) = F_0 \sum_{0}^{\infty} \left[\operatorname{erf}\left(\frac{2n + 1 + x}{2\sqrt{t}}\right) - \operatorname{erf}\left(\frac{2n + 1 - x}{2\sqrt{t}}\right) \right].$$

Details are left to the problems.

50   A COOLING FIN OR EVAPORATION PLATE

A thin semi-infinite plate occupies the space $x \geq 0, 0 \leq z \leq z_0, -\infty < y < \infty$. Heat transfer takes place at the faces $z = 0$ and $z = z_0$, into a medium at temperature zero, according to the linear law of surface heat transfer (Sec. 45);

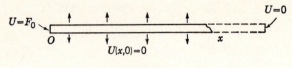

Fig. 59

but the thickness z_0 of the plate is small enough that variations of temperature with z can be neglected. If the initial temperature is zero and the end $x = 0$ is kept at temperature F_0 (Fig. 59), the boundary value problem for the temperature function becomes

(1)
$$U_t(x,t) = kU_{xx}(x,t) - hU(x,t) \qquad (x > 0, t > 0),$$
$$U(x,0) = 0, \qquad U(0,t) = F_0, \qquad \lim_{x \to \infty} U(x,t) = 0.$$

Here $U(x,t)$ may also represent the concentration of moisture in the plate when the plate is initially dry and evaporation takes place into a dry medium, if the concentration at the end $x = 0$ is a constant F_0. In that case $k = K$, the coefficient of diffusion (Sec. 45), and the positive constant h is a coefficient of evaporation.

In terms of transforms, our problem (1) becomes

$$ku_{xx}(x,s) - (s + h)u(x,s) = 0 \qquad (x > 0),$$

$$u(0,s) = \frac{F_0}{s}, \qquad \lim_{x \to \infty} u(x,s) = 0.$$

The solution of this problem is

(2)
$$u(x,s) = \frac{F_0}{s} \exp\left(-x\sqrt{\frac{s + h}{k}}\right).$$

Knowing the inverse transform of $e^{-x\sqrt{s/k}}$, we can write, with the aid of our property on substitution of $s + h$ for s (Sec. 7),

$$L^{-1}\left\{\exp\left(-x\sqrt{\frac{s + h}{k}}\right)\right\} = \frac{xe^{-ht}}{2\sqrt{\pi kt^3}} \exp\left(-\frac{x^2}{4kt}\right).$$

It follows from formula (2) that

(3)
$$U(x,t) = \frac{F_0 x}{2\sqrt{\pi k}} \int_0^t e^{-h\tau} \exp\left(-\frac{x^2}{4k\tau}\right) \frac{d\tau}{\tau^{\frac{3}{2}}}$$

$$= \frac{2F_0}{\sqrt{\pi}} \int_{\frac{1}{2}x/\sqrt{kt}}^{\infty} \exp\left(-\lambda^2 - \frac{hx^2}{4k\lambda^2}\right) d\lambda,$$

where the second integral is obtained by the substitution $\lambda = \frac{1}{2}x/\sqrt{k\tau}$.

The formula (3) can be changed to a more useful form with the aid of the integration formula (see Prob. 14 below)

(4) $\qquad \dfrac{4}{\sqrt{\pi}} \displaystyle\int_\tau^\infty \exp\left(-\lambda^2 - \dfrac{a^2}{\lambda^2}\right) d\lambda = e^{2a}\,\mathrm{erfc}\left(r + \dfrac{a}{r}\right) + e^{-2a}\,\mathrm{erfc}\left(r - \dfrac{a}{r}\right).$

This formula can be verified by noting that its two members have the same derivative with respect to the parameter r and that both members vanish as r tends to infinity. With the aid of formula (4) the solution (3) of our problem becomes

(5) $\qquad U(x,t) = \dfrac{F_0}{2}\left[e^{bx}\,\mathrm{erfc}\left(\dfrac{x}{2\sqrt{kt}} + \sqrt{ht}\right) + e^{-bx}\,\mathrm{erfc}\left(\dfrac{x}{2\sqrt{kt}} - \sqrt{ht}\right)\right],$

where $b = \sqrt{h/k}$. Note that $U(x,t) > 0$ for each x whenever $t > 0$, if $F_0 > 0$.

In obtaining Eq. (5) we have found a desirable form of the inverse transform of the function (2). The verification of our formal solution of problem (1) is left to the problems.

51 TEMPERATURES IN A COMPOSITE SOLID

Let us write a formula for temperatures $U(x,t)$ in a solid $x \geq 0$, composed of a layer $0 < x < a$ of material initially at uniform temperature A in perfect thermal contact with a semi-infinite solid $x > a$ of another material initially at temperature zero, when the face $x = 0$ is kept insulated (Fig. 60).

If the thermal conductivity and diffusivity are K_1 and k_1, respectively, in the first part and K_2 and k_2 in the second part, the boundary value problem is the following one:

(1) $\qquad\qquad U_t(x,t) = k_1 U_{xx}(x,t) \qquad\qquad (0 < x < a,\ t > 0),$

(2) $\qquad\qquad U_t(x,t) = k_2 U_{xx}(x,t) \qquad\qquad (x > a,\ t > 0),$

(3) $\qquad U(x,0) = A\,(0 < x < a), \qquad U(x,0) = 0 \qquad\qquad (x > a),$

(4) $\qquad U_x(0,t) = 0, \qquad \lim_{x\to\infty} U(x,t) = 0 \qquad\qquad (t > 0),$

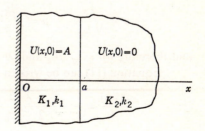

Fig. 60

(5)
$$U(a - 0, t) = U(a + 0, t) \qquad (t > 0),$$

(6)
$$K_1 U_x(a - 0, t) = K_2 U_x(a + 0, t) \qquad (t > 0).$$

Condition (5) states that the temperature at the interface $x = a$ is the same after $t = 0$ when the point approaches the interface from either direction. Condition (6) states that the flux of heat out of the first part through the interface must be equal to the flux into the second part, at each instant.

The problem in the transform of $U(x,t)$ is then

(7)
$$xu(x,s) - A = k_1 u_{xx}(x,s) \qquad (0 < x < a),$$

(8)
$$su(x,s) = k_2 u_{xx}(x,s) \qquad (x > a),$$

(9)
$$u_x(0,s) = 0, \qquad \lim_{x \to \infty} u(x,s) = 0,$$

(10)
$$u(a - 0, s) = u(a + 0, s),$$
$$K_1 u_x(a - 0, s) = K_2 u_x(a + 0, s).$$

The solution of Eq. (7) that satisfies the first of conditions (9) is

$$u(x,s) = C_1 \cosh x \sqrt{\frac{s}{k_1}} + \frac{A}{s} \qquad (0 \leq x < a),$$

and the solution of (8) that satisfies the second of conditions (9) is

$$u(x,s) = C_2 \exp\left(-x\sqrt{\frac{s}{k_2}}\right) \qquad (x > a).$$

By applying the conditions (10) to these functions, the values of C_1 and C_2 are easily found. The formulas for $u(x,s)$ can then be written

(11)
$$u(x,s) = \frac{A}{s}\left(1 - \frac{1 - \lambda\, e^{-\sigma(a-x)} + e^{-\sigma(a+x)}}{2} \frac{}{1 - \lambda e^{-2\sigma a}}\right) \qquad (0 < x < a),$$

(12)
$$u(x,s) = \frac{A(1 + \lambda)\, e^{-\sigma\mu(x-a)} - e^{-\sigma(2a+\mu x - \mu a)}}{2s} \frac{}{1 - \lambda e^{-2\sigma a}} \qquad (x > a),$$

where

$$\sigma = \sqrt{\frac{s}{k_1}}, \qquad \mu = \sqrt{\frac{k_1}{k_2}}, \qquad \lambda = \frac{K_1\sqrt{k_2} - K_2\sqrt{k_1}}{K_1\sqrt{k_2} + K_2\sqrt{k_1}},$$

and therefore $\sigma\mu = \sqrt{s/k_2}$ and $|\lambda| < 1$.

Equation (11) can be written in the form

$$\frac{u(x,s)}{A} = \frac{1}{s} - \frac{1 - \lambda}{2} \sum_{n=0}^{\infty} \frac{\lambda^n}{s}(e^{-\sigma(ma-x)} + e^{-\sigma(ma+x)}) \qquad (0 < x < a),$$

where $m = 2n + 1$, and (12) can be written

$$\frac{u(x,s)}{A} = \frac{1 + \lambda}{2} \sum_{n=0}^{\infty} \frac{\lambda^n}{s} (e^{-\sigma(2na + \mu x - \mu a)} - e^{-\sigma(2na + 2a + \mu x - \mu a)}) \qquad (x > a).$$

It therefore follows that

$$(13) \qquad U(x,t) = A - A\frac{1 - \lambda}{2} \sum_{0}^{\infty} \lambda^n \left\{ \text{erfc}\left[\frac{(2n + 1)a - x}{2\sqrt{k_1 t}} \right] \right.$$

$$\left. + \text{erfc}\left[\frac{(2n + 1)a + x}{2\sqrt{k_1 t}} \right] \right\} \qquad (0 < x < a),$$

$$(14) \qquad U(x,t) = A\frac{1 + \lambda}{2} \sum_{0}^{\infty} \lambda^n \left\{ \text{erfc}\left[\frac{2na + \mu(x - a)}{2\sqrt{k_1 t}} \right] \right.$$

$$\left. - \text{erfc}\left[\frac{(2n + 2)a + \mu(x - a)}{2\sqrt{k_1 t}} \right] \right\} \qquad (x > a).$$

In such problems if $U(x,t)$ represents the concentration of a substance diffusing in a composite porous medium, the concentrations in adjacent layers of different media may maintain some ratio α ($\alpha \neq 1$), the ratio of equilibrium concentrations in the two media (see Prob. 13 below). Interface condition (5) would then be replaced by the condition

$$(15) \qquad U(a + 0, t) = \alpha U(a - 0, t).$$

PROBLEMS

1. Complete the derivation of formula (1), Sec. 49.

2. When $x = 1$, show that the general term of the series in formula (3), Sec. 49, becomes

$$\frac{2}{\sqrt{\pi}} \int_{n/\sqrt{t}}^{(n+1)/\sqrt{t}} e^{-\lambda^2} \, d\lambda,$$

and hence that $U(1,t) = F_0$.

3. In Sec. 50, verify the solution (5) of problem (1).

4. In the problem (1) of Sec. 50, show that the flux into the plate at the end $x = 0$ is given by the formula

$$\Phi(0,t) = KF_0\left(\frac{1}{\sqrt{\pi k t}} e^{-ht} + \sqrt{\frac{h}{k}} \, \text{erf} \sqrt{ht} \right).$$

5. In the problem (1) of Sec. 50, make the substitution

$$V(x,t) = e^{ht}U(x,t)$$

and show that the resulting problem in $V(x,t)$ is a special case of one solved in Sec. 47.

6.   The faces $x = 0$ and $x = 1$ of a slab of material for which $k = 1$ are kept at temperatures $U = 0$ and $U = 1$, respectively, until the temperature distribution becomes $U = x$. After time $t = 0$ both faces are held at temperature $U = 0$. Derive the temperature formula

$$U(x,t) = x - \sum_{n=0}^{\infty} \left(\operatorname{erfc} \frac{2n + 1 - x}{2\sqrt{t}} - \operatorname{erfc} \frac{2n + 1 + x}{2\sqrt{t}} \right) \qquad (t > 0).$$

7.   In Prob. 6, let the face $x = 1$ be insulated when $t > 0$ while $U = 0$ as before at the face $x = 0$. Find the flux of heat outward through the face $x = 0$.

$$Ans. \quad K \left[1 - 2 \sum_{n=0}^{\infty} (-1)^n \operatorname{erfc} \frac{2n + 1}{2\sqrt{t}} \right].$$

8.   Heat is generated in a long bar $x \geq 0$ at a rate $R(t)$ units per unit volume [Sec. 45, Eq. (7)], and the lateral surface of the bar is insulated. If the initial temperature is zero and the end $x = 0$ is kept at that temperature while the infinite end is insulated (Fig. 61), derive the temperature formula

$$U(x,t) = Q(t) - \frac{2}{\sqrt{\pi}} \int_{\frac{1}{2}x/\sqrt{kt}}^{\infty} Q\left(t - \frac{x^2}{4k\lambda^2} \right) \exp(-\lambda^2) \, d\lambda,$$

where

$$Q(t) = \frac{1}{c} \int_0^t R(\tau) \, d\tau.$$

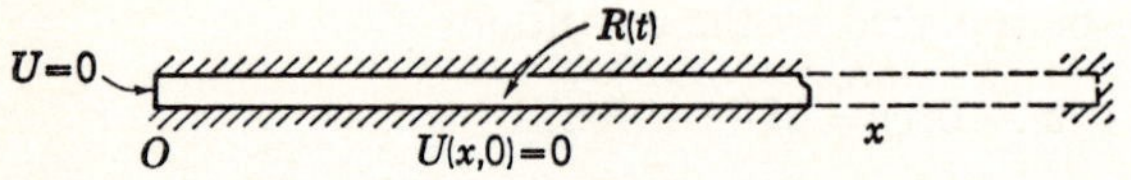

Fig. 61

9.   In Prob. 8, let the rate of generation of heat be a constant B and show that

$$U(x,t) = \frac{B}{c} \int_0^t \operatorname{erf} \frac{x}{2\sqrt{k\tau}} \, d\tau.$$

10.   The lateral surface of a bar of unit length is insulated while its ends $x = 0$ and $x = 1$ are kept at temperature zero. Heat is generated throughout the bar at a constant rate of B units per unit volume. If the initial temperature is $B(x - x^2)/(2K)$, show that the bar retains that initial temperature distribution when $t > 0$.

11.   Units of time and distance are chosen so that $k = 1$ and a given bar has unit length. The lateral surface and the end $x = 0$ are insulated. The initial temperature of the bar is zero, and the end $x = 1$ is kept at that temperature. If heat is generated throughout the bar at a rate of $R(t)$ units per unit volume, derive the temperature formula

$$cU(x,t) = \int_0^t R(\tau) \, d\tau - \int_0^t R(t - \tau) E(x,\tau) \, d\tau,$$

where

$$E(x,t) = \sum_0^{\infty} (-1)^n \left[\operatorname{erfc} \left(\frac{2n + 1 + x}{2\sqrt{t}} \right) + \operatorname{erfc} \left(\frac{2n + 1 - x}{2\sqrt{t}} \right) \right].$$

12. Let the semi-infinite evaporation plate (Sec. 50, Fig. 59) have a uniform initial concentration of moisture $U(x,0) = u_0$, and let the end $x = 0$ be kept dry, $U(0,t) = 0$, while the infinite end is impervious to moisture. Derive, and verify, this formula for the concentration:

$$U(x,t) = u_0 e^{-ht} \operatorname{erf} \frac{x}{2\sqrt{Kt}}.$$

13. The entire surface of a long porous cylinder is impervious to moisture. The cylinder consists of two semi-infinite parts, the first $(-\infty < x < 0)$ of material with a coefficient of diffusion of moisture K_1 and the second $(0 < x < \infty)$ with a coefficient K_2 (Fig. 62). Initially, the concentration of moisture is zero in the part $x < 0$ and u_0 in the part $x > 0$. Let α be the ratio of equilibrium concentrations in the two parts so that, as indicated in Sec. 51, the concentration $U(x,t)$ satisfies the conditions

$$U(+0,t) = \alpha U(-0,t), \qquad K_2 U_x(+0,t) = K_1 U_x(-0,t)$$

at the interface $x = 0$. Derive the formula

$$U(x,t) = \frac{u_0}{\alpha + \beta} \operatorname{erfc}\left(\frac{-x}{2\sqrt{K_1 t}}\right) \qquad \text{when } x < 0,$$

$$= \frac{u_0}{\alpha + \beta}\left(\alpha + \beta \operatorname{erf} \frac{x}{2\sqrt{K_2 t}}\right) \qquad \text{when } x > 0,$$

where $\beta = \sqrt{K_1/K_2}$. Show that the ratio of the limiting concentrations, as $t \to \infty$, in two parts is α.

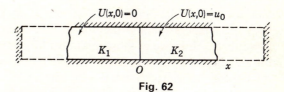

Fig. 62

14. Derive the integration formula (4), Sec. 50. *Suggestion*: Write

$$z(x) = \frac{4}{\sqrt{\pi}} \int_r^\infty \exp\left[-\left(\lambda - \frac{x}{\lambda}\right)^2\right] d\lambda;$$

then the integral (4), Sec. 50, is $e^{-2a}z(a)$. Show that $z'(x) = 4e^{4x} \operatorname{erfc}(r + x/r)$ and $z(0) = 2 \operatorname{erfc} r$, then use integration by parts to find $z(x)$.

52   OBSERVATIONS ON THE METHOD

All the problems treated in this chapter involve partial differential equations and boundary conditions that are *linear*, that is, of first degree in the unknown function and its derivatives. The limitation of our operational method to the treatment of such linear boundary value problems is a natural one, since we have presented no formula giving the Laplace transform of the product of two functions in terms of the transforms of the individual functions. It is known that the transform of the product of two arbitrary functions can be expressed by a convolution integral of the two transforms, where the integration is one in the complex plane of s. But it is safe to say that no advantage can be anticipated in replacing nonlinear differential forms by complex nonlinear integral forms.

We have solved problems with constant coefficients. If the coefficients are functions of t, the variable with respect to which the transformation is made, the transformed problem is not likely to be simpler than the original one. For even when the coefficients are polynomials in t, the transformed problem involves derivatives with respect to s in place of the derivatives with respect to t present in the original one. If the coefficients are not functions of t, the transformed problem will be simpler.

We have made the transformation with respect to time t in all our problems in partial differential equations. If the physical problem involved the first derivative U_t, the initial value $U(x,0)$ was prescribed; if it involved Y_{tt}, then $Y(x,0)$ and $Y_t(x,0)$ were both prescribed. Consequently, when we applied the formula for the transformation of those derivatives, $u(x,s)$ or $y(x,s)$ was the only unknown function arising. But suppose the Laplace transformation with respect to x had been applied in the temperature problems. Then if $L\{U(x,t)\} = u(z,t)$,

$$L\{U_{xx}(x,t)\} = z^2 u(z,t) - zU(0,t) - U_x(0,t),$$

and not both of the functions $U(0,t)$ and $U_x(0,t)$ could be prescribed, since both the temperature and the flux of heat at the surface $x = 0$ cannot be prescribed. Thus one of those unknown functions must be determined with the aid of other boundary conditions, and the procedure becomes unwieldy; the Laplace transformation is not the integral transformation that is adapted to the boundary conditions with respect to x here.

The operational method of solving partial differential equations is of course not limited to equations of the second order. We shall soon take up a more powerful method of obtaining inverse transforms, and then we can attack problems whose solutions depend upon more involved inverse transformations than those in this chapter.

As we have illustrated here and in the preceding chapter, the operational method is well adapted to the solution of many problems in differential equations in which some of the given functions or their derivatives are discontinuous. That is one of the remarkable features of the method.

In case the boundary value problem involves more than two independent variables, say x, y, and t, a Laplace transformation with respect to t still leaves us with a partial differential equation with independent variables x and y. The new problem may be adapted to solution by one of the Fourier transformations or another integral transformation with respect to x or y, depending on the differential forms and boundary conditions involved there. That procedure of using *successive transformations* will be illustrated in Chap. 11.

As the problems in the present chapter have indicated, transformation methods help to show whether the number and type of boundary conditions accompanying a given partial differential equation are adequate to determine a definite solution of the boundary value problem. A physical interpretation of the problem serves as another guide for setting boundary conditions that lead to just one solution.

A complete treatment or rigorous solution of a boundary value problem, however, consists first of proving that the formal solution does satisfy the differential equation and all boundary conditions and continuity requirements, then showing that the function is the only one that satisfies all those conditions. That procedure of establishing a unique solution is apt to be lengthy. We shall illustrate it in later chapters.

5
Functions of a Complex Variable

We give now a synopsis of theorems and definitions from the basic theory of functions of a complex variable, material needed to develop further the theory of the Laplace transformation. For complete treatments of the topics, including proofs of statements and theorems, the reader may refer to books on the subject.[1] Derivations of extentions of the Cauchy integral formulas are given in the final section of this chapter.

53 COMPLEX NUMBERS

We shall not review the algebra of complex numbers except to remind the reader of properties of absolute values and conjugates. We write

$$z = x + iy = r(\cos \theta + i \sin \theta),$$

[1] See, for instance, the author's "Complex Variables and Applications," 2d ed., 1960, and references listed in the Bibliography in that book.

$$x = \text{Re } z, \qquad y = \text{Im } z, r = |z| \qquad \text{and} \qquad \theta = \arg z.$$

If z_1 and z_2 are complex numbers, then

$$|z_1 z_2| = |z_1| |z_2|, \qquad \left| \frac{z_1}{z_2} \right| = \frac{|z_1|}{|z_2|} \qquad \text{if } z_2 \neq 0.$$

The triangle inequalities are

$$|z_1 \pm z_2| \leq |z_1| + |z_2|, \qquad |z_1 \pm z_2| \geq ||z_1| - |z_2||.$$

Neither the statement $z_1 > z_2$ nor $z_1 < z_2$ has meaning unless z_1 and z_2 are both real numbers.

If m and n are integers and $z \neq 0$, then

$$z^{m/n} = r^{m/n} \left(\cos \frac{m\theta}{n} + i \sin \frac{m\theta}{n} \right), \qquad |z^{m/n}| = |z|^{m/n}.$$

The conjugate operation on z, $\bar{z} = x - iy$, has these distributive properties:

$$\overline{z_1 + z_2} = \bar{z}_1 + \bar{z}_2, \qquad \overline{z_1 z_2} = \bar{z}_1 \bar{z}_2, \qquad \overline{(z_1/z_2)} = \bar{z}_1/\bar{z}_2 \qquad (z_2 \neq 0).$$

Also $z\bar{z} = |z|^2$.

A *neighborhood* of a point z_0 in the complex plane is a two-dimensional circular domain, or disk, consisting of all points z such that $|z - z_0| < \epsilon$, where ϵ is some positive number.

54 ANALYTIC FUNCTIONS

Let w denote the values of a single-valued function f of the complex variable z, defined on some set of points in the z plane. We write

$$w = f(z) = u(x,y) + iv(x,y)$$

where u and v are real-valued functions of two real variables. If $w = z^2$ for all z, for example, then

$$f(z) = (x + iy)^2 = x^2 - y^2 + 2xyi;$$

thus $u(x,y) = x^2 - y^2$ and $v(x,y) = 2xy$. If $w = \text{Re } z$ then $u(x,y) = x$ and $v(x,y) = 0$.

If w is a single-valued function $z^{\frac{1}{2}}$ such that

$$z^{\frac{1}{2}} = \sqrt{r} \left(\cos \frac{\theta}{2} + i \sin \frac{\theta}{2} \right) \qquad (r > 0, -\pi < \theta < \pi),$$

then it is convenient to write the components u and v as functions of r and θ:

$$u(r,\theta) = \sqrt{r} \cos \tfrac{1}{2}\theta, \qquad v(r,\theta) = \sqrt{r} \sin \tfrac{1}{2}\theta$$
$$(r > 0, -\pi < \theta < \pi).$$

Here u and v are single-valued in the domain consisting of all points of the complex plane except those for which $y = 0$ and $x \leq 0$.

A function f has a limit w_0 at a point z_0,

$$\lim_{z \to z_0} f(z) = w_0,$$

if and only if for each positive number ϵ there is some number δ such that

$$|f(z) - w_0| < \epsilon \qquad \text{whenever } |z - z_0| < \delta \text{ and } z \neq z_0.$$

Thus the value of f is arbitrarily close to w_0 throughout some deleted neighborhood of z_0, a two-dimensional neighborhood with z_0 excluded. The definition requires f to be defined throughout some neighborhood of the point z_0, except possibly at the point itself.

If we write $\operatorname{Re} w_0 = u_0$, $\operatorname{Im} w_0 = v_0$, and $z_0 = x_0 + iy_0$, then it turns out that the condition

$$\lim_{z \to z_0} f(z) = \lim_{z \to z_0} [u(x,y) + iv(x,y)] = u_0 + iv_0$$

is satisfied if and only if the two-dimensional limits of the real-valued functions u and v at the point (x_0, y_0) are u_0 and v_0, respectively. Since f is *continuous* at z_0 only if its limit w_0 is the same as its value $f(z_0)$, then it is continuous there if and only if both u and v are continuous at the point (x_0, y_0).

The derivative of f at a point z is defined as

$$f'(z) = \lim_{\Delta z \to 0} \frac{f(z + \Delta z) - f(z)}{\Delta z}$$

provided the limit exists, that is, provided that for each positive number ϵ there is a number δ such that

$$\left| \frac{f(z + \Delta z) - f(z)}{\Delta z} - f'(z) \right| < \epsilon$$

$$\text{whenever } |\Delta z| < \delta (\Delta z \neq 0).$$

The two-dimensional limit here with respect to the variable Δz (Fig. 63) requires that, if $f'(z)$ exists, the partial derivatives of first order of the functions u and v exist at the point (x,y) and satisfy the *Cauchy-Riemann conditions*

$$\frac{\partial u}{\partial x} = \frac{\partial v}{\partial y}, \qquad \frac{\partial u}{\partial y} = -\frac{\partial v}{\partial x},$$

at the point. Also, when $f'(z)$ exists, we find that

$$f'(z) = u_x(x,y) + iv_x(x,y) = v_y(x,y) - iu_y(x,y).$$

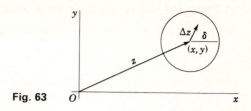

Fig. 63

Suppose that u and v are real-valued functions of x and y whose partial derivatives of first-order are continuous at a point (x,y).[1] Then in order for the function $f = u + iv$ to have a derivative $f'(z)$ at the point $z = x + iy$, it is *necessary* and *sufficient* that u and v satisfy the Cauchy-Riemann conditions at the point.

For example, the components u and v of the function z^2 satisfy those conditions everywhere; hence

$$\frac{d}{dz}(z^2) = \frac{\partial}{\partial x}(x^2 - y^2) + i\frac{\partial}{\partial x}(2xy) = 2(x + iy) = 2z.$$

In contrast, the components of the function $|z|^2 = x^2 + y^2$, whose partial derivatives are continuous everywhere, satisfy the Cauchy-Riemann conditions at the origin only, where $2x = 0$ and $2y = 0$. Thus the derivative of $|z|^2$ exists at $z = 0$ only, where its value is zero.

A function f is *analytic* at a point z_0 if its derivative $f'(z)$ *exists at every point in some neighborhood* of z_0. Analyticity implies continuity, but not conversely. The functions $\text{Re } z, \bar{z},$ and $|z|^2$, for instance, are everywhere continuous, but nowhere analytic because their derivatives do not exist throughout any neighborhood.

Formulas for derivatives of sums, products, quotients, and powers of functions and of composite functions $g[f(z)]$ correspond to those in calculus. Consequently the sum, product, or quotient of two analytic functions is analytic except, in the case of the quotient, at those points where the denominator vanishes. An analytic function $g(w)$ of an analytic function $w = f(z)$ is analytic for values of z in a *domain* (a connected open region) where w is analytic and where the values of w are interior to a domain where $g(w)$ is analytic.

The function $1/z$ is an example of a quotient of two functions that are everywhere analytic. The function is analytic at all points except the origin. Its derivative is $-1/z^2$ if $z \neq 0$.

An *entire function* is one that is analytic for all finite z. Every polynomial in z is an entire function.

---

[1] The continuity of those partial derivatives ensures the continuity of the functions u and v themselves, at the point.

55 EXPONENTIAL AND TRIGONOMETRIC FUNCTIONS

The exponential function, written exp z or e^z, can be defined by the equation

$$(1) \qquad\qquad \exp z = e^x(\cos y + i \sin y) \qquad\qquad (z = x + iy),$$

where y is the radian measure of the argument of the trigonometric functions. Its components u and v have partial derivatives which are continuous and satisfy the Cauchy-Riemann conditions everywhere. Therefore exp z *is an entire function.* Its derivative is the function itself:

$$(2) \qquad\qquad \frac{d}{dz} \exp z = u_x(x,y) + iv_x(x,y) = \exp z.$$

We note that $|\exp z| = e^x > 0$ for all z; thus e^z is never zero. When $z = x$, the function exp z becomes the real-valued function e^x.

The polar representation of z can now be written $re^{i\theta}$:

$$z = r(\cos \theta + i \sin \theta) = r \exp (i\theta).$$

The hyperbolic cosine and sine are these linear combinations of the entire functions e^z and e^{-z}:

$$(3) \qquad\qquad \cosh z = \tfrac{1}{2}(e^z + e^{-z}), \qquad \sinh z = \tfrac{1}{2}(e^z - e^{-z});$$

hence they are entire. Their components are shown by the formulas

$$\cosh z = \cosh x \cos y + i \sinh x \sin y,$$

$$\sinh z = \sinh x \cos y + i \cosh x \sin y.$$

Each of the remaining hyperbolic functions of z,

$$\tanh z = \frac{\sinh z}{\cosh z}, \qquad \coth z = \frac{1}{\tanh z},$$

$$\operatorname{sech} z = \frac{1}{\cosh z}, \qquad \operatorname{csch} z = \frac{1}{\sinh z},$$

is analytic except at those points where the denominator on the right-hand side vanishes.

The circular functions of z can be defined as follows:

$$(4) \qquad \cos z = \frac{e^{iz} + e^{-iz}}{2}, \qquad \sin z = \frac{e^{iz} - e^{-iz}}{2i}, \qquad \tan z = \frac{\sin z}{\cos z},$$

the remaining three being the reciprocals of these. The functions $\cos z$ and $\sin z$ are entire. In view of these definitions, it follows that

$$\cos z = \cosh iz = \cos x \cosh y - i \sin x \sinh y,$$

$$\sin z = -i \sinh iz = \sin x \cosh y + i \cosh x \sinh y.$$

Also, $\sin z$ and $\cos z$ are periodic with period 2π, while $\sinh z$ and $\cosh z$ are periodic with period $2\pi i$.

The forms of the trigonometric identities are the same as those for functions with real arguments; for instance,

$$\sin^2 z + \cos^2 z = 1, \qquad \cos 2z = \cos^2 z - \sin^2 z,$$

$$\sin(z_1 + z_2) = \sin z_1 \cos z_2 + \cos z_1 \sin z_2.$$

The same is true for the relations between the six hyperbolic functions. Furthermore, the formulas for the derivatives of all those functions retain the same form when the variable is real. Thus for all z in a region where $w(z)$ is analytic, the composite functions $\sin w(z)$ and $\cos w(z)$ are analytic functions of z and

$$\frac{d}{dz} \sin w = \cos w \frac{dw}{dz}, \qquad \frac{d}{dz} \cos w = -\sin w \frac{dw}{dz}.$$

The absolute values of $\sin z$ and $\cos z$ are *not bounded* as $|y| \to \infty$. It is left to the problems to show that

(5) $|\sin z|^2 = \sin^2 x + \sinh^2 y,$ $|\cos z|^2 = \cos^2 x + \sinh^2 y,$

(6) $|\sinh z|^2 = \sinh^2 x + \sin^2 y,$ $|\cosh z|^2 = \sinh^2 x + \cos^2 y.$

From those formulas we can see that the zeros of $\sin z$ and $\cos z$ are just the real zeros of those functions, while the zeros of $\sinh z$ and $\cosh z$ are pure imaginary numbers.

PROBLEMS

1. (a) Interpret the triangle inequalities (Sec. 53) in terms of lengths of sides of triangles.
(b) Extend the first triangle inequality to apply to n complex numbers as follows:

$$|z_1 + z_2 + \cdots + z_n| \leq |z_1| + |z_2| + \cdots + |z_n|.$$

2. If $|z_2| \neq |z_3|$, prove that

$$\left| \frac{z_1}{z_2 + z_3} \right| \leq \frac{z_1}{||z_2| - |z_3||}.$$

3. If $n = 2, 3, \ldots$, show that $z^{1/n}$ has n distinct values.

4. If a_n and b_n are real numbers, prove that the function

$$f(z) = a_1 x + b_1 y + c + i(a_2 x + b_2 y) \qquad (z = x + iy)$$

is an entire function if and only if the coefficients are such that $f(z) = (a_1 - ib_1)z + c$.

5. Prove that the function $f(z) = x^2 + iy^2$ is nowhere analytic.

6. State why the functions $\sin(e^z)$ and $\exp(\cos z)$ are entire.

7.   Show that (a) $e^{-z} = 1/e^z$; (b) $\exp(z_1 + z_2) = (\exp z_1)(\exp z_2)$; (c) $|\exp(-z^2)|$ is not bounded for all z.

8.   In Sec. 55, prove (a) identities (5); (b) identities (6).

9.   Prove that (a) $\sin z = 0$ if and only if $z = \pm n\pi$; (b) $\sinh z = 0$ if and only if $z = \pm n\pi i$, where $n = 0, 1, 2, \ldots$.

10.   Prove that $\tanh z$ is analytic except at the points $z = \pm(n - \frac{1}{2})\pi i$, where $n = 1, 2, \ldots$.

11.   Show that (a) $\tan(z + \pi) = \tan z$; (b) $\tanh(z + \pi i) = \tanh z$.

12.   Establish the inequalities $\sinh |x| \le |\sinh z| \le \cosh x$.

13.   Use the definition of $f'(z)$ when $\Delta z = \Delta x$, then when $\Delta z = i\Delta y$, to show that $f' = u_x + iv_x = v_y - iu_y$ when $f'(z)$ exists. Hence deduce that the components u and v necessarily satisfy the Cauchy-Riemann conditions at points where $f'(z)$ exists.

14.   In polar coordinates, where $z = re^{i\theta}$, $r > 0$, verify that the Cauchy-Riemann conditions take the form

$$\frac{\partial u}{\partial r} = \frac{1}{r}\frac{\partial v}{\partial \theta}, \qquad \frac{1}{r}\frac{\partial u}{\partial \theta} = -\frac{\partial v}{\partial r},$$

and that $f'(z) = e^{-i\theta}[u_r(r,\theta) + iv_r(r,\theta)]$.

56   CONTOUR INTEGRALS

Consider first a complex-valued function G of a real variable t, with components U and V,

$$G(t) = U(t) + iV(t).$$

We call it sectionally continuous over a bounded interval if both the real-valued functions U and V have that property. Its derivative is $U'(t) + iV'(t)$ wherever U' and V' exist. Its integral

$$(1) \qquad \int_a^b G(t)\,dt = \int_a^b U(t)\,dt + i\int_a^b V(t)\,dt$$

has the properties of integrals of real-value functions as to integrals of sums, multiplication by a constant, reversal of limits of integration, and integration by parts. Also, when G is sectionally continuous on the interval $a < t < b$, it is absolutely integrable over the interval and

$$(2) \qquad \left| \int_a^b G(t)\,dt \right| \le \int_a^b |G(t)|\,dt.$$

The limit b of the integral (1) can be replaced by ∞ if the corresponding integrals of U and V exist. The comparison test for convergence of integrals (Prob. 14, Sec. 5) is easily extended to the improper integral of G. Thus if

G is sectionally continuous over each bounded subinterval of the half line $t > a$ and if $|G(t)| \leq M(t)$ where the real-valued function M is integrable from a to ∞, then G is integrable from a to ∞. In particular, G is integrable there if $|G|$ is integrable, and a proof[1] of inequality (2) can be extended to the improper integral to write

$$(3) \qquad \left| \int_a^\infty G(t)\,dt \right| \leq \int_a^\infty |G(t)|\,dt.$$

Now let C denote either an arc of a smooth curve

$$(4) \qquad x = \phi(t), \qquad y = \psi(t),$$

where ϕ, ψ, ϕ', and ψ' are continuous and the derivatives do not vanish simultaneously for any of the values of t, or a curve formed by joining a finite number of such arcs endwise. Then C is called a *contour*. If C is closed and does not intersect itself, it is called a *closed contour*; boundaries of triangles, rectangles, and circles are examples.

If the parametric equations (4) represent a contour C over which $a \leq t \leq b$ and if $f(z)$ is sectionally continuous over C, that is, if the function $f[\phi(t) + i\psi(t)]$, $a < t < b$, is sectionally continuous, then the *contour integral* of f over C can be defined in terms of an integral of type (1), namely

$$(5) \qquad \int_C f(z)\,dz = \int_a^b f[\phi(t) + i\psi(t)][\phi'(t) + i\psi'(t)]\,dt.$$

The integrand is a sectionally continuous function of t. Inequality (2) enables us to write

$$(6) \qquad \left| \int_C f(z)\,dz \right| \leq \int_C |f(z)|\,|dz| \leq ML,$$

if $|f(z)| \leq M$ for all z on C, L is the length of C, and $|dz|$ is the element of arc length of C: $|dz| = |\phi'(t) + i\psi'(t)|\,dt = ds$.

Using the components $u(x,y)$ and $v(x,y)$ of $f(z)$, we can write the contour integral (5) in terms of real-line integrals over C as follows:

$$(7) \qquad \int_C f(z)\,dz = \int_C (u\,dx - v\,dy) + i \int_C (u\,dy + v\,dx),$$

a result that is arrived at formally by writing $f = u + iv$ and $dz = dx + i\,dy$, then expanding the product $f(z)\,dz$.

---

[1] Churchill, R. V., "Complex Variables and Applications," 2d ed., p. 96, 1960.

57   INTEGRAL THEOREMS

We now state basic theorems on contour integrals of analytic functions.

Cauchy integral theorem   *If a function f is analytic at all points interior to and on a closed contour C, then*

(1)
$$\int_C f(z)\,dz = 0.$$

A *simply connected* domain D is an open-connected region such that every closed contour within it encloses only points of D. The interior of a closed contour and the entire z plane are examples, but the exterior of a closed contour is not simply connected, nor is the domain between two concentric circles.

The Cauchy integral theorem permits C to be any closed contour within a simply connected domain throughout which f is analytic. The theorem is easily extended so as to permit C to be replaced by the boundary B of certain multiply connected regions, as follows.

Let C_0 denote a closed contour, and $C_1, C_2, \ldots, C_n$ a finite number of closed contours within C_0 with no interior points in common (Fig. 64). Let R be the closed multiply connected region consisting of all points on and within C_0 except for points interior to $C_j\,(j = 1, 2, \ldots, n)$. Let B denote the boundary of R, consisting of C_0 and all C_j described in such a direction that the region R lies on the left of the boundary. Then if f is analytic at all points of R, it is true that

(2)
$$\int_B f(z)\,dz = 0.$$

The value of f at each point interior to a closed contour C is determined by the values on C by the following theorem.

Cauchy's integral formula   *Let f be analytic everywhere within and on a closed contour C. Then if z_0 is a point interior to C,*

(3)
$$f(z_0) = \frac{1}{2\pi i}\int_C \frac{f(z)}{z - z_0}\,dz,$$

where the integration is counterclockwise around C.

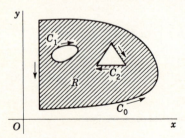

Fig. 64

Here, too, C can be replaced by the boundary B of a multiply connected region under the conditions prescribed for formula (2).

Derivatives of the integral (3) with respect to z_0 exist and are given by the integral of the derivative of the integrand. In particular,

$$(4) \qquad f''(z_0) = \frac{1}{\pi i} \int_C \frac{f(z)}{(z - z_0)^3} \, dz.$$

Consequently *the derivative f' is analytic wherever f is analytic*; thus the derivative $f^{(n)}$ of any order is analytic there.

If z_0 and z are points of a simply connected domain D over which f is analytic, then a contour integral between the points,

$$(5) \qquad \int_{z_0}^{z} f(\xi) \, d\xi = F(z),$$

has a value $F(z)$ independent of the contour C as long as C lies entirely within D. Moreover that single-valued function F is analytic over D. It is an antiderivative, or indefinite integral, of f:

$$(6) \qquad F'(z) = f(z) \qquad\qquad \text{for all } z \text{ in } D.$$

Any two antiderivatives of f differ by a constant. In terms of an antiderivative F of the analytic function f, we can write

$$\int_{z_0}^{z_1} f(z) \, dz = F(z_1) - F(z_0)$$

over any contour interior to D from point z_0 to point z_1.

58 POWER SERIES

Let f be analytic over the interior of a circle $|z - z_0| = r_0$. Then at each interior point z it is true that

$$(1) \qquad f(z) = f(z_0) + \sum_{n=1}^{\infty} \frac{1}{n!} f^{(n)}(z_0)(z - z_0)^n;$$

that is, *Taylor's series* converges whenever $|z - z_0| < r_0$ and its sum is $f(z)$. That theorem is a consequence of Cauchy's integral formula.

The region of convergence of any power series

$$(2) \qquad \sum_{n=0}^{\infty} a_n(z - z_0)^n,$$

where a_n are complex constants, is always a disk about the point z_0, $|z - z_0| < r_0$, unless the series converges only when $z = z_0$ or for all z. Let $f(z)$ denote the sum of the series. The series can be differentiated term by term to give a power series that converges over that disk to $f'(z)$. Thus series (2) represents an analytic function over that disk; in fact, it is Taylor's series representation, about the point z_0, of its sum $f(z)$. Thus

$$a_0 = f(z_0), \qquad a_n = \frac{1}{n!} f^{(n)}(z_0) \qquad (n = 1, 2, \ldots).$$

The power series (2) converges absolutely and uniformly with respect to z over each interior disk $|z - z_0| \leq r_1$, where $r_1 < r_0$.

Since e^z is an entire function, its *Maclaurin series* $(z_0 = 0)$,

$$(3) \qquad e^z = 1 + \sum_{n=1}^{\infty} \frac{1}{n!} z^n \qquad (|z| < \infty),$$

represents that function for all z. Consequently, the function $g(z) = \exp(z^2)$ has the power series representation

$$g(z) = \exp(z^2) = 1 + \sum_{n=1}^{\infty} \frac{1}{n!} z^{2n} \qquad (|z| < \infty),$$

which must be Maclaurin's series for g, so it follows that $g^{(2n)}(0) = (2n)!/n!$ and $g^{(2n-1)}(0) = 0$.

With the aid of Taylor's series we find that *the zeros of an analytic function are isolated.* More precisely, if a function f is analytic at a point z_0 and not identically zero over some neighborhood of that point, then there is a neighborhood of z_0 such that $f(z) \neq 0$ at any of its points except possibly z_0 itself.

A generalization of Taylor's series also follows from Cauchy's integral formula. If f is analytic over an annular region R, $r_1 < |z - z_0| < r_0$, it is represented there by *Laurent's series*:

$$(4) \qquad f(z) = \sum_{n=-\infty}^{\infty} A_n(z - z_0)^n \qquad (r_1 < |z - z_0| < r_0)$$

where, if C is any closed contour described counterclockwise around the annulus R (Fig. 65),

$$(5) \qquad A_n = \frac{1}{2\pi i} \int_C \frac{f(z)}{(z - z_0)^{n+1}} \, dz \qquad (n = 0, \pm 1, \pm 2, \ldots).$$

We note in particular that

$$A_{-1} = \frac{1}{2\pi i} \int_C f(z) \, dz.$$

Taylor's and Laurent's series can be integrated term by term within their domains of convergence. It follows that a representation of f by a

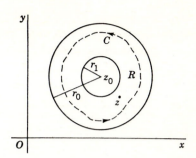

Fig. 65

convergent series of form (4) is necessarily the Laurent series, and hence the values of the integrals (5) can be seen from such a representation.

For example, from Maclaurin's series (3) for e^z we can write this Laurent series representation:

$$(6) \qquad \exp\left(\frac{1}{z}\right) = 1 + \frac{1}{z} + \frac{1}{2!}\frac{1}{z^2} + \frac{1}{3!}\frac{1}{z^3} + \cdots \qquad (|z| > 0).$$

Here $A_{-1} = 1$, $A_{-2} = \frac{1}{2}$, and $z_0 = 0$. According to formula (5), if C is a closed contour described counterclockwise around the origin, then

$$\int_C \exp\left(\frac{1}{z}\right) dz = 2\pi i, \qquad \int_C z\exp\left(\frac{1}{z}\right) dz = \pi i.$$

The region of convergence of a series of type (4) is always an annulus with center at z_0. The series represents f as a sum of two functions: the sum of the series of nonnegative powers of $z - z_0$ and the sum of the series of negative powers. The first function is analytic when $|z - z_0| < r_0$ and the second one when $|z - z_0| > r_1$.

Two power series of type (2) in which z_0 is the same for both can be added, multiplied, or divided to produce series of that type which converge to the corresponding combinations of the sums of the series, under natural restrictions on the regions. Those operations apply to some types of Laurent series too, of course, as we can see by replacing $z - z_0$ in series (2) by $(z - z_0)^{-1}$ or by multiplying the series by a negative power of $z - z_0$. But in the general case the product of two Laurent series is a double series.[1]

59 SINGULAR POINTS AND RESIDUES

If a function is analytic at some point in every neighborhood of a point z_0, but not at z_0 itself, then z_0 is called a *singular point* of the function. If the function is analytic at all points except z_0, in some neighborhood of z_0, then

---

[1] For properties of double series see, for example, E. Hille, "Analytic Function Theory," vol. 1, pp. 114 ff., 1959.

z_0 is an *isolated singular point*. The function $(z^2 + 1)^{-1}$, for example, is analytic everywhere except for the two isolated singular points $z = \pm i$.

About an isolated singular point z_0 a function always has a Laurent series representation:

$$
(1) \qquad f(z) = \frac{A_{-1}}{z - z_0} + \frac{A_{-2}}{(z - z_0)^2} + \cdots + A_0 + A_1(z - z_0) + \cdots
$$

$$
(0 < |z - z_0| < r_0),
$$

where r_0 is the radius of the neighborhood in which f is analytic except at z_0. The coefficients A_n are given by formula (5), Sec. 58, where the inner radius r_1 of the annulus, around which C is described, is now zero. In particular,

$$
(2) \qquad A_{-1} = \frac{1}{2\pi i} \int_C f(z)\, dz,
$$

where C can be any closed contour, described counterclockwise, containing z_0 in its interior and such that, except for z_0, f is analytic within and on C.

The complex number A_{-1}, the coefficient of $(z - z_0)^{-1}$ in the expansion (1), is called the *residue* of f at the isolated singular point z_0; $2\pi i A_{-1}$ is the value of the integral of f in the positive direction around a contour that encloses no other singular points but z_0. The expansion (6), Sec. 58, for instance, shows that the residue of the function $\exp(1/z)$ at the point $z = 0$ has the value $A_{-1} = 1$.

If C is a closed contour within and on which f is analytic except for a finite number of singular points $z_1, z_2, \ldots, z_m$, interior to the region bounded by C, the *residue theorem* states that

$$
(3) \qquad \int_C f(z)\, dz = 2\pi i (\rho_1 + \rho_2 + \cdots + \rho_m),
$$

where ρ_n denotes the residue of f at z_n and where the integration is in the positive direction around C. Note that the singular points z_n are necessarily isolated because of their finite number.

In the representation (1) the series of negative powers of $(z - z_0)$ is called the *principal part* of $f(z)$ about the isolated singular point z_0. The point z_0 is an *essential singular point* of f if the principal part has an infinite number of nonvanishing terms. It is a *pole of order m* if $A_{-m} \neq 0$ and $A_{-n} = 0$ when $n > m$. It is called a *simple pole* when $m = 1$; thus, if z_0 is a simple pole, a number r_0 exists such that

$$
(4) \qquad f(z) = \frac{A_{-1}}{z - z_0} + \sum_{n=0}^{\infty} A_n(z - z_0)^n \qquad (A_{-1} \neq 0,\, 0 < |z - z_0| < r_0).
$$

The function $\exp{(1/z)}$ has an essential singular point at the origin, in view of its representation (6), Sec. 58. The function

$$\frac{\cos z}{z^2} = \frac{1}{z^2} \sum_{n=0}^{\infty} \frac{(-1)^n}{(2n)!} z^{2n} = \frac{1}{z^2} - \frac{1}{2!} + \frac{z^2}{4!} + \cdots \qquad (|z| > 0)$$

has a pole of order 2 at $z = 0$, where the residue of the function is zero.

If a function is not analytic at z_0 but can be made so by merely assigning a suitable value to the function at that point, then z_0 is a *removable singular point* of the function. Thus the function

$$\frac{\sin z}{z} = 1 - \frac{z^2}{3!} + \frac{z^4}{5!} - \cdots \qquad (|z| > 0)$$

is analytic when $z \neq 0$. If we define its value to be unity when $z = 0$, the function is represented by the above convergent power series for all z and is therefore entire. Consequently $z = 0$ is a removable singular point of $z^{-1} \sin z$.

When f has a pole of order m at z_0, let us write

$$(5) \qquad\qquad \phi(z) = (z - z_0)^m f(z) \qquad (0 < |z - z_0| < r_0).$$

From the Laurent expansion (1) of f about z_0 it follows that z_0 is a removable singular point of ϕ, that ϕ is analytic at z_0 if we make the definition

$$(6) \qquad\qquad \phi(z_0) = A_{-m}.$$

Then ϕ is represented by Taylor's series when $|z - z_0| < r_0$, and it can be seen from Eq. (5) that

$$(7) \qquad\qquad A_{-1} = \frac{\phi^{(m-1)}(z_0)}{(m-1)!}.$$

This is a useful formula for the residue of f at a pole. For a simple pole ($m = 1$) it becomes

$$(8) \qquad\qquad A_{-1} = \phi(z_0) = \lim_{z \to z_0} (z - z_0)f(z).$$

We also note that $|f(z)| \to \infty$ as $z \to z_0$ whenever z_0 is a pole.

Conversely, for a given f, suppose that a positive integer m exists such that the function ϕ, defined by Eq. (5), is analytic at z_0 when $\phi(z_0)$ is suitably defined and that this value $\phi(z_0) \neq 0$. Then Taylor's series for ϕ shows that f has a pole of order m at z_0 and that the residue of f there is given by formula (7), or by (8) if $m = 1$. For example, if

$$f(z) = \frac{e^{-z}}{z^2 + \pi^2} = \frac{e^{-z}}{z + \pi i} \frac{1}{z - \pi i}$$

then corresponding to the singular point $z = \pi i$,

$$\phi(z) = \frac{e^{-z}}{z + \pi i}$$

and $m = 1$, so that πi is a simple pole. Here $\phi(\pi i) = e^{-\pi i}/(2\pi i)$; thus the residue of f at this pole is $i/(2\pi)$.

Let functions p and q be analytic at z_0, where $p(z_0) \neq 0$. Then the function

(9) $$f(z) = \frac{p(z)}{q(z)} = \frac{p(z)}{q(z_0) + q'(z_0)(z - z_0) + \cdots}$$

has a simple pole at z_0 if and only if $q(z_0) = 0$ and $q'(z_0) \neq 0$. The residue of f at the simple pole is given by the formula

(10) $$A_{-1} = \frac{p(z_0)}{q'(z_0)}.$$

If $q(z_0) = q'(z_0) = 0$ and $q''(z_0) \neq 0$, then z_0 is a pole of f of order 2, and conversely. Similarly for poles of higher order. Since formulas for residues in terms of p and q are awkward when $m > 1$, we rely on formula (7) or the Laurent expansion instead.

PROBLEMS

1. When C is the boundary of a square with opposite vertices at the points $z = 0$ and $z = 1 + i$, evaluate the integral $\int_C z^2 \, dz$ directly in terms of real line integrals to show that its value is zero.

2. When C is the circle $|z| = 2$ described counterclockwise, use Cauchy's integral theorem or Cauchy's integral formula to prove that

(a) $\displaystyle\int_C \frac{z}{z^2 + 9} \, dz = 0$;   (b) $\displaystyle\int_C \sec\frac{z}{2} \, dz = 0$;

(c) $\displaystyle\int_C \frac{z^2 + 4}{z - 1} \, dz = 10\pi i$;   (d) $\displaystyle\int_C \frac{\sinh z}{2z + \pi i} \, dz = \pi$.

3. Establish these expansions in the regions indicated:

(a) $\displaystyle\frac{1}{1 + z} = \sum_{n=0}^{\infty} (-1)^n z^n$ $(|z| < 1)$;

(b) $\displaystyle\frac{1}{1 - z^2} = \sum_{n=0}^{\infty} z^{2n}$ $(|z| < 1)$;

(c) $\displaystyle\sinh z = \sum_{n=1}^{\infty} \frac{z^{2n-1}}{(2n - 1)!}$ $(|z| < \infty)$;

(d) $\displaystyle\frac{z}{z - 1} = \sum_{n=0}^{\infty} \frac{1}{z^n}$ $(|z| > 1)$.

4. Show that the function $\tan z$ is analytic except for simple poles at $z = \pm(n - \tfrac{1}{2})\pi$ and that its integral around the square bounded by the lines $x = \pm 2, y = \pm 2$ has the value $-4\pi i$.

5. Show that the integral of $\tanh z$ around the circle $|z| = 2\pi$ has the value $8\pi i$.

6. Show that the functions $\sin(1/z)$ and $\cos(1/z)$ are analytic except at $z = 0$ and that $z = 0$ is an essential singular point. Find the residues there. *Ans.* 1; 0.

7. Find the residues of these functions at their singular points:

$$(a)\ \frac{z + 1}{z - 1}; \qquad (b)\ \frac{1}{\sin z}; \qquad (c)\ \frac{\cos z}{(z - z_0)^2}; \qquad (d)\ z^3 \cosh \frac{1}{z^2}; \qquad (e)\ \frac{\sin(z^2)}{z^5}.$$

Ans. $(a)\ 2; (b)\ \pm 1; (c)\ -\sin z_0; (d)\ \tfrac{1}{2}; (e)\ 0.$

8. If C is the circle $|z| = 8$ described counterclockwise, show that

$$\frac{1}{2\pi i} \int_C \frac{e^{zt}}{\sinh z} \, dz = 1 - 2 \cos \pi t + 2 \cos 2\pi t.$$

9. If C is the circle $|z| = 4$ described counterclockwise, show that

$$\int_C \frac{e^z}{(z^2 + \pi^2)^2} \, dz = \frac{i}{\pi}.$$

10. Show that the singular point $z = 0$ of the function

$$f(z) = \frac{1}{\sin(\pi/z)}$$

is not isolated.

11. With the aid of the Cauchy-Riemann conditions prove that the components u and v of an analytic function f are *harmonic functions*, that is, they are continuous with continuous partial derivatives up to the second order and satisfy *Laplace's equation*

$$u_{xx}(x,y) + u_{yy}(x,y) = 0.$$

60 BRANCHES OF MULTIPLE-VALUED FUNCTIONS

When $z \neq 0$, the function

$$(1) \qquad z^{\frac{1}{2}} = \sqrt{r}\left(\cos\frac{\theta}{2} + i \sin\frac{\theta}{2}\right)$$

has one value corresponding to a particular choice of θ and a second value when that argument is increased by 2π. Those two values of $z^{\frac{1}{2}}$, which differ only in algebraic sign, are the only possible values for a given z; thus the function (1) is double-valued.

A *branch* of a multiple-valued function f is a single-valued function that is analytic in some region and whose value at each point there coincides with one of the values of f at the point. The function

(2)
$$f_1(z) = \sqrt{r}\left(\cos\frac{\theta}{2} + i\sin\frac{\theta}{2}\right) \quad (-\pi < \theta < \pi, r > 0),$$

for example, is a branch of the double-valued function (1). The definition of the derivative can be used to show that f_1 is analytic everywhere except at the origin and points on the negative real axis. Since $f_1(z)$ tends to $i\sqrt{r}$ when $\theta \to \pi$ and to $-i\sqrt{r}$ when $\theta \to -\pi$, the function has no limit as $z \to -r \, (r > 0)$. Thus f_1 cannot be defined on the negative real axis so as to make the function continuous there, and the ray $\theta = \pi$ must be excluded from the region of analyticity (Fig. 66).

The negative real axis is called a *branch cut* of f_1, a boundary that is needed to define the branch in the greatest possible region. Note that each point of the branch cut is a singular point of f_1 that is not isolated. The origin is a *branch point*, a point associated with the multiple-valued function in this way: each branch of the function has a branch cut running from that point. Since the function (1) is double-valued unless the range of θ is limited to the value 2π or less, its branches must have cuts running out from the origin.

A second branch of the function (1) with the same cut as f_1 is

(3)
$$f_2(z) = \sqrt{r}\left(\cos\frac{\theta}{2} + i\sin\frac{\theta}{2}\right) \quad (\pi < \theta < 3\pi, r > 0).$$

Here $f_2(z) = -f_1(z)$. The number of branches is unlimited. A branch with the positive real axis as its branch cut is

(4)
$$f_3(z) = \sqrt{r}\left(\cos\frac{\theta}{2} + i\sin\frac{\theta}{2}\right) \quad (0 < \theta < 2\pi, r > 0);$$

one that has the positive y axis as a cut is the function (1) with the restriction $\pi/2 < \theta < 5\pi/2, r > 0$; etc.

A branch of the n-valued function $z^{1/n}$ $(n = 2, 3, \ldots)$ with a ray $\theta = \theta_0$ as the branch cut is described by the conditions

(5)
$$z^{1/n} = \sqrt[n]{r}\left(\cos\frac{\theta}{n} + i\sin\frac{\theta}{n}\right) \quad (\theta_0 < \theta < \theta_0 + 2\pi, r > 0).$$

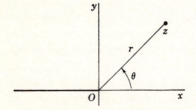

Fig. 66

For this function it can be shown that

$$\frac{d}{dz}(z^{1/n}) = \frac{1}{n}\frac{z^{1/n}}{z}.$$

The *logarithmic function* log z, the inverse of the exponential function, is infinitely multiple-valued. It can be written

(6) $$\log z = \log(re^{i\theta}) = \text{Log } r + i\theta,$$

where Log r denotes the real natural logarithm of the positive number r. The branch known as the principal value of log z is the function

(7) $$\text{Log } z = \text{Log } r + i\theta \qquad (-\pi < \theta < \pi, r > 0),$$

whose branch cut is the negative real axis. Its derivative is $1/z$.

When c is a complex number, we define z^c thus:

(8) $$z^c = \exp(c \log z) \qquad (z \neq 0).$$

Unless c is an integer, the function z^c is multiple-valued. If $c = m/n$ where m and n are integers and $n \neq 0, \pm 1$, z^c becomes the n-valued function $z^{m/n}$. Branches of z^c can be written by using branches of log z in definition (8).

The inverse trigonometric and hyperbolic functions, which can be written in terms of logarithms, are further examples of multiple-valued functions.

61   ANALYTIC CONTINUATION

If a function is single-valued and analytic throughout a region, it is uniquely determined throughout the region by its values over an arc, or over a sub-region, within the given region. That theorem can be proved with the aid of Taylor's series.

Let f_1 denote a given function analytic in a region R_1, and let R be a greater region containing R_1. A function f, analytic throughout R and equal to f_1 whenever z is in R_1, may exist. If so, there is only one such function in view of the foregoing theorem. The function f is called the *analytic continuation* of f_1 into the larger region R.

As an example, let f_1 be defined by this power series:

$$f_1(z) = \sum_{n=0}^{\infty} z^n \qquad (|z| < 1).$$

Then f_1 is analytic in the region $|z| < 1$, the region of convergence of the series; but the function is undefined for all other values of z because the series diverges whenever $|z| \geq 1$. The series is the Maclaurin series representing the function $(1-z)^{-1}$ in the region; thus $f_1(z) = (1-z)^{-1}$ when $|z| < 1$.

This second representation of f_1 discloses the analytic continuation

$$f(z) = \frac{1}{1 - z} \qquad (z \neq 1)$$

of f_1 into the entire z plane, excluding the point $z = 1$, because f is everywhere analytic except at that point and $f(z) = f_1(z)$ when $|z| < 1$.

As another example, consider the Laplace transform

$$f_1(z) = \int_0^\infty e^{-zt}\, dt = \int_0^\infty e^{-xt} \cos yt\, dt - i \int_0^\infty e^{-xt} \sin yt\, dt.$$

The Laplace integral here exists only if $x > 0$, where its value is $1/z$. Thus $f_1(z) = 1/z\ (x > 0)$, and the function is analytic in that right half plane. But since $1/z$ is analytic everywhere except at the origin, the analytic continuation of f_1 out of the half plane is the function

$$f(z) = \frac{1}{z} \qquad (z \neq 0).$$

We also note that if the Laplace integral were known to represent an analytic function in the half plane and to have the value $1/x$ when $z = x$ there, then the integral represents $1/z$ there because the latter function is analytic and has the values $1/x$ on the x axis.

Finally, consider the branch (2), Sec. 60, of the function $z^{\frac{1}{2}}$. In the half plane $y > 0$ it is given by the conditions

$$f_1(z) = \sqrt{r}\left(\cos\frac{\theta}{2} + i \sin\frac{\theta}{2}\right) \qquad (0 < \theta < \pi, r > 0).$$

It is not analytic on the negative real axis. The function

$$f(z) = \sqrt{r}\left(\cos\frac{\theta}{2} + i \sin\frac{\theta}{2}\right) \qquad (0 < \theta < 2\pi, r > 0)$$

is analytic except on the positive real axis, and its values coincide with those of f_1 in the half plane $y > 0$. Thus f is the analytic continuation of f_1 downward across the negative real axis.

The following *principle of reflection* is easily established with the aid of the Cauchy-Riemann conditions. Let a function $w = f(z)$ be analytic in some region R that includes a segment of the x axis and is symmetric with respect to that axis. If $f(x)$ is real whenever x is a point on that segment, then

(1) $$f(\bar{z}) = \bar{w}$$

whenever z is in R. Conversely, if condition (1) is satisfied, then $f(x)$ is real. Condition (1) can be written $f(\bar{z}) = \overline{f(z)}$, or $\overline{f(\bar{z})} = f(z)$.

As examples, the entire functions z^2, e^z, and $\sin z$ are real when z is real, and the complex conjugate of each function is the same as the function of $\bar{z}$. But the entire functions iz and $z^2 + i$ are not real when $z = x$; their conjugates are different from $i\bar{z}$ and $\bar{z}^2 + i$.

62 IMPROPER CAUCHY INTEGRALS

In the following chapter we shall use extensions of the Cauchy integral formula and Cauchy's integral theorem in which the closed contour C is replaced by a straight line $x = \gamma$, the boundary of a half plane $x \geqq \gamma$ over which the function f is analytic. The following proof of those extensions will serve to illustrate a prominent type of application of contour integration.

First we introduce the notion of order of magnitude of a function of z for large $|z|$. Order conditions serve our purpose more fully than the concept of analyticity at the infinite point of the complex plane.

A function f is *of the order of z^k*, written $\mathcal{O}(z^k)$, as $|z| \to \infty$ in a specified part of the z plane, if positive numbers M and r_0 exist such that $|z^{-k}f(z)| < M$ there whenever $|z| > r_0$; that is, when $|z|$ is sufficiently large,

$$|f(z)| < M|z|^k.$$

Theorem 1   *Let f be analytic over a right half plane $\mathrm{Re}\, z \geqq \gamma$ and $\mathcal{O}(z^{-k})$ there as $|z| \to \infty$, where $k > 0$. Then if $\mathrm{Re}\, z_0 > \gamma$, $f(z_0)$ is given by the improper Cauchy integral formula*

$$(1) \qquad f(z_0) = -\frac{1}{2\pi i}\int_{\gamma - i\infty}^{\gamma + i\infty} \frac{f(z)}{z - z_0}\, dz = -\frac{1}{2\pi}\int_{-\infty}^{\infty} \frac{f(z)}{z - z_0}\, dy,$$

where the integration is along the line $x = \gamma$; thus $z = \gamma + iy$.

To prove the theorem, let C_R denote the arc $x \geqq \gamma$ of a circle $|z| = R$ where $R > |\gamma|$ and $R > |z_0|$, so that the point z_0 is interior to the region bounded by the arc and the line $x = \gamma$ (Fig. 67). If we write $\beta = \sqrt{R^2 - \gamma^2}$,

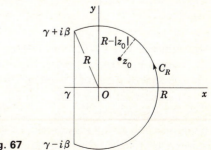

Fig. 67

then according to Cauchy's integral formula,

$$(2) \qquad f(z_0) = \frac{1}{2\pi i}\left[\int_{C_R} \frac{f(z)}{z - z_0}\,dz - \int_{\gamma - i\beta}^{\gamma + i\beta} \frac{f(z)}{z - z_0}\,dz\right],$$

where C_R is described in the positive sense.

When point z is on C_R, then $|z - z_0| \geq R - |z_0|$ and, since $|f(z)| < M|z|^{-k}$ when $|z| = R$ and R is sufficiently large, it follows that

$$\left|\frac{f(z)}{z - z_0}\right| < \frac{M}{R^k}\frac{1}{R - |z_0|} \qquad\qquad \text{when } R > r_0.$$

The integrands of the integrals in formula (2) are continuous. The length of C_R is less than $2\pi R$; therefore (Sec. 56)

$$\left|\int_{C_R} \frac{f(z)}{z - z_0}\,dz\right| < \frac{2\pi RM}{R^k(R - |z_0|)} = \frac{2\pi M}{R^k(1 - |z_0|R^{-1})}$$

and the last member clearly tends to zero as $R \to \infty$, since $k > 0$. It also tends to zero as $\beta \to \infty$, because $R^2 = \beta^2 + \gamma^2$. Since $f(z_0)$ is independent of β, we see from formula (2) that

$$(3) \qquad f(z_0) = -\frac{1}{2\pi i}\lim_{\beta \to \infty}\int_{\gamma - i\beta}^{\gamma + i\beta} \frac{f(z)}{z - z_0}\,dz;$$

$$= -\frac{1}{2\pi}\lim_{\beta \to \infty}\left[\int_0^\beta \frac{f(\gamma + iy)}{\gamma + iy - z_0}\,dy + \int_0^\beta \frac{f(\gamma - iy)}{\gamma - iy - z_0}\,dy\right].$$

The integrands of the last two integrals are of the order of $1/y^{k+1}$ as $y \to \infty$, and since $k + 1 > 1$, the improper integrals, as $\beta \to \infty$, exist and

$$(4) \qquad f(z_0) = -\frac{1}{2\pi}\left[\int_0^\infty \frac{f(\gamma + iy)}{\gamma + iy - z_0}\,dy + \int_{-\infty}^0 \frac{f(\gamma + iy)}{\gamma + iy - z_0}\,dy\right].$$

This is the extension (1) of Cauchy's integral formula.

Formula (1) states that the integral of the function

$$\phi(z) = \frac{f(z)}{z - z_0}$$

along the line $x = \gamma$ is the residue $f(z_0)$ of ϕ at its singular point z_0. The proof was based upon the analyticity of $\phi(z)(z - z_0)$ throughout the half plane and the fact that ϕ is $\mathcal{O}(1/z^{k+1})$ there. But Taylor's series representation of the analytic function f about the point z_0 shows that z_0 is a removable singular point of ϕ if and only if $f(z_0) = 0$. Thus if ϕ is also analytic at z_0, its integral along the line $x = \gamma$ is zero, the value $f(z_0)$, where $f(z) = (z - z_0)\phi(z)$, and we have established the following extension of the Cauchy integral theorem.

Theorem 2 *Let ϕ be analytic over a half plane $x \geq \gamma$ and $\mathcal{O}(z^{-k})$ as $|z| \to \infty$ there, where $k > 1$. Then*

$$(5) \qquad \int_{\gamma - i\infty}^{\gamma + i\infty} \phi(z)\, dz = i \int_{-\infty}^{\infty} \phi(\gamma + iy)\, dy = 0.$$

The proof of formula (5) applies as well when ϕ is analytic and $\mathcal{O}(z^{-k})$ in the *left* half plane $x \leq \gamma$, if $k > 1$. Also, the residue theorem (Sec. 59) can be extended in the same manner to the case in which ϕ has a *finite* number of singular points interior to either half plane. For the left half plane that extension of the residue theorem can be stated as follows.

Theorem 3 *Let ϕ be analytic when $x \leq \gamma$ except for n singular points $z_1, z_2, \ldots, z_n$ interior to that half plane, where ϕ has residues $\rho_1, \rho_2, \ldots,$ ρ_n. Then if for some real numbers M, k, and r_0, where $k > 1$ and r_0 exceeds the greatest of the numbers $|z_1|, |z_2|, \ldots, |z_n|$, ϕ satisfies the order condition $|\phi(z)| < M|z|^{-k}$ when $|z| > r_0$ in that half plane, it is true that*

$$(6) \qquad \int_{\gamma - i\infty}^{\gamma + i\infty} \phi(z)\, dz = 2\pi i(\rho_1 + \rho_2 + \cdots + \rho_n).$$

As examples, consider first the function $f(z) = e^{-z}/z$ which satisfies the conditions in Theorem 1 when $\gamma = 1$, for it is analytic over the half plane $x \geq 1$ and $\mathcal{O}(1/z)$ there because

$$|f(z)| = \frac{1}{|z|}e^{-x} \leq \frac{e^{-1}}{|z|} \qquad \text{when } x \geq 1.$$

Thus formula (1) applies to f, giving the integration formula

$$\int_{1 - i\infty}^{1 + i\infty} \frac{e^{-z}}{z(z - z_0)}\, dz = -\frac{2\pi i}{z_0} e^{-z_0} \qquad \text{when Re } z_0 > 1.$$

The function $\phi(z) = z^{-2}$ satisfies the conditions in Theorem 2 when $\gamma = 1$; therefore

$$(7) \qquad \int_{1 - i\infty}^{1 + i\infty} \frac{dz}{z^2} = i \int_{-\infty}^{\infty} \frac{dy}{(1 + iy)^2} = 0.$$

When t is real and $t \geq 0$ and $\gamma = 1$, the conditions in Theorem 3 are satisfied by the function $e^{zt}/(z^2 + 1)$ with simple poles $\pm i$; therefore

$$(8) \qquad \frac{1}{2\pi i} \int_{1 - \infty}^{1 + \infty} \frac{e^{zt}}{z^2 + 1}\, dz = \frac{e^{it}}{2i} - \frac{e^{-it}}{2i} = \sin t \qquad (t \geq 0).$$

As we shall see in the following chapter, the left-hand member of formula (8) is an integral representation of the inverse Laplace transform of $1/(s^2 + 1)$.

PROBLEMS

1.   Use Maclaurin's series for cos Z to prove that the function $\cos \sqrt{z}$ is single-valued and entire when defined as 1 at $z = 0$.

2.   Point out why the function

$$g_1(z) = \exp\left(-\sqrt{r}\cos\frac{\theta}{2}\right)\left[\cos\left(\sqrt{r}\sin\frac{\theta}{2}\right) - i\sin\left(\sqrt{r}\sin\frac{\theta}{2}\right)\right],$$

where $r > 0$ and $0 < \theta < 2\pi$, is a branch (Sec. 60) of the double-valued function $\exp(-\sqrt{z})$ that is analytic everywhere except at points on the real axis where $x \geq 0$.

3.   The branch g_1 of $\exp(-\sqrt{z})$ described in Prob. 2 is not analytic along the positive real axis. Show why its analytic continuation from the first quadrant into the fourth quadrant is the branch

$$g_2(z) = \exp\left(-\sqrt{r}\cos\frac{\theta}{2}\right)\left[\cos\left(\sqrt{r}\sin\frac{\theta}{2}\right) - i\sin\left(\sqrt{r}\sin\frac{\theta}{2}\right)\right]$$

where $r > 0$ and $-\frac{1}{2}\pi \leq \theta \leq \frac{1}{2}\pi$.

4.   (a) If $\gamma > 0$, prove that the branch g_2 of $\exp(-\sqrt{z})$ described in Prob. 3 is $\mathcal{O}(z^{-k})$ as $|z| \to \infty$ in the half plane $x \geq \gamma$ for each fixed positive number k. [Note that $|g_2(z)| < \exp(-\sqrt{r/2})$ there.] (b) Apply Theorem 2, Sec. 62, to show that

$$\int_{\gamma-i\infty}^{\gamma+i\infty} z^n g_2(z)\, dz = 0 \qquad\qquad (n = 0, 1, 2, \ldots; \gamma > 0).$$

5.   Show that the complex-valued Laplace transform of sin t,

$$f_1(z) = \int_0^\infty e^{-zt}\sin t\, dt \qquad\qquad (\text{Re } z > 0),$$

is analytic over the half plane $x > 0$ and that its analytic continuation into the rest of the z plane, excluding the points $\pm i$, is the function $f(z) = 1/(z^2 + 1)$.

6.   Use Theorem 1, Sec. 62, to prove that, if k is real and positive,

$$\int_{-i\infty}^{i\infty} \frac{dz}{z^2 - k^2} = -\frac{\pi i}{k} \qquad\qquad (k > 0).$$

Verify that result by integrating with respect to y.

7.   In the half plane $x \geq 1$ show that

$$|\tanh z| \leq \frac{\cosh x}{\sinh x} \leq \frac{1}{\tanh 1},$$

and that the function $z^{-1}\tanh z$ satisfies the conditions in Theorem 1, so that

$$-\frac{1}{2\pi i}\int_{1-i\infty}^{1+i\infty} \frac{\tanh z}{z(z - z_0)}\, dz = \frac{\tanh z_0}{z_0} \qquad\qquad \text{if Re } z_0 > 1.$$

8.   Apply Theorem 2 to prove that

$$\int_{1-i\infty}^{1+i\infty} \frac{e^{zt}}{z^2}\, dz = 0 \qquad\qquad \text{if } t \leq 0.$$

9. Show that the integral (8), Sec. 62, vanishes if $t \leq 0$.

10. Use Theorems 2 and 3 to prove that, if $\gamma > 0$,

$$\frac{1}{2\pi i}\int_{\gamma-i\infty}^{\gamma+i\infty} \frac{e^{zt}}{(z^2 + 1)^2}\,dz = \begin{cases} \frac{1}{2}\sin t - \frac{1}{2}t\cos t & \text{if } t \geq 0, \\ 0 & \text{if } t \leq 0. \end{cases}$$

Thus when $t \geq 0$, the left-hand member represents $L^{-1}\{(s^2 + 1)^{-2}\}$.

11. Prove Theorem 1, Sec. 62, by using open rectangular contours C_β with sides along the lines $x = \beta$, $y = \pm\beta$ (Fig. 68) in place of the circular arcs C_R.

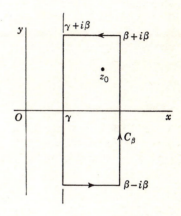

Fig. 68

12. Use the residue theorem (Sec. 59) and give details of the proof of Theorem 3, Sec. 62.

6

The Inversion Integral

We shall now extend our theory of the Laplace transformation by letting the letter s in the transform $f(s)$ represent a complex variable. As before $F(t)$ represents a real-valued function of the positive real variable t; but the transform $f(s)$ can assume complex values. We shall see that properties of the transformation already obtained by assuming that s is real carry over to the case in which s is complex.

63 ANALYTIC TRANSFORMS

When s is a complex variable,

$$s = x + iy,$$

the Laplace transform of a real-valued function $F(t)$,

(1)
$$f(s) = \int_0^\infty e^{-st}F(t)\,dt = \int_0^\infty e^{-xt}e^{-iyt}F(t)\,dt,$$

can be written in the form

(2) $$f(s) = u(x,y) + iv(x,y),$$

where the components u and v are real integrals:

(3)
$$u(x,y) = \int_0^\infty e^{-xt} \cos ytF(t)\,dt,$$

$$v(x,y) = -\int_0^\infty e^{-xt} \sin ytF(t)\,dt.$$

Since the integrands of the integrals (3) are the real and imaginary coefficients of the integrand $e^{-st}F(t)$ of the Laplace integral (1), we note that

(4)
$$|e^{-xt} \cos ytF(t)| \leq |e^{-st}F(t)|,$$

$$|-e^{-xt} \sin ytF(t)| \leq |e^{-st}F(t)|.$$

These inequalities will be useful in discussing the convergence of the Laplace integral.

Under broad conditions on F its transform $f(s)$ is analytic over a right half plane and bounded there. Let F be sectionally continuous over each interval $0 < t < T$ and of exponential order as $t \to \infty$. Then constants M and α exist such that $|F(t)| < Me^{\alpha t}$ when $t \geq 0$. Consequently whenever $x \geq x_0$, where $x_0 > \alpha$,

(5) $$|e^{-st}F(t)| = e^{-xt}|F(t)| < M \exp[-(x_0 - \alpha)t].$$

The integrands of the real-valued integrals (3) are continuous functions of x, y, and t except for possible finite jumps of $F(t)$. Their absolute values are less than $M \exp[-(x_0 - \alpha)t]$ when $x \geq x_0$, in view of inequalities (4) and (5). According to the Weierstrass test, then, the integrals (3) converge uniformly with respect to x and y on the half plane $x \geq x_0$. Thus they represent continuous functions $u(x,y)$ and $v(x,y)$ there (Sec. 15).

When F is $\mathcal{O}(e^{\alpha t})$, we noted earlier (Sec. 19) that $tF(t)$ is $\mathcal{O}(\exp x_1 t)$ whenever $x_1 > \alpha$. It follows that

(6) $$u_x(x,y) = \int_0^\infty e^{-xt}(-t) \cos ytF(t)\,dt \qquad (x > \alpha),$$

because the integral here converges uniformly with respect to x and y when $x \geq x_0(x_0 > \alpha)$; also u_x is continuous there. From the second of Eqs. (3) we see that the integral (6) also represents $v_y(x,y)$; hence u and v satisfy the first Cauchy-Riemann condition over the half plane $x > \alpha$. In the same manner we find that the second condition $u_y = -v_x$ is satisfied there. Those conditions along with the continuity of the first-order partial derivatives of u and v establish the analyticity of f over the half plane $x > \alpha$ (Sec. 54).

Now $f' = u_x + iv_x$ wherever f is analytic; thus when $x > \alpha$

$$f'(s) = -\int_0^\infty te^{-xt}(\cos yt - i\sin yt)F(t)\,dt = L\{-tF(t)\}.$$

Since $tF(t)$ is also sectionally continuous and of exponential order, our result applies as well to that function; thus

$$f''(s) = L\{t^2F(t)\} \qquad\qquad (x > \alpha),$$

and so on for $f^{(n)}(s)$.

With the aid of condition (5) we find that f is bounded over the half plane $x \geqq x_0$ since (Sec. 56)

$$|f(s)| \leqq \int_0^\infty e^{-xt}|F(t)|\,dt < M\int_0^\infty e^{-(x_0-\alpha)t}\,dt = \frac{M}{x_0 - \alpha}.$$

Similarly, we find that $|f'(s)| < M/(x_0 - \alpha)^2$, etc., so the derivatives of f are bounded over the half plane.

Finally we note that $f(s)$ is real when $s = x$ because $F(t)$ is real, and it follows from the principle of reflection (Sec. 61), or directly from Eqs. (2) and (3), that $f(s) = f(\bar{s})$. Our results can be stated as follows.

Theorem 1 *Let a real-valued function F be sectionally continuous over each interval $0 < t < T$ and $\mathcal{O}(e^{\alpha t})$ as $t \to \infty$. Then its transform*

$$f(s) = \int_0^\infty e^{-st}F(t)\,dt = L\{F\} \qquad\qquad (s = x + iy)$$

is an analytic function of s over the half plane $x > \alpha$. Its Laplace integral is absolutely and uniformly convergent over each half plane $x \geqq x_0(x_0 > \alpha)$, and f and its derivatives are bounded there. Also,

(7) $$f^{(n)}(s) = L\{(-t)^nF(t)\} \qquad (n = 1, 2, \ldots; x > \alpha),$$

(8) $$\overline{f(s)} = f(\bar{s}) \qquad\qquad (x > \alpha).$$

Formula (7) was derived in Sec. 19 when s is real.

The conditions in Theorem 1 can be relaxed. For instance F may become infinite at $t = t_0$ in such a way that $(t - t_0)^kF(t)$ remains bounded as $t \to t_0$, where $k < 1$. Then the conclusions in the theorem are still valid. As an example, the transform of $t^{-\frac{1}{2}}$ is analytic in the half plane $x > 0$, and formula (7) applies to it.

64   PERMANENCE OF FORMS

We have seen that the Laplace integral of $F(t)$ represents a function $f(s)$ that is analytic in a half plane $x > \alpha$, where $s = x + iy$. If the integration is performed when $s = x$, a real function $\phi(x)$ is obtained that is identical with $f(s)$ to the right of point $s = \alpha$ along the real axis; that is, $\phi(x) = f(x)$ when $x > \alpha$. If $\phi(s)$ is an analytic function in the half plane, then it must be identical with $f(s)$, since two different analytic functions cannot be identical along a line (Sec. 61).

It follows that transforms can be found by carrying out the integration as if s were a real variable. That the function $f(s)$ so found is analytic when $\mathrm{Re}\, s > \alpha$ can be seen in the particular cases; but it is true in general because the integration formulas are the same whether the parameter in the integral is complex or real. The transform of t^2, for instance, was found to be $2s^{-3}$ when s is real. Now t^2 is $\mathcal{O}(e^{\alpha t})$ for any positive α, and $2s^{-3}$ is analytic when $\mathrm{Re}\, s > 0$. Therefore, $L\{t^2\} = 2s^{-3}$ for all complex s in the half plane $x > 0$.

All our transforms of particular functions, tabulated in Appendix A, Table A.2, are valid when s is complex. We seldom need the value of α which determines the half plane in which s lies; the existence of the number α usually suffices.

The operational properties of the transformation developed in the first two chapters and tabulated in Appendix A, Table A.1, are likewise valid when s is a complex variable in some half plane $x > \alpha$. For the sake of simplicity, we may make an exception of operation 10, Appendix A, Table A.1:

$$L\left\{\frac{1}{t}F(t)\right\} = \int_s^\infty f(\lambda)\,d\lambda,$$

and agree that s is real here.

The permanence of the forms of those properties is again a consequence of the fact that the steps used in their derivations are independent of the real or complex character of the parameter s. However, the derivations could be rewritten when $s = x + iy$ by taking the corresponding steps with the real integrals (3), Sec. 63, that represent $u(x,y)$ and $v(x,y)$.

65   ORDER PROPERTIES OF TRANSFORMS

When $|s| \to \infty$ in a half plane of convergence of the Laplace integral, the behavior of the transform $f(s)$ is governed by regularity properties of the object function F.

Theorem 2   *Let a function F as well as its derivative F' be sectionally continuous over each bounded interval $0 < t < T$ and let F itself be $\mathcal{O}(e^{\alpha t})$. Then in any half plane $\mathrm{Re}\, s \geqq x_0$ where $x_0 > \alpha$, its transform $f(s)$ tends to zero as $|s| \to \infty$:*

$$(1) \qquad\qquad \lim_{|s| \to \infty} f(s) = 0 \qquad\qquad (\mathrm{Re}\, s \geqq x_0 > \alpha).$$

For points $s = x + iy$ in the half plane $x \geqq x_0$ we are to prove that for each positive number ϵ there is a number r_ϵ such that

(2)
$$|f(s)| < \epsilon \quad \text{whenever } |s| > r_\epsilon \text{ and } x \geqq x_0.$$

The Laplace integral of F converges uniformly with respect to s in the half plane $x \geqq x_0$ (Theorem 1), so the remainder for that improper integral can be made uniformly small in absolute value for all s there. That is, to the given number ϵ there corresponds a number T_ϵ independent of s such that

(3)
$$\left| \int_0^\infty e^{-st} F(t) \, dt \right| \leqq \left| \int_0^{T_\epsilon} e^{-st} F(t) \, dt \right| + |R(s, T_\epsilon)|$$

and $|R(s, T_\epsilon)| < \frac{1}{2}\epsilon$ for all s in the half plane. Now we shall determine r_ϵ such that the first term on the right of condition (3) is less than $\frac{1}{2}\epsilon$ when $|s| > r_\epsilon$.

The interval $(0, T_\epsilon)$ consists of a finite number of subintervals interior to each of which F and F' are continuous and have limits from the interior at the end points. Consider a subinterval $(0, T_1)$ as a sample. Over it we can apply integration by parts to the complex-valued integral (Sec. 56); thus

$$-s \int_0^{T_1} e^{-st} F(t) \, dt = e^{-st} F(t) \Big]_0^{T_1} - \int_0^{T_1} e^{-st} F'(t) \, dt.$$

The absolute value of the right-hand member does not exceed the function

$$|F(0)| + \exp(-xT_1)|F(T_1)| + \int_0^{T_1} e^{-xt} |F'(t)| \, dt$$

and, since $x \geqq x_0$ where x_0 may be negative, and F' is bounded, $|F'(t)| < N_1$, this function is less than M_1, where

$$M_1 = |F(0)| + \exp(|x_0| T_1)[|F(T_1)| + N_1 T_1].$$

Similarly for the remaining subintervals of $(0, T_\epsilon)$. Thus a number M exists such that

(4)
$$|s| \left| \int_0^T e^{-st} F(t) \, dt \right| < M \qquad \text{when } x \geqq x_0.$$

Then from condition (3) we conclude that

$$f(s) < \frac{M}{|s|} + \frac{\epsilon}{2} < \epsilon \qquad\qquad \text{if } |s| > \frac{2M}{\epsilon},$$

that is, condition (2) is established by selecting r_ϵ as $2M/\epsilon$.

Weaker-order properties on $f(s)$, such as the boundedness of $xf(x + iy)$, can be derived by assuming only that F itself is sectionally continuous and of exponential order (Probs. 7 and 8). Stronger ones will now be noted in case F or its derivatives are continuous and of exponential order.

Theorem 3 *If F is continuous and F′ and F″ are sectionally continuous over each interval $0 \leqq t \leqq T$ while both F and F′ are $\mathcal{O}(e^{\alpha t})$, then*

(5)
$$\lim_{|s| \to \infty} sf(s) = F(0) \qquad \text{when } x \geqq x_0 > \alpha;$$

in particular, f is $\mathcal{O}(1/s)$ in that half plane:

(6)
$$|s|\,|f(s)| < M \qquad (x \geqq x_0 > \alpha).$$

Under the conditions stated we know that

(7)
$$L\{F'(t)\} = sf(s) - F(0).$$

Since F' satisfies the conditions on F in Theorems 1 and 2, then $L\{F'\}$ is bounded over the half plane $x \geqq x_1$ and tends to zero as $|s| \to \infty$ there. Therefore $sf(s)$ is bounded and, since it equals $L\{F'\} + F(0)$, its limit as $|s| \to \infty$ exists and is given by formula (5).

As illustrations the functions 1, $\cos t$ and t satisfy the conditions in Theorem 3 whenever $\alpha > 0$. Their transforms $1/s$, $s/(s^2 + 1)$, and $1/s^2$ are $\mathcal{O}(1/s)$ and satisfy condition (5). In fact the last of those transforms is $\mathcal{O}(1/s^2)$, a conclusion that follows from properties of the function t with the aid of the following extension of Theorem 3.

Theorem 4 *If F and F′ are continuous and F″ and F‴ are sectionally continuous over each interval $0 \leqq t \leqq T$ while F, F′, and F″ are $\mathcal{O}(e^{\alpha t})$, then $s^2f(s) - sF(0)$ is bounded over any half plane $x \geqq x_0$ where $x_0 > \alpha$; in fact*

(8)
$$\lim_{|s| \to \infty} [s^2 f(s) - sF(0)] = F'(0) \qquad (x \geqq x_0 > \alpha),$$

and the additional condition $F(0) = 0$ is then adequate and necessary for $f(s)$ to be $\mathcal{O}(1/s^2)$ in the half plane.

Since our formula for $L\{F''\}$ is valid here, then

$$L\{F''(t)\} + F'(0) = s^2 f(s) - sF(0) \qquad (x \geqq x_0).$$

In view of Theorems 1 and 2 the function on the left is bounded and has the limit $F'(0)$ as $|s| \to \infty$, so the same is true for the function $s^2f(s) - sF(0)$. The second component $-sF(0)$ of that function is bounded if and only if $F(0) = 0$, in which case $s^2f(s)$ is bounded over the half plane.

Extensions of Theorem 4 are evident. For instance, $f(s)$ is $\mathcal{O}(1/s^3)$ over the half plane if F, F', and F'' are continuous and the next two derivatives are sectionally continuous, while F and its first three derivatives are $\mathcal{O}(e^{\alpha t})$ and $F(0) = F'(0) = 0$; also $s^3f(s) \to F''(0)$ as $|s| \to \infty$ in the half plane.

PROBLEMS

1. State why none of the following functions can be Laplace transforms of real-valued sectionally continuous functions of exponential order (Theorem 1).

$$(a)\ s; \quad (b)\ \frac{\sin s}{s + 1}; \quad (c)\ \exp(-s^2); \quad (d)\ \frac{1}{s}; \quad (e)\ \frac{1}{s + i}.$$

2. Use Theorem 2 to show that none of the following functions can be Laplace transforms of a function F of exponential order such that F and F' are sectionally continuous.

$$(a)\ 1; \quad (b)\ \frac{s}{s - 2}; \quad (c)\ e^{-s}; \quad (d)\ \frac{\exp(-s^2)}{s^2}.$$

3. Determine an order property of the transform $f(s)$ from the character of the function F in each of the following cases, and verify by writing $f(s)$. (a) $F(t) = \sin t$; (b) $F(t) = \cosh t$; (c) $F(t) = t \sin t$; (d) $F(t) = (t - 1)S_0(t - 1)$.

$$Ans.\ (a)\ \mathcal{O}\frac{1}{s^2}, x \geqq x_0 > 0;\ (b)\ \mathcal{O}\frac{1}{s}, x \geqq x_0 > 1;\ (c)\ \mathcal{O}\frac{1}{s^3}, x \geqq x_0 > 0.$$

4. If a function F is sectionally continuous over a bounded interval $0 < t < T$ and $F(t) = 0$ whenever $t > T$, state why $F(t)$ is $\mathcal{O}(e^{\alpha t})$ *for every real* α, and why its transform $f(s)$ is an entire function of the complex variable s.

5. The step function $1 - S_0(t - 1)$ satisfies the conditions on F in Prob. 4. Show directly that its transform can be written

$$f(s) = \frac{1 - e^{-s}}{s} \qquad \text{whenever } s \neq 0, \text{ and } f(0) = 1;$$

then represent f as a power series in s convergent for all s to verify that f is entire.

6. If a function F is sectionally continuous over some interval $0 < t < T$ but continuous when $t \geqq T$, and if F' is sectionally continuous over every such interval and both F and F' are $\mathcal{O}(e^{\alpha t})$, use the formula for $L\{F'\}$ to prove that the transform $f(s)$ is $\mathcal{O}(1/s)$ over the half plane $\operatorname{Re} s \geqq x_0$ when $x_0 > \alpha$.

7. If F is sectionally continuous over each interval $0 < t < T$ and $\mathcal{O}(e^{\alpha t})$, prove that its transform $f(s)$ satisfies the condition that $xf(x + iy)$ is bounded over the half plane $x \geqq x_0$ when $x_0 > \alpha$; consequently

$$\lim_{x \to \infty} f(x + iy) = 0.$$

8. According to the *Riemann-Lebesgue theorem* in the theory of convergence of Fourier series,[1] if a function G is sectionally continuous over an interval $a < t < b$, then

$$\lim_{y \to \infty} \int_a^b G(t) \cos yt\, dt = \lim_{y \to \infty} \int_a^b G(t) \sin yt\, dt = 0,$$

a result that is plausible if graphs of the integrands for large values of y are considered.

[1] See the author's "Fourier Series and Boundary Value Problems," 2d ed., p. 87, 1963.

(*a*) If G is sectionally continuous over each interval $0 < t < T$, and absolutely integrable from 0 to ∞, extend the above result to improper integrals by showing that

$$\lim_{|y|\to\infty} \int_0^\infty e^{-iyt}G(t)\,dt = 0.$$

(*b*) Write $G(t) = e^{-xt}F(t)$ where F is sectionally continuous over each interval $(0,T)$ and $\mathcal{O}(e^{\alpha t})$ to prove that the transform $f(s)$ of F satisfies this condition when x is fixed:

$$\lim_{|y|\to\infty} f(x + iy) = 0 \qquad\qquad (x > \alpha).$$

66   THE INVERSION INTEGRAL

According to our extension of Cauchy's integral formula (Sec. 62), a function $f(s)$ that is analytic and $\mathcal{O}(s^{-k})$ over a half plane $\operatorname{Re} s \geqq \gamma$, where $k > 0$, can be represented in terms of its values on the boundary of the half plane by a line integral:

$$(1) \qquad\qquad f(s) = \frac{1}{2\pi i} \int_{\gamma - i\infty}^{\gamma + i\infty} \frac{f(z)}{s - z}\,dz \qquad\qquad (\operatorname{Re} s > \gamma).$$

Suppose $f(s) = L\{F(t)\}$. If we formally apply the inverse Laplace transformation to the functions of s in formula (1) and interchange the order of the operation L^{-1} and the integration, we find that

$$(2) \qquad F(t) = \frac{1}{2\pi i} \int_{\gamma - i\infty}^{\gamma + i\infty} e^{tz}f(z)\,dz = \frac{e^{\gamma t}}{2\pi} \int_{-\infty}^{\infty} e^{iyt}f(\gamma + iy)\,dy.$$

This is an integral transformation of $f(s)$ that may represent in a direct manner the inverse Laplace transform of that function.

The improper integral (2) exists if and only if the integrals of $e^{iyt}f(\gamma + iy)$ from $-\infty$ to 0 and from 0 to ∞ both exist. For some transforms $f(s)$ those improper integrals fail to exist, at least for certain values of t, while the *Cauchy principal value* of the integral (2),

$$(3) \qquad\qquad \frac{e^{\gamma t}}{2\pi} \lim_{\beta\to\infty} \int_{-\beta}^{\beta} e^{iyt}f(\gamma + iy)\,dy,$$

does exist. This is illustrated in Prob. 7, Sec. 70. If the improper integral (2) does exist, its principal value exists and is equal to the improper integral.

The principal value (3) is called the *complex inversion integral* for the Laplace transformation. We use the symbol L_i^{-1} for that linear integral transformation of $f(s)$, a symbol that is intended to suggest an integration as well as an inverse Laplace transformation; thus

$$(4) \qquad\qquad L_i^{-1}\{f(s)\} = \frac{1}{2\pi i} \lim_{\beta\to\infty} \int_{\gamma - i\beta}^{\gamma + i\beta} e^{tz}f(z)\,dz.$$

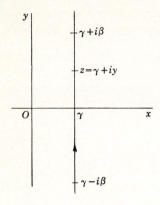

Fig. 69

We assume that $f(\bar{s}) = \overline{f(s)}$, a condition that is satisfied by transforms of real-valued functions $F(t)$ (Sec. 63). Then although the inversion integral is an integral in the complex plane along the line $x = \gamma$ (Fig. 69), we can write it as a *real improper integral*. Since $z = \gamma + iy$, where γ is fixed, the integral in Eq. (4) can be written as

$$ie^{\gamma t}\left[\int_{-\beta}^{0} e^{iyt}f(\gamma + iy)\,dy + \int_{0}^{\beta} e^{iyt}f(\gamma + iy)\,dy\right].$$

When the variable of integration in the first integral here is replaced by $-y$, the expression inside the brackets becomes

$$\int_{0}^{\beta} [e^{-iyt}f(\gamma - iy) + e^{iyt}f(\gamma + iy)]\,dy.$$

The integrand of this last integral is $2\,\mathrm{Re}\,e^{iyt}f(\gamma + iy)$, because the complex conjugate of $e^{iyt}f(\gamma + iy)$ is $e^{-iyt}f(\gamma - iy)$. We write

$$f(x + iy) = u(x,y) + iv(x,y);$$

then

$$\mathrm{Re}\,[e^{iyt}f(\gamma + iy)] = u(\gamma,y)\cos yt - v(\gamma,y)\sin yt$$

and therefore

(5) $$L_i^{-1}\{f(s)\} = \frac{e^{\gamma t}}{\pi}\int_{0}^{\infty} [u(\gamma,y)\cos yt - v(\gamma,y)\sin yt]\,dy.$$

When $f(s)$ satisfies certain conditions, we shall see that the value of the real integral (5) is independent of γ for all values of γ greater than a prescribed number. Even for simple functions f, however, the integration involved in the real form (5) is generally difficult. To evaluate the inversion integral, we shall use the complex form (4) in conjunction with the theory of residues or processes of changing the path of integration.

In the following sections conditions on either f or F will be established under which the inversion integral represents the inverse transform $F(t)$ of $f(s)$.

Other formulas for the inverse transformation are known;[1] but the inversion integral (4) has greater general utility than the others.

67 CONDITIONS ON $f(s)$

The following theorem gives conditions on a function f that are sufficient to ensure that $L_i^{-1}\{f(s)\}$ exists and represents a function F whose transform is $f(s)$. Useful properties of $F(t)$ also follow from the conditions on $f(s)$, properties that give the initial value $F(0)$, the continuity and exponential order of F.

Theorem 5 *Let f be any function of the complex variable s that is analytic and $\mathcal{O}(s^{-k})$ for all $s(s = x + iy)$ over a half plane $x \geqq \alpha$, where $k > 1$; also let $f(x)$ be real when $x \geqq \alpha$. Then for all real t the inversion integral of $f(s)$ along any line $x = \gamma$, where $\gamma \geqq \alpha$, converges to a real-valued function F that is independent of γ,*

$$(1) \qquad F(t) = L_i^{-1}\{f(s)\} = \frac{1}{2\pi i}\int_{\gamma - i\infty}^{\gamma + i\infty} e^{tz}f(z)\,dz \qquad (|t| < \infty),$$

whose Laplace transform is the given function $f(s)$:

$$(2) \qquad L\{F(t)\} = f(s) \qquad (\text{Re } s > \alpha).$$

Furthermore $F(t)$ is $\mathcal{O}(e^{\alpha t})$, it is continuous $(-\infty < t < \infty)$, and

$$(3) \qquad F(t) = 0 \qquad \text{when } t \leqq 0.$$

The integrand $e^{tz}f(z)$ of the inversion integral, where $z = \gamma + iy$ and $\gamma \geqq \alpha$, is everywhere continuous in y and t since f is analytic when $\text{Re } z \geqq \alpha$. Thus the integral along each bounded segment $|y| \leqq y_0$ of the line $x = \gamma$ exists as a continuous function of t. According to the order condition on f, positive constants M and y_0 exist such that

$$(4) \qquad |f(\gamma + iy)| < \frac{M}{(\gamma^2 + y^2)^{k/2}} \leqq \frac{M}{|y|^k} \qquad \text{when } |y| > y_0.$$

Since $k > 1$, the improper integrals of $M|y|^{-k}$ from y_0 to ∞ and from $-\infty$ to $-y_0$ exist, so the inversion integral exists as an improper integral, and it is equal to its principal value.

---

[1] See books by Doetsch and by Widder listed in the Bibliography.

Also, since $f(x)$ is real, then, according to the reflection principle for analytic functions, $f(\bar{z}) = \overline{f(z)}$, and the inversion integral takes its real form

$$(5) \qquad L_i^{-1}\{f(s)\} = \frac{e^{\gamma t}}{\pi} \int_0^{\infty} [u(\gamma,y) \cos ty - v(\gamma,y) \sin ty]\, dy.$$

Now the components $u(\gamma,y)$ and $v(\gamma,y)$ of $f(\gamma + iy)$, as well as f itself, satisfy condition (4). Thus the absolute value of the integrand of integral (5) is less than $2My^{-k}$ when $y > y_0$, a function that is integrable from y_0 to ∞ and independent of t. Consequently, integral (5) converges uniformly with respect to t, and it follows that $L_i^{-1}\{f\}$ is a continuous real-valued function $F(t)$ for all real t.

To show that $F(t)$ is independent of γ when $\gamma \geq \alpha$, we use a second path of integration $x = \gamma'$, where $\gamma' > \gamma$. Since $e^{tz}f(z)$ is analytic when $x \geq \gamma$, the integral of that function around the boundary of the rectangle $ABCD$ in Fig. 70 vanishes, according to Cauchy's integral theorem. On sides BC and AD, $|z| \geq \beta$ and, since $|f(z)| < M|z|^{-k}$,

$$|e^{tz}f(z)| < e^{tx}\frac{M}{\beta^k} \leq \frac{N(t)}{\beta^k},$$

where $N(t)$ is the greater of the two numbers $Me^{t\gamma}$ and $Me^{t\gamma'}$. The length of those sides is $\gamma' - \gamma$; hence the absolute value of the integral of $e^{tz}f(z)$ along either side BC or AD does not exceed $N(t)(\gamma' - \gamma)\beta^{-k}$, a quantity that vanishes as $\beta \to \infty$.

The vanishing of the contour integral around the boundary can be expressed in the form

$$\int_{AB} e^{tz}f(z)\, dz + \int_{CD} e^{tz}f(z)\, dz = \int_{AD} e^{tz}f(z)\, dz - \int_{BC} e^{tz}f(z)\, dz.$$

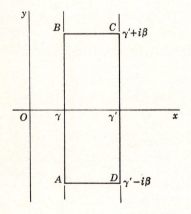

Fig. 70

The left-hand member tends to zero as $\beta \rightarrow \infty$ because the member on the right does so. But we have seen that the limit as $\beta \rightarrow \infty$ of each of the integrals on the left exists when $k > 1$. It follows that

$$\lim_{\beta \to \infty} \int_{\gamma - i\beta}^{\gamma + i\beta} e^{tz} f(z)\, dz = \lim_{\beta \to \infty} \int_{\gamma' - i\beta}^{\gamma' + i\beta} e^{tz} f(z)\, dz,$$

so the function F defined by Eq. (1) is independent of γ.

Let us write $\gamma = \alpha$ in Eq. (1). Then

$$e^{-\alpha t}|F(t)| = \frac{1}{2\pi} \left| \int_{-\infty}^{\infty} e^{ity} f(\alpha + iy)\, dy \right| \leq \frac{1}{2\pi} \int_{-\infty}^{\infty} |f(\alpha + iy)|\, dy.$$

The last integral exists because f is $\mathcal{O}(y^{-k})$ as $|y| \rightarrow \infty$ where $k > 1$, and f is continuous. Its value is independent of t. Therefore $F(t)$ is $\mathcal{O}(e^{\alpha t})$ as $t \rightarrow \infty$.

To evaluate F when $t \leq 0$, it is convenient to select a positive number for γ; this can be done since γ can be any number in the range $\gamma \geqq \alpha$. Then over the half plane $x \geqq \gamma$, $|e^{tz}| = e^{xt} \leq 1$ when $t \leq 0$ and

$$|e^{tz} f(z)| < M|z|^{-k} \qquad \text{where } k > 1.$$

It follows from our extension of Cauchy's integral theorem (Theorem 2, Sec. 62) that

$$\int_{\gamma - i\infty}^{\gamma + i\infty} e^{tz} f(z)\, dz = 0 \qquad\qquad (t \leq 0);$$

that is, $F(t) = 0$ when $t \leq 0$. In particular, $F(0) = 0$.

Finally, the transform of $F(t)$ must exist when $\operatorname{Re} s > \alpha$ because F is continuous and $\mathcal{O}(e^{\alpha t})$, so

$$L\{F(t)\} = \frac{1}{2\pi} \lim_{T \to \infty} \int_0^T e^{-st} \int_{-\infty}^{\infty} e^{tz} f(z)\, dy\, dt \qquad (\operatorname{Re} s > \alpha),$$

when $z = \alpha + iy$. Since the convergence of the inversion integral is uniform with respect to t, the order of integration can be interchanged:

$$(6) \qquad L\{F\} = \frac{1}{2\pi} \lim_{T \to \infty} \int_{-\infty}^{\infty} f(z) \int_0^T e^{-(s-z)t}\, dt\, dy$$

$$= \frac{1}{2\pi} \lim_{T \to \infty} \int_{-\infty}^{\infty} \left[\frac{f(z)}{s - z} - \frac{f(z)}{s - z} e^{-(s-z)T} \right] dy.$$

The first term inside the brackets here is independent of T and, in view of the extended Cauchy integral formula (Theorem 1, Sec. 62), its integral exists and has the value $2\pi f(s)$:

$$(7) \qquad \frac{1}{2\pi} \int_{-\infty}^{\infty} \frac{f(\alpha + iy)}{s - \alpha - iy}\, dy = f(s) \qquad (\operatorname{Re} s > \alpha).$$

The limit as $T \to \infty$ of the integral of the second term inside the brackets in Eq. (6) is zero. For if we write $s = a + ib$, where $a > \alpha$, and note that $|s - z| \geq a - \alpha$ and $|e^{-(s-z)T}| = e^{-(a-\alpha)T}$, it follows that

$$\left| \int_{-\infty}^{\infty} \frac{f(z)}{s - z} e^{-(s-z)T} \, dy \right| \leqq \frac{e^{-(a-\alpha)T}}{a - \alpha} \int_{-\infty}^{\infty} |f(z)| \, dy.$$

The member on the right vanishes as $T \to \infty$ because f is absolutely integrable along the line $x = \alpha$. Thus

(8) $$L\{F(t)\} = L\{L_i^{-1}[f]\} = f(s) \qquad (\text{Re } s > \alpha).$$

The proof of Theorem 5 is now complete.

68 CONDITIONS ON $F(t)$

The foregoing conditions under which the inversion integral formula is valid are severe. They are not satisfied, for example, by the function $1/s$, which is $\mathcal{O}(1/s^k)$ where $k = 1$, nor by transforms of functions F that are discontinuous or for which $F(0) \neq 0$. By using a Fourier integral theorem and qualifying the function F instead of f, we can relax the conditions so that the inversion integral formula can be seen to apply in nearly all cases of interest to us. In fact we shall see that our formula is only a modified form of the Fourier integral formula.

Let G be a function defined for all real t, sectionally continuous over each bounded interval of the t axis, and absolutely integrable from $-\infty$ to ∞. We agree to define $G(t)$ at each point t_0 where G is discontinuous as the mean value of its limits from the right and left at t_0:

(1) $$G(t_0) = \tfrac{1}{2}[G(t_0 + 0) + G(t_0 - 0)].$$

Also let the derivative G' be sectionally continuous over each bounded interval; this condition actually implies that G itself is sectionally continuous (Prob. 10, Sec. 70).

Then $G(t)$ is represented in terms of trigonometric functions of t by the *Fourier integral formula*[1]

(2) $$G(t) = \frac{1}{\pi} \int_0^{\infty} \int_{-\infty}^{\infty} G(\tau) \cos y(t - \tau) \, d\tau \, dy \qquad (-\infty < t < \infty)$$

$$= \frac{1}{2\pi} \lim_{\beta \to \infty} \int_0^{\beta} \int_{-\infty}^{\infty} [G(\tau) e^{iy(t-\tau)} + G(\tau) e^{-iy(t-\tau)}] \, d\tau \, dy.$$

Since the integral of $|G(\tau)|$ from $-\infty$ to ∞ exists, the inner integral of each term within the brackets converges uniformly with respect to y to a con-

[1] See, for instance, the author's "Fourier Series and Boundary Value Problems," 2d ed., pp. 83, 115 ff., 1963.

tinuous function of y whose integral from $y = 0$ to $y = \beta$ exists. Thus the iterated integral in the second line of formula (2) can be written as the sum

$$\int_0^\beta e^{iyt} \int_{-\infty}^\infty G(\tau)e^{-iy\tau}\, d\tau\, dy + \int_{-\beta}^0 e^{iyt} \int_{-\infty}^\infty G(\tau)e^{-iy\tau}\, d\tau\, dy.$$

Hence the Fourier integral formula has the *exponential form*

$$(3) \qquad G(t) = \frac{1}{2\pi} \lim_{\beta \to \infty} \int_{-\beta}^\beta e^{iyt} \int_{-\infty}^\infty G(\tau)e^{-iy\tau}\, d\tau\, dy \qquad (-\infty < t < \infty).$$

Note that the existence of the principal value of the improper integral with respect to y from $-\infty$ to ∞ is ensured here, but not the convergence of the improper integral itself.

Now let F denote a function defined when $t \geqq 0$ and $\mathcal{O}(e^{\alpha t})$ as $t \to \infty$, such that F', and therefore F itself, is sectionally continuous over each interval $0 < t < T$. If we write

$$(4) \qquad G(t) = e^{-\gamma t}F(t) \qquad \text{when } t > 0, \qquad G(t) = 0 \qquad \text{when } t < 0,$$

where $\gamma > \alpha$, then G satisfies the sufficient conditions stated above for the representation (3), so

$$(5) \qquad G(t) = \frac{1}{2\pi} \lim_{\beta \to \infty} \int_{-\beta}^\beta e^{iyt} \int_0^\infty e^{-(\gamma + iy)\tau}F(\tau)\, d\tau\, dy \qquad (|t| < \infty).$$

The inner integral represents $f(\gamma + iy)$, where $f(s)$ is the Laplace transform of $F(t)$. If $z = \gamma + iy$, it follows that, for all real t,

$$(6) \qquad e^{\gamma t}G(t) = \frac{1}{2\pi i} \lim_{\beta \to \infty} \int_{\gamma - i\beta}^{\gamma + i\beta} e^{tz}f(z)\, dz = L_i^{-1}\{f(s)\}.$$

When $t < 0$, $G(t) = 0$, and therefore $L_i^{-1}\{f(s)\} = 0$.

When $t > 0$, $e^{\gamma t}G(t) = F(t)$, and Eq. (6) represents $F(t)$ as the inversion integral $L_i^{-1}\{f(s)\}$. But since the function G represented by formula (5) has the mean value (1) at each point t_0 of discontinuity, the inversion integral converges to $F(t_0)$ at a point t_0 where F has a jump, provided we agree that

$$(7) \qquad F(t_0) = \tfrac{1}{2}[F(t_0 + 0) + F(t_0 - 0)] \qquad (t_0 > 0).$$

When $t = 0$, the value of the inversion integral is $\tfrac{1}{2}[F(+0) + 0]$.

The existence of the inversion integral (5) in the form of the principal value of the improper integral along the line $x = \gamma$ was established here, but the improper integral itself may not exist under the conditions stated. The following theorem is now established.

Theorem 6   *If $f(s)$ is the Laplace transform of a function F of exponential order $\mathcal{O}(e^{\alpha t})$ whose derivative F' is sectionally continuous over each* *interval* $0 < t < T$, *then the inversion integral of f along any line* Re $s = \gamma$, *where* $\gamma > \alpha$, *exists and represents* $F(t)$:

$$(8) \qquad\qquad L_i^{-1}\{f(s)\} = F(t) \qquad\qquad (t > 0).$$

At a point $t_0(t_0 > 0)$ *where F is discontinuous, the inversion integral has the mean value* (7) *of* $F(t)$; *when* $t = 0$, *it has the value* $\frac{1}{2}F(+0)$, *and when* $t < 0$, *it has the value zero.*

The conditions here on F can be relaxed because the conditions on G for the validity of the Fourier integral formula (2) can be modified in various ways.[1] For instance, on each bounded interval let the function G have at most a finite number of *infinite* discontinuities but still be absolutely integrable over the entire t axis, where each interval can be subdivided into a finite number of open intervals such that G is monotonic (nondecreasing or nonincreasing) over each one. Then the Fourier integral expression (2) has the value $\frac{1}{2}[G(t + 0) + G(t - 0)]$ at each point t where $G(t + 0)$ and $G(t - 0)$ exist.

As a consequence of those modified conditions, for instance,

$$L_i^{-1}\left\{\sqrt{\frac{\pi}{s}}\right\} = \frac{1}{\sqrt{t}} \qquad\qquad (\gamma > 0, t > 0),$$

because the function $G(t) = e^{-\gamma t}t^{-\frac{1}{2}}$ when $t > 0$, $G(t) = 0$ when $t < 0$, is represented by its Fourier integral formula when $t > 0$ and $\gamma > 0$, and because $L\{t^{-\frac{1}{2}}\} = (\pi/s)^{\frac{1}{2}}$ when Re $s > 0$.

Even though its conditions are on F, the function usually sought, rather than f, Theorem 6 serves as a useful supplement to Theorem 5. The following example and corollary illustrate this.

Example   Let $\sqrt{s}$ denote the branch $\sqrt{r}e^{i\theta/2}$, $r > 0$, $|\theta| < \pi/2$, where $s = re^{i\theta}$. Show that the inversion integral formula applies, when $\gamma > 0$, to the function

$$f(s) = \frac{\sqrt{s}}{s\sqrt{s} + 1}.$$

On a half plane Re $s \geq \alpha > 0$, the branch $\sqrt{s}$ is analytic and $s\sqrt{s} + 1$ never vanishes since Re $(s + 1/\sqrt{s}) > 0$. Thus f is analytic there but only $\mathcal{O}(1/s)$, which fails to satisfy the order condition in Theorem 5. By carrying out one step of a division, however, we can write f as a sum of two components,

$$(9) \qquad\qquad f(s) = \frac{1}{s} - \frac{1}{s^2\sqrt{s} + s}.$$

[1] See, for instance, H. S. Carslaw, "Fourier's Series and Integrals," chap. 10, 1930, and E. C. Titchmarsh, "Theory of Fourier Integrals," p. 13, 1937.

such that Theorem 5 applies to the second, which is $\mathcal{O}(s^{-5/2})$ and analytic over the half plane, and real when s is real, while Theorem 6 applies to the first to give $L_i^{-1}\{1/s\} = 1$ when $t > 0$. Thus $L_i^{-1}\{f\} = F(t)$ where

$$F(t) = L_i^{-1}\left\{\frac{1}{s}\right\} + L_i^{-1}\left\{\frac{-1}{s^2\sqrt{s+s}}\right\} = 1 + F_1(t) \qquad (t > 0),$$

and F_1 is continuous $(t \geq 0)$; also $F_1(0) = 0$ and F_1 is $\mathcal{O}(e^{\alpha t})$ if $\alpha > 0$. Consequently F is continuous when $t > 0$ and $\mathcal{O}(e^{\alpha t})$, and $F(+0) = 1$. At the end of this chapter we present methods for simplifying such representations as $L_i^{-1}\{f\}$ of our function F.

The example suggests a corollary to Theorems 5 and 6.

Corollary *If $f(s)$ can be written in the form*

(10) $$f(s) = f_1(s) + f_2(s) \qquad (\text{Re } s \geq \alpha)$$

where f_1 satisfies the conditions in Theorem 5 and f_2 is the transform of a function F_2 satisfying the conditions in Theorem 6, then $L_i^{-1}\{f\}$ is the inverse transform $F(t)$ of $f(s)$. Moreover F is $\mathcal{O}(e^{\alpha t})$, F is sectionally continuous on each bounded interval, and $F(+0) = F_2(+0)$.

69 UNIQUENESS OF INVERSE TRANSFORMS

In order to make a simple analysis of uniqueness of the inverse Laplace transformation, we consider *the class $\mathcal{E}$ of real-valued functions on the half line $t > 0$*, defined in Sec. 6. The class consists of *all functions F that are sectionally continuous over each bounded interval, where $F(t)$ is defined as the mean value of its limits from right and left at each point of discontinuity, and such that each function F is $\mathcal{O}(e^{\alpha t})$ for some α.* Then F has a transform $f(s)$ when Re $s > \alpha$. Our theorem on uniqueness of inverse transforms can be stated as follows.

Theorem 7 *Among all functions of class $\mathcal{E}$ no two distinct functions can have the same Laplace transform over a right half plane Re $s > \alpha$, or for all real s on a half line $s > \alpha$.*

Since transforms of functions of class $\mathcal{E}$ are analytic functions of s over some right half plane, any two transforms are the same over a half plane if they are identical for all real s on a half line $s > \alpha$ (Sec. 64).

In particular, the theorem shows that no two functions of exponential order that are continuous when $t \geq 0$ can have identical transforms for all real s on a half line $s > \alpha$.

To prove the theorem, first let $\mathcal{E}'$ denote the subclass of all functions F in class $\mathcal{E}$ for which F' is also sectionally continuous over each bounded interval on the half line $t > 0$. According to Theorem 6 if F belongs to $\mathscr{E}'$, it is represented by the inversion integral of its transform f. That is, when $F(t)$ is $\mathcal{O}(e^{\alpha t})$, then $F(t) = L_i^{-1}\{f(s)\}$ when $t > 0$ regardless of the value of γ in the inversion integral as long as $\gamma > \alpha$. In case a function F in $\mathscr{E}'$ has the vanishing transform $f(s) = 0$ for all s in some right half plane, it follows that $F(t) = 0$ *is the only function of class $\mathscr{E}'$ whose transform vanishes over a right half plane.*

Next let G and H be two functions of class $\mathscr{E}$, so that no conditions are imposed on their derivatives. Let both be $\mathcal{O}(e^{\alpha t})$ and suppose that they have identical transforms $g(s)$:

(1) $$L\{G(t)\} = L\{H(t)\} = g(s) \qquad \text{(Re } s > \alpha).$$

Now the function

(2) $$F(t) = \int_0^t [G(\tau) - H(\tau)]\,d\tau \qquad (t \geqq 0)$$

is continuous and $\mathcal{O}(e^{\beta t})$ whenever $\beta > \alpha$ (Sec. 17). Moreover, on each interval where G and H are continuous,

(3) $$F'(t) = G(t) - H(t);$$

thus F' is sectionally continuous over each bounded interval, and F is therefore in class $\mathscr{E}'$.

We can choose β as a positive number. Then from condition (1) and the representation (2) of F as a special convolution, it follows that

(4) $$L\{F(t)\} = \frac{1}{s}[g(s) - g(s)] = 0 \qquad \text{(Re } s > \beta),$$

and since F is in $\mathscr{E}'$, we conclude that $F(t) = 0$ when $t > 0$. Therefore $F'(t) = 0 (t > 0)$; that is, wherever G and H are continuous,

(5) $$G(t) - H(t) = 0.$$

At each point where G or H has a jump, it follows that the one-sided limits of $G(t) - H(t)$ vanish; thus

$$\tfrac{1}{2}[G(t + 0) + G(t - 0)] = \tfrac{1}{2}[H(t + 0) + H(t - 0)] \qquad (t > 0).$$

Consequently $G(t) = H(t)$ whenever $t > 0$, and Theorem 7 is proved.[1]

70 DERIVATIVES OF THE INVERSION INTEGRAL

When the solution of a boundary value problem is found in the form of an inversion integral $L_i^{-1}\{f\}$, it is often possible to verify the solution by noting

---

[1] For theorems on uniqueness that permit $L^{-1}\{f\}$ to have certain infinite discontinuities, see G. Doetsch, "Handuch der Laplace-Transformation," vol. 1, pp. 72 ff., 1950.

properties of the function f. The two theorems in this section are useful for that purpose. Their proofs follow from Theorem 5 and properties of uniformly convergent integrals (Sec. 15).

When the inversion integral in Theorem 5 is differentiated with respect to t under the integral sign, we obtain $L_i^{-1}\{sf(s)\}$:

(1)
$$\frac{1}{2\pi}\int_{-\infty}^{\infty}\frac{\partial}{\partial t}[e^{tz}f(z)]\,dy = \frac{1}{2\pi}\int_{-\infty}^{\infty}e^{tz}zf(z)\,dy,$$

where $z = \gamma + iy$. If $sf(s)$ as well as $f(s)$ satisfies the conditions imposed on f in that theorem, the integral (1) converges uniformly with respect to t and represents $F'(t)$, the derivative with respect to t of $L_i^{-1}\{f(s)\}$. Then F' satisfies the conditions stated for the function F in the theorem. The additional condition needed here is that $sf(s)$ is $\mathcal{O}(s^{-k})$ where $k > 1$; that is, that $f(s)$ is $\mathcal{O}(s^{-k-1})$.

If we replace $f(s)$ above by $sf(s)$, we see that $F''(t)$ is represented by $L_i^{-1}\{s^2f(s)\}$, and so on, giving the following theorem.

Theorem 8 *Let f be a function of s that is analytic and $\mathcal{O}(s^{-k-m})$ over a half plane $x \geq \alpha$, where $s = x + iy$, $k > 1$, and m is a positive integer. Also let $f(x)$ be real $(x \geq \alpha)$. Then the inversion integral of f along any line $x = \gamma$, where $\gamma \geq \alpha$, converges to the inverse transform of f, a real-valued function F,*

(2)
$$L_i^{-1}\{f(s)\} = F(t) \qquad\qquad (t \geq 0),$$

and the derivatives of F are given by the formula

(3)
$$F^{(n)}(t) = L_i^{-1}\{s^nf(s)\} \qquad (n = 1, 2, \ldots, m);$$

furthermore, F and each of its derivatives (3) are continuous when $t \geq 0$ and $\mathcal{O}(e^{\alpha t})$, and their initial values are zero:

(4)
$$F(0) = F'(0) = \cdots = F^{(m)}(0) = 0.$$

Consider again, as an example, the inverse transform F_1 of the second function on the right in Eq. (9), Sec. 68;

$$F_1(t) = L_i^{-1}\left\{\frac{-1}{s^2\sqrt{s}+s}\right\}.$$

Its transform is $\mathcal{O}(s^{-5/2})$, so $m = 1$, and it follows from Theorem 8 that both F_1 and F_1' are continuous when $t \geq 0$ and both are $\mathcal{O}(e^{\alpha t})$ when $\alpha > 0$, and that

$$F_1(0) = F_1'(0) = 0.$$

It will be recalled that formula (3) cannot be valid unless either F or f is subjected to rather severe restrictions. For according to our basic property of transforming derivatives,

$$F^{(n)}(t) = L^{-1}\{s^n f(s) - s^{n-1}F(0) - \cdots - F^{(n-1)}(0)\}.$$

If $F^{(n)}(t)$ is to be the inverse transform of $s^n f(s)$, then it is necessary that $F(0), \ldots, F^{(n-1)}(0)$ all vanish.

Of course the function f in the above theorem may involve variables independent of s provided the statements in the theorem are understood to apply for fixed values of those parameters. Concerning differentiation and continuity with respect to such a parameter ρ, the following theorem can be stated from properties of uniformly convergent integrals.

Theorem 9 *Let f be a continuous function of two variables ρ and s ($s = x + iy$) when $x \geq \alpha$, analytic with respect to s over that half plane for each fixed ρ, and such that $|f(\rho,s)| < M|s|^{-k}$ there where $k > 1$ and the constants M, k, and α are independent of ρ. Also let $f(\rho,x)$ be real when $x \geq \alpha$. Even if ρ has an unbounded range of values, let $f(\rho, \alpha + iy)$ be bounded for all ρ and y. Then the inverse transform of f with respect to s, which can be written*

$$F(\rho,t) = L_i^{-1}\{f(\rho,s)\} \qquad\qquad (\gamma \geq \alpha, t \geq 0),$$

is a continuous real-valued function of its two variables ρ and t ($t \geq 0$) for which a constant N independent of ρ exists such that $|F(\rho,t)| < Ne^{\alpha t}$; in particular, F is a bounded function of ρ for each fixed t. If the partial derivative $f_\rho(\rho,t)$ also satisfies the conditions imposed here on f, then

$$(5) \qquad\qquad \frac{\partial}{\partial \rho}F(\rho,t) = L_i^{-1}\left\{\frac{\partial}{\partial \rho}f(\rho,s)\right\},$$

and $|F_\rho(\rho,t)| < Ne^{\alpha t}$.

The theorem can be applied to $f_\rho(\rho,s)$ to obtain corresponding results for $F_{\rho\rho}(\rho,t)$, and so on for derivatives of higher order. Applications of Theorems 8 and 9 are made in the following chapters.

PROBLEMS

1. Apply Theorem 5 to the functions (a) $f(s) = (s^2 - 1)^{-1}$ and (b) $g(s) = s^{-2}e^{-s}$ to show that their inverse transforms $F(t)$ and $G(t)$ are represented by inversion integrals, also to find order and continuity properties of F and G and to show that $F(0) = G(0) = 0$. Verify those properties by writing F and G as known inverse transforms.

2. Apply Theorem 8 to the function $f(s) = s^{-4}$ to deduce that the inverse transform $F(t)$ and its derivatives of first and second order at least are continuous when $t \geq 0$, and $\mathcal{O}(e^{\alpha t})$ if $\alpha > 0$, and that $F(0) = F'(0) = F''(0) = 0$. Verify those properties by recalling that $F(t) = t^3/6$.

3. Apply Theorem 6 to the transform of the function e^{-t} and simplify the inversion integral, when $\gamma = 0$, to establish the integration formula

$$\int_0^\infty \frac{\cos ty + y \sin ty}{y^2 + 1} \, dy = \pi e^{-t} \qquad\qquad (t > 0).$$

4. When $f(s)$ in Theorem 5 is the transform $(s + 1)^{-2}$ of te^{-t}, simplify the inversion integral, when $\gamma = 0$, to prove the integration formula

$$\int_0^\infty \frac{(1 - y^2) \cos ty + 2y \sin ty}{(y^2 + 1)^2} \, dy = \pi te^{-t} \qquad\qquad (t \geq 0).$$

5. Let $\sqrt{s}$ denote the branch $\sqrt{r}e^{i\theta/2}$, $-\pi < \theta < \pi, r > 0$, where $s = re^{i\theta}$, an analytic function over each half plane Re $s \geq \alpha > 0$. If $f(s) = \exp(-\sqrt{s})$, prove that $|f(s)| < \exp(-\sqrt{r/2})$ over the half plane and that $f(s)$ is $\mathcal{O}(s^{-m})$ there for *every* positive integer m. Show that Theorem 8 applies to establish $f(s)$ as the transform of a function $F(t)$ with continuous derivatives of all order $(t \geq 0)$ such that

$$F(0) = 0 \qquad \text{and} \qquad F^{(n)}(0) = 0 \qquad\qquad (n = 1, 2, \ldots).$$

[Compare the transformation (6), Sec. 27.]

6. Use power series to prove that the function g, where

$$g(z) = \frac{1 - e^{-z}}{z} \qquad\qquad \text{if } z \neq 0, g(0) = 1,$$

is entire and hence bounded over a neighborhood of the origin in the z plane. Then apply our corollary, Sec. 68, to the function

$$f(s) = \frac{1}{s} e^{-1/s} = \frac{1}{s} - \frac{s(1 - e^{-1/s})}{s^2}$$

to prove that $L^{-1}\{f\}$ is a continuous function $F(t)$ when $t > 0$, $\mathcal{O}(e^{\alpha t})$ if $\alpha > 0$, and that $F(+0) = 1$.

7. State why the inversion integral formula applies when $\gamma = 1$ to the known transform $f(s) = s^{-1}e^{-s}$ for all t; but show that when $t = 1$ the Cauchy principal value $L_i^{-1}\{f\}$ must be used because the improper integral itself fails to exist for that value of t. Also verify that, when $t = 1$, $L_i^{-1}\{f(s)\} = \frac{1}{2}$.

8. The entire function $g(s) = \exp(s^2)$ is not $\mathcal{O}(s^{-k})$ over a right half plane for any positive k. But prove that $L_i^{-1}\{g\}$, integrated along the imaginary axis, is the function

$$F(t) = \frac{1}{\pi} \int_0^\infty \exp(-y^2) \cos ty \, dy.$$

A known integration formula [Eq. (7), Sec. 74] shows that $F(t) = \frac{1}{2}\pi^{-\frac{1}{2}} \exp(-t^2/4)$, and thus (transformation 109, Appendix A, Table A.2)

$$L\{F\} = L\left\{\frac{1}{2\sqrt{\pi}} \exp\left(-\frac{t^2}{4}\right)\right\} = \frac{1}{2} \exp(s^2) \text{ erfc } s \neq g(s).$$

Thus the inversion integral may exist without representing the inverse transform.

9. Show why the function G, where

$$G(t) = e^{-t} \quad \text{when } t > 0, \qquad G(0) = \tfrac{1}{2}, \qquad G(t) = 0 \quad \text{when } t < 0,$$

satisfies all conditions for its representation by the exponential form of the Fourier integral formula. When $t = 0$, however, show that the Cauchy principal value (3), Sec. 68, must be used in that representation, for the improper integral itself, with respect to y, does not exist.

10. Prove that $F(t)$ itself is sectionally continuous over an interval if its derivative $F'(t)$ is sectionally continuous there. Note that F' is continuous over each of a finite number of subintervals, such as $a \leqq t \leqq b$, when properly defined at the end points, so F is continuous when $a < t < b$ and, when $a < t_0 < b$, then

$$\int_{t_0}^{t} F'(\tau)\,d\tau = F(t) - F(t_0) \qquad\qquad (a < t < b).$$

Deduce that $F(a + 0)$ and $F(b - 0)$ exist.

71   REPRESENTATION BY SERIES OF RESIDUES

Throughout this and the following two sections, $f(s)$ denotes a function that is analytic for all finite values of the complex variable s except for a set of *isolated* singular points

$$s_1, s_2, \ldots, s_n, \ldots$$

confined to some left half plane $\operatorname{Re} s < \gamma$. We assume also that $f(s)$ satisfies conditions under which its inversion integral along the line $x = \gamma$ converges to the inverse transform $F(t)$, say the conditions in either Theorem 5, Sec. 67, or Theorem 6, Sec. 68. Then $F(t)$ can be represented *formally* by a series, finite or infinite, depending on the number of singular points, and we shall establish practical conditions under which the representation is valid.

Since e^{tz} is an entire function of z, the singular points $z = s_n$ of $f(z)$ are the singular points of the integrand $e^{tz}f(z)$ of the inversion integral. Let $\rho_n(t)$ denote the residue of the integrand, for any fixed t, at the isolated singular point s_n:

(1) $\rho_n(t) = $ residue of $e^{tz}f(z)$ at $z = s_n$.

According to the residue theorem, the integral of $e^{tz}f(z)$ around a path enclosing the points $s_1, s_2, \ldots, s_N$ has the value

$$2\pi i[\rho_1(t) + \rho_2(t) + \cdots + \rho_N(t)].$$

Let the path be made up of the line segment joining the points $\gamma - i\beta_N$, $\gamma + i\beta_N$, and some contour C_N beginning at the second and ending at the

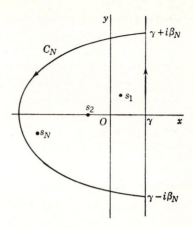

Fig. 71

first of these two points and lying in the half plane $x \leqq \gamma$ (Fig. 71). Then

$$(2) \qquad \frac{1}{2\pi i} \int_{\gamma - i\beta_N}^{\gamma + i\beta_N} e^{tz} f(z) \, dz + \frac{1}{2\pi i} \int_{C_N} e^{tz} f(z) \, dz = \sum_{n=1}^{N} \rho_n(t).$$

As $\beta_N \to \infty$, the value of the first integral here tends to $L_i^{-1}\{f(s)\}$, since the inversion integral is the limit of the corresponding integral involving β, as $\beta \to \infty$ in any manner. Let the numbers β_N ($N = 1, 2, \ldots$) be selected so that $\beta_N \to \infty$, and let the curves C_N together with the line $x = \gamma$ enclose the points $s_1, s_2, \ldots, s_N$, if the number of singular points is infinite. If the number is finite, let all of them be enclosed when N is greater than some fixed number. Then if $f(z)$ and C_N satisfy *additional conditions* under which

$$(3) \qquad \frac{1}{2\pi i} \lim_{N \to \infty} \int_{C_N} e^{tz} f(z) \, dz = 0,$$

it follows, by letting $N \to \infty$ in Eq. (2), that

$$L_i^{-1}\{f(s)\} = \sum_{n=1}^{\infty} \rho_n(t),$$

or by a finite series in case the number of singular points is finite. The series on the right is necessarily convergent because the limit, as $N \to \infty$, of the left-hand member of Eq. (2) exists.

Since the inversion integral represents $F(t)$ by hypothesis, the inverse transform of $f(s)$ is represented as the series of residues of $e^{tz} f(z)$:

$$(4) \qquad F(t) = \sum_{n=1}^{\infty} \rho_n(t).$$

It is not essential that the line $x = \gamma$ and the curve C_N enclose exactly N of the poles, of course. For example, if two poles are included in the ring between C_N and C_{N+1}, the residues at these two poles are simply grouped as a single term in the series.

It is sometimes convenient to use the series representation (4) directly and formally, without regard to conditions under which the inversion integral represents $F(t)$ and the integral over C_N tends to zero. The function F so obtained may be such that its transform can be shown to be $f(s)$, or such that it satisfies all conditions in a problem whose solution was sought by the transformation method. As we shall see in the problems at the end of this chapter, however, there are simple cases in which such a formal procedure leads to incorrect results.

72   RESIDUES AT POLES

At singular points s_n, which are poles of the function f, useful formulas for the residues $\rho_n(t)$ of $e^{tz}f(z)$ can be found.

Let s_n be a pole of f of order m. Over a neighborhood of that pole with the point s_n itself excluded $(0 < |z - s_n| < c_n)$, f is represented by a Laurent series

$$(1) \qquad f(z) = \frac{A_{-1}}{z - s_n} + \frac{A_{-2}}{(z - s_n)^2} + \cdots + \frac{A_{-m}}{(z - s_n)^m} + \sum_{j=0}^{\infty} A_j(z - s_n)^j,$$

where $A_{-m} \neq 0$ and the A's depend on s_n. Thus $(z - s_n)^m f(z)$ is represented there by the power series

$$(2) \qquad A_{-1}(z - s_n)^{m-1} + A_{-2}(z - s_n)^{m-2} + \cdots + A_{-m} + \sum_{j=0}^{\infty} A_j(z - s_n)^{j+m}.$$

Taylor's series expansion of the entire function e^{tz} about the point s_n is

$$(3) \qquad e^{tz} = \exp(s_n t)\left[1 + t(z - s_n) + \cdots + \frac{t^{m-1}}{(m-1)!}(z - s_n)^{m-1} + \cdots\right].$$

The coefficient of $(z - s_n)^{-1}$ in the Laurent expansion of $e^{tz}f(z)$ over the domain $0 < |z - s_n| < c_n$, the residue $\rho_n(t)$ of that function at s_n, is the coefficient of $(z - s_n)^{m-1}$ in the product of the power series (2) and (3). Thus we find that

$$(4) \qquad \rho_n(t) = \exp(s_n t)\left[A_{-1} + tA_{-2} + \cdots + \frac{t^{m-1}}{(m-1)!}A_{-m}\right].$$

Formula (4) has a useful feature. It presents the residue of $e^{tz}f(z)$ at a pole s_n of order m as *the inverse Laplace transform*, with respect to z, of the *principal part of the Laurent expansion* (1) *of the function f.*

When s_n is a *simple pole* $(m = 1)$ of f

$$(5) \qquad \rho_n(t) = \exp(s_n t)A_{-1} = \exp(s_n t)\lim_{z \to s_n}[(z - s_n)f(z)].$$

Note that A_{-1} is the residue of f itself at s_n. In particular, if

(6)
$$f(z) = \frac{p(z)}{q(z)}$$

where p and q are analytic at s_n and $q(s_n) = 0$ while $q'(s_n) \neq 0$ and $p(s_n) \neq 0$, then

(7)
$$\rho_n(t) = \frac{p(s_n)}{q'(s_n)} \exp(s_n t).$$

If all singular points of f are simple poles and f has the fractional form (6), then formula (4), Sec. 71, can be written

(8)
$$F(t) = L^{-1}\left\{\frac{p(s)}{q(s)}\right\} = \sum_{n=1}^{\infty} \frac{p(s_n)}{q'(s_n)} \exp(s_n t).$$

In case p and q are polynomials, the number of poles of f is finite, and formula (8) becomes Heaviside's expansion (4), Sec. 24.

Assume now that $f(\bar{z}) = \overline{f(z)}$ except at the singular points of f, all of which are isolated, a condition that is satisfied if $f(z)$ is real over the real axis except at singular points (Sec. 61). Let $s_1 = a + ib$, where $b \neq 0$, be a pole of f of order m. Then over a domain $0 < |z - s_1| < c_1$,

(9)
$$f(z) = \frac{A_{-1}}{z - s_1} + \cdots + \frac{A_{-m}}{(z - s_1)^m} + \sum_{j=0}^{\infty} A_j(z - s_1)^j \qquad (A_{-m} \neq 0).$$

The operation of taking complex conjugates is distributive to the terms of a convergent series; that is, if an infinite series Σz_n converges to a sum S, then the series $\Sigma \bar{z}_n$ converges to $\bar{S}$, as we can see by considering remainders. It follows from representation (9) and the condition $\overline{f(z)} = f(\bar{z})$ that

(10)
$$f(\bar{z}) = \frac{\bar{A}_{-1}}{\bar{z} - \bar{s}_1} + \cdots + \frac{\bar{A}_{-m}}{(\bar{z} - \bar{s}_1)^m} + \sum_{j=0}^{\infty} \bar{A}_j(\bar{z} - \bar{s}_1)^j$$

when $0 < |z - s_1| < c_1$, that is, when $0 < |\bar{z} - \bar{s}_1| < c_1$. By writing s for $\bar{z}$ here, we see that $f(s)$ has a Laurent series expansion in the domain $0 < |s - \bar{s}_1| < c_1$ with the $\bar{A}$'s as coefficients, and, since $\bar{A}_{-m} \neq 0$, $\bar{s}_1$ is also a pole of f of order m. According to formula (4) then, the sum of the residues of $e^{tz}f(z)$ at the poles $s_1 = a + ib$ and $\bar{s}_1$ is $\rho_1(t) + \bar{\rho}_1(t)$, where

$$\rho_1(t) = e^{(a+ib)t}\left[A_{-1} + tA_{-2} + \cdots + \frac{t^{m-1}}{(m-1)!}A_{-m}\right].$$

Therefore *the component of $F(t)$ corresponding to the pair of poles $a \pm ib$ ($b \neq 0$) of $f(s)$ is the real-valued function*

(11) $$\rho_1(t) + \overline{\rho_1(t)} = 2e^{at}\,\text{Re}\left\{e^{ibt}\left[A_{-1} + tA_{-2} + \cdots + \frac{t^{m-1}}{(m-1)!}A_{-m}\right]\right\}.$$

Let us write the residue A_{-1} of f at a pole $s_1 = a + ib$ in its polar form as

$$(12) \qquad\qquad\qquad A_{-1} = r_1 \exp(i\theta_1).$$

If s_1 is a *simple pole*, formula (11) then gives the component of F corresponding to the poles $a \pm ib$ as

$$(13) \qquad\qquad \rho_1(t) + \overline{\rho_1(t)} = 2e^{at}r_1 \cos(bt + \theta_1) \qquad (b \neq 0, r > 0),$$

a result that includes the component $G(t)$ obtained in Sec. 26, where f was a quotient of polynomials.

In case the singular points of f consist only of the simple poles $0, \pm in\omega$ $(n = 1, 2, \ldots)$, where f has residue r_0 at $z = 0$ and residues $r_n \exp(i\theta_n)$ at $in\omega$, formula (4), Sec. 71, becomes

$$(14) \qquad\qquad F(t) = r_0 + 2\sum_{n=1}^{\infty} r_n \cos(n\omega t + \theta_n);$$

then F is a periodic function of t with period $2\pi/\omega$.

If $s_1 = a + ib$ is a *second-order pole* of f, we write

$$(15) \qquad\qquad A_{-2} = \lim_{z \to s_1} [(z - s_1)^2 f(z)] = \rho_1 \exp(i\psi_1) \qquad (\rho_1 > 0).$$

Then if $b \neq 0$, according to formulas (11) and (12),

$$(16) \qquad \rho_1(t) + \overline{\rho_1(t)} = 2e^{at}[r_1 \cos(bt + \theta_1) + \rho_1 t \cos(bt + \psi_1)].$$

When $a = 0$, the term $2\rho_1 t \cos(bt + \psi_1)$ is one of *resonance type* representing an *unstable* component of $F(t)$, since $\rho_1 > 0$.

Similarly if ib is a pole of order m of $f(z)$ then $F(t)$ contains an unstable component of the type $Ct^{m-1}\cos(bt + \theta)$ where $C \neq 0$. In fact we can see from the above formulas that $F(t)$ *contains an unstable component if $f(z)$ has a real or imaginary pole $a + ib$ of any order such that $a > 0$, or a pole $z = 0$ or $z = ib$ of order $m = 2, 3, \ldots$* This is a generalization of results found in Sec. 26 where partial fractions could be used.

73   VALIDITY OF THE REPRESENTATION BY SERIES

We now establish sufficient conditions on f under which the integrals of $e^{tz}f(z)$ over certain useful contours C_N tend to zero as $N \to \infty$ as assumed in Sec. 71. Since the integrals must exist, the contours should not pass through singular points of f, nor should they become arbitrarily close to such points as N increases if the limit of the integrals is to exist. The ease of finding an order of magnitude of $|f(z)|$ when z is on C_N depends on the particular function f and the type of paths chosen for C_N.

We treat three types of paths C_N that are especially useful. In each case we assume that $f(z)$ is $\mathcal{O}(z^{-k})$ on C_N where $k > 0$; that is, constants M

and k, independent of N, exist such that

(1) $$|f(z)| < \frac{M}{|z|^k} \qquad \text{for points } z \text{ on } C_N \qquad (k > 0),$$

and we shall prove that, since $\beta_N \to \infty$ as $N \to \infty$,

(2) $$\lim_{N \to \infty} \int_{C_N} e^{tz} f(z)\, dz = 0 \qquad (t > 0).$$

If $k > 1$, statement (2) holds also when $t = 0$.

RECTANGULAR PATHS

Order properties of many functions on horizontal and vertical lines are easily determined. Let C_N be an open rectangle with sides along lines $y = \pm\beta_N$ and $x = -\beta_N$ as indicated in Fig. 72, where $\beta_N \to \infty$ as $N \to \infty$. Then if f satisfies the order condition (1) on C_N, the integrand of the inversion integral satisfies the condition

$$|e^{tz} f(z)| < \frac{M}{|z|^k} e^{tx} \qquad \text{when } z \text{ is on } C_N.$$

The integrals over the sides $y = \pm\beta_N$ are then such that

(3) $$\left| \int_{-\beta_N}^{\gamma} e^{tz} f(z)\, dx \right| < \frac{M}{\beta_N{}^k} \int_{-\beta_N}^{\gamma} e^{tx}\, dx = \frac{M}{t} \frac{e^{t\gamma} - e^{-t\beta_N}}{\beta_N{}^k}$$

when $t > 0$. Thus they tend to zero as $N \to \infty$.

The length of the third side of C_N is $2\beta_N$, and the absolute value of the integrand $e^{tz} f(z)$ is less than $M \exp(-\beta_N t)\beta_N{}^{-k}$ there, so when $t > 0$, the integrals over those sides vanish as $N \to \infty$.

Thus condition (2), where $t > 0$, is established. Moreover, if $k > 1$, then $\int_{C_N} f(z)\, dz \to 0$ as $N \to \infty$, so that condition (2) holds when $t = 0$.

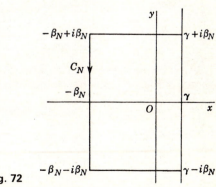

Fig. 72

CIRCULAR ARCS

Let C_N be arcs $|z| = R_N$, $x \leqq \gamma$. Then $R_N{}^2 = \beta_N{}^2 + \gamma^2$ and $R_N \to \infty$ as $N \to \infty$. If $\gamma > 0$, the length of each of the two arcs of C_N in the strip $0 \leqq x \leqq \gamma$ tends to γ as $N \to \infty$. Since f satisfies condition (1), then $|e^{tz}f(z)| < Me^{t\gamma}R_N{}^{-k}$ on those arcs, and the integrals over them tend to zero as $N \to \infty$.

Regardless of the sign of γ, the integral over the arc C_N' in the region $x \leqq 0$ satisfies the condition

(4)
$$\left| \int_{C_N'} e^{tz}f(z)\,dz \right| < \frac{M}{R_N{}^k} \int_{\pi/2}^{3\pi/2} \exp\,(tR_N \cos\theta)R_N\,d\theta$$

$$= \frac{2M}{R_N{}^k} \int_0^{\pi/2} \exp\,(-tR_N \sin\phi)R_N\,d\phi.$$

The graph of $\sin\phi$ shows that $\sin\phi > 2\phi/\pi$ when $0 < \phi < \pi/2$. Therefore the final member of condition (4) is less than

$$\frac{2M}{R_N{}^k} \int_0^{\pi/2} \exp\left(-2tR_N\frac{\phi}{\pi}\right)R_N\,d\phi = \frac{\pi M}{t}\frac{1 - \exp\,(-tR_N)}{R_N{}^k}$$

when $t > 0$, and this quantity vanishes as $N \to \infty$.

Therefore condition (2) is satisfied when C_N are those circular arcs. Moreover, if $k > 1$, condition (2) follows when $t = 0$ because

(5)
$$\left| \int_{C_N} f(z)\,dz \right| < \frac{M}{R_N{}^k}2\pi R_N \to 0 \qquad \text{as} \qquad N \to \infty.$$

PARABOLIC ARCS

Transforms involving $\sqrt{s}$ arise in solutions of heat equations. It is often desirable to select expanding paths C_N on which $\sqrt{z}$ is bounded away from zero. But $\mathrm{Re}\sqrt{z} = \sqrt{r}\cos\theta/2$ which vanishes on the negative real axis $(\theta = \pi)$. We may choose paths on which $\mathrm{Im}\sqrt{z}$ never vanishes by taking C_N as arcs of curves $\sqrt{r}\sin\theta/2 = \sqrt{a_N}$, where $a_N \to \infty$ as $N \to \infty$.

Then C_N is a parabolic arc

(6)
$$r = \frac{2a_N}{1 - \cos\theta} \qquad\qquad (x \leqq \gamma)$$

with vertex at $x = -a_N, y = 0$ and focus at the origin, resembling the arc shown in Fig. 71. Its equation can be written in the form

(7)
$$y^2 = 4a_N(x + a_N) \qquad [x \leqq \gamma, \beta_N{}^2 = 4a_N(\gamma + a_N)]$$

from which we find that

$$1 + \left(\frac{dy}{dx}\right)^2 = \frac{x + 2a_N}{x + a_N}.$$

Since $|z| \geq a_N$ and f satisfy condition (1) on C_N,

$$
(8) \qquad \left| \int_{C_N} e^{tz} f(z)\, dz \right| < \frac{2M}{a_N{}^k} \int_{-a_N}^{\gamma} e^{tx} \sqrt{\frac{x + 2a_N}{x + a_N}}\, dx.
$$

Let N be large enough that $a_N > 2|\gamma|$. Then when $t > 0$,

$$
(9) \qquad \int_{-a_N}^{-\frac{1}{2}a_N} e^{tx} \sqrt{\frac{x + 2a_N}{x + a_N}}\, dx < \exp\left(-\frac{ta_N}{2}\right) \sqrt{2a_N} \int_{-a_N}^{-\frac{1}{2}a_N} \frac{dx}{\sqrt{x + a_N}}
$$

and when we evaluate the last integral, we find that the right-hand member of condition (9) vanishes as $N \to \infty$. Also,

$$
(10) \qquad \frac{1}{a_N{}^k} \int_{-\frac{1}{2}a_N}^{\gamma} e^{tx} \sqrt{\frac{x + 2a_N}{x + a_N}}\, dx < \frac{\sqrt{\gamma + 2a_N}}{a_N{}^k \sqrt{a_N/2}} \int_{-\frac{1}{2}a_N}^{\gamma} e^{tx}\, dx
$$

which tends to zero as $N \to \infty$ if $t > 0$, so condition (2) is satisfied on the parabolic arcs. Again we find that the right-hand member of condition (8) vanishes as $N \to \infty$ when $t = 0$ in case $k > 1$. Our results can be summarized as follows.

Theorem 10 *Let f be analytic everywhere except for isolated singular points $s_n(n = 1, 2, \ldots)$ in a half plane $x < \gamma$ and such that its inversion integral along the line $x = \gamma$ represents the inverse Laplace transform $F(t)$ of $f(s)$. On contours $C_N(N = 1, 2, \ldots)$ in the left hand plane $x \leq \gamma$, let f satisfy the order condition $|f(z)| < M|z|^{-k}$ where M and k are positive constants independent of N, and C_N is either (a) the rectangular path shown in Fig. 72, (b) circular arcs $|z| = (\beta_N{}^2 + \gamma^2)^{\frac{1}{2}}, x \leq \gamma$, or (c) the parabolic arcs (7), where $\beta_N \to \infty$ as $N \to \infty$. Then whenever $t > 0$, the series of residues $\rho_n(t)$ of $e^{tz} f(z)$ at s_n converges to $F(t)$,*

$$
(11) \qquad F(t) = \sum_{n=1}^{\infty} \rho_n(t) \qquad\qquad (t > 0),
$$

if its terms corresponding to points s_n within the ring between successive paths C_N and C_{N+1} are grouped as a single terms of the series.
 Also, if $k > 1$ and $L_i^{-1}\{f\}$ represents $\frac{1}{2}F(+0)$ when $t = 0$, then

$$
(12) \qquad \tfrac{1}{2}F(+0) = \sum_{n=1}^{\infty} \rho_n(0).
$$

74 ALTERATIONS OF THE INVERSION INTEGRAL

When the singular points of $f(s)$ are not all isolated, it is often possible to reduce the inversion integral to a desirable form by introducing a new path of integration. We illustrate the procedure by finding an inverse transform that was obtained by another method in Sec. 27.

Let us find $F(t)$ when

(1) $$f(s) = \frac{1}{s}\exp\left(-s^{\frac{1}{2}}\right),$$

where $s = re^{i\theta}$ and $s^{\frac{1}{2}}$ is the single-valued function

(2) $$s^{\frac{1}{2}} = \sqrt{re^{\frac{1}{2}i\theta}} = \sqrt{r}\left(\cos\frac{\theta}{2} + i\sin\frac{\theta}{2}\right) \qquad (r > 0, -\pi < \theta < \pi).$$

Since the function (2) is analytic everywhere except on the branch cut $\theta = \pi$, $r \geq 0$, f is analytic in the same region. In the half plane $\operatorname{Re} s \geq \gamma$, where γ is a positive constant, $-\frac{1}{2}\pi < \theta < \frac{1}{2}\pi$ and $\cos\frac{1}{2}\theta > 1/\sqrt{2}$; therefore

$$|s^k f(s)| = r^{k-1}\exp\left(-\sqrt{r}\cos\frac{\theta}{2}\right) < r^{k-1}e^{-\sqrt{r}/2}.$$

This last function of r is continuous when $r \geq \gamma$ and vanishes as $r \to \infty$; hence it is bounded, and therefore f is $\mathcal{O}(s^{-k})$ in the half plane for *every* value of the constant k, and in particular if $k > 1$. According to Theorem 5, the inversion integral along the line $x = \gamma$ therefore represents the inverse transform $F(t)$,

(3) $$F(t) = \frac{1}{2\pi i}\lim_{\beta \to \infty}\int_{\gamma-i\beta}^{\gamma+i\beta} e^{tz}\exp\left(-z^{\frac{1}{2}}\right)\frac{dz}{z} \qquad (t \geq 0).$$

The sum of the integral in formula (3) and the integral along the path $ACDD'C'A'$, consisting of the circular arcs and line segments shown in Fig. 73, is zero, since the integrand is analytic except on the nonpositive real axis. Thus if I_{AC} denotes the integral of $e^{tz}f(z)$ over the arc AC, etc., we can write

(4) $$-\int_{\gamma-i\beta}^{\gamma+i\beta} e^{tz}f(z)\,dz = I_{AC} + I_{CD} + I_{DD'} + I_{D'C'} + I_{C'A'}.$$

Let R and r_0 denote the radii of the large and small cicular arcs; thus $R^2 = \gamma^2 + \beta^2$ so that $\beta \to \infty$ when $R \to \infty$. Along the arc AC, $z = Re^{i\theta}$, $dz = iRe^{i\theta}\,d\theta$, and $z^{\frac{1}{2}} = \sqrt{R}e^{i\theta/2}$. Hence the integrand of the integral is a continuous function of θ whenever $\epsilon \geq 0$, where ϵ is the angle between DC or $D'C'$ and the negative real axis. For any fixed R, the limit of the integrals I_{AC} and $I_{A'C'}$, as $\epsilon \to 0$, therefore exists. Likewise for any fixed positive r_0 the limits of the integrals over the other parts of the path exist. Since formula (4) is true for every positive ϵ and the integral on the left is independent of ϵ, it follows that we can let each of the integrals on the right have their limiting values as $\epsilon \to 0$, and consider hereafter the path in Fig. 74.

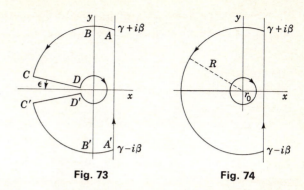

Fig. 73   **Fig. 74**

Write $J_{AC} = \lim_{\epsilon \to 0} I_{AC}$, and so on, for the integrals over the other arcs and lines. We shall now let r_0 approach zero and R tend to infinity, so that the left-hand member of Eq. (4), which is incidentally independent of r_0, becomes the inversion integral or $F(t)$, except for a factor $-2\pi i$.

When z is on the circle $r = r_0$, $z = r_0 e^{i\theta}$, and the integral over that circle can be written

$$J_{DD'} = i \int_{\pi}^{-\pi} \exp\left(t r_0 e^{i\theta} - \sqrt{r_0} e^{\frac{1}{2}i\theta}\right) d\theta,$$

and the integrand is a continuous function of θ and r_0 when $r_0 \geq 0$. Therefore

$$\lim_{r_0 \to 0} J_{DD'} = i \int_{\pi}^{-\pi} d\theta = -2\pi i.$$

On the line CD, $z = r e^{i(\pi - \epsilon)}$ and $z^{\frac{1}{2}} = \sqrt{r} e^{i(\pi - \epsilon)/2}$; thus as $\epsilon \to 0$, $z \to -r$ and $z^{\frac{1}{2}} \to i\sqrt{r}$. On the limiting position of $D'C'$, however, $z = -r$ and $z^{\frac{1}{2}} = -i\sqrt{r}$. Therefore

$$J_{CD} + J_{D'C'} = \int_{R}^{r_0} e^{-tr} e^{-i\sqrt{r}} \frac{dr}{r} + \int_{r_0}^{R} e^{-tr} e^{i\sqrt{r}} \frac{dr}{r}$$

$$= 2i \int_{r_0}^{R} e^{-tr} \frac{\sin \sqrt{r}}{r} dr,$$

and

$$\lim_{r_0 \to 0} (J_{CD} + J_{D'C'}) = 2i \int_{0}^{R} e^{-tr} \frac{\sin \sqrt{r}}{r} dr$$

$$= 4i \int_{0}^{\sqrt{R}} \exp(-t\mu^2) \frac{\sin \mu}{\mu} d\mu.$$

Consequently we can write, in view of formula (4),

$$(5) \quad F(t) = -\frac{1}{2\pi i} \lim_{R \to \infty} (J_{AC} + J_{C'A'}) + 1 - \frac{2}{\pi} \int_{0}^{\infty} \exp(-t\mu^2) \frac{\sin \mu}{\mu} d\mu.$$

Now on arcs AB and $A'B'$, $\text{Re}\,(tz - z^{\frac{1}{2}}) \leqq t\gamma$; thus the integrand of the integrals J_{AB} and $J_{B'A'}$ satisfies the condition

$$|z^{-1}\exp(tz - z^{\frac{1}{2}})| \leqq R^{-1}e^{t\gamma}.$$

Since the length of each arc tends to γ as $R \to \infty$, those integrals tend to zero as $R \to \infty$.

Finally, on arcs BC and $B'C'$, $\text{Re}\,(tz - z^{\frac{1}{2}}) \leqq tR\cos\theta$ so that

$$|J_{BC}| \leqq \int_{\pi/2}^{\pi} \exp(tR\cos\theta)\,d\theta = \int_{0}^{\pi/2} \exp(-tR\sin\phi)\,d\phi,$$

and $J_{C'B'}$ satisfies the same condition. The value of the last integrand does not exceed $\exp(-2tR\phi/\pi)$, as pointed out under Circular Arcs, Sec. 73, and an elementary integration then shows that J_{BC} and $J_{C'B'}$ tend to zero as $R \to \infty$, if $t > 0$.

The limit in formula (5) therefore vanishes, and

$$(6) \qquad F(t) = 1 - \frac{2}{\pi}\int_{0}^{\infty} \exp(-t\mu^2)\frac{\sin\mu}{\mu}\,d\mu \qquad (t > 0).$$

When we integrate both members of the known[1] formula

$$(7) \qquad \int_{0}^{\infty} \exp(-t\mu^2)\cos\alpha\mu\,d\mu = \frac{1}{2}\sqrt{\frac{\pi}{t}}\exp\left(-\frac{\alpha^2}{4t}\right) \qquad (t > 0),$$

with respect to α, from zero to 1, we find that

$$\frac{2}{\pi}\int_{0}^{\infty} \exp(-t\mu^2)\frac{\sin\mu}{\mu}\,d\mu = \frac{1}{\sqrt{\pi t}}\int_{0}^{1} \exp\left(\frac{\alpha^2}{4t}\right)d\alpha$$

$$= \frac{2}{\sqrt{\pi}}\int_{0}^{1/(2\sqrt{t})} \exp(-\lambda^2)\,d\lambda.$$

Thus we can write our result in the form

$$(8) \qquad F(t) = 1 - \text{erf}\left(\frac{1}{2\sqrt{t}}\right) = \text{erfc}\left(\frac{1}{2\sqrt{t}}\right).$$

---

[1] See, for instance, R. V. Churchill, "Complex Variables and Applications," 2d ed., p. 171, exercise 7, 1960. Formula (7) also follows from the result in Prob. 7, Sec. 35.

PROBLEMS

1. Instead of simply using partial fractions, show that Theorems 5 and 10 apply to $f(s)$ when

$$(a)\ f(s) = \frac{2s + 1}{s(s^2 + 1)}, \qquad (b)\ f(s) = \frac{1}{s^2(s + 1)},$$

and find $F(t)$ as a sum of residues.   *Ans.* (a) $F(t) = 1 - \cos t + 2 \sin t$.

2. Use identities (6), Sec. 55, to prove that the function $\tanh z$ is bounded over each of the half planes $x \geq \gamma$ and $x \leq -\gamma$ when $\gamma > 0$, and on the lines $y = \pm N\pi (N = 1, 2, \ldots)$ uniformly with respect to N. Thus show that the function

$$f(s) = \frac{\tanh s}{s^2}$$

satisfies all conditions in Theorems 5 and 10 such that $F(t)$ is represented by a series of residues as follows:

$$F(t) = 1 - \frac{8}{\pi^2} \sum_{n=1}^{\infty} \frac{1}{(2n-1)^2} \cos \frac{(2n-1)\pi t}{2} \qquad (t \geq 0).$$

Note that, according to Sec. 23, F is the triangular-wave function $H(2,t)$ shown graphically in Fig. 10.

3. In Prob. 4, Sec. 24, we found that the function

$$f(s) = \frac{1}{s^2} - \frac{1}{s \sinh s}$$

is the transform of the periodic function F for which

$$F(t) = t \qquad \text{when } -1 < t < 1, F(t + 2) = F(t).$$

Show that $s = 0$ is a removable singular point of f, that f satisfies all conditions in Theorems 6 and 10, and thus that F has the sine series representation

$$F(t) = \frac{2}{\pi} \sum_{n=1}^{\infty} \frac{(-1)^{n+1}}{n} \sin n\pi t.$$

4. Use Maclaurin's series for $\cosh z$ to show that the function $g(s) = \cosh \sqrt{s}$, $s \neq 0$, $g(0) = 1$, is an entire function of s for any branch of $\sqrt{s}$. Then show that the function

$$f(s) = \frac{1}{s \cosh \sqrt{s}}$$

is analytic except for simple poles $s = 0$, $s = -(n - \frac{1}{2})^2\pi^2$ and, with the aid of one of identities (6), Sec. 55, that f satisfies all conditions in Theorems 5 and 10 when C_N are arcs of parabolas $\sqrt{r} \sin \theta/2 = N\pi$. Thus find $L^{-1}\{f\}$ in the form

$$F(t) = 1 + \frac{4}{\pi} \sum_{n=1}^{\infty} \frac{(-1)^n}{2n - 1} \exp\left[-\frac{(2n-1)^2\pi^2}{4} t\right] \qquad (t > 0).$$

5. Write $F(t)$ formally as a series of residues when

$$f(s) = \frac{2\tanh s}{4s^2 + \pi^2}.$$

$$Ans.\ \pi^2 F(t) = \pi t \sin\frac{\pi t}{2} + \cos\frac{\pi t}{2} - \sum_{n=2}^{\infty}\frac{1}{n(n-1)}\cos\frac{(2n-1)\pi t}{2}.$$

6. When $f(s) = s^{-2}e^{-s}$, show that $F(t)$ is represented by the inversion integral of f, where $\gamma > 0$, but that the residue of $e^{tz}f(z)$ at the lone singular point of f represents $F(t)$ only when $t \geq 1$. This example and those in Probs. 7 and 8 below illustrate how the formal procedure of writing the sum of residues $\rho_n(t)$ may fail to give the inverse transform. Note that in this case $|e^{tz}f(z)| \to \infty$ as $x \to -\infty$ if $t < 1$.

7. State why $L_i^{-1}\{s^{-1}e^{-s}\} = S_0(t-1)$ if $\gamma > 0$, but show that the sum of residues $\rho_n(t)$, consisting of just one term, fails to represent $S_0(t-1)$ when $t < 1$.

8. Apply the real form of the inversion integral, formula (5), Sec. 66, to the entire function

$$f(s) = \frac{1 - e^{-s}}{s} \qquad\qquad [s \neq 0, f(0) = 1],$$

to find $F(t)$ directly. Here we can write $\gamma = 0$ and use the integration formula (8), Sec. 34. Note that the residue procedure can represent $F(t)$ only where $F(t) = 0$ because $e^{tz}f(z)$ is entire.

9. From properties of the transform of the function erfc $(\frac{1}{2}t^{-\frac{1}{2}})$ noted in Sec. 74, prove that the function and each of its derivatives tends to zero as $t \to +0$.

10. When $\sqrt{s} = \sqrt{r}e^{i\theta/2}(-\pi < \theta < \pi, r > 0)$, use the procedure in Sec. 74 to prove that

$$L_i^{-1}\left\{\frac{1}{\sqrt{s}}\right\} = \frac{2}{\pi\sqrt{t}}\int_0^{\infty}\exp(-\lambda^2)\,d\lambda = \frac{1}{\sqrt{\pi t}} \qquad (t > 0)$$

11. Generalize Theorem 10 for rectangular paths by letting the sides of C_N lie on lines $x = -c\beta_N$ and $y = \pm\beta_N$ and using only a condition of boundedness $|f(z)| < B$ on the sides $x = -c\beta_N$ when $t > 0$, where c and B are positive constants independent of N.

12. Prove Theorem 10 when C_N are semicircles $|z - \gamma| = \beta_N, x \leq \gamma$, and $|f(z)| < M|z|^{-k}$ on C_N.

7

Problems in Heat Conduction

We shall now illustrate the use of the theory just developed in solving further boundary value problems in the conduction of heat in solids. We present examples of problems that cannot be fully treated with the more elementary theory used in Chap. 4.[1]

The formal solution of the problem in the next section is followed by a complete mathematical treatment of that problem. The purpose is to illustrate a means of rigorously establishing the solutions of such problems. Since the procedure is lengthy, the reader is advised to use it sparingly in his work on the sets of problems that follow. A clear understanding of the formal method of solution is of primary importance.

---

[1] Alternative methods for solving a few of the problems in this and the following chapter are presented in chap. 7 of the author's "Fourier Series and Boundary Value Problems," 2d ed., 1963.

75 TEMPERATURES IN A BAR WITH ENDS AT FIXED TEMPERATURES

Let $U(x,t)$ denote the temperature at any point in a bar (Fig. 75) with insulated lateral surface and with its ends $x = 0$ and $x = 1$ kept at temperatures zero and F_0, respectively, when the initial temperature is zero throughout.

In Sec. 49 we obtained a formula for $U(x,t)$ in the form of a series of error functions, a series that converges rapidly when t is small. We shall now obtain another series representation of this temperature function. This series will converge rapidly for large t. Let us proceed formally to the solution here, leaving the full justification of our result to the following sections.

We have taken the unit of length as the length of the bar, and we observed earlier that, by a proper choice of the unit of time, we can write $k = 1$ in the heat equation, where k is the diffusivity. The boundary value problem in $U(x,t)$ is then

$$U_t(x,t) = U_{xx}(x,t) \qquad\qquad (0 < x < 1, t > 0),$$

$$U(x,+0) = 0 \qquad\qquad (0 < x < 1),$$

$$U(+0,t) = 0, \qquad U(1 - 0, t) = F_0 \qquad\qquad (t > 0),$$

where F_0 is a constant.

The problem in the transform of $U(x,t)$ is

$$su(x,s) = u_{xx}(x,s) \qquad\qquad (0 < x < 1),$$

$$u(+0,s) = 0, \qquad u(1 - 0, s) = \frac{F_0}{s}.$$

Since this problem in ordinary differential equations has a solution that is continuous at $x = 0$ and $x = 1$, $u(+0,s) = u(0,s)$ and $u(1 - 0, s) = u(1,s)$. The solution is

(1)
$$u(x,s) = F_0 \frac{\sinh x\sqrt{s}}{s \sinh \sqrt{s}},$$

where the symbol $\sqrt{s}$ denotes some branch of the double-valued function $s^{\frac{1}{2}}$. As long as $\sqrt{s}$ represents the same branch, regardless of which one, in both

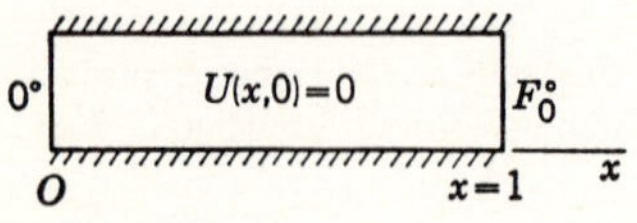

Fig. 75

numerator and denominator, the quotient of hyperbolic sines can be written

(2) $$\frac{\sinh x\sqrt{s}}{\sinh \sqrt{s}} = \frac{x\sqrt{s} + (x\sqrt{s})^3/3! + \cdots}{\sqrt{s} + (\sqrt{s})^3/3! + \cdots} = \frac{x + (x^3 s/3!) + \cdots}{1 + s/3! + \cdots},$$

except at those points where $\sinh \sqrt{s} = 0$, namely,

(3) $$s = 0, \qquad s = -n^2\pi^2 \qquad (n = 1, 2, \ldots).$$

The final member of Eqs. (2) is the quotient of two power series in s that converge for every s. It follows that $u(x,s)$ is an analytic function of s except for isolated singular points (3). It also follows from Eqs. (1) and (2) that

(4) $$\lim_{s \to 0} su(x,s) = F_0 x;$$

therefore when $x \neq 0$, the function $u(x,s)$ has a simple pole at $s = 0$ with residue $F_0 x$. We note that u is real-valued when s is real.

Now let the branch cut of $\sqrt{s}$ be taken as the positive real axis, so that $\sqrt{s}$ is analytic on the negative real axis. Then $u(x,s)$ has the fractional form $p(x,s)/q(s)$, where the functions

$$p(x,s) = \frac{F_0}{s} \sinh x\sqrt{s}, \qquad q(s) = \sinh \sqrt{s}$$

are analytic at $s = -n^2\pi^2$, and

$$q'(-n^2\pi^2) = \frac{1}{2\sqrt{-n^2\pi^2}} \cosh \sqrt{-n^2\pi^2} = \frac{\cos n\pi}{2n\pi i} \neq 0.$$

Hence $s = -n^2\pi^2$ are simple poles, where the residues of $e^{st}u(x,s)$ are

(5) $$\frac{p(x, -n^2\pi^2)}{q'(-n^2\pi^2)} e^{-n^2\pi^2 t} = \frac{F_0 i \sin n\pi x}{-n^2\pi^2}(-1)^n 2n\pi i \exp(-n^2\pi^2 t).$$

At least formally then (Sec. 71) the inverse transform of $u(x,s)$ is the sum of the residues (4) and (5) of $e^{st}u(x,s)$ at the poles (3); that is,

(6) $$U(x,t) = F_0\left[x + \frac{2}{\pi} \sum_{n=1}^{\infty} \frac{(-1)^n}{n} \exp(-n^2\pi^2 t) \sin n\pi x \right].$$

This formal solution can be verified by showing that the function defined by formula (6) satisfies all the conditions of our boundary value problem (see Probs. 5 and 6, Sec. 81). But we shall now see that the theory in the preceding chapter enables us to make the verification in another way that has some advantages.

76   THE SOLUTION ESTABLISHED

We have seen that the function

$$u(x,s) = F_0 \frac{\sinh x\sqrt{s}}{s \sinh \sqrt{s}}$$

is analytic with respect to s in any half plane $\operatorname{Re} s \geq \gamma$ where $\gamma > 0$. To examine its order in that half plane, let us write

$$s = re^{i\theta}, \qquad \sqrt{s} = \sqrt{r}\, e^{i\theta/2} \qquad \left(-\frac{\pi}{2} < \theta < \frac{\pi}{2} \right);$$

then $\operatorname{Re} \sqrt{s} = \sqrt{r} \cos(\theta/2) > \sqrt{r/2} \geq \sqrt{\gamma/2}$. Thus

$$\left| \frac{\sinh x\sqrt{s}}{\sinh \sqrt{s}} \right| = \left| \exp\left[(x-1)\sqrt{s}\right] \left[\frac{1 - \exp(-2x\sqrt{s})}{1 - \exp(-2\sqrt{s})} \right] \right|$$

$$< \frac{1 + \exp(-2x\sqrt{\gamma/2})}{1 - \exp(-2\sqrt{\gamma/2})} \exp\left[-(1-x)\sqrt{r/2} \right] \leq M \exp\left[-(1-x)\sqrt{r/2} \right],$$

where $M = 2/[1 - \exp(-2\sqrt{\gamma/2})]$. Thus when $x_1 < 1$, constants M_m, independent of x over the interval $0 \leq x \leq x_1$, exist such that for each integer $m = 2, 3, \ldots$

$$r^m |u(x,s)| < F_0 M r^{m-1} \exp\left[-(1-x_1)\sqrt{r/2} \right] < M_m \qquad (\operatorname{Re} s \geq \gamma);$$

that is, u satisfies the order condition

(1) $$|u(x,s)| < \frac{M_m}{|s|^m} \qquad (m = 2, 3, \ldots, \operatorname{Re} s \geq \gamma, 0 \leq x \leq x_1).$$

According to Theorem 8, Sec. 70, the inversion integral of u along the line $\operatorname{Re} s = \gamma$, therefore, represents the inverse transform $U(x,t)$ of $u(x,s)$,

(2) $$U(x,t) = L_i^{-1}\{u(x,s)\} \qquad (t \geq 0, 0 \leq x < 1),$$

and U is a real-valued continuous function of t that satisfies the required initial condition

(3) $$U(x,+0) = U(x,0) = 0 \qquad (0 \leq x < 1)$$

and also the condition

(4) $$U_t(x,t) = L_i^{-1}\{su(x,s)\} \qquad (0 \leq x < 1).$$

Now in view of Theorem 9, Sec. 70, U is a continuous function of x and t together when $t \geq 0$ and $0 \leq x \leq x_1$; thus

(5) $$U(+0, t) = U(0,t) = L_i^{-1}\{u(0,s)\} = 0$$

since $u(0,s) = 0$. Likewise, the derivatives

$$u_x(x,s) = F_0 \frac{\cosh x\sqrt{s}}{\sqrt{s}\, \sinh\sqrt{s}}, \qquad u_{xx}(x,s) = su(x,s),$$

are $\mathcal{O}(s^{-m})$ in the half plane, for each integer m, uniformly with respect to x when $0 \leq x \leq x_1 < 1$. This is evident by comparing those functions with u itself. Therefore

$$U_{xx}(x,t) = L_i^{-1}\{u_{xx}(x,s)\} = L_i^{-1}\{su(x,s)\} \qquad (0 < x < 1),$$

and it follows from Eq. (4) that U satisfies the heat equation

(6) $$U_t(x,t) = U_{xx}(x,t) \qquad (0 < x < 1).$$

We have now shown that our function (2) satisfies all conditions of our boundary value problem except the end condition

(7) $$U(1 - 0, t) = F_0 \qquad (t > 0).$$

It is evident that U satisfies the condition

$$U(1,t) = L_i^{-1}\{u(1,s)\} = L_i^{-1}\left\{\frac{F_0}{s}\right\} = F_0 \qquad (t > 0);$$

but this does not assure us that $U(x,t) \to F_0$ as $x \to 1$, which is the condition the temperature function should satisfy.

By writing hyperbolic sines in terms of exponential functions and carrying out one step of a division, we find that

(8) $$\frac{\sinh x\sqrt{s}}{\sinh\sqrt{s}} = \exp[-(1 - x)\sqrt{s}] + \frac{\exp[-(2 - x)\sqrt{s}] - \exp(-x\sqrt{s})}{\exp(\sqrt{s}) - \exp(-\sqrt{s})}.$$

It is the first term on the right that has a weak order with respect to s when $x = 1$. Now we can write

(9) $$u(x,s) = \frac{F_0}{s} \exp[-(1 - x)\sqrt{s}] + g(x,s)$$

where

(10) $$g(x,s) = \frac{F_0}{s} \exp[-(1 + x)\sqrt{s}]\frac{\exp[-2(1 - x)\sqrt{s}] - 1}{1 - \exp(-2\sqrt{s})}.$$

On the half plane $\operatorname{Re} s \geq \gamma$ we can see that

$$|g(x,s)| < \frac{|F_0|}{r} \exp(-\sqrt{r/2})\frac{2}{1 - \exp(-\sqrt{2\gamma})} \qquad (0 \leq x \leq 1)$$

so that g is $\mathcal{O}(s^{-m})$ there for each integer m, uniformly with respect to

x $(0 \leq x \leq 1)$. Consequently, $L_i^{-1}\{g\}$ represents a function $G(x,t)$ that is continuous in x and t and, since $g(1,s) = 0$,

$$G(1 - 0, t) = G(1,t) = L_i^{-1}\{g(1,s)\} = 0 \qquad (t \geq 0).$$

With the aid of the inversion integral evaluated in Sec. 74, we find that

$$(11)\ L_i^{-1}\left\{\frac{F_0}{s}\exp\left[-(1 - x)\sqrt{s}\right]\right\} = F_0\,\text{erfc}\left(\frac{1 - x}{2\sqrt{t}}\right) \qquad (0 \leq x \leq 1, t > 0).$$

It now follows from Eq. (9) that

$$(12) \qquad U(x,t) = F_0\,\text{erfc}\left(\frac{1 - x}{2\sqrt{t}}\right) + G(x,t) \qquad (0 \leq x \leq 1, t > 0),$$

and $U(1 - 0, t) = F_0 + G(1,t) = F_0$ when $t > 0$.

Our function (2) thus satisfies the end condition (7), and it is therefore completely established as a solution of our boundary value problem. Moreover, we have shown that the transform of our temperature function is the function $u(x,s)$ from which we obtained $U(x,t)$. Some interesting properties of U will follow from the order properties of u.

We still have to prove that the series obtained in the last section represents our solution (2).

77 THE SERIES FORM ESTABLISHED

We have seen that the function

$$u(x,s) = F_0\frac{\sinh x\sqrt{s}}{s\sinh\sqrt{s}}$$

is analytic except for the poles $s = 0$ and $s = -n^2\pi^2$ and that its inversion integral converges to a function $U(x,t)$ that is a solution of our boundary value problem. The series representation of $U(x,t)$, given in Sec. 75, is valid provided the integral

$$\int_{C_n} e^{tz}u(x,z)\,dz,$$

taken along a curve C_n of a family of curves $(n = 1, 2, \ldots)$ between the poles, tends to zero as n tends to infinity (Sec. 71).

To select curves C_n so that $|u(x,z)|$ has a suitable order property whenever z is on C_n, we first note that

$$|\sinh\sqrt{z}|^2 = \sinh^2\left(\sqrt{r}\cos\frac{\theta}{2}\right) + \sin^2\left(\sqrt{r}\sin\frac{\theta}{2}\right).$$

Hence if $\sqrt{r}\sin\frac{1}{2}\theta = a_n$, where

(1)
$$a_n = \left(n - \frac{1}{2}\right)\pi \qquad (n = 1, 2, \ldots),$$

then

(2)
$$|\sinh\sqrt{z}|^2 = \sinh^2\left(\sqrt{r}\cos\frac{\theta}{2}\right) + 1 \geq 1.$$

Let us therefore take C_n as the arc of the parabola

(3)
$$r = \frac{a_n{}^2}{\sin^2\theta/2} = \frac{2a_n{}^2}{1 - \cos\theta}$$

that lies to the left of the line $\xi = \gamma$, where γ is any positive constant and $z = \xi + i\eta$ (Fig. 76). Then when z is on C_n

(4)
$$\left|\frac{zu(x,z)}{F_0}\right|^2 = \frac{\sinh^2\left(x\sqrt{r}\cos\theta/2\right) + \sin^2\left(x\sqrt{r}\sin\theta/2\right)}{\sinh^2\left(\sqrt{r}\cos\theta/2\right) + 1} \leq 1,$$

because $0 \leq x \leq 1$ and $\sinh^2 y$ increases when $|y|$ increases. Thus $u(x,z)$ is $O(z^{-1})$ on C_n and, according to Theorem 10, Sec. 73, the series of residues of $e^{zt}u(x,z)$ converges to the inversion integral for all positive values of t. The series (6), Sec. 75, therefore does represent a solution of our problem; that is, our solution can be written

(5)
$$U(x,t) = F_0\left[x + \frac{2}{\pi}\sum_{n=1}^{\infty}\frac{(-1)^n}{n}\exp\left(-n^2\pi^2 t\right)\sin n\pi x\right] \qquad (t > 0).$$

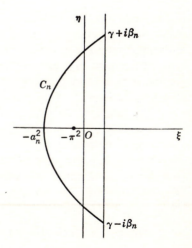

Fig. 76

We have established the sum of series (5) as a solution of our boundary value problem by identifying it with our solution $L_i^{-1}\{u\}$. Owing to the uniform convergence of such series when $t \geq t_0 > 0$ (Prob. 5, Sec. 81) however, we find easily that series (5) represents a function that does satisfy the end conditions $U(1 - 0, t) = F_0$, $U(+0, t) = 0$ and the heat equation. Hence a simpler verification consists of those observations coupled with our brief proof that $L_i^{-1}\{u\}$ satisfies the initial condition (3), Sec. 76, after establishing the series representation of the inversion integral.

We found in Sec. 76 that $L_i^{-1}\{u\} = 0$ when $t = 0$ and $0 \leq x < 1$. Certain order conditions satisfied by u on C_n are actually adequate to show that the series representation (5) of $L_i^{-1}\{u\}$ is valid when $t = 0$ and consequently that

$$(6) \qquad\qquad x = -\frac{2}{\pi} \sum_{n=1}^{\infty} \frac{(-1)^n}{n} \sin n\pi x \qquad\qquad (0 \leq x < 1).$$

For if ϵ is a small positive angle and z is on either part of any of the parabolic paths C_n on which $|\theta| < \pi - \epsilon$, it can be seen from the Eq. (4) that for each fixed x $(x < 1)$ a number M_ϵ independent of n exists such that $|z^k u(x,z)| < M_\epsilon$, for any constant k. This order property, together with the inequality (4) satisfied on the remaining part of C_n, is sufficient to show that the representation (5) is valid when $t = 0$ and $x < 1$. The proof, made by first selecting ϵ small and then n large, is left to the problems. It establishes the Fourier sine series representation (6) of the function x on the interval $0 \leq x < 1$. We shall use a similar procedure in Chap. 9 to establish a generalization of Fourier series representations of arbitrary functions.

78   PROPERTIES OF THE TEMPERATURE FUNCTION

It was shown in Sec. 76 that for any integer m the transform $u(x,s)$ of our temperature function $U(x,t)$ is $\mathcal{O}(s^{-m})$ in a right half plane of the variable s, uniformly for all x in any interval $0 \leq x \leq x_1$ where $x_1 < 1$. The derivatives of u with respect to x also satisfy that order property. As a consequence, our temperature function possesses the following properties, according to Theorems 5, 8, and 9 of the preceding chapter.

The function $U(x,t)$ is a continuous function of both x and t when $t \geq 0$ and $0 \leq x < 1$, and each of its derivatives with respect to x or t has this continuity property.

At any interior point of the bar, the temperature begins to change very slowly at the time $t = 0$, since $U(x,0) = 0$ and

$$(1) \qquad\qquad U_t(x,0) = U_{tt}(x,0) = U_{ttt}(x,0) = \cdots = 0 \qquad\qquad (0 \leq x < 1).$$

Of course, $U(x,t)$ is not identically zero because $u(x,s)$ is not. In fact, from the representation (3), Sec. 49, of $U(x,t)$ as a series of error functions, we can see that

(2) $$\frac{1}{F_0} U(x,t) > 0 \qquad \text{whenever } t > 0 \text{ and } 0 < x < 1.$$

The flux of heat through any section $x = x_0$,

$$\Phi(x_0,t) = -KU_x(x_0,t),$$

where K is thermal conductivity, has the transform

$$\phi(x_0,s) = -KF_0 \frac{\cosh x_0 \sqrt{s}}{\sqrt{s} \sinh \sqrt{s}} \qquad \text{if } 0 \leq x_0 < 1.$$

From the order of ϕ we can see that the flux also changes very slowly from its initial value $\Phi(x_0,0) = 0$, since

(3) $$\Phi_t(x_0,0) = \Phi_{tt}(x_0,0) = \cdots = 0 \qquad \text{when } 0 \leq x_0 < 1.$$

In Sec. 76 we found that

$$U(x,t) = F_0 \operatorname{erfc}\left(\frac{1-x}{2\sqrt{t}}\right) + G(x,t) \qquad (0 \leq x \leq 1, t > 0),$$

where G and its derivatives vanish as $t \to 0$. Therefore, the flux through the right-hand face of the bar can be written

(4) $$\Phi(1,t) = -\frac{F_0 K}{\sqrt{\pi t}} - KG_x(1,t)$$

and

(5) $$\lim_{t \to 0}\left[\Phi(1,t) + \frac{F_0 K}{\sqrt{\pi t}} \right] = 0.$$

That is, the flux at that face is $\mathcal{O}(1/\sqrt{t})$ as $t \to 0$; it becomes infinite as $t \to 0$ like $-F_0 K/\sqrt{\pi t}$.

The total quantity of heat that has passed through a unit area of the face $x = 1$ up to time t can be written

$$Q(1,t) = \int_0^t \Phi(1,\tau) \, d\tau = -\frac{F_0 K}{\sqrt{\pi}} \int_0^t \frac{d\tau}{\sqrt{\tau}} - K \int_0^t G_x(1,\tau) \, d\tau,$$

in view of formula (4). Thus

$$Q(1,t) = -\frac{2F_0 K}{\sqrt{\pi}} \sqrt{t} - K \int_0^t G_x(1,\tau) \, d\tau.$$

Since $G_x(1,t)$ is continuous when $t \geqq 0$,

(6) $$\lim_{t \to 0} Q(1,t) = 0,$$

a condition that would not be satisfied if there were an instantaneous source of heat over the surface $x = 1$ at $t = 0$. Such a source is an idealization of an actual situation in which a large quantity of heat is generated over a surface in a very short time interval, by combustion, for instance.

79 UNIQUENESS OF THE SOLUTION

Our treatment of the boundary value problem is not strictly complete until we have shown that our solution is the only one possible. The physical problem of the temperatures in a bar with prescribed initial temperature and prescribed thermal conditions at the boundary must have just one solution. If we have completely stated the problem as one in mathematics, that problem must also have a unique solution.

The conditions we have imposed on $U(x,t)$, namely,

(1) $$U_t(x,t) = U_{xx}(x,t) \qquad\qquad (0 < x < 1, t > 0),$$

(2) $$U(x,+0) = 0 \qquad\qquad (0 < x < 1),$$

(3) $$U(+0,t) = 0, \qquad U(1 - 0, t) = F_0 \qquad\qquad (t > 0),$$

are not sufficient to ensure just one solution. They do not exclude the possibility of instantaneous sources of heat at the ends of the bar at $t = 0$. The equation of conduction (1) is the statement that heat distributes itself interior to the bar after the time $t = 0$, by conduction. In the derivation of that equation, it is assumed that the functions U, U_t, U_x, and U_{xx} are continuous with respect to the two variables x and t, interior to the solid after conduction begins. We shall therefore require our solution to have those properties of continuity. Physically, the presence of heat sources interior to the bar when $t > 0$ is then prohibited.

Let the required temperature function satisfy the conditions (1), (2), and (3) and the following continuity and order conditions.

1. U is a continuous function of x and t when $t > 0$ and $0 \leqq x \leqq 1$, and U, U_t, and U_x are continuous when $t \geqq 0$ and $0 \leqq x < 1$.

2. Constants α and M exist such that $|U(x,t)| < Me^{\alpha t}$ over the strip $t > 0$, $0 < x < 1$; also for each $x_1 (0 < x_1 < 1)$, a constant M_1 exists such that the functions U_t and U_x are less in absolute value than $M_1 e^{\alpha t}$ over the strip $t > 0$, $0 < x < x_1$.

We could write $\alpha = 0$ here, simply requiring the functions to be bounded over the regions specified; but it is easier to show that our functions are

$\mathcal{O}(e^{\alpha t})$ than to prove they are bounded for all t. Note that U_{xx} satisfies the conditions imposed on U_t because $U_t = U_{xx}$.

We have seen that the function

(4)
$$U(x,t) = L_i^{-1} \left\{ F_0 \frac{\sinh x\sqrt{s}}{s \sinh \sqrt{s}} \right\}$$

satisfies conditions (1), (2). and (3) and that it is continuous, together with U_t and U_x, when $t \geq 0$ and $0 \leq x < 1$. We saw also that

$$U(x,t) = F_0 \operatorname{erfc}\left(\frac{1-x}{2\sqrt{t}}\right) + G(x,t) \qquad (0 \leq x \leq 1, t > 0),$$

where G is continuous and $\mathcal{O}(e^{\alpha t})$ when $t \geq 0$ and $0 \leq x \leq 1$, if $\alpha > 0$, according to the character of $g(x,s)$. The error function here is continuous in x and t when $t > 0$ and $0 \leq x \leq 1$. Hence the function (4) satisfies our continuity requirements 1.

The first of conditions 2 is satisfied because that error function is bounded over the strip $t > 0$, $0 < x < 1$, and G is $\mathcal{O}(e^{\alpha t})$ there. The order condition there on U_t and U_x when $0 < x < x_1$ is satisfied according to the representation (4).

Our function (4) therefore satisfies conditions (1), (2), (3), and the continuity and order conditions. If a second function V does so, then the function

(5)
$$W(x,t) = U(x,t) - V(x,t)$$

also satisfies those continuity and order conditions, and also these homogeneous conditions:

(6)
$$W_t(x,t) = W_{xx}(x,t) \qquad\qquad (0 < x < 1, t > 0),$$

(7)
$$W(x,0) = 0 \qquad\qquad (0 < x < 1),$$

(8)
$$W(0,t) = 0, \qquad W(1,t) = 0 \qquad\qquad (t > 0).$$

In writing conditions (8), we have used the fact that W is continuous when $t > 0$ and $0 \leq x \leq 1$. We shall prove that this problem in W has only the trivial solution $W(x,t) \equiv 0$ in the class of functions that satisfy the continuity and order conditions above.

Since W satisfies those conditions, its transform $w(x,s)$ exists when $\operatorname{Re} s > \alpha$ as well as the transforms $w'(x,s)$ and $w''(x,s)$ of W_x and W_{xx}, where the primes denote derivatives with respect to x. Moreover, those transforms are continuous functions of x and s when $0 \leq x < 1$. In view of condition (7) the transform of W_t is sw. It follows from conditions (6) and (8) that for every s in the right half plane $\operatorname{Re} s > \alpha$

(9)
$$sw(x,s) = w''(x,s) \qquad\qquad (0 < x < 1),$$

(10) $w(0,s) = 0,$ $w(1,s) = 0.$

The continuity of w with respect to x actually extends to the point $x = 1$. To prove it, we write

$$w(x,s) = \int_0^{t_0} e^{-st} W(x,t)\, dt + \int_{t_0}^{\infty} e^{-st} W(x,t)\, dt \qquad (t_0 > 0),$$

where the second integral is a continuous function of x when $0 \le x \le 1$ that vanishes when $x = 1$. Since $|W(x,t)| < M e^{\alpha t}$ and $\mathrm{Re}\, s > \alpha$, then $|e^{-st} W(x,t)| < M$ throughout the strip $t > 0,\ 0 < x < 1$. Thus we can take t_0 small enough to make the absolute value of the first integral arbitrarily small independent of x when $0 < x < 1$. Then for that t_0 the absolute value of the second integral is arbitrarily small when x is sufficiently close to 1. Hence $w(x,s) \to 0$ when $x \to 1$; that is, $w(1 - 0, s) = 0 = w(1,s)$.

We have now shown that for each s in a right half plane, w is a continuous function of $x\ (0 \le x \le 1)$ that satisfies the linear ordinary differential equation (9) with constant coefficients, also that w' is continuous when $0 \le x < 1$, and $w(0,s) = 0$. In the theory of linear differential equations it is shown that, when the value $w'(0,s)$ of the derivative at $x = 0$ is also prescribed, such an equation has one and only one solution satisfying those continuity conditions. In our case that solution is

$$w(x,s) = \frac{w'(0,s)}{\sqrt{s}} \sinh x\sqrt{s} \qquad (0 \le x \le 1).$$

But $w(1,s) = 0$, and since $\sinh \sqrt{s} = 0$, only when $s = -n^2\pi^2\ (n = 0, 1, 2 \ldots)$ and not for all s in a right half plane, then $w'(0,s) = 0$; hence

(11) $w(x,s) = 0.$

Since $w(x,s)$ is the transform of a continuous function W of exponential order and W_t is continuous, our Theorem 7, Sec. 69, on uniqueness of inverse transforms applies to show that $W(x,0) = 0$; that is,

(12) $V(x,t) = U(x,t)$ $(0 \le x < 1).$

The proof that the problem consisting of conditions (1), (2), (3), and the continuity and order conditions has just one solution is now complete.

80   ARBITRARY END TEMPERATURES

Let the temperature of the end $x = 1$ of the bar be a prescribed function $F(t)$ (Fig. 58). The temperature function U then satisfies the heat equation $U_t = U_{xx}$, the initial condition $U(x,+0) = 0$, and the end conditions

$$U(+0,t) = 0, \qquad U(1 - 0,t) = F(t) \qquad (t > 0).$$

As noted in Sec. 49, the solution of the transformed problem is

(1)
$$u(x,s) = f(s)\frac{\sinh x\sqrt{s}}{\sinh \sqrt{s}}.$$

Let $V(x,t)$ denote the temperature function found in the preceding sections when $F(t) = 1$. Then

(2)
$$u(x,s) = sf(s)v(x,s)$$

since
$$v(x,s) = \frac{\sinh x\sqrt{s}}{s \sinh \sqrt{s}}.$$

Now $sv(x,s)$ is the transform of $V_t(x,t)$ when $0 \le x < 1$. In view of the convolution property, it follows from Eq. (2) that

(3)
$$U(x,t) = \int_0^t F(t - \tau)V_t(x,\tau)\, d\tau.$$

It was shown that $V(x,t)$ is represented by a series:

$$V(x,t) = x + \frac{2}{\pi} \sum_{n=1}^{\infty} \frac{(-1)^n}{n} \exp(-n^2\pi^2 t)\sin n\pi x.$$

The series obtained by differentiating this series term by term with respect to t does not converge when $t = 0$; but it was shown that the function $V_t(x,t)$ is continuous when $t \ge 0$ and $0 \le x < 1$ and that

$$V_t(x,0) = 0 \qquad\qquad (0 \le x < 1).$$

The differentiated series simply fails to represent $V_t(x,t)$ at $t = 0$.

To arrive at another form of the temperature function U, we assume F continuous, F' sectionally continuous, and F of exponential order. Then

$$L\{F'(t)\} = sf(s) - F(+0),$$

and
$$u(x,s) = F(+0)v(x,s) + L\{F'(t)\}v(x,s).$$

Consequently we have the formula

(4)
$$U(x,t) = F(+0)V(x,t) + \int_0^t F'(t - \tau)V(x,\tau)\, d\tau.$$

The two formulas (3) and (4) give the temperature $U(x,t)$ in terms of the temperature $V(x,t)$ corresponding to a fixed surface temperature. They are two forms of Duhamel's formula (Sec. 85). The above series for $V(x,t)$ can be substituted into formula (4), and it can be shown that the temperature function can be written

$$(5) \quad U(x,t) = xF(t) + \frac{2F(+0)}{\pi} \sum_{n=1}^{\infty} \frac{(-1)^n}{n} e^{-n^2\pi^2 t} \sin n\pi x$$

$$+ \frac{2}{\pi} \sum_{n=1}^{\infty} \frac{(-1)^n}{n} \sin n\pi x \int_0^t F'(t-\tau)e^{-n^2\pi^2\tau} d\tau.$$

81   SPECIAL END TEMPERATURES

When the end temperature $F(t)$ is a specific function, a convenient formula for $U(x,t)$ may be found directly from the transform $u(x,s)$.

For example, let

$$(1) \qquad\qquad\qquad F(t) = At$$

in the problem of the last section, where A is a constant. Then

$$u(x,s) = A\frac{\sinh x\sqrt{s}}{s^2 \sinh \sqrt{s}},$$

a function with a pole of the second order at $s = 0$. We noted in Sec. 75 that

$$\frac{\sinh x\sqrt{s}}{\sinh \sqrt{s}} = \frac{x + (x^3 s/3!) + (x^5 s^2/5!) + \cdots}{1 + (s/3!) + (s^2/5!) + \cdots}.$$

By carrying out the indicated division here, the first two terms are found to be $x + x(x^2 - 1)s3!$; hence $u(x,s)$ has the following representation in a neighborhood of $s = 0$:

$$u(x,s) = A\left[\frac{x}{s^2} + \frac{x(x^2 - 1)}{3!s} + \sum_{n=0}^{\infty} a_n(x)s^n\right].$$

The residue of $e^{zt}u(x,z)$ at $z = 0$ is therefore (Sec. 72)

$$A\left[xt + \frac{x(x^2 - 1)}{3!}\right].$$

The residue of $e^{zt}u(x,z)$ at the simple pole $z = -n^2\pi^2$ is

$$2A\frac{e^{zt} \sinh x\sqrt{z}}{z\sqrt{z} \cosh \sqrt{z}}\bigg]_{z=-n^2\pi^2} = \frac{2A(-1)^{n-1}}{\pi^3 n^3} \sin(n\pi x) \exp(-n^2\pi^2 t).$$

Consequently the formula for the temperatures can be written

$$(2) \qquad U(x,t) = A\left[\frac{x^3 - x}{6} + xt + \frac{2}{\pi^3} \sum_{n=1}^{\infty} \frac{(-1)^{n-1}}{n^3} \exp(-n^2\pi^2 t) \sin n\pi x\right].$$

That function can be verified completely as a solution of the boundary value problem by just the same procedure that was used in Secs. 76 and 77. But the procedure can be simplified in this case in view of the fact that $u(x,s)$ is $\mathcal{O}(s^{-2})$ in a right half plane and on the parabolas C_n, uniformly with respect to x when $0 \leq x \leq 1$. Consequently U is a continuous function of its two variables for all x and t $(0 \leq x \leq 1, t \geq 0)$; also the series representation (2) is valid at $t = 0$. Since $U(x,0) = 0$, it follows from Eq. (2) that

$$x - x^3 = \frac{12}{\pi^3} \sum_{n=1}^{\infty} \frac{(-1)^{n-1}}{n^3} \sin n\pi x \qquad (0 \leq x \leq 1).$$

which is the Fourier sine series expansion of the function $x - x^3$ on the interval $0 \leq x \leq 1$.

PROBLEMS

1. A bar with its lateral surface insulated is initially at uniform temperature A (Fig. 77). Its ends $x = 0$ and $x = l$ are then kept at constant temperatures B and C, respectively. Show formally that its temperature distribution can be written

$$U(x,t) = B + \frac{C - B}{l} x + \frac{2}{\pi} \sum_{n=1}^{\infty} \frac{A - B + (-1)^n(C - A)}{n} \sin \frac{n\pi x}{l} \exp\left(-\frac{n^2\pi^2 kt}{l^2}\right).$$

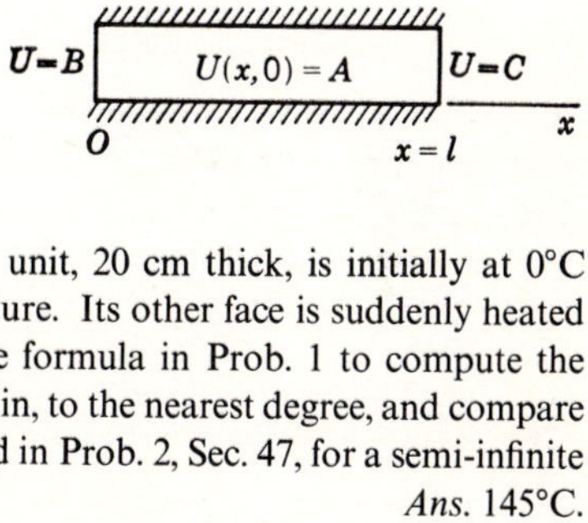

$U=B$ | $U(x,0) = A$ | $U=C$

Fig. 77 $\quad O \qquad x = l$

2. A slab of iron with diffusivity $k = 0.15$ cgs unit, 20 cm thick, is initially at $0°C$ throughout, and one face is kept at that temperature. Its other face is suddenly heated to $500°C$ and kept at that temperature. Use the formula in Prob. 1 to compute the temperature of the midsection of the slab after 5 min, to the nearest degree, and compare the result with a corresponding temperature found in Prob. 2, Sec. 47, for a semi-infinite solid of the same material. *Ans.* $145°C$.

3. For fixed values of A, B, C, and k, note that the temperature $U(x,t)$ in Prob. 1 is determined by the two numbers x/l and t/l^2. For two slabs with those same fixed values and widths l_1 and l_2, show why the temperature of a section $x = x_1$ in the first at time t_1 is the same as the temperature of the corresponding section $x = x_2(x_2/l_2 = x_1/l_1)$ of the second at time t_2 if

$$t_2 = t_1 \frac{l_2^2}{l_1^2} = t_1 \frac{x_2^2}{x_1^2}.$$

Thus the slowness of heating varies as the square of the width.

4. When $k = l = 1$ for the bar in Prob. 1, let $U_0(x,t)$ denote the temperatures in case $C = 0$ and $W(x,t)$ the temperatures when C is replaced by $F(t)$, so that $W(1,t) = F(t)$,

$W(0,t) = B$, and $W(x,0) = A$. If $U(x,t)$ is the function (5), Sec. 80, so that $U_t = U_{xx}$, $U(x,0) = U(0,t) = 0$ and $U(1,t) = F(t)$, show by superposition that

$$W(x,t) = U_0(x,t) + U(x,t).$$

5.   When $t \geq t_0 > 0$ and $0 \leq x \leq 1$, prove that the series (6), Sec. 75, and the series obtained by differentiating that series once or twice with respect to x or once with respect to t are uniformly convergent with respect to x and t. The series then represent continuous functions of x and t, and termwise differentiation is valid, since the terms themselves are continuous functions. As a consequence, show that the function $U(x,t)$ defined by the series satisfies the heat equation $U_t = U_{xx}$ when $0 < x < 1$ and $t > 0$, and end conditions $U(+0,t) = 0$, $U(1 - 0, t) = F_0$, when $t > 0$.

6.   According to *Abel's test for uniform convergence*,[1] a series $\Sigma_{n=1}^{\infty} A_n T_n(t)$ converges uniformly with respect to t if the series $\Sigma_{n=1}^{\infty} A_n$ converges, and if functions $T_n(t)$ are bounded uniformly with respect to t and n and such that $T_{n+1}(t) \leq T_n(t)$. Given that the Fourier series representation (6), Sec. 77, is valid, use Abel's test when $0 \leq t \leq t_0$ to show that the function defined by Eq. (6), Sec. 75, is continuous with respect to t when $t \geq 0$ and that it satisfies the initial condition $U(x,+0) = 0$ for each fixed x such that $0 < x < 1$. This, together with the result in Prob. 5, again establishes the solution (6), Sec. 75.

7.   Give details of the proof outlined in Sec. 77 of the Fourier sine series representation (6) of the function $x(0 \leq x < 1)$. *Suggestion*: Show first that when z is on the parabolic arc C_n and $|\theta| < \pi - \epsilon$, then

$$\left| \frac{zu(x,z)}{F_0} \right| \leq \frac{\cosh{(x\sqrt{r}\cos\theta/2)}}{\cosh{(\sqrt{r}\cos\theta/2)}} < \frac{2\exp{[-(1-x)\sqrt{r}\sin\epsilon/2]}}{1 + \exp{(-2\sqrt{r}\sin\epsilon/2)}}.$$

8.   As in Sec. 80 let $V(x,t)$ denote temperatures in the slab with faces $x = 0$ and $x = 1$ when $V(x,0) = V(0,t) = 0$, $V(1,t) = 1$, and $k = 1$, and write $V(x,t) = 0$ when $t < 0$. Let $U(x,t)$ be the temperatures under the same conditions except that the face $x = 1$ is kept at temperature A from $t = 0$ to $t = t_0$ and thereafter at temperature zero; thus

$$U(1,t) = A - AS_0(t - t_0) \qquad\qquad (t > 0).$$

(*a*) Show that

$$U(x,t) = A[V(x,t) - V(x, t - t_0)].$$

(*b*) Note that the total quantity of heat conducted across a unit area of a section $x = x_0$, $Q = -K\int_0^{\infty} U_x(x_0,t)\,dt$, is the value of $L\{U_x(x_0,t)\}$ when $s = 0$ if that transform exists when $s = 0$. Thus for the above slab with temperatures U show formally that

$$Q = -AKt_0.$$

9.   Obtain the solution of the temperature problem

$$U_t(x,t) = U_{xx}(x,t) \qquad\qquad (0 < x < 1, t > 0),$$

$$U(x,0) = 1, \qquad U(0,t) = U(1,t) = 0,$$

[1] For a proof see the author's "Fourier Series and Boundary Value Problems," 2d ed., chap. 10, 1963, which also treats uniqueness of solutions.

in the form

$$U(x,t) = \frac{4}{\pi} \sum_{n=1}^{\infty} \frac{\sin (2n - 1)\pi x}{2n - 1} \exp[-(2n - 1)^2\pi^2 t].$$

10. Derive the formula

$$U(x,t) = 1 - \frac{4}{\pi} \sum_{n=1}^{\infty} \frac{(-1)^{n-1}}{2n - 1} \cos \frac{(2n - 1)\pi x}{2l} \exp\left[-\frac{(2n - 1)^2\pi^2 t}{4l^2}\right]$$

for the temperatures in a wall with its face $x = 0$ insulated and its face $x = l$ kept at temperature $U = 1$, if the initial temperature is zero (Fig. 78) and if $k = 1$.

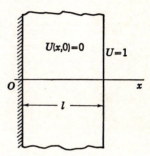

Fig. 78

11. Establish the formula in Prob. 10 as a solution of the boundary value problem.
12. Find U in Prob. 10 as a series of error functions.
  Ans. $U = 1 - U_0(x,t)$, where U_0 is the function (4), Sec. 48, when $u_0 = k = 1$.
13. Obtain the solution of the problem in Sec. 48 in the form

$$U(x,t) = \frac{4u_0}{\pi} \sum_{n=1}^{\infty} \frac{(-1)^{n-1}}{2n - 1} \cos \frac{(2n - 1)\pi x}{2l} \exp\left[-\frac{(2n - 1)^2\pi^2 kt}{4l^2}\right].$$

14. Let the temperature of the face $x = l$ of the wall in Prob. 10 be $F(t)$, where F is continuous, F' is sectionally continuous, and $F(0) = 0$. Derive the temperature formula

$$U(x,t) = F(t) - \frac{4}{\pi} \sum_{n=1}^{\infty} \frac{(-1)^{n-1}}{2n - 1} \cos \frac{(2n - 1)\pi x}{2l} G_n(x,t),$$

where

$$G_n(x,t) = \int_0^t F'(t - r) \exp\left[-\frac{(2n - 1)^2\pi^2 r}{4l^2}\right] dr.$$

15. The face $x = 0$ of a slab is insulated. The temperature of the face $x = \pi$ is $U = t(t \geq 0)$. If the slab is initially at temperature zero throughout and $k = 1$.
  (*a*) derive the formula

$$U(x,t) = t - \frac{\pi^2 - x^2}{2}$$

$$- \frac{16}{\pi} \sum_{n=1}^{\infty} (-1)^n \frac{\cos (n - \tfrac{1}{2})x}{(2n - 1)^3} \exp\left[-(2n - 1)^2\frac{t}{4}\right] \qquad (0 \leq x \leq \pi, t \geq 0).$$

  (*b*) Verify this solution of the problem.

16.   Solve the boundary value problem

$$cU_t(x,t) = KU_{xx}(x,t) + R(t), \qquad U_x(0,t) = U(1,t) = U(x,0) = 0,$$

for the temperatures in an internally heated bar (cf. Prob. 11, Sec. 51) in the form

$$U(x,t) = \int_0^t R(t - \tau)G(x,\tau)\,d\tau,$$

$$G(x,t) = \frac{4}{\pi c}\sum_{n=1}^{\infty}\frac{(-1)^{n+1}}{2n-1}\cos\frac{(2n-1)\pi x}{2}\exp\left[-\frac{(2n-1)^2\pi^2 kt}{4}\right]$$

where $k = K/c$. Note that $G(x,t)$ is formally the temperature distribution when $R(t) = \delta(t)$.

17.   If heat is introduced through the face $x = 1$ of a slab at a uniform rate A per unit area while the face $x = 0$ is kept at the initial temperature zero of the slab, the temperature function U satisfies the conditions

$$U_t(x,t) = kU_{xx}(x,t) \qquad\qquad (0 < x < 1, t > 0),$$

$$U(x,0) = U(0,t) = 0, \qquad KU_x(1,t) = A.$$

Write $m = n - \frac{1}{2}$ and derive the solution

$$U(x,t) = \frac{A}{K}\left[x + \frac{2}{\pi^2}\sum_{n=1}^{\infty}\frac{(-1)^n}{m^2}\sin(m\pi x)\exp(-m^2\pi^2 kt)\right].$$

18.   At the face $x = 0$ of a wall, heat transfer takes place into a medium at temperature zero according to the linear law of surface heat transfer, so that

$$U_x(0,t) = hU(0,t) \qquad\qquad (h > 0).$$

Units are chosen so that $k = 1$, and the wall has unit thickness. If the other conditions are those indicated in Fig. 79, set up the boundary value problem for temperatures $U(x,t)$ and show that

$$u(x,s) = \frac{1}{s}\frac{h\sinh x\sqrt{s} + \sqrt{s}\cosh x\sqrt{s}}{h\sinh\sqrt{s} + \sqrt{s}\cosh\sqrt{s}}.$$

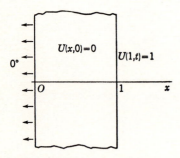

Fig. 79

Obtain formally the solution

$$U(x,t) = \frac{hx+1}{h+1} - 4\sum_{n=1}^{\infty}\frac{\sin\alpha_n(1-x)}{2\alpha_n - \sin 2\alpha_n}\exp(-\alpha_n^2 t),$$

where $\alpha_1, \alpha_2, \ldots$ are the positive roots of the equation

$$\tan \alpha = -\frac{\alpha}{h}.$$

Show how those roots can be approximated graphically when the value of h is known, and note that α_n is only slightly greater than $(n - \frac{1}{2})\pi$ when n is large. To show that $s = 0$ and $s = -\alpha_n^2$ are the only singular points of $u(x,s)$, write $p(z) = h \sinh z + z \cosh z$ and $z = \lambda + i\mu$ and prove that $p(\sqrt{s}) = 0$ only if $s = 0$ or s is real and negative by showing that

$$2|p(z)| \geq |e^\lambda \sqrt{(h + \lambda)^2 + \mu^2} - e^{-\lambda} \sqrt{(h - \lambda)^2 + \mu^2}| > 0$$

if $\lambda > 0$ or if $\lambda < 0$. Hence $\lambda = 0$ and $z = i\mu$ when $p(z) = 0$.

82 ARBITRARY INITIAL TEMPERATURES

Let the initial temperature distribution in a bar or slab be a prescribed function $g(x)$ of the distance from one face. When the lateral surface of the bar is insulated and the ends are kept at temperature zero (Fig. 80), units can be selected so that the boundary value problem for the temperature in the bar becomes

$$U_t(x,t) = U_{xx}(x,t) \qquad\qquad (0 < x < 1, t > 0),$$

$$U(x,+0) = g(x) \qquad\qquad (0 < x < 1),$$

$$U(+0,t) = U(1 - 0, t) = 0 \qquad\qquad (t > 0).$$

The problem in the transform of $U(x,t)$,

(1) $$u''(x,s) - su(x,s) = -g(x) \qquad\qquad (0 < x < 1),$$

(2) $$u(0,s) = u(1,s) = 0,$$

can be solved by any one of several methods, including a Laplace transformation with respect to x, or the process of using a Green's function to be described in Chap. 9. Regardless of the method used, the solution can be written in terms of Green's function R for the problem, in the form (see Probs. 9 to 11, Sec. 83)

(3) $$u(x,s) = -\int_0^1 R(x,\xi,s)g(\xi)\,d\xi,$$

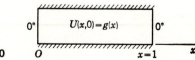

Fig. 80

where R is this symmetric function of x and ξ:

$$(4) \qquad R(x,\xi,s) = -\frac{\sinh[(1-x)\sqrt{s}]\sinh\xi\sqrt{s}}{\sqrt{s}\sinh\sqrt{s}} \qquad (0 \leqq \xi \leqq x),$$

$$= -\frac{\sinh[(1-\xi)\sqrt{s}]\sinh x\sqrt{s}}{\sqrt{s}\sinh\sqrt{s}} \qquad (x \leqq \xi \leqq 1).$$

By writing Maclaurin's series in powers of $\sqrt{s}$ for the hyperbolic sines and properly defining $R(x,\xi,0)$, we see that, regardless of which branch of $s^{\frac{1}{2}}$ is represented by $\sqrt{s}$, R is an analytic function of s except for the singular points

$$(5) \qquad\qquad\qquad s = -n^2\pi^2 \qquad\qquad (n = 1, 2, \ldots).$$

If g is sectionally continuous, the order of integration with respect to ξ in Eq. (3) and differentiation with respect to s can be interchanged, and hence $u(x,s)$ is analytic except for the singular points (5). By taking the branch cut of $\sqrt{s}$ along the positive real axis and selecting $q(s)$ as $\sinh\sqrt{s}$ in formula (7), Sec. 72, we find that the points (5) are simple poles of $u(x,s)$, where the residues of $e^{st}u(x,s)$ are

$$-\frac{\exp(-n^2\pi^2t)}{\frac{1}{2}\cos n\pi}\left[\int_0^x g(\xi)\sin n\pi(1-x)\sin n\pi\xi\, d\xi\right.$$

$$\left.+\int_x^1 g(\xi)\sin n\pi x\sin n\pi(1-\xi)\, d\xi\right].$$

Since $\sin n\pi(1-x) = -\cos n\pi\sin n\pi x$, the residues can be written

$$2\exp(-n^2\pi^2t)\sin n\pi x\int_0^1 g(\xi)\sin n\pi\xi\, d\xi.$$

Formally, therefore, the solution of our problem is

$$(6) \qquad U(x,t) = 2\sum_{n=1}^{\infty}\exp(-n^2\pi^2t)\sin n\pi x\int_0^1 g(\xi)\sin n\pi\xi\, d\xi.$$

When $t = 0$, the series here is the Fourier sine series for the function $g(x)$, on the interval $0 < x < 1$. In fact, the boundary value problem in $U(x,t)$ here is especially well adapted to the classical method of solution by using separation of variables and Fourier series, a method we shall discuss in Chap. 9.

VERIFICATION

Solution (6) can be established by the procedure used in Secs. 76 and 77. Since some variations are needed, we outline the method here. Over any half plane $\mathrm{Re}\, s \geq \gamma > 0$ we find that $|R(x,\xi,s)| < ME(x,\xi,r)/\sqrt{r}$, where $E(x,\xi,r) = \exp(-|x - \xi|\sqrt{r/2})$, $r = |s|$, and M is a constant. Since g is bounded and $\int_0^x E \, d\xi$ and $\int_x^1 E \, d\xi$ are $\mathcal{O}(r^{-\frac{1}{2}})$, it follows from Eq. (3) that u is $\mathcal{O}(r^{-1})$. This is not sufficient to show that the inversion integral applies to u.

For the sake of brevity we assume g'' continuous and g''' sectionally continuous ($0 \leq x \leq 1$). We integrate both integrals by parts in the expression

$$u(x,s) = -\int_0^x R(x,\xi,s)g(\xi)\, d\xi - \int_x^1 R(x,\xi,s)g(\xi)\, d\xi,$$

and introduce an integral P of R with respect to ξ, where

(7)
$$P(x,\xi,s) = -\frac{\sinh[(1-x)\sqrt{s}]\cosh\xi\sqrt{s}}{s\sinh\sqrt{s}} \qquad (0 \leq \xi < x),$$

$$= \frac{\cosh[(1-\xi)\sqrt{s}]\sinh x\sqrt{s}}{s\sinh\sqrt{s}} \qquad (x < \xi \leq 1),$$

and thus isolate the term in u of weak order by writing

(8)
$$u = \frac{g(x)}{s} + g(0)P(x,0,s) - g(1)P(x,1,s) + \int_0^1 Pg'(\xi)\, d\xi.$$

The integral in Eq. (8) is $\mathcal{O}(r^{-3/2})$ in the half plane $\mathrm{Re}\, s \geq \gamma$. From Sec. 76 we find that the inversion integrals of $P(x,0,s)$ and $P(x,1,s)$ exist and satisfy certain boundary conditions. Since $L_i^{-1}\{g(x)/s\} = g(x)$, we can conclude that $L_i^{-1}\{u\}$ represents a function $U(x,t)$ that satisfies the conditions $U(x, +0) = g(x)$ and $U(+0, t) = U(1 - 0, t) = 0$.

From Eq. (8) we can show that

$$u' = g(0)P_x(x,0,s) - g(1)P_x(x,1,s) + \int_0^1 g'(\xi)P_x(x,\xi,s)\, d\xi$$

and, by an integration by parts here, that $u'(x,s)$ is $\mathcal{O}(r^{-3/2})$ when $0 < x < 1$. That $u''(x,s)$ also has this same order follows from the fact that $u'' = su - g$ and integration by parts twice in Eq. (8). Thus $U_{xx} = L_i^{-1}\{u''\}$ and, since $s(u - g/s)$ is $\mathcal{O}(r^{-3/2})$, then

$$L_i^{-1}\left\{s\left(u - \frac{g}{s}\right)\right\} = [U(x,t) - g(x)]_t = U_t(x,t) = U_{xx}(x,t) \qquad (0 < x < 1).$$

The function

(9)
$$U(x,t) = L_i^{-1}\{u(x,s)\} \qquad (\gamma > 0)$$

is therefore established as a solution of our problem.

From Eqs. (3) and (4) it can be shown that $u(x,s)$ is $O(r^{-\frac{1}{2}})$ when s is on the parabolas $\sqrt{r}\sin \theta/2 = (n - \frac{1}{2})\pi$. Therefore the series (6) represents the solution (9) when $t > 0$.

In addition to this, we can see from Eq. (8) that the inversion integral of the function $u(x,s) - g(x)/s$ converges to zero when $t = 0$ and that it is represented by its series of residues when $t = 0$, provided $0 < x < 1$. It follows that

$$(10) \qquad 2 \sum_{n=1}^{\infty} \sin n\pi x \int_0^1 g(\xi) \sin n\pi\xi \, d\xi - g(x) = 0 \qquad (0 < x < 1),$$

which is the Fourier series expansion mentioned above.

The results hold true if g or its derivatives are sectionally continuous instead of continuous. The proof is longer, since it involves the writing of each integral as the sum of integrals over intervals on which the functions are continuous.

83 TEMPERATURES IN A CYLINDER

Let us derive formally the temperature function $U(r,t)$ for a solid circular cylinder of infinite length whose initial temperature is zero and whose surface is kept at unit temperature (Fig. 81). Units of time and length can be chosen so that the boundary value problem becomes

$$U_t(r,t) = U_{rr}(r,t) + \frac{1}{r}U_r(r,t) \qquad (0 \leqq r < 1, t > 0),$$

$$(1) \qquad\qquad U(r,0) = 0 \qquad\qquad (0 \leqq r < 1)$$

$$U(1,t) = 1 \qquad\qquad (t > 0),$$

where the cylindrical coordinate r is the distance from the axis of the cylinder.

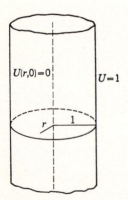

$U(r,0)=0$ $U=1$

Fig. 81

The function U is to be continuous interior to the cylinder, at $r = 0$ in particular. We can assume that U is bounded over the domain $0 < r < 1$, $t > 0$. Then the formal problem in the transform $u(r,s)$ is

$$(2) \qquad\qquad su = \frac{d^2u}{dr^2} + \frac{1}{r}\frac{du}{dr} \qquad (0 \leqq r < 1, \operatorname{Re} s > 0),$$

$$(3) \qquad\qquad u(1,s) = \frac{1}{s}, \qquad u(r,s) \text{ continuous} \qquad (0 \leqq r \leqq 1, \operatorname{Re} s > 0).$$

Let μ denote a complex constant. The substitutions $t = \mu z$ and $y(z) = Y(\mu z)$ in Bessel's equation (7), Sec. 21, leads to the form

$$(4) \qquad z^2 y''(z) + zy'(z) + (\mu^2 z^2 - n^2)y(z) = 0 \qquad (n = 0, 1, 2, \ldots).$$

The solution of Bessel's equation (4), which is continuous at $z = 0$ as well as elsewhere, is $y = CJ_n(\mu z)$, where C is any constant. Equation (2) is a special case of that equation in which $n = 0$, $\mu^2 = -s$, and $z = r$, so its solution that satisfies our continuity requirement is

$$(5) \qquad\qquad u(r,s) = CJ_0(ir\sqrt{s})$$

where (Sec. 21) for each branch of $\sqrt{s}$

$$(6) \qquad\qquad J_0(ir\sqrt{s}) = \sum_{k=0}^{\infty} \frac{1}{(k!)^2}\left(\frac{r}{2}\right)^{2k} s^k.$$

Since this power series in s converges for all complex s, it represents an entire function of s for each fixed value of r.

The condition $u(1,s) = 1/s$ determines C, so that

$$(7) \qquad\qquad u(r,s) = \frac{J_0(ir\sqrt{s})}{sJ_0(i\sqrt{s})}.$$

Now the zeros of Bessel's function $J_0(z)$ are all real and simple; they consist of an infinite sequence of positive numbers $\alpha_n (n = 1, 2, \ldots)$ together with their negatives $-\alpha_n$, such that $\alpha_n \to \infty$ as $n \to \infty$.[1] Their values are tabulated; in particular, to four significant figures.

$$\alpha_1 = 2.405, \qquad \alpha_2 = 5.520, \qquad \alpha_3 = 8.654, \qquad \alpha_4 = 11.79.$$

Thus $J_0(\pm\alpha_n) = 0$ and $J_0'(\pm\alpha_n) \neq 0$. In view of formula (7), the singular points of $u(r,s)$ in the plane of the complex variable s consist of the simple poles

$$(8) \qquad\qquad s = 0, \qquad s = -\alpha_n^2 \qquad (n = 1, 2, \ldots),$$

---

[1] For such properties of Bessel functions see, for instance, chap. 8 of the author's "Fourier Series and Boundary Value Problems," 2d ed., 1963. The function $J_0(iz)$ is also written as $I_0(z)$.

since $s = -\alpha_n^2$ whenever $i\sqrt{s} = \pm\alpha_n$, for a branch of $\sqrt{s}$ that is analytic over the negative real axis so that $\sqrt{-\alpha_n^2}$ is defined, say the branch $\sqrt{s} = \sqrt{|s|}e^{i\theta/2}$ where $0 < \theta < 2\pi$.

The residue of $u(r,s)e^{st}$ at $s = 0$ is 1, and at $s = -\alpha_n^2$ it is

$$\frac{e^{st}}{s}\frac{J_0(ir\sqrt{s})}{d\,J_0(i\sqrt{s})/ds}\Bigg]_{s=-\alpha_n^2} = \frac{2}{-\alpha_n}\frac{J_0(-\alpha_n r)}{J_0'(-\alpha_n)}\exp(-\alpha_n^2 t),$$

where $J_0'(z) = dJ_0(z)/dz = -J_1(z) = J_1(-z)$. Formally, then,

$$(9) \qquad U(r,t) = 1 - 2\sum_{n=1}^{\infty}\frac{J_0(\alpha_n r)}{\alpha_n J_1(\alpha_n)}\exp(-\alpha_n^2 t).$$

To write this formula in terms of standard units of length and time, centimeters and seconds, for example, let ρ denote the radial distance and τ the time in such units. If the radius of the cylinder is ρ_0 and the thermal diffusivity of the material is k, then to transform the heat equation in problem (1) into $U_\tau = k(U_{\rho\rho} + U_\rho/\rho)$ $(0 \le \rho < \rho_0)$, we put

$$r = \frac{\rho}{\rho_0}, \qquad t = \frac{k\tau}{\rho_0^2},$$

where r and t are the variables used in formula (9). Also write $V(\rho,\tau) = AU(r,t)$ so that the constant surface temperature is arbitrary:

$$V(\rho_0,\tau) = A.$$

Our temperature formula then takes the form

$$(10) \qquad V(\rho,\tau) = A\left[1 - 2\sum_{n=1}^{\infty}\frac{J_0(\alpha_n\rho/\rho_0)}{\alpha_n J_1(\alpha_n)}\exp\left(-\frac{\alpha_n^2 k\tau}{\rho_0^2}\right)\right].$$

PROBLEMS

1. The initial temperature of a slab is $U(x,0) = Ax$. If the faces $x = 0$ and $x = l$ are kept at temperature zero, derive the temperature formula

$$U(x,t) = \frac{2Al}{\pi}\sum_{n=1}^{\infty}\frac{(-1)^{n-1}}{n}\exp\left(-\frac{n^2\pi^2 kt}{l^2}\right)\sin\frac{n\pi x}{l}.$$

2. Derive the following formula for the temperature function in Prob. 1:

$$U(x,t) = Ax - Al\sum_{n=0}^{\infty}\left\{\mathrm{erf}\left[\frac{(2n+1)l + x}{2\sqrt{kt}}\right] - \mathrm{erf}\left[\frac{(2n+1)l - x}{2\sqrt{kt}}\right]\right\}.$$

3. If the slab in Prob. 1 is 20 cm thick and is made of iron for which $k = 0.15$ cgs unit, and if the initial temperature varies uniformly through the slab from 0 to 100°C, find to the nearest degree the temperature at the center after the faces have been kept at 0°C (a) for 1 min, (b) for 100 min.   Ans. (a) 48°C; (b) 0°C.

4. The faces $x = 0$ and $x = l$ of a slab are insulated (Fig. 82) and the initial temperature distribution is prescribed: $U(x,0) = g(x)$. Find the transform of the temperature function $U(x,t)$ in the form

$$u(x,s) = -\frac{1}{k}\int_0^l g(\xi)R(x,\xi,s)\,d\xi,$$

where

$$R(x,\xi,s) = -\frac{\cosh x\sqrt{s/k}\cosh (l - \xi)\sqrt{s/k}}{\sqrt{s/k}\sinh l\sqrt{s/k}}$$

when $x \leq \xi$ and $R(x,\xi,s) = R(\xi,x,s)$. Then obtain the formula

$$U(x,t) = \frac{1}{2}a_0 + \sum_{n=1}^{\infty} a_n \cos\frac{n\pi x}{l}\exp\left(-\frac{n^2\pi^2 kt}{l^2}\right)$$

where

$$a_n = \frac{2}{l}\int_0^l g(x)\cos\frac{n\pi x}{l}\,dx \qquad (n = 0, 1, 2, \ldots).$$

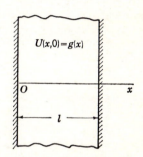

$U(x,0) = g(x)$

O

x

l

Fig. 82

5. In Prob. 4, if $g(x) = A$ when $0 < x < l/2$ and $g(x) = 0$ when $l/2 < x < l$, show that

$$U(x,t) = \frac{A}{2} + \frac{2A}{\pi}\sum_{n=1}^{\infty}\frac{(-1)^{n-1}}{2n-1}\cos\frac{(2n-1)\pi x}{l}\exp\left[-\frac{(2n-1)^2\pi^2 kt}{l^2}\right].$$

6. The face $x = 0$ of a slab is kept at temperature zero while the face $x = 1$ is insulated. If $k = 1$ and the initial temperature is $U(x,0) = x$, derive the temperature formula

$$U(x,t) = \frac{8}{\pi^2}\sum_{n=1}^{\infty}\frac{(-1)^{n+1}}{(2n-1)^2}\sin\frac{(2n-1)\pi x}{2}\exp\left[-\frac{(2n-1)^2\pi^2 t}{4}\right].$$

7. The initial temperature of a cylinder of infinite length is zero. If the surface $r = 1$ is kept at temperature A from $t = 0$ to $t = t_0$ and at temperature zero thereafter, and if $k = 1$, derive the following formula for the temperatures in the cylinder:

$$W(r,t) = A[U(r,t) - U(r, t - t_0)],$$

where $U(r,t)$ is the function defined by formula (9), Sec. 83, when $t \geq 0$ and $U(r,t) = 0$ when $t < 0$.

8. The flux of heat into a cylinder of infinite length through its surface $r = 1$ is a constant, so that $U_r(1,t) = A$. If $k = 1$ and the initial temperature is zero, derive the temperature formula

$$U(r,t) = \frac{A}{4}\left[2r^2 - 1 + 8t - 8 \sum_{n=1}^{\infty} \frac{J_0(\beta_n r)}{\beta_n^2 J_0(\beta_n)} \exp(-\beta_n^2 t) \right],$$

where $\beta_1, \beta_2, \ldots$ are the positive roots of the equation $J_1(\beta) = 0$, given that the equation has only real roots and that $J_1'(\beta_n) \neq 0$. Note that according to Bessel's equation $-J_0''(x) = J_0(x) - J_1(x)/x$, and since $-J_0''(x) = J_1'(x)$ it follows that $J_1'(\beta_n) = J_0(\beta_n)$.

9. Use symbolic transforms, as in Sec. 13, to indicate formally that the solution of the problem

$$\frac{d^2 R}{dx^2} - sR = \delta(x - \xi), \qquad R(0,\xi,s) = R(1,\xi,s) = 0,$$

where $0 < \xi < 1$, is Green's function (4), Sec. 82.

10. Multiply the members of all equations in Prob. 9 by $-g(\xi)$ and integrate from $\xi = 0$ to $\xi = 1$. Formally interchange the operations of differentiation with respect to x and integration with respect to ξ, as might be justified if $\delta(x - \xi)$ in Prob. 9 were replaced by $I(h, x - \xi)$, to indicate that the function

$$u(x,s) = -\int_0^1 R(x,\xi,s)g(\xi)\, d\xi$$

is the solution of the problem (Sec. 82)

$$\frac{d^2 u}{dx^2} - su = -g(x), \qquad u(0,s) = u(1,s) = 0.$$

Write R in the form (4), Sec. 82, and verify the solution.

11. (*a*) Note that Green's function R defined by Eqs. (4), Sec. 82, is a continuous function of x ($0 \leq x \leq 1$) when $0 < \xi < 1$, but show that *its derivative* R_x *has a unit jump at* $x = \xi$:

$$R_x(\xi + 0, \xi,s) - R_x(\xi - 0, \xi,s) = 1 \qquad (0 < \xi < 1).$$

(*b*) Integrate the members of the equation

$$\frac{d^2 R}{dx^2} - sR = \delta(x - \xi) \qquad (0 < \xi < 1)$$

with respect to x from $\xi - \epsilon$ to $\xi + \epsilon$ and let ϵ tend to zero to indicate formally that R_x has a unit jump with respect to x at $x = \xi$.

12. Show that the derivative $R_x(x,\xi,s)$ of Green's function R in Prob. 4 has a unit jump with respect to x at the point $x = \xi$ ($0 < \xi < l$); also show that R itself is a continuous function of x ($0 \leq x \leq l$) that satisfies these conditions:

$$k\frac{d^2 R}{dx^2} - sR = 0 \qquad (x \neq \zeta), \qquad R_x(0,\zeta,s) = R_x(l,\zeta,s) = 0.$$

84 EVAPORATION FROM A THICK SLAB

We now illustrate a type of problem that is especially well adapted to treatment by the Laplace transformation in contrast with other methods.

Let $C(x,\tau)$ denote the concentration of moisture at time τ in a porous semi-infinite solid ($x \geq 0$) with a uniform initial concentration C_0 (Fig. 83). Evaporation of moisture takes place at the face $x = 0$ into a dry adjacent medium according to the linear law that the outward flux is $E[C(0,\tau) - 0]$, where the positive constant E is a coefficient of evaporation. A boundedness condition on C will take the place of a condition of diffusion at a right-hand boundary. The condition that $C(x,\tau)$ is $\mathcal{O}(e^{\alpha\tau})$ uniformly with respect to x, whenever $\alpha > 0$, will be sufficient. Then if K is the coefficient of diffusion of moisture in the solid, $C(x,\tau)$ satisfies the conditions

(1)
$$C_\tau = KC_{xx}, \qquad (x > 0, \tau > 0); \qquad C(x,0) = C_0, \qquad (x > 0);$$
$$KC_x(0,\tau) = EC(0,\tau), \qquad |C(x,\tau)| < Me^{\alpha\tau}.$$

In terms of new variables t, U, and a positive constant h, where

(2)
$$t = K\tau, \qquad U(x,t) = \frac{C(x,\tau)}{C_0}, \qquad h = \frac{E}{K},$$

the problem in the relative concentration $U(x,t)$ becomes

(3)
$$U_t = U_{xx}, \qquad (x > 0, t > 0); \qquad U(x,0) = 1, \qquad (x > 0);$$
$$U_x(0,t) = hU(0,t), \qquad |U(x,t)| < Ne^{\beta t}$$

for some constant N and each positive constant β. The problem in the transform of U becomes, formally,

(4)
$$su(x,s) - 1 = u''(x,s) \qquad (x > 0, \operatorname{Re} s \geq \gamma > \beta),$$

$$u'(0,s) = hu(0,s), \qquad |u(x,s)| < \frac{N}{\gamma - \beta}.$$

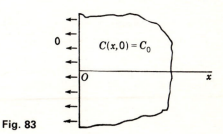

$$0$$

$$C(x,0) = C_0$$

$$0 \qquad\qquad x$$

Fig. 83

If $\sqrt{s} = \sqrt{r}\, e^{i\theta/2}$ where $-\pi < \theta < \pi$ and $s = re^{i\theta}$, the solution of problem (4) can be written as

$$(5) \qquad u(x,s) = \frac{1}{s} - \frac{h}{s(h + \sqrt{s})} \exp\left(-x\sqrt{s}\right).$$

Since $\operatorname{Re}\sqrt{s} < 0$ when $\theta \ne \pi$, then $h + \sqrt{s} \ne 0$ and u is an analytic function of s everywhere except on the nonpositive real axis. The ray $\theta = \pi$ is the branch cut of $\sqrt{s}$ and of u. The inversion integral on the line $\operatorname{Re} s = \gamma$ applies to both terms on the right in Eq. (5) since the second term is $\mathcal{O}(s^{-3/2})$, uniformly with respect to x. According to our theorems in Chap. 6, it represents a function

$$(6) \qquad U(x,t) = 1 - L_i^{-1}\left\{\frac{h}{s(h + \sqrt{s})}\exp\left(-x\sqrt{s}\right)\right\} \qquad (t > 0, x \geqq 0)$$

that is $\mathcal{O}(e^{\gamma t})$ uniformly in x where γ can be any positive number. Moreover that function is continuous in x and t, and $U(x,+0) = 1$. When $x \geqq x_0 > 0$, the function inside the braces is $\mathcal{O}(s^{-m})$ when $\operatorname{Re} s \geqq \gamma$, for each integer m, and it follows readily that $U_t = U_{xx}$ when $t > 0$ and $x > 0$. Thus the function (6) satisfies all the conditions (3) except possibly the condition $U_x(0,t) = hU(0,t)$.

The procedure in Sec. 74 of altering the path of integration of the inversion integral can be applied to the last term in Eq. (6) with no essential change. In particular we find that the integral around the small circle $|s| = r_0$ in Figs. 73 and 74,

$$\frac{1}{2\pi i}\int_{\theta=-\pi}^{\theta=\pi} \frac{e^{-x\sqrt{s}}}{s(h + \sqrt{s})} e^{st}\, ds \qquad \text{where } s = r_0 e^{i\theta},$$

tends to 1 as $r_0 \to 0$. The integrals over the rays $s = re^{-i\pi}$ and $s = re^{i\pi}$ constitute the rest of the value of $U(x,t) - 1$, and the result is this more useful representation of the function (6):

$$(7) \qquad U(x,t) = \frac{h}{\pi}\int_0^\infty \frac{h\sin x\sqrt{r} + \sqrt{r}\cos x\sqrt{r}}{r(h^2 + r)} e^{-rt}\, dr \qquad (t > 0, x \geqq 0).$$

Since $|y^{-1}\sin y| < 1$ when $y > 0$, we can see that the integrand in formula (7) satisfies the condition

$$\left|\frac{xh\sin x\sqrt{r}/(x\sqrt{r}) + \cos x\sqrt{r}}{\sqrt{r}(h^2 + r)} e^{-rt}\right| < \frac{x_0 h + 1}{h^2\sqrt{r}} e^{-rt}$$

when $0 \leqq x < x_0$. Thus the integral (7) converges uniformly with respect to $x(0 \leqq x < x_0)$ when $t > 0$ and

$$U(+0,t) = U(0,t) = \frac{h}{\pi}\int_0^\infty \frac{e^{-rt}}{h^2 + r}\frac{dr}{\sqrt{r}} \qquad (t > 0).$$

The integral of the derivative with respect to x is uniformly convergent in x when $t > 0$ and $x \geq 0$, so that

$$U_x(+0,t) = U_x(0,t) = \frac{h^2}{\pi} \int_0^\infty \frac{e^{-rt}}{h^2 + r} \frac{dr}{\sqrt{r}} \qquad (t > 0).$$

Thus $U_x(0,t) = hU(0,t)$ when $t > 0$, so the function U with representations (6) and (7) satisfies all conditions in boundary value problem (3).

That function has the transform (5); thus

$$u(0,s) = \frac{1}{s} - \frac{h}{s(h + \sqrt{s})} = \frac{1}{s - h^2} - \frac{h}{\sqrt{s}(s - h^2)}$$

and it follows from Prob. 4, Sec. 17, or transform 40, Appendix A, Table A.2, that

$$(8) \quad U(0,t) = \exp(h^2 t)\,\mathrm{erfc}(h\sqrt{t}) = \frac{2}{\sqrt{\pi}} \int_0^\infty \exp(-y^2 - 2h\sqrt{t}\,y)\,dy \lesseqgtr 1.$$

Using formulas (7) and (8) and the substitutions (2), we can write the concentration C in the slab and at the face as

$$(9) \qquad C(x,\tau) = \frac{2hC_0}{\pi} \int_0^\infty \frac{h\sin xy + y\cos xy}{y(y^2 + h^2)} \exp(-K\tau y^2)\,dy,$$

$$(10) \qquad C(0,\tau) = C_0 \exp(h^2 K\tau)\,\mathrm{erfc}(h\sqrt{K\tau}).$$

Another representation of $C(x,\tau)$, in terms of error functions, can be written from our formula for $u(x,s)$ (Prob. 1, Sec. 85).

85   DUHAMEL'S FORMULA

In Sec. 80 we obtained a formula for the temperatures in a bar with variable end temperature, in terms of the temperature function when the end temperature is constant. We now obtain a more general formula that simplifies heat conduction problems in the same way.

Let $U(x,y,z,t)$ be the temperatures in any solid, filling a region R, that is initially at temperature zero throughout. Let the temperature at every point of some part S of the boundary be a prescribed function $F(t)$ of time, and let the remainder S' of the boundary be kept at temperature zero (Fig. 84).

If $\Lambda(U)$ represents the linear differential form

$$(1) \qquad \Lambda(U) = \frac{1}{c}\left[\frac{\partial}{\partial x}\left(K\frac{\partial U}{\partial x}\right) + \frac{\partial}{\partial y}\left(K\frac{\partial U}{\partial y}\right) + \frac{\partial}{\partial z}\left(K\frac{\partial U}{\partial z}\right)\right],$$

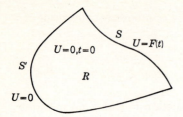

Fig. 84

the equation of conduction can be written

(2) $$U_t = \Lambda(U) \qquad [(x,y,z) \text{ in } R, t > 0].$$

We assume that the thermal coefficients K and c are either constants or independent of t. The boundary value problem in U consists of Eq. (2) and the conditions

(3) $$U(x,y,z,0) = 0 \qquad\qquad \text{interior to } R,$$

(4) $$U = 0 \qquad\qquad \text{on } S'$$
$$= F(t) \qquad\qquad \text{on } S.$$

The transform $u(x,y,z,s)$ then satisfies the conditions

(5) $$su = \Lambda(u) \qquad\qquad \text{in } R,$$

(6) $$u = 0 \qquad\qquad \text{on } S'$$
$$= f(s) \qquad\qquad \text{on } S.$$

Let $V(x,y,z,t)$ be the temperature function U when $F(t) = 1$; that is, V satisfies the heat equation (2), the initial condition (3), and the surface conditions

(7) $$V = 0 \qquad\qquad \text{on } S'$$
$$= 1 \qquad\qquad \text{on } S.$$

Then the transform $v(x,y,z,s)$ satisfies the differential equation (5) and the conditions

(8) $$v = 0 \qquad\qquad \text{on } S'$$
$$= \frac{1}{s} \qquad\qquad \text{on } S.$$

Since s is a parameter in the linear homogeneous differential equation (5), the product of $sf(s)$ by the solution v is also a solution. But according to conditions (8),

$$sf(s)v = 0 \qquad\qquad \text{on } S'$$
$$= f(s) \qquad\qquad \text{on } S;$$

thus the function $sf(s)v$ also satisfies all the conditions (5) and (6). If the problem consisting of conditions (5) and (6) is to have but one solution, then $sf(s)v$ must be the same as the solution u:

(9) $$u(x,y,z,s) = sf(s)v(x,y,z,s).$$

Since sv is the transform of V_t, it follows from Eq. (9) with the aid of the convolution that

(10) $$U(x,y,z,t) = \int_0^t F(t - \tau)V_t(x,y,z,\tau)\, d\tau$$

$$= \frac{\partial}{\partial t} \int_0^t F(t - \tau)V(x,y,z,\tau)\, d\tau.$$

These are two cases of Duhamel's formula representing temperatures U corresponding to variable surface temperatures in terms of temperatures V in the simpler problem where surface temperatures are constant. A more general case is given in Prob. 14.

Given that V is the solution of the simpler problem, the solution (10) of the problem in U can be verified directly under reasonable conditions on F.

PROBLEMS

1. From formula (5), Sec. 84, for $u(x,s)$ and transform 86, Appendix A, Table A.2, obtain this formula for the concentration of moisture in the slab $x \geq 0$ with evaporation at the surface:

$$C(x,\tau) = C_0\left[\operatorname{erf}\left(\frac{x}{2\sqrt{K\tau}}\right) + \exp\left(hx + h^2 K\tau\right)\operatorname{erfc}\left(\frac{x}{2\sqrt{K\tau}} + h\sqrt{K\tau}\right)\right].$$

2. Identify the two expressions in formula (8), Sec. 85, for the concentration $U(0,t)$ at the face of the semi-infinite solid and thus show that $0 < U(0,t) \leq 1$ when $t \geq 0$.

3. A layer of finite thickness extending from the face of the slab in Sec. 84 is initially dry. The initial concentration of the rest of the semi-infinite slab is uniform. Choose units so that the relative concentration $U(x,t)$ satisfies the conditions

$$U_t = U_{xx}, x > 0, t > 0, \qquad U(x,0) = S_0(x - 1), \qquad U_x(0,t) = hU(0,t),$$

and a boundedness condition; also U and U_x are to be continuous when $t > 0$, at $x = 1$ in particular. Show formally that

$$2su(x,s) = \exp\left[-(1 - x)\sqrt{s}\,\right] - \exp\left[-(1 + x)\sqrt{s}\,\right]\frac{h - \sqrt{s}}{h + \sqrt{s}} \qquad (0 \leq x \leq 1),$$

and use the tables of transforms to get these particular results:

$$U(0,t) = \exp\left(h + h^2 t\right)\operatorname{erfc}\left(\frac{1}{2\sqrt{t}} + h\sqrt{t}\right),$$

$$U(1,t) = \frac{1}{2}\operatorname{erf}\left(\frac{1}{\sqrt{t}}\right) + \exp\left(2h + h^2 t\right)\operatorname{erfc}\left(\frac{1}{\sqrt{t}} + h\sqrt{t}\right).$$

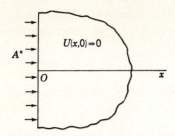

Fig. 85

4. The face $x = 0$ of a semi-infinite solid (Fig. 85) is exposed to a medium at constant temperature A. Heat is transferred from that medium to the face of the solid according to the law that the flux of heat is $E[A - U(0,t)]$, where $U(x,t)$ is the temperature in the solid. Thus the boundary condition at the face becomes

$$U_x(0,t) = h[U(0,t) - A],$$

where $h = E/K$. If the initial temperature is zero, derive the following formula with the aid of the tables in Appendix A:

$$U(x,t) = A\left[\operatorname{erfc}\left(\frac{x}{2\sqrt{kt}}\right) - \exp\left(hx + h^2 kt\right)\operatorname{erfc}\left(h\sqrt{kt} + \frac{x}{2\sqrt{kt}}\right)\right].$$

5. Let $V(x,t)$ be the temperature function for the solid in Prob. 4 when $A = 1$. Let $W(x,t)$ be the temperature of the solid when the constant A is replaced by a function $\Phi(t)$, so that the medium to which the face is exposed has a variable temperature. Derive the formula

$$W(x,t) = \int_0^t \Phi(t - \tau)V_t(x,\tau)\,d\tau.$$

6. The temperature of the face of a semi-infinite solid $x \geq 0$ varies in the following manner:

$$U(0,t) = A \sin \omega t.$$

Taking the initial temperature as zero, for convenience, show that, when t is large, the temperature $U(x,t)$ is approximately $V(x,t)$, where

$$V(x,t) = A \sin\left(\omega t - x\sqrt{\frac{\omega}{2k}}\right)\exp\left(-x\sqrt{\frac{\omega}{2k}}\right),$$

a simple periodic function of time. Note that the closed contour corresponding to the one shown in Fig. 73 will enclose two simple poles $s = \pm i\omega$ of $u(x,s)$ in this case. (Also, see Prob. 7.)

7. Use the following elementary method to obtain the formula in Prob. 6 for the undamped component V of the temperature function U. Assume V periodic in t with frequency ω, for each x, and write $V(x,t) = \operatorname{Im} W(x,t)$ where $W(x,t) = F(x)e^{i\omega t}$, F is complex-valued, and $W_t = kW_{xx}$, $W(\infty,t) = 0$, $W(0,t) = Ae^{i\omega t}$. Then $V_t = kV_{xx}$, $V(\infty,t) = 0$, and $V(0,t) = A \sin \omega t$. Thus write the problem in ordinary differential equations in the function F and show that $F(x) = A \exp\left[-(1 + i)x\sqrt{\omega}/\sqrt{2k}\right]$.

8. The diffusivity k of the earth's soil in a certain locality is 0.005 cgs unit. The temperature of the surface of the soil has an annual variation from -8 to $22°C$. Assuming the variation is approximately sinusoidal (Prob. 6), show that the freezing temperature will penetrate to a depth of approximately 170 cm (considerably less, because of the latent heat of freezing).

9. In Prob. 8 find the approximate depth at which the variation of temperature with time is 6 months out of phase with the variation of the surface temperature. Show that the amplitude of the variation at that depth is less than $1°C$.

Ans. $x = 705$ cm $= 23$ ft, approximately.

10. Let the functions $V(x,t)$ and $W(y,t)$ satisfy the heat equations $V_t = kV_{xx}$ and $W_t = kW_{yy}$, respectively. Prove by direct substitution that the *product of those temperature functions*,

$$U(x,y,t) = V(x,t)W(y,t),$$

satisfies the heat equation

$$U_t = k(U_{xx} + U_{yy}).$$

If in addition $V(0,t) = V(a,t) = 0$ and $W(0,t) = W(b,t) = 0$, and if $V(x,0) = f(x)$ and $W(y,0) = g(y)$, then show that $U(x,y,t)$ represents the temperatures in a rectangular plate (Fig. 86) with insulated faces, if the edges are at temperature zero and the initial temperature is

$$U(x,y,0) = f(x)g(y).$$

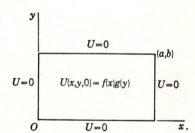

Fig. 86

11. Use the product of solutions (Prob. 10) to obtain the following formula for the temperatures in an infinite prism with a square cross section, if the initial temperature is A and the surface temperature is zero, if the unit of length is the side of the square, and $k = 1$:

$$V(x,y,t) = AU(x,t)U(y,t),$$

where U is the temperature function found in Prob. 9 of Sec. 81.

12. With the aid of Prob. 10, derive the formula

$$U(x,y,t) = \frac{4}{\pi} \text{erf}\left(\frac{x}{2\sqrt{t}}\right) \sum_{n=1}^{\infty} \frac{\sin(2n-1)\pi y}{2n-1} \exp[-(2n-1)^2\pi^2 t]$$

for the temperatures in the semi-infinite slab $x \geq 0, 0 \leq y \leq 1$ with its boundary at temperature zero, if $U(x,y,0) = 1$ (Fig. 87) and $k = 1$.

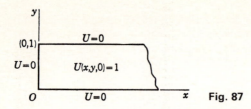

Fig. 87

13. Generalize the method of Prob. 10 to the case of three dimensions, and give an illustration of its use in finding temperatures in a cube.

14. Use the generalized convolution property found in Prob. 20, Sec. 27, to make a formal derivation of the following more general Duhamel formula. Let $\Lambda(U)$ denote the differential form (1), Sec. 85, P the point (x,y,z), n the distance normal to the boundary surface S or S', and let $a(P)$, $b(P)$, H, F, and G represent prescribed functions. The temperature function $U(P,t)$ in the region R is to satisfy the conditions

$$U_t = \Lambda(U) + H(P,t), \qquad U(P,0) = G(P) \qquad\qquad (P \text{ in } R);$$

$$a(P)U + b(P)\frac{dU}{dn} = F(P,t) \qquad\qquad (P \text{ on } S \text{ or } S').$$

When the functions $H(P,t)$ and $F(P,t)$ are replaced by $H(P,t')$ and $F(P,t')$ respectively, where t' is any fixed value of t, $V(P,t,t')$ denotes the solution of the problem. Write the problem in the transform $v(P,s,t')$ and transform its members with respect to t', using the same parameter s, to obtain a problem in the iterated transform $\hat{v}(P,s,s)$ and thus derive this generalization of formula (10):

$$U(P,t) = \frac{\partial}{\partial t}\int_0^t V(P, t - \tau, \tau)\, d\tau.$$

Note that if $a(P) = F(P,t) = 0$ when P is on S', then the surface S' is insulated. Thus the generalized formula could be applied to Prob. 14, Sec. 81.

8
Problems in Mechanical Vibrations

This chapter contains further illustrations of the uses of those properties of the Laplace transformation that involve complex variables. The problems taken as illustrations deal with vibrations and resonance in continuous mechanical systems—systems in which the mass and elastic characteristics are distributed over the system. Consequently these problems are boundary value problems in partial differential equations, of the type treated in Chap. 4. It is the intention here to present fairly simple physical problems in their mathematical form, although the mathematical problems may have more important physical interpretations. Some electrical analogs of mechanical vibrations, involving transmission lines, are included among the exercises.

86   A BAR WITH A CONSTANT FORCE ON ONE END

In Sec. 44 we derived a formula for the longitudinal displacements in an elastic bar in the form of a prism, when one end of the bar is fixed and a

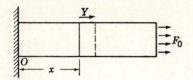

Fig. 88

constant force F_0 per unit area acts parallel to the bar on the other end (Fig. 88). Let all parts of the bar be initially at rest and unstrained. The displacements $Y(x,t)$ then satisfy the conditions in the boundary value problem

$$Y_{tt}(x,t) = a^2 Y_{xx}(x,t) \qquad (0 < x < c, t > 0),$$

(1)
$$Y(x,0) = Y_t(x,0) = 0,$$

$$Y(0,t) = 0, \qquad E Y_x(c,t) = F_0,$$

where $a^2 = E/\rho$, E is Young's modulus of elasticity, and ρ is the mass per unit volume of the material.

Let us obtain another formula for $Y(x,t)$ here. From problem (1) we found this transform of $Y(x,t)$:

(2)
$$y(x,s) = \frac{aF_0}{E} \frac{\sinh(sx/a)}{s^2 \cosh(sc/a)},$$

an analytic function of s everywhere except at the origin and at the zeros $\pm s_n$ of $\cosh(sc/a)$, where

(3)
$$s_n = \frac{(2n-1)\pi a}{2c} i \qquad (n = 1, 2, \dots).$$

When $s \neq 0$ and $s \neq \pm s_n$, y is continuous in x and s, real-valued when s is real, and $y(0,s) = 0$.

Now the quotient of hyperbolic functions in formula (2),

$$Q(x,s) = \frac{\sinh(sx/a)}{\cosh(sc/a)}$$

fails to enhance the order property of y over a right or left half plane of the variable s. For if $s = \lambda + i\mu$, then (Sec. 55)

(4)
$$|Q(x,s)|^2 = \frac{\sinh^2(\lambda x/a) + \sin^2(\mu x/a)}{\sinh^2(\lambda c/a) + \cos^2(\mu c/a)}$$

$$\leq \frac{\sinh^2(\lambda x/a)}{\sinh^2(\lambda c/a)} + \frac{1}{\sinh^2(\lambda c/a)} \qquad (\lambda \neq 0).$$

Thus Q is bounded over either half plane $|\lambda| \geq \gamma$ if $\gamma > 0$, since $|Q|^2 \leq 1 + 1/\sinh^2(\gamma c/a)$. But Q does not always vanish there as $|s| \to \infty$ because

$|Q|^2 \geqq \sinh^2(\lambda x/a)/\cosh^2(\lambda c/a)$, and this quotient remains fixed when λ is constant and $\mu \to \pm \infty$.

Therefore $y(x,s)$ is $\mathcal{O}(1/s^2)$ over half planes $\lambda \geqq \gamma$ and $\lambda \leqq -\gamma$, when $\gamma > 0$, uniformly with respect to x, and $L_i^{-1}\{y\}$ is a continuous function $Y(x,t)$ with transform $y(x,s)$. Also, $Y(x,0) = Y(0,t) = 0$. But the order properties of the derivatives y_x and y_{xx} are not adequate for the representation of derivatives of Y by corresponding inversion integrals.

We can see from Eq. (4) that Q is bounded on horizontal lines $\mu = \pm n\pi a/c$ between the singular points $\pm s_n$ since

$$|Q|^2 \leqq \frac{\cosh^2(\lambda x/a)}{\cosh^2(\lambda c/a)} \leqq 1$$

for all points s on those lines. Thus y is $\mathcal{O}(1/s^2)$ on those lines, and $L_i^{-1}\{y\}$ is represented by the infinite series of residues of $e^{st}y(x,s)$ (Theorem 10, Sec. 73).

We can write

$$\frac{1}{s}\sinh\frac{sx}{a} = \frac{x}{a} + \frac{s^2 x^3}{3!a^3} + \frac{s^4 x^5}{5!a^5} + \cdots,$$

a function that is analytic at the point $s = 0$. In view of Eq. (2) then the origin is a simple pole of $e^{st}y(x,s)$ when $x \neq 0$, and the residue there is $(aF_0/E)(x/a)$, or $F_0 x/E$.

The residue of y itself at the simple pole s_n can be written

(5) $$\frac{F_0 a^2}{Ecs_n^2}\frac{\sinh(s_n x/a)}{\sinh(s_n c/a)} = \frac{4F_0 c}{\pi^2 E}\frac{(-1)^n}{(2n-1)^2}\sin\frac{(2n-1)\pi x}{2c}.$$

It is real-valued. Thus according to formula (13), Sec. 72, in which $\theta_1 = 0$ and $a = 0$, the sum of the residues of $e^{st}y(x,s)$ at the pair of poles $\pm s_n$, or s_n and $\bar{s}_n$, is

(6) $$\rho_n + \bar{\rho}_n = \frac{8F_0 c}{\pi^2 E}\frac{(-1)^n}{(2n-1)^2}\sin\frac{(2n-1)\pi x}{2c}\cos\frac{(2n-1)\pi at}{2c}.$$

Finally, then, the inversion integral $L_i^{-1}\{y(x,s)\}$ is a function $Y(x,t)$ represented by this infinite series:

(7) $$Y = \frac{F_0}{E}\left[x + \frac{8c}{\pi^2}\sum_{n=1}^{\infty}\frac{(-1)^n}{(2n-1)^2}\sin\frac{(2n-1)\pi x}{2c}\cos\frac{(2n-1)\pi at}{2c}\right].$$

Every term in that convergent series is periodic in t with period

(8) $$T_0 = \frac{4c}{a} = 4c\sqrt{\frac{\rho}{E}},$$

so the function Y itself has that periodicity: $Y(x, t + T_0) = Y(x,t)$. Note, however, that the series is not twice differentiable with respect to x or t. Neither the series nor the inversion integral representation can be used directly to verify that our function Y satisfies the wave equation. We shall sum the series (7) in order to represent Y in a form that enables us to complete the verification of our solution of problem (1).

87 ANOTHER FORM OF THE SOLUTION

In establishing the inversion integral representation of our function Y, we noted that $Y(x,0) = 0$. Since the function is also represented by series (7), Sec. 86, it follows that

$$(1) \qquad x = -\frac{8c}{\pi^2} \sum_{n=1}^{\infty} \frac{(-1)^n}{(2n-1)^2} \sin\frac{(2n-1)\pi x}{2c} \qquad (0 \le x \le c).$$

Each term of series (1) is an odd function of x, as is x itself, so the series represents the function x over the interval $-c \le x \le c$. When x is replaced by $x + 2c$, each term merely changes sign so the sum of the series is an antiperiodic function for all x, with period $2c$; consequently it is periodic with period $4c$. Thus for all real x the series represents the triangular-wave function H shown in Fig. 89, described by the two conditions

$$(2) \quad H(x) = x \quad (-c \le x \le c), \qquad H(x + 2c) = -H(x) \quad (-\infty < x < \infty).$$

The function H also has these properties for all x:

$$(3) \qquad H(x) = -H(-x) = H(x + 4c) = H(2c - x).$$

Now the sum of the series in Eq. (1) can be written

$$(4) \qquad \frac{8c}{\pi^2} \sum_{n=1}^{\infty} \frac{(-1)^{n-1}}{(2n-1)^2} \sin\frac{(2n-1)\pi x}{2c} = H(x) \qquad (-\infty < x < \infty).$$

The series is, in fact, the Fourier series representing the odd periodic function H for all x.

We write $m = (2n-1)\pi/(2c)$ and note that

$$2 \sin mx \cos mat = \sin m(x + at) + \sin m(x - at).$$

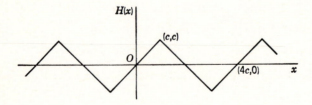

Fig. 89

Then our series representation (7), Sec. 86, of $Y(x,t)$ can be written

$$Y = \frac{F_0}{E}\left\{x - \frac{4c}{\pi^2}\sum_{n=1}^{\infty}\frac{(-1)^{n-1}}{(2n-1)^2}[\sin m(x + at) + \sin m(x - at)]\right\}$$

and, in view of formula (4), Y has this representation:

(5) $$Y(x,t) = \frac{F_0}{E}[x - \tfrac{1}{2}H(x + at) - \tfrac{1}{2}H(x - at)].$$

We can use formula (5) to complete the verification of the solution of our boundary value problem (1), Sec. 86. We first note that $H(x)$ has derivatives everywhere except at the points $x = \pm(2n - 1)c$ $(n = 1, 2, \ldots)$. Since any twice-differentiable function of $x + at$ or $x - at$ satisfies the linear homogeneous-wave equation $Y_{tt} = a^2 Y_{xx}$, and x itself satisfies it, our function (5) satisfies that equation except when x and t are such that $x \pm at = \pm(2n - 1)c$, for which values the partial derivatives of Y do not exist.

To see that the condition $EY_x(c,t) = F_0$ is satisfied by our function (5), we note that

(6) $$EY_x(c,t) = F_0[1 - \tfrac{1}{2}H'(c + at) - \tfrac{1}{2}H'(c - at)].$$

But since $H(x) = H(2c - x)$, then $H'(x) = -H'(2c - x)$. Therefore $H'(c + at) = -H'(2c - c - at) = -H'(c - at)$ and, except for the values of t such that $c \pm at = \pm(2n - 1)c$, Eq. (6) reduces to $EY_x(c,t) = F_0$.

The condition $Y_t(x,0) = 0$ is easily verified from formula (5). This completes the verification of our solution.

Formula (5) is well adapted to graphical descriptions of the variation of Y with either x or t. The graph of $H(x + at)$ for a fixed value of t, for example, is obtained by translating the graph of $H(x)$ to the left through a distance at.

The displacement at the end $x = c$ can be written

(7) $$Y(c,t) = \frac{F_0}{E}[c - \tfrac{1}{2}H(at + c) + \tfrac{1}{2}H(at - c)]$$

$$= \frac{F_0}{E}[c + H(at - c)]$$

since $H(at + c) = -H(at + c - 2c)$. $Y(c,t)$ is shown in Fig. 45.

The reader can show that the force per unit area on the fixed support, $EY_x(0,t)$, assumes the values $2F_0$ and zero periodically.

Incidentally, we have established this transformation here:

(8) $$L^{-1}\left\{\frac{2a \sinh (sx/a)}{s^2 \cosh (sc/a)}\right\} = 2x - H(x + at) - H(x - at).$$

88   RESONANCE IN THE BAR WITH A FIXED END

Let the end $x = 0$ of the bar again be fixed while a simple periodic force

$$F(t) = A \sin \omega t \qquad\qquad (\omega > 0)$$

per unit area acts lengthwise at the end $x = c$. If the bar is initially unstrained and at rest, only the end condition at $x = c$ in problem (1), Sec. 86, need be changed to read

$$(1) \qquad\qquad EY_x(c,t) = A \sin \omega t.$$

The transform of the displacement now becomes

$$(2) \qquad\qquad y(x,s) = \frac{A a \omega}{E(s^2 + \omega^2)} \frac{\sinh (sx/a)}{s \cosh (sc/a)}.$$

Here y is an analytic function of s except at the origin $s = 0$, a removable singular point, and the points $s = \pm i\omega$ and $s = \pm s_n$ where

$$s_n = \frac{(2n - 1)\pi a}{2c} i \qquad\qquad (n = 1, 2, \ldots).$$

If $i\omega$ is not equal to any one of the numbers s_n, that is, if

$$\omega \neq \frac{(n - \tfrac{1}{2})\pi a}{c} \qquad\qquad (n = 1, 2, \ldots),$$

the singular points $\pm i\omega$ and $\pm s_n$ are all simple poles unless the value of x is such that some are removable, in which case they serve as simple poles with residue zero.

According to formula (13), Sec. 72, the component of Y corresponding to the pair of poles $\pm i\omega$ of y is a term of the type $b \cos(\omega t + \theta)$; here the real numbers b and θ depend on x. Similar components correspond to a pair of simple poles $\pm s_n$. Thus Y is represented formally by a series of this type:

$$(3) \qquad Y(x,t) = b_0(x) \cos [\omega t + \theta_0(x)] + \sum_{n=1}^{\infty} b_n(x) \cos [\omega_n t + \theta_n(x)],$$

where $\omega_n = (n - \tfrac{1}{2})\pi a/c$. We shall find the b's and θ's in Sec. 89 and establish formula (3) as a solution of our boundary value problem in Y. The series shows that each section of the bar moves as a superposition of two periodic motions, one with frequency ω and the other with frequency $\omega_1 = \pi a/(2c)$, or period $T_1 = 4c/a$.

But if the frequency ω of the external force coincides with one of the frequencies ω_n, say $\omega = \omega_r$ for some positive integer r, then the function

$$q(s) = (s^2 + \omega_r^2) \cosh \frac{sc}{a}$$

in the denominator of expression (2) for $y(x,s)$ is such that $q(i\omega_r) = q'(i\omega_r) = 0$ while $q''(i\omega_r) \neq 0$. Consequently the points $s = \pm i\omega_r$ are poles of y of the second order, and corresponding to that pair of poles, according to formula (16), Sec. 72, $Y(x,t)$ contains an *unstable* component of the type

(4) $$tb_r(x) \cos [\omega_r t + \psi_r(x)], \qquad \text{where } b_r(x) \neq 0.$$

The remaining component of Y is periodic with frequency ω_1, the common frequency of all its periodic terms.

That unstable oscillation of sections of the bar is called *resonance*. The periodic external force is in resonance with the bar when its frequency ω coincides with any one of the resonance frequencies

(5) $$\omega_r = \frac{(r - \frac{1}{2})\pi a}{c} \qquad (r = 1, 2, \ldots).$$

The frequencies ω_r depend on the physical properties of the bar and the manner in which the bar is supported. If the end $x = 0$ of the bar is free or elastically supported, for instance, we find that the set of resonance frequencies is different from the set (5).

To produce resonance, the external force need not be restricted to the simple form $A \sin \omega_r t$. For any prescribed force $F(t)$ at the end $x = c$, the transform of $Y(x,t)$ is

(6) $$y(x,s) = \frac{a}{E} f(s) \frac{\sinh (sx/a)}{s \cosh (sc/a)}.$$

Consequently if $F(t)$ contains a term $A_r \sin \omega_r t$ or $B_r \cos \omega_r t$, where ω_r is a number of the set (5), then $y(x,s)$ will contain a term with the product $(s^2 + \omega_r^2) \cosh (sc/a)$ in the denominator, and $Y(x,t)$ will have an unstable component of type (4). In fact, if F is any periodic function with a frequency ω_r, then resonance will occur, as can be seen from the form in Sec. 23 of the transform of a periodic function.

89   VERIFICATION OF SOLUTIONS

When the formal solution of a boundary value problem can be written in terms of a finite number of simple functions, that form is usually best for verifying the solution, as we illustrated in Sec. 87. But to establish a solution represented only by an infinite series or by an improper integral, the procedure based on properties of its transform, used in Chap. 7, may be useful. Let us illustrate a combination of the two methods for the problem in the preceding section.

The following transform of displacements $Y(x,t)$ in a bar with its end $x = 0$ fixed and with pressure $A \sin \omega t$ applied at the end $x = c$ was found formally:

$$(1) \qquad y(x,s) = \frac{B}{s^2 + \omega^2} \frac{\sinh(sx/a)}{s \cosh(sc/a)} \qquad \left(B = \frac{Aa\omega}{E} \right).$$

The quotient of hyperbolic functions here was shown (Sec. 86) to be bounded uniformly in x over half planes $\operatorname{Re} s \geq \gamma$ and $\operatorname{Re} s \leq -\gamma$ when $\gamma > 0$; $\cosh(sx/a)/\cosh(sc/a)$ is also bounded there. Hence y is $\mathcal{O}(s^{-3})$ and y_x is $\mathcal{O}(s^{-2})$ over those half planes, uniformly in x, and $L_i^{-1}\{y\}$ along a line $\operatorname{Re} s = \gamma$ represents a continuous function $Y(x,t)$ such that $Y(x,0) = Y(0,t) = 0$; also Y has continuous derivatives of first order,

$$Y_t(x,t) = L_i^{-1}\{sy(x,s)\}, \qquad Y_x(x,t) = L_i^{-1}\{y_x(x,s)\},$$

such that $Y_t(x,0) = 0$ and $EY_x(c,t) = A \sin \omega t$. Hence Y satisfies all boundary conditions in the problem, but the order of y is not adequate to show directly that $Y_{tt} = a^2 Y_{xx}$.

To prove that Y does satisfy the wave equation, we write

$$\frac{1}{s^2 + \omega^2} = \frac{1}{s^2} - \frac{\omega^2}{s^2(s^2 + \omega^2)},$$

$$(2) \qquad u(x,s) = \frac{B}{s} \frac{\sinh(sx/a)}{s^2 \cosh(sc/a)}, \qquad v(x,s) = \frac{-B\omega^2 \sinh(sx/a)}{s^3(s^2 + \omega^2)\cosh(sc/a)}.$$

Then

$$y(x,s) = u(x,s) + v(x,s)$$

where v is $\mathcal{O}(s^{-5})$ and v_{xx} is $\mathcal{O}(s^{-3})$. It follows readily that $L_i^{-1}\{v\}$ satisfies the wave equation.

The function $L_i^{-1}\{u\}$ can be written as a linear combination of known solutions of the wave equation since (Sec. 87)

$$2au(x,s) = Bs^{-1}L\{2x - H(x + at) - H(x - at)\}$$

where H is the triangle-wave function shown in Fig. 89. Thus $L_i^{-1}\{u\}$ is the function

$$U(x,t) = \frac{B}{2a} \int_0^t [2x - H(x + a\tau) - H(x - a\tau)]\, d\tau$$

$$= \frac{B}{2a}\left[2xt - \frac{1}{a}\int_{x-at}^{x+at} H(\xi)\, d\xi \right].$$

In terms of the periodic function

$$(3) \qquad\qquad G(r) = \int_0^r H(\xi)\, d\xi \qquad\qquad (-\infty < r < \infty)$$

we can write U in the form

$$(4) \qquad\qquad U(x,t) = \frac{B}{2a^2}[2axt - G(x + at) + G(x - at)],$$

from which we see that $U_{tt} = a^2 U_{xx}$. Therefore our function

$$(5) \qquad Y(x,t) = L_i^{-1}\{y(x,s)\} = L_i^{-1}\left\{\frac{B\sinh(sx/a)}{s(s^2 + \omega^2)\cosh(sc/a)}\right\}$$

satisfies the wave equation and all the boundary conditions.

We shall use Theorem 10, Sec. 73, to prove that the series of residues of $e^{st}y(x,s)$ converges to the inversion integral for each frequency ω. Then we shall have a series representation of our solution (5).

Our function y is analytic in s except for the poles

$$(6) \qquad \pm i\omega \quad \text{and} \quad \pm iam_n, \quad \text{where } m_n = \frac{(n - \frac{1}{2})\pi}{c} \quad (n = 1, 2, \ldots).$$

As noted in Sec. 86, the quotient $\sinh(sx/a)/\cosh(sc/a)$ is bounded on the horizontal lines $\text{Im } s = \pm an\pi/c$. Thus $y(x,s)$ is $\mathcal{O}(s^{-3})$ on all such lines for which $an\pi/c > \omega$. Since y has that order over the left half plane $\text{Re } s < -\gamma$, the infinite series of residues of $e^{st}y(x,s)$ converges to $L_i^{-1}\{y\}$ when $t \geq 0$ and $0 \leq x \leq c$.

Let us write that series explicitly when ω is not one of the resonance frequencies am_n. Then the poles of y are all simple, and formula (13), Sec. 72, gives the sum of the residues of $e^{st}y$ at each of its pairs of conjugate poles. The residue of y itself at the poles $s = i\omega$ and $s = iam_n$ are easily found to be

$$(7) \qquad -i\frac{B\sin(\omega x/a)}{2\omega^2\cos(\omega c/a)} \quad \text{and} \quad -i\frac{B(-1)^n\sin m_n x}{cm_n(am_n^2 - \omega^2)},$$

respectively. The sum of the residues of $e^{st}y$ at the pair of poles $\pm i\omega$ is therefore

$$\frac{B\sin(\omega x/a)}{\omega^2\cos(\omega c/a)}\cos\left(\omega t - \frac{\pi}{2}\right);$$

at the pair $\pm iam_n$ the sum is

$$\frac{2B(-1)^n\sin m_n x}{cm_n(am_n^2 - \omega^2)}\cos\left(am_n t - \frac{\pi}{2}\right).$$

Thus we have this representation of our solution (5):

$$(8) \qquad Y(x,t) = B\frac{\sin(\omega x/a)\sin\omega t}{\omega^2\cos(\omega c/a)}$$

$$+ \frac{2B}{c}\sum_{n=1}^{\infty}\frac{(-1)^n}{m_n}\frac{\sin m_n x \sin am_n t}{am_n^2 - \omega^2} \quad \left(m_n = \frac{2n-1}{2c}\pi\right).$$

When ω coincides with one of the resonance frequencies am_r, then y has a conjugate pair of poles of order two, and the series for Y contains a resonance term of type (4), Sec. 88.

PROBLEMS

1.   The end $x = 0$ of an elastic bar is free while a constant longitudinal force F_0 per unit area acts longitudinally at the end $x = c$ (Fig. 48). The bar is initially at rest and unstrained. One formula for the longitudinal displacements $Y(x,t)$ in the bar was found in Prob. 8, Sec. 44. Derive this alternate formula:

$$Y = \frac{F_0}{6Ec}\left[3x^2 + 3a^2t^2 - c^2 - \frac{12c^2}{\pi^2}\sum_{n=1}^{\infty}\frac{(-1)^n}{n^2}\cos\frac{n\pi x}{c}\cos\frac{n\pi at}{c}\right].$$

2.   Reduce the solution of Prob. 1 to the form

$$Y(x,t) = \frac{F_0}{12Ec}[6x^2 + 6a^2t^2 - 2c^2 + Q(x + at) + Q(x - at)]$$

where Q is this periodic function:

$$Q(x) = c^2 - 3x^2 \quad (-c < x < c), \qquad Q(x + 2c) = Q(x) \quad (-\infty < x < \infty).$$

Use that form to verify the solution of the boundary value problem.

3.   An unstrained elastic bar or heavy coiled spring is clamped along its length c so as to prevent longitudinal displacements, then hung from its end $x = 0$ (Fig. 90). At the instant $t = 0$, the clamp is released and the bar or coil vibrates longitudinally because of its own weight. Thus

$$Y_{tt}(x,t) = a^2 Y_{xx}(x,t) + g \quad (0 < x < c, t > 0), \qquad Y_x(c,t) = 0,$$

where g is the acceleration of gravity. Complete the boundary value problem, write $m = n - \frac{1}{2}$, and derive the formula

$$Y(x,t) = \frac{g}{2a^2}\left[x(2c - x) - \frac{4c^2}{\pi^3}\sum_{n=1}^{\infty}\frac{1}{m^3}\sin\frac{m\pi x}{c}\cos\frac{m\pi at}{c}\right].$$

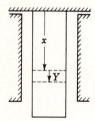

Fig. 90

4.   Reduce the solution of Prob. 3 to the form

$$Y(x,t) = \frac{g}{4a^2}[2P(x) - P(x + at) - P(x - at)],$$

where the function P is described by the conditions

$$P(x) = x(2c - x) \quad (0 \le x \le 2c), \qquad P(x + 2c) = -P(x) \quad (-\infty < x < \infty).$$

Use that form to verify the solution of the problem.

5. The ends $x = \pm c$ of an unstrained shaft spinning about its axis with angular velocity ω are brought to rest at the instant $t = 0$ (Fig. 52, Prob. 13, Sec. 44). Derive this formula for the angular displacements of its cross sections:

$$\Theta(x,t) = \frac{8c\omega}{\pi^2 a} \sum_{n=1}^{\infty} \frac{(-1)^{n-1}}{(2n-1)^2} \cos \frac{(2n-1)\pi x}{2c} \sin \frac{(2n-1)\pi a t}{2c}.$$

6. Reduce the formula for Θ in Prob. 5 to the form

$$\Theta(x,t) = \frac{\omega}{2a}[H(x+at) - H(x-at)],$$

where H is the triangular-wave function shown in Fig. 89 and represented by series (4), Sec. 87. Use this simple representation of Θ to verify the solution of the boundary value problem in Prob. 5.

7. The end $x = 0$ of an elastic bar is fixed while a force proportional to t^2 acts longitudinally at the end $x = 1$ so that $Y_x(1,t) = At^2$, where $Y(x,t)$ denotes longitudinal displacements. The unit of time is such that $a = 1$. If the bar is initially at rest and unstrained, derive the formula

$$Y(x,t) = A\left[x(t^2 - 1) + \frac{x^3}{3} - \frac{4}{\pi^4} \sum_{n=1}^{\infty} \frac{(-1)^n}{m^4} \sin m\pi x \cos m\pi t \right]$$

where $m = n - \frac{1}{2}$. Verify the solution fully with the aid of the inversion integral representation of Y. (Also, see Probs. 8 and 9, below.)

8. Reduce the solution of Prob. 7 to the form

$$Y(x,t) = A\left[x(t^2 - 1) + \frac{x^3}{3} + \frac{1}{2}R(x + t) + \frac{1}{2}R(x - t) \right]$$

where

$$R(x) = x - \frac{x^3}{3} \quad (-1 \leq x \leq 1), \qquad R(x+2) = -R(x) \quad (-\infty < x < \infty).$$

Show R graphically. Use this simple representation of Y to verify the solution of Prob. 7.

9. Show that the function Y found in Prob. 7 and its derivatives of first and second order are everywhere continuous in x and t and $\mathcal{O}(e^{\alpha t})$ uniformly with respect to x ($0 \leq x \leq 1$). Prove that no other function of that type can satisfy all conditions in the boundary value problem. (Cf. Sec. 79; but here the proof of uniqueness is simpler because of the favorable continuity conditions.)

10. The end $x = 0$ of an elastic bar is fixed while the end $x = c$ is displaced longitudinally in the manner $Y(c,t) = A \sin \omega t$. Show that the resonance frequencies are $\omega = \omega_r$, where

$$\omega_r = \frac{r\pi a}{c} \qquad\qquad (r = 1, 2, \ldots).$$

Note that the lowest frequency ω_1 is twice as great as the lowest in the set (5), Sec. 88.

11. If the end $x = 0$ of the bar in Prob. 10 is free, rather than fixed, show that the resonance frequencies are the same as the set (5), Sec. 88.

12. A simple periodic longitudinal force acts uniformly on all elements of an elastic bar with its end $x = 0$ fixed and its end $x = c$ free, so that the displacements $Y(x,t)$ satisfy the conditions

$$Y_{tt}(x,t) = a^2 Y_{xx}(x,t) + B \sin \omega t,$$

$$Y(0,t) = Y_x(c,t) = Y(x,0) = Y_t(x,0) = 0.$$

Show that resonance occurs when $\omega = \omega_r$, where

$$\omega_r = \frac{(r - \frac{1}{2})\pi a}{c} \qquad\qquad (r = 1, 2, \dots).$$

13. A simple periodic transverse force acts uniformly on all elements of a stretched string of length c with fixed ends, so that the transverse displacements $Y(x,t)$ satisfy the equation $Y_{tt} = a^2 Y_{xx} + B \sin \omega t$. Let the string be initially at rest in its position of equilibrium. Show that resonance occurs only when $\omega = \omega_r$ where

$$\omega_r = \frac{(2r - 1)\pi a}{c} \qquad\qquad (r = 1, 2, \dots).$$

14. The supports of the ends of a stretched string vibrate in a transverse direction so that

$$Y(0,t) = A \sin \alpha t, \qquad Y(c,t) = B \sin \beta t.$$

The string is initially at rest on the x axis. In general, show that resonance takes place if either α or β has one of the values $n\pi a/c$. But if $\beta = \alpha$, show that the resonance frequencies are $2n\pi a/c$ when $B = -A$, and $(2n - 1)\pi a/c$ when $B = A$, where $n = 1, 2, \dots$. *Hint:* When $\beta = \alpha$ and $B = -A$, show that

$$y(x,s) = \frac{A\alpha}{s^2 + \alpha^2} \frac{\sinh \left[s(c - 2x)/(2a)\right]}{\sinh \left[sc/(2a)\right]}.$$

90   FREE VIBRATIONS OF A STRING

A string stretched between the origin and a point $(c,0)$ is given a prescribed initial displacement $Y(x,0) = g(x)$ and released from rest in that position. To find its transverse displacements $Y(x,t)$, we solve the problem

$$Y_{tt}(x,t) = a^2 Y_{xx}(x,t) \qquad \left(0 < x < c, t > 0; a^2 = \frac{H}{\rho}\right),$$

$$Y(x,0) = g(x), \qquad Y_t(x,0) = Y(0,t) = Y(c,t) = 0.$$

The function g must naturally be continuous $(0 \leqq x \leqq c)$, and vanish when $x = 0$ and when $x = c$. Let g' be sectionally continuous.

The solution of the transformed problem in $y(x,s)$,

$$a^2 y'' - s^2 y = -sg(x), \qquad y(0,s) = y(c,s) = 0,$$

can be written (cf. Sec. 82)

$$y(x,s) = \frac{p(x,s)}{a \sinh (sc/a)}$$

when $s \neq 0$, where

$$p(x,s) = \sinh \frac{(c - x)s}{a} \int_0^x g(\xi) \sinh \frac{\xi s}{a} \, d\xi$$

$$+ \sinh \frac{xs}{a} \int_x^c g(\xi) \sinh \frac{(c - \xi)s}{a} \, d\xi.$$

When $s = 0$, the solution is $y(x,0) = 0$.

Now p is an entire function of s, and the singular points of y are the simple poles $s = \pm s_n$, where

$$s_n = \frac{in\pi a}{c} \qquad\qquad (n = 1, 2, \dots).$$

If we write

(1)
$$b_n = \frac{2}{c} \int_0^c g(\xi) \sin \frac{n\pi\xi}{c} \, d\xi,$$

then $p(x,s_n) = \frac{1}{2}b_n c \cos n\pi \sin (n\pi x/c)$, and the residue of y at s_n is $\frac{1}{2}b_n \sin (n\pi x/c)$. Thus the sum of the residues of $e^{st}y(x,s)$ at the pair of poles $\pm s_n$ is (Sec. 72)

$$b_n \sin \frac{n\pi x}{c} \cos \frac{n\pi at}{c}.$$

The formal solution of the boundary value problem is then

(2)
$$Y(x,t) = \sum_{n=1}^{\infty} b_n \sin \frac{n\pi x}{c} \cos \frac{n\pi at}{c},$$

where the coefficients b_n are given by formula (1).

Since $Y(x,0) = g(x)$, formula (2) implies that

$$\sum_{n=1}^{\infty} b_n \sin \frac{n\pi x}{c} = g(x) \qquad\qquad (0 \leq x \leq c).$$

Each term of this series is an odd periodic function of x with period $2c$. Consequently, for all x the series should represent the odd periodic extension G of g defined by the conditions

(3)
$$G(x) = g(x) \qquad\qquad \text{when } 0 \leq x \leq c,$$
$$G(-x) = -G(x), \qquad G(x + 2c) = G(x) \text{ for all } x.$$

Then G is everywhere continuous, and

(4)
$$\sum_{n=1}^{\infty} b_n \sin \frac{n\pi x}{c} = G(x) \qquad (-\infty < x < \infty).$$

Since the terms in series (2) can be written as

$$\frac{1}{2}\left[b_n \sin \frac{n\pi(x + at)}{c} + b_n \sin \frac{n\pi(x - at)}{2} \right],$$

our formal solution has the simple representation

(5)
$$Y(x,t) = \tfrac{1}{2}[G(x + at) + G(x - at)].$$

The function (5) is easily verified as a solution of our boundary value problem. Its form is a convenient one to use in studying the motion of the string.

91 RESONANCE IN A BAR WITH A MASS ATTACHED

The end $x = 0$ of an elastic bar is fixed. To the end $x = c$ a rigid mass M is attached, and a periodic longitudinal force $B \sin \omega t$ acts on that mass (Fig. 91). The elastic displacements within the mass itself are assumed negligible, and the bar is assumed to be too heavy to be considered as a coiled spring with negligible mass. We shall find the frequencies ω for which resonance takes place.

Let A denote the area of the cross section of the bar, m the mass of the bar, and $Y(x,t)$ the longitudinal displacements in the bar. The force exerted by the bar on the attached mass M is then $-AEY_x(c,t)$, and so the end conditions are

(1)
$$Y(0,t) = 0, \qquad MY_{tt}(c,t) = -AEY_x(c,t) + B \sin \omega t.$$

If the bar is initially at rest and unstrained, the remaining conditions in the boundary value problem are

(2)
$$Y_{tt}(x,t) = a^2 Y_{xx}(x,t) \qquad \left(0 < x < c, t > 0, a^2 = \frac{E}{\rho}\right),$$

(3)
$$Y(x,0) = Y_t(x,0) = 0.$$

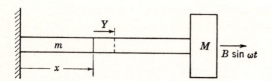

Fig. 91

This is also the problem of the torsional vibrations in a shaft with one end fixed and with a flywheel, on which a periodic torque acts, attached to the other end. Note that one end condition here involves a derivative of the second order. An interpretation of the problem in terms of torsional vibrations in a propeller shaft is given in the problems at the end of the chapter.

The problem in the transform y is

(4)
$$a^2 y''(x,s) - s^2 y(x,s) = 0,$$

$$y(0,s) = 0, \qquad AEy'(c,s) + Ms^2 y(c,s) = \frac{B\omega}{s^2 + \omega^2}.$$

Since $m = Ac\rho = AEc/a^2$, we can write the solution of problem (4) in terms of a function q, where

(5)
$$q(z) = Mz \sinh z + m \cosh z,$$

in this form:

(6)
$$y(x,s) = \frac{Bc\omega \sinh (sx/a)}{as(s^2 + \omega^2)q(sc/a)}.$$

Thus y is analytic in s except at the points $s = \pm i\omega$ and at the zeros of $q(sc/a)$. But if we write $z = \lambda + i\mu$, we find from Eq. (5) that

$$2|q(z)| = |(Mz + m)e^z - (Mz - m)e^{-z}|$$

$$\geq ||Mz + m|e^{\lambda} - |Mz - m|e^{-\lambda}|$$

$$= |[(M\lambda + m)^2 + M^2\mu^2]^{\frac{1}{2}}e^{\lambda} - [(M\lambda - m)^2 + M^2\mu^2]^{\frac{1}{2}}e^{-\lambda}|,$$

from which we can see that $|q(z)| > 0$ if $\lambda > 0$ or if $\lambda < 0$. Therefore $q(z)$ can vanish only if $\lambda = 0$, and thus $z = i\mu$, where μ is a root of the equation $-M\mu \sin \mu + m \cos \mu = 0$; that is,

(7)
$$\tan \mu = \frac{m}{M} \frac{1}{\mu}.$$

The graphs of the functions $\tan \mu$ and $m/(M\mu)$ show that Eq. (7) has an infinite sequence of roots $\pm\mu_n$ where $0 < \mu_1 < \pi/2$ and $\mu_n - (n - 1)\pi \to 0$ as $n \to \infty$ (Fig. 92). If $\omega = a\mu_n/c$, where n is some positive integer, then the

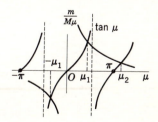

Fig. 92

two points $s = \pm ia\mu_n/c$ are poles of y of the second order and Y contains a term of resonance type $b_n(x)t \cos [\omega_n t + \theta_n(x)]$ with the resonance frequency

(8)
$$\omega_n = \frac{a\mu_n}{c} \qquad \left(\tan \mu_n = \frac{m}{M\mu_n}, n = 1, 2, \dots\right).$$

The representation of $Y(x,t)$ as a series of residues can be written in the usual way. Except for a term of resonance type in case $\omega = \omega_n$, each term of the series is periodic in t, but the terms have no common period.

92 TRANSVERSE VIBRATIONS OF BEAMS

Let $Y(x,t)$ denote the transverse displacement of a point at distance x from one end of a bar or beam, at time t. As in the case of static displacements discussed in Sec. 36, the instantaneous bending moment transmitted through a cross section is $EIY_{xx}(x,t)$, where I is the moment of inertia of the cross section with respect to its neutral axis. The shearing force at a cross section is $EI\partial^3 Y/\partial x^3$. When the cross section is uniform and no external force acts along the beam, the displacement function $Y(x,t)$ satisfies the equation

(1)
$$\frac{\partial^2 Y}{\partial t^2} + a^2\frac{\partial^4 Y}{\partial x^4} = 0 \qquad \left(a^2 = \frac{EI}{A\rho}\right)$$

under certain idealizing assumptions, where $A\rho$ is the mass of the beam per unit length.

Let the end $x = 0$ be hinged, so that no bending moment is transmitted across the section at $x = 0$ and the displacement is zero there. Let the end $x = c$ be hinged on a support which moves parallel to the Y axis in a simple harmonic manner (Fig. 93). If the beam is initially at rest along the x axis, the boundary conditions that accompany Eq. (1) are then

$$Y(x,0) = Y_t(x,0) = 0,$$

$$Y(0,t) = Y_{xx}(0,t) = 0,$$

$$Y(c,t) = B \sin \omega t, \qquad Y_{xx}(c,t) = 0.$$

Let us find the frequencies ω at which resonance will occur.

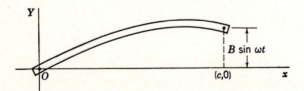

Fig. 93

The problem in the transform $y(x,s)$ is

(2)
$$a^2\frac{d^4y}{dx^4} + s^2y = 0,$$

$$y(0,s) = y_{xx}(0,s) = y_{xx}(c,s) = 0, \qquad y(c,s) = \frac{B\omega}{s^2 + \omega^2}.$$

It will be convenient to write

(3)
$$s = iaq^2.$$

Then $s^2 = -a^2q^4$, and the general solution of Eq. (2) can be written

(4)
$$y(x,s) = C_1 \sin qx + C_2 \cos qx + C_3 \sinh qx + C_4 \cosh qx,$$

where the C's can be functions of the parameter s. When those constants are determined so that the boundary conditions on $y(x,s)$ are satisfied, the solution (4) becomes

(5)
$$y(x,s) = \frac{B\omega}{s^2 + \omega^2}\frac{\sin qx \sinh qc + \sinh qx \sin qc}{2 \sin qc \sinh qc}.$$

The Maclaurin series representations of the sine and hyperbolic sine functions show that the final fraction in Eq. (5) is an analytic function of q^2, and therefore of s, except at points where the denominator vanishes. The point $s = 0$ is a removable singular point. The denominator vanishes when $qc = \pm n\pi$ and $qc = \pm in\pi$; that is, whenever $q^2c^2 = \pm n^2\pi^2$. In view of Eqs. (3) and (5) then, the singular points of $y(x,s)$ are

(6)
$$s = \pm i\omega, \qquad s = \pm\frac{in^2\pi^2 a}{c^2} \qquad (n = 1, 2, \dots).$$

If the value of ω is distinct from all the numbers

(7)
$$\omega_n = \frac{n^2\pi^2 a}{c^2} \qquad (n = 1, 2, \dots),$$

the derivative of the entire denominator in the expression (5) for $y(x,s)$ does not vanish at any of the points (6), nor does the numerator except for particular values of x. Those points are therefore simple poles of $y(x,s)$. In this case the formula for the displacements has the form

$$Y(x,t) = b_0(x) \cos[\omega t + \theta_0(x)] + \sum_{n=1}^{\infty} b_n(x) \cos[\omega_n t + \theta_n(x)].$$

If the value of ω coincides with one of the numbers ω_n given by formula (7), then y has poles of the second order at the points $s = \pm i\omega_n$, and a term of resonance type appears in $Y(x,t)$. Hence the resonance frequencies are the frequencies (7).

In case the hinge at $x = c$ is kept fixed and a simple harmonic bending moment acts on that end of the beam, the conditions at $x = c$ have the form

$$Y(c,t) = 0, \qquad Y_{xx}(c,t) = B \sin \omega t.$$

The reader can show that the expression for $y(x,s)$ then has the same denominator as it does in Eq. (5). The resonance frequencies are again those given by formula (7). Other cases are included in the problems at the end of the chapter.

93   DUHAMEL'S FORMULA FOR VIBRATION PROBLEMS

As in the case of problems in heat conduction (Sec. 85), the convolution property of the transform displays a relation between the solutions of problems in vibrations with variable boundary conditions and corresponding problems with fixed conditions. Consider for example the transverse displacements $Y(x,t)$ in a string. If both damping and an elastic support are present (Fig. 94), the equation of motion has the form

$$(1) \qquad Y_{tt}(x,t) = a^2 Y_{xx}(x,t) - b Y_t(x,t) - h Y(x,t).$$

To permit the end $x = 0$ to be elastically supported or kept fixed or to slide freely along the Y axis, we can write the condition

$$(2) \qquad\qquad\qquad \lambda_1(Y) = 0 \qquad\qquad\qquad \text{at } x = 0,$$

where $\lambda_1(Y) = h_1 Y - k_1 Y_x$. If a prescribed force $F(t)$ acts on the end $x = c$, a fairly general boundary condition is

$$(3) \qquad\qquad\qquad \lambda_2(Y) = F(t) \qquad\qquad\qquad \text{at } x = c,$$

where $\lambda_2(Y) = h_2 Y + k_2 Y_x$. Then if

$$(4) \qquad\qquad\qquad Y(x,0) = Y_t(x,0) = 0,$$

the transform $y(x,s)$ satisfies the conditions

$$(5) \qquad\qquad\qquad (s^2 + bs + h)y(x,s) = a^2 y_{xx}(x,s),$$

$$(6) \qquad\qquad \lambda_1[y(0,s)] = 0, \qquad \lambda_2[y(c,s)] = f(s).$$

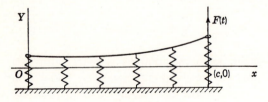

Fig. 94

Let $Z(x,t)$ represent the displacement $Y(x,t)$ in the special case in which $F(t) = 1$.

Then $z(x,s)$ satisfies Eq. (5) and the boundary conditions

$$\lambda_1[z(0,s)] = 0, \qquad \lambda_2[z(c,s)] = \frac{1}{s}.$$

It follows that the product $sf(s)z(x,s)$ satisfies the conditions (5) and (6), and therefore

(7) $$y(x,s) = sf(s)z(x,s).$$

In view of the convolution property then,

(8) $$Y(x,t) = \frac{\partial}{\partial t} \int_0^t F(t - \tau)Z(x,\tau)\,d\tau,$$

or

(9) $$Y(x,t) = \int_0^t F(t - \tau)Z_t(x,\tau)\,d\tau.$$

These are two forms of Duhamel's formula for the resolution of the problem in $Y(x,t)$ with a variable end condition into one with a fixed end condition.

The derivation of these formulas can be extended easily to other problems. In the case of transverse vibrations of bars, for instance, the derivative of the fourth order with respect to x replaces $Y_{xx}(x,t)$, and additional boundary conditions are involved. In the case of transverse displacements in a membrane, the Laplacian of Y replaces Y_{xx} in our problem. In these cases, the steps in the derivation of Duhamel's formula are the same as in the case treated above.

PROBLEMS

1. An elastic bar with its end $x = 0$ kept fixed is initially stretched so that its longitudinal displacements are $Y(x,0) = Ax$ $(0 \leq x \leq c)$, then released from rest in that position at the instant $t = 0$ with its end $x = c$ kept free. Derive these two formulas for its displacements:

$$Y(x,t) = \frac{8Ac}{\pi^2} \sum_{n=1}^{\infty} \frac{(-1)^{n-1}}{(2n-1)^2} \sin\frac{(2n-1)\pi x}{2c} \cos\frac{(2n-1)\pi at}{2c}$$

$$= \tfrac{1}{2}A[H(x + at) + H(x - at)],$$

where H is the triangular-wave function (2), Sec. 87 (Fig. 89). Verify the solution.

2. The end $x = 0$ of an elastic bar is elastically supported (Fig. 95) so that the longitudinal force exerted on that end is proportional to the longitudinal displacement:

$$Y_x(0,t) = hY(0,t) \qquad\qquad (h > 0).$$

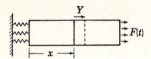

Fig. 95

When a longitudinal periodic force $F(t) = A \sin \omega t$ acts at the end $x = c$, show that resonance takes place if ω has any one of the values $a\alpha_n/c$ $(n = 1, 2, \ldots)$, where α_n is a positive root of the equation $\tan \alpha = hc/\alpha$. Show those roots graphically (cf. Fig. 92).

3. A string stretched from the origin to a point on the x axis is initially straight with a uniform velocity in the direction of the Y axis, as if a moving frame supporting the end points is brought to rest at the instant $t = 0$. There is a small damping force per unit length proportional to the velocity. Select units so that the transverse displacements satisfy the conditions

$$Y_{tt}(x,t) = Y_{xx}(x,t) - 2bY_t(x,t) \qquad (0 < x < \pi, t > 0, 0 < b < 1),$$

$$Y(0,t) = Y(\pi,t) = Y(x,0) = 0, \qquad Y_t(x,0) = v_0.$$

If p is a branch of the function $(s^2 + 2bs)^{\frac{1}{2}}$, show that

$$L\{Y\} = y(x,s) = \frac{v_0}{p^2} \frac{\cosh (\pi p/2) - \cosh (x - \pi/2)p}{\cosh (\pi p/2)}$$

and obtain formally this formula for displacements:

$$Y(x,t) = \frac{4v_0}{\pi} e^{-bt} \sum_{n=1}^{\infty} \frac{\sin (2n - 1)x \sin mt}{(2n - 1)m} \qquad [m^2 = (2n - 1)^2 - b^2].$$

4. A string is stretched from the origin to the point $(c,0)$ and given an initial velocity $Y_t(x,0) = g(x)$ but no initial displacement. Derive this formula for its displacements:

$$Y(x,t) = \frac{2}{\pi a} \sum_{n=1}^{\infty} \frac{1}{n} \sin \frac{n\pi x}{c} \sin \frac{n\pi at}{c} \int_0^c g(\xi) \sin \frac{n\pi \xi}{c} d\xi.$$

5. In Prob. 4 let $G(x)$ be the odd periodic extension, with period $2c$, of the velocity function $g(x)$ and show formally that

$$2Y_t(x,t) = G(x + at) + G(x - at)$$

and
$$Y(x,t) = \frac{1}{2a} \int_{x-at}^{x+at} G(\xi) d\xi.$$

Verify that integral form of Y as a solution of the problem.

6. Let $I(x,t)$ denote the current and $V(x,t)$ the voltage at distance x from one end of a transmission line, at time t. Following the notation used in the telegraph equation (11), Sec. 40, and indicated in Fig. 96, let R, L, S, and K represent resistance, inductance, leakage conductance, and capacitance to ground, respectively, per unit length of line. Then the functions $I(x,t)$ and $V(x,t)$ satisfy the system of partial differential equations

(a) \qquad\qquad $-V_x = RI + LI_t, \qquad -I_x = SV + KV_t,$

called the *transmission line equations*. By differentiating the members of equations (a) and eliminating one of the functions and its derivatives, show that either I or V satisfies

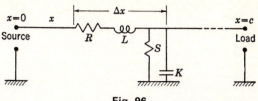

Fig. 96

the telegraph equation. Note that equations (a) are better adapted to initial and end conditions that involve values of both I and V, however, than is the telegraph equation (see Probs. 7 to 9).

7. Suppose that $R = S = 0$ in the transmission line shown in Fig. 96, and that the current and voltage are zero initially. If one end is kept grounded, $V(0,t) = 0$, and the current at the end $x = c$ is proportional to t, find the transform $v(x,s)$ of $V(x,t)$ using the transmission line equations (a), Prob. 6. Thus show that $V(x,t)$ is an electrical analog of the displacements $Y(x,t)$ in the bar in Sec. 86, where a^2 corresponds to $(KL)^{-1}$.

8. In the transmission line shown in Fig. 96, $R = S = 0$ and $I(x,0) = V(x,0) = 0$. If $V(0,t) = A \sin \omega t$ and the end $x = c$ is grounded, find the values of ω for which the voltage and current become unstable, using the transmission line equations (a), Prob. 6.   *Ans.* $\omega = n\pi c^{-1}(KL)^{-\frac{1}{2}}$ $(n = 1, 2, \ldots)$.

9. Solve Prob. 8 when the condition that the end $x = c$ is grounded is replaced by the condition that the circuit is open at that end.
$$Ans. \quad \omega = (n - \tfrac{1}{2})\pi c^{-1}(KL)^{-\frac{1}{2}} \quad (n = 1, 2, \ldots).$$

10. One end of a beam is built into a rigid support (Fig. 97); thus if $Y(x,t)$ denotes transverse displacements,

$$Y(0,t) = Y_x(0,t) = 0.$$

Fig. 97

The pin-support of the other end $x = c$ vibrates, so that

$$Y(c,t) = B \sin \omega t, \qquad Y_{xx}(c,t) = 0.$$

Initially the beam is at rest along the x axis. If a^2 is the coefficient in Eq. (1), Sec. 92, show that resonance occurs in the vibration of this beam if $\omega = a\alpha_n^2/c^2$, where α_n is any positive root of the equation $\tan \alpha = \tanh \alpha$. Show how the roots α_n can be approximated graphically.

11. One end of a bar of length c is free. The other end is built into a support that undergoes a transverse vibration $Y = B \sin \omega t$ (Fig. 98). The bar is initially at rest with no displacements. Show that resonance takes place in the transverse vibrations of the bar when $\omega = a\alpha_n^2/c^2$, where α_n is any positive root of the equation $\cos \alpha = -\operatorname{sech} \alpha$ and a^2 is the coefficient in Eq. (1), Sec. 92. Also show how the roots α_n can be approximated graphically.

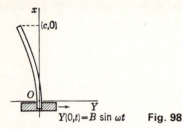

$$Y(0,t) = B \sin \omega t \qquad \textbf{Fig. 98}$$

12. Both ends of a beam of length c are built into rigid supports. The beam is initially at rest with no displacements. A simple periodic force per unit length acts along the entire span, in a direction perpendicular to the beam, so that the transverse displacements $Y(x,t)$ satisfy the equation

$$\frac{\partial^2 Y}{\partial t^2} + a^2 \frac{\partial^4 Y}{\partial x^4} = B \sin \omega t \qquad \left(a^2 = \frac{EI}{A\rho}\right).$$

Show that resonance occurs when $\omega = 4a\alpha_n^2/c^2$, where α_n is any positive root of the equation $\tan \alpha = -\tanh \alpha$.

13. A membrane, stretched across a fixed circular frame $r = c$, is initially at rest in its position of equilibrium. If a simple periodic force per unit area in a direction perpendicular to the membrane acts at all points, the transverse displacements $Z(r,t)$ satisfy an equation of the type

$$Z_{tt} = b^2 \left(Z_{rr} + \frac{1}{r}Z_r\right) + A \sin \omega t.$$

Show that the resonance frequencies are $\omega = b\alpha_n/c$, where α_n is any positive root of the equation $J_0(\alpha) = 0$.

14. The end $x = 0$ of the propeller shaft for a ship is connected rigidly to a massive flywheel that turns with uniform angular velocity ω. The end $x = c$ supports a propeller whose moment of inertia with respect to the axis of the shaft is I. In addition to the steady-state torque transmitted between shaft and propeller, the action in the water induces a perturbation in the nature of a torque proportional to $\sin 4\omega t$ exerted on the propeller. Let $\Theta(x,t)$ denote the corresponding perturbation in the angular displacements of the circular cross sections of the shaft. Thus $\Theta(x,t) = 0$ when the shaft operates under steady-state conditions with angular velocity ω. If I_p is the polar moment of inertia of the cross section of the shaft, of radius r_0 $(I_p = \frac{1}{2}r_0^4)$, and if E_s is the modulus of elasticity in shear for the material in the shaft, show that

$$I\Theta_{tt}(c,t) = -E_s I_p \Theta_x(c,t) + B \sin 4\omega t,$$

and that the boundary value problem in $\Theta(x,t)$ is an analog of the problem in Sec. 91, where a^2 now represents E_s/ρ and ρ is mass of the material per unit volume. Show that torsional resonance takes place when $\omega = a\lambda_n/(4c)$, where λ_n is any positive root of the equation $\tan \lambda = k/\lambda$ and where $k = c\rho I_p/I$.

15. The propeller shaft in Prob. 14 is 150 ft long with a diameter of 10 in., and made of steel. The moment of inertia I of the propeller equals that of a steel disk 4 in. thick and 4 ft in diameter. The steel weighs 0.28 lb/in.^{3} ($g\rho = 0.28$), and $E_s = 12 \times 10^6$ lb/in.^{2} Show that the lowest flywheel speed at which torsional resonance takes place is approximately 135 revolutions per minute.

9

Generalized Fourier Series

We shall use theory developed in Chap. 6, on the inversion integral for the Laplace transformation, to assist us in establishing some basic theory of eigenvalue problems in differential equations. The theory shows that the solutions of certain types of homogeneous boundary value problems in ordinary differential equations form orthogonal sets of functions, and that an arbitrary function can be represented by a series of functions of such a set, a generalized Fourier series.

A classical method of solving some types of linear boundary value problems in partial differential equations begins with a separation of variables, then requires a representation of a prescribed function by a generalized Fourier series. The orthogonal functions in the series are the solutions of an eigenvalue problem that arises from the separation of variables. The method will be illustrated at the end of this chapter.

In the following chapter we shall see how such eigenvalue problems are encountered in a process of designing an integral transformation to fit the requirements of a given boundary value problem. Then the inverse transformation is represented by the corresponding generalized Fourier series.

94 SELF-ADJOINT DIFFERENTIAL EQUATIONS

We begin with some concepts and an existence theorem from the theory of linear ordinary differential equations.

The *adjoint* D_2^*y of the linear differential form

$$(1) \qquad D_2y = \alpha_2(x)y''(x) + \alpha_1(x)y'(x) + \alpha_0(x)y(x)$$

of the second order is this related form:

$$(2) \qquad D_2^*y = [\alpha_2(x)y(x)]'' - [\alpha_1(x)y(x)]' + \alpha_0(x)y(x).$$

It is useful in constructing forms that represent derivatives of other forms. With the aid of a second function $v(x)$, for instance, a form that is an exact derivative can be written

$$(3) \qquad vD_2y - yD_2^*v = \frac{d}{dx}[\alpha_2vy' - (\alpha_2v)'y + \alpha_1vy].$$

The *Lagrange identity* (3) is easily verified.

We find that the form

$$(4) \qquad \mathscr{D}_2y = [r(x)y'(x)]' - Q(x)y(x)$$

is *self-adjoint*, $\mathscr{D}_2^*y = \mathscr{D}y$. For it, Lagrange's identity becomes

$$(5) \qquad v\mathscr{D}_2y - y\mathscr{D}_2v = \frac{d}{dx}[(vy' - v'y)r].$$

When both members of the differential equation

$$(6) \qquad y''(x) + \beta_1(x)y'(x) + \beta_0(x)y(x) = f(x)$$

are multiplied by an integrating factor

$$(7) \qquad r(x) = \exp\left[\int^x \beta_1(x)\,dx\right],$$

the *equation* assumes its *self-adjoint* form

$$(8) \qquad [r(x)y'(x)]' - Q(x)y(x) = F(x),$$

where $Q(x) = -r(x)\beta_0(x)$ and $F(x) = r(x)f(x)$.

Adjoints of differential forms of order higher than 2 are defined in a manner that corresponds to the definition (2) (Prob. 3, Sec. 97). But linear differential equations of higher order do not always have self-adjoint forms.

An existence theorem for solutions of the homogeneous form of Eq. (6), in which the coefficient $\beta_0(x)$ may involve a parameter λ in a linear manner, will now be cited. The theorem can be proved by a method of successive approximations.[1]

THEOREM

Let A, B, and C be continuous functions of x over some closed interval I and let x_0 be a fixed point on I. Let λ denote a complex-valued parameter and y_0 and y_1 two prescribed constants independent of λ. Then the *initial-value problem*

(9)
$$y'' + A(x)y' + [B(x) + \lambda C(x)]y = 0,$$
$$y(x_0,\lambda) - y_0, \qquad y'(x_0,\lambda) - y_1$$

in the complex-valued function $y(x,\lambda)$ has one and only one solution $y = u(x,\lambda)$ such that u and u' are continuous functions of x and λ for all λ and all x on the interval I and, for each x on I, u and u' are *entire* functions of the complex parameter λ.

In Sec. 96 we shall see that further restrictions are needed to ensure a unique solution of a *two-point boundary value problem* in linear ordinary differential equations. This fact is illustrated by the simple problem

(10)
$$y'' + \lambda y = 0, \qquad y(0,\lambda) = 0, \qquad y(1,\lambda) = 0,$$

which clearly has a solution $y(x,\lambda) \equiv 0$. But when the parameter λ has any one of the values $n^2\pi^2$ $(n = 1, 2, \ldots)$, problem (10) has the additional solutions

(11)
$$y = k \sin n\pi x,$$

where the constant k is arbitrary. Also, note that the problem

(12)
$$y''(x) + \pi^2 y(x) = 0, \qquad y(0) = 0, \qquad y(1) = 1,$$

has no solution, since the function $k \sin \pi x$ that satisfies the first two conditions fails to satisfy the condition $y(1) = 1$ for any value of the constant k.

95 GREEN'S FUNCTIONS

On an interval $a \leqq x \leqq b$ let r, Q, and F denote prescribed continuous functions of x such that r' is also continuous and $r(x) > 0$. In the boundary value problem

(1)
$$[r(x)y'(x)]' - Q(x)y(x) = F(x), \qquad y(a) = y(b) = 0,$$

---

[1] See, for instance, E. L. Ince, "Ordinary Differential Equations," pp. 72 ff., 1927, also, p. 123 for more on adjoints, or E. C. Titchmarsh, "Eigenfunction Expansions," p. 6, 1946.

the unknown function y can be interpreted as the *static* transverse displacements in a stretched string attached to an elastic foundation. The elastic support may have variable properties so that it contributes a transverse force $-Q(x)y(x)$ per unit length as well as a variable tension $r(x)$ in the string. An external transverse force $-F(x)$, per unit length, acts along the string. (Cf. Secs. 40 and 93.) The ends are fixed at the points $(a,0)$ and $(b,0)$. We make the interpretation to assist us in writing the solution of problem (1), at first formally, in terms of solutions of simpler problems.

Let $F(x)$ be replaced formally by the unit impulse symbol $\delta(x - \xi)$, where $a < \xi < b$. This corresponds to the application of a negative unit transverse force at the point $x = \xi$ (Fig. 99). When the string has that concentrated load, we write $G(x,\xi)$ for the displacements $y(x)$. Since the vertical force from left to right is $-r(x)y'(x)$, we note that the value of $-ry'$ must jump by the amount -1 at $x = \xi$; that is,

$$(2) \qquad G_x(\xi + 0, \xi) - G_x(\xi - 0, \xi) = \frac{1}{r(\xi)}.$$

The displacement function $G(x,\xi)$ itself should be continuous in the square region $a \leqq x \leqq b$, $a \leqq \xi \leqq b$, and its derivative $G_x(x,\xi)$ should be continuous there except for the jump prescribed by Eq. (2) at the diagonal $x = \xi$. Since $\delta(x - \xi)$ is zero except when $x = \xi$,

$$(3) \qquad [r(x)G_x(x,\xi)]_x - Q(x)G(x,\xi) = 0 \qquad\qquad (x \neq \xi);$$

therefore, in view of the conditions imposed on r and Q, the function $G_{xx}(x,\xi)$ is continuous except at the diagonal $x = \xi$. Also, $G(x,\xi)$ satisfies the boundary conditions

$$(4) \qquad G(a,\xi) = G(b,\xi) = 0.$$

A function that satisfies conditions (2) to (4) and the stated continuity conditions is called *Green's function* for problem (1). Specifically, and for a problem with general linear homogeneous boundary conditions, we describe the function as follows.

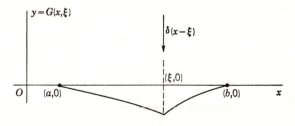

Fig. 99

DEFINITION

Green's function $G(x,\xi)$ for this boundary value problem with self-adjoint differential equation:

(5) $$[r(x)y'(x)]' - Q(x)y(x) = F(x),$$

(6) $$A_1 y(a) + A_2 y'(a) = 0, \qquad B_1 y(b) + B_2 y'(b) = 0,$$

is a solution of the homogeneous differential equation (3) when $a < x < \xi$ and when $\xi < x < b$ that satisfies the following conditions. It is, continuous in x and ξ together over the square $a \leq x \leq b$, $a \leq \xi \leq b$ (Fig. 100). Its derivative G_x is continuous over the closed triangle T_1 ($a \leq x \leq b$, $a \leq \xi \leq x$) when properly defined on the diagonal $x = \xi$, and continuous over the closed triangle T_2 when again properly defined on the diagonal, but G_x has the jump $1/r(\xi)$ from T_1 to T_2 at $x = \xi$ as specified by condition (2). Also, when ξ is fixed ($a < \xi < b$), G satisfies boundary conditions (6):

(7) $$A_1 G(a,\xi) + A_2 G_x(a,\xi) = B_1 G(b,\xi) + B_2 G_x(b,\xi) = 0.$$

Of course, one of the A's and one of the B's here must not vanish. A convenient way to specify this is to write

(8) $$A_1 = \cos\alpha, \qquad A_2 = \sin\alpha, \qquad B_1 = \cos\beta, \qquad B_2 = \sin\beta,$$

where α and β are constants. Note that G_{xx} has the continuity properties of G_x, in view of Eq. (3).

We now indicate how the solution of problem (1) may be written as a weighted superposition of solutions G of the problem in which the load $F(x)$ is replaced by concentrated loads $\delta(x - \xi)$.

The impulse symbol has the formal properties $\delta(-x) = \delta(x)$ and

$$F(x) = \int_a^b F(\xi)\delta(\xi - x)\,d\xi = \int_a^b F(\xi)\delta(x - \xi)\,d\xi;$$

that is, $F(x)$ is a superposition of impulses. If all members of the symbolic equations

$$[r(x)G_x(x,\xi)]_x - Q(x)G(x,\xi) = \delta(x - \xi),$$

$$G(a,\xi) = G(b,\xi) = 0,$$

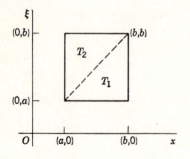

Fig. 100

are multiplied by $F(\xi)$ and integrated with respect to ξ from $\xi = a$ to $\xi = b$, the result is the three statements in problem (1) when we write

$$(9) \qquad\qquad y(x) = \int_a^b F(\xi)G(x,\xi)\,d\xi$$

and identify $(ry')'$ with the symbol $\int_a^b F(\xi)[r(x)G_x(x,\xi)]_x\,d\xi$.

In order to verify the formal solution (9), we first note that $y(a) = y(b) = 0$ because $G(x,\xi)$ satisfies the boundary conditions (4). When we write

$$y(x) = \int_a^x F(\xi)G(x,\xi)\,d\xi + \int_x^b F(\xi)G(x,\xi)\,d\xi,$$

the integrand of each integral and its first two partial derivatives with respect to x are continuous functions of (x,ξ) in the triangular regions T_1 and T_2, respectively, in Fig. 100. Differentiation by the Leibnitz formula (Sec. 14) is therefore valid. We find that

$$y'(x) = \int_a^x F(\xi)G_x(x,\xi)\,d\xi + \int_x^b F(\xi)G_x(x,\xi)\,d\xi$$

since G is continuous, and that

$$(10) \qquad\qquad (ry')' = \int_a^b F(\xi)[r(x)G_x(x,\xi)]_x\,d\xi + F(x)r(x)J(x),$$

where $J(x) = G_x(x, x - 0) - G_x(x, x + 0)$. Thus $J(x)$ is the jump in the value of G_x as the point (x,ξ) moves across the diagonal from triangle T_1 into triangle T_2 (Fig. 100). According to Eq. (2), $J(x) = 1/r(x)$. In view of Eq. (3), Eq. (10) can now be written

$$(ry')' = Q(x)\int_a^b F(\xi)G(x,\xi)\,d\xi + F(x) = Qy + F\,;$$

that is, the function (9) also satisfies the differential equation in problem (1).

The solution (9) of problem (1) is now established if Green's function G exists. In fact the solution is valid if F is a sectionally continuous function. For modifications of the procedure when the boundary conditions are type (6), see Probs. 8 and 9, Sec. 97.

96   CONSTRUCTION OF GREEN'S FUNCTION

A function $f(x)$ is of *class C' on an interval* $a \leq x \leq b$ if both f and f' are continuous over the interval.

Let r be of class C' over that interval, where $r(x) > 0$, and let Q be continuous there. Then, according to the existence theorem stated in Sec. 94, the initial-value problem

(1) $[r(x)u'(x)]' - Q(x)u(x) = 0,$ $u(a) = 0,$ $u'(a) = 1,$

has a unique solution $u(x)$ of class C' on that interval. Let $v(x)$ be the unique solution, of that class, of the initial-value problem

(2) $[r(x)v'(x)]' - Q(x)v(x) = 0,$ $v(b) = 0,$ $v'(b) = 1.$

Then on the square region $a \leq x \leq b, a \leq \xi \leq b$, the function

(3) $$G(x,\xi) = \begin{cases} A(\xi)u(x) & (x \leq \xi) \\ B(\xi)v(x) & (x \geq \xi), \end{cases}$$

where A and B are continuous functions, satisfies the self-adjoint differential equation (3), Sec. 95, and the boundary conditions

$$G(a,\xi) = 0,   G(b,\xi) = 0.$$

It is continuous there, on the diagonal $x = \xi$ in particular, and its derivative G_x has the jump prescribed by Eq. (2), Sec. 95, at $x = \xi$ if A and B satisfy the simultaneous equations

(4) $B(\xi)v(\xi) - A(\xi)u(\xi) = 0,$ $B(\xi)v'(\xi) - A(\xi)u'(\xi) = \dfrac{1}{r(\xi)}.$

According to Lagrange's identity (5), Sec. 94, the determinant $-v(\xi)u'(\xi) + u(\xi)v'(\xi)$ of system (4) satisfies the condition

(5) $-\dfrac{d}{d\xi}[(vu' - uv')r] = -[(ru')' - Qu]v + [(rv')' - Qv]u.$

The right-hand member vanishes since u and v satisfy the differential equations in problems (1) and (2). Hence $(vu' - uv')r$ is constant and, since $v(b) = 0$ and $v'(b) = 1$,

$$-[v(\xi)u'(\xi) - u(\xi)v'(\xi)]r(\xi) = u(b)r(b).$$

Then in case $u(b) \neq 0$, we find that the solution of system (4) is

$$B(\xi) = \frac{u(\xi)}{u(b)r(b)},   A(\xi) = \frac{v(\xi)}{u(b)r(b)},$$

and formula (3) for Green's function becomes

(6) $G(x,\xi) = \dfrac{u(x)v(\xi)}{u(b)r(b)}$ when $x \leq \xi$,

                          $= \dfrac{u(\xi)v(x)}{u(b)r(b)}$ when $x \geq \xi$.

Note that G is *symmetric* in its two arguments:

(7) $$G(x,\xi) = G(\xi,x),$$

a consequence of the self-adjoint character of the differential equation.

In case $u(b) = 0$ then, since $u(a) = 0$ and $u'(a) = 1$, the function $w = u(x)$ is not identically zero, and it is a solution of class C' of the *homogeneous problem*

(8) $$(rw')' - Qw = 0, \qquad w(a) = w(b) = 0.$$

Since v satisfies the same differential equation and $v(b) = 0$, it follows that $v(x) = ku(x)$ when $u(b) = 0$, where $ku'(b) = 1$. The following theorem is now established.

Theorem 1 *If the homogeneous problem* (8) *has no solution of class C' on the interval $a \leq x \leq b$ other than $w(x) \equiv 0$, then Green's function $G(x,\xi)$ exists for the boundary value problem*

(9) $$[r(x)y'(x)]' - Q(x)y(x) = F(x), \qquad y(a) = y(b) = 0,$$

where r is of class C', Q is continuous, F is sectionally continuous, and $r(x) > 0$, on the interval $a \leq x \leq b$. Green's function is expressed in terms of solutions of the initial-value problems (1) *and* (2) *by formula* (6); *it is symmetric in x and ξ. The solution, of class C', of problem* (9) *is*

(10) $$y(x) = \int_a^b F(\xi)G(x,\xi)\,d\xi.$$

Although the initial-value problems (1) and (2) are simpler than the boundary value problem (9), they can be solved explicitly only for limited types of coefficients r and Q.

Under the conditions assumed in Theorem 1, suppose that there is another solution $z(x)$ of problem (9), in addition to $y(x)$. Then the function $w = y - z$ satisfies the homogeneous problem (8) which has only the solution $w(x) \equiv 0$. Thus $z = y$, and we have the following theorem on uniqueness.

Theorem 2 *Under the conditions stated in Theorem 1, the solution* (10) *of problem* (9) *is the only one of class C'.*

Examples and additional material on Green's functions are given in the problems following Sec. 97.[1]

97 ORTHOGONAL SETS OF FUNCTIONS

Consider now real-valued functions of x that are *continuous* over an interval $a \leq x \leq b$. A function f may be called a generalized vector, with one component $f(x)$ corresponding to each value of x on the interval (Sec. 10).

[1] For further theory on Green's functions for ordinary and partial differential equations see, for instance, R. Courant and D. Hilbert, "Methods of Mathematical Physics," vol. 1, 1953.

A generalized scalar product, the *inner product* (f,g), of two functions f and g is defined as the sum of products of corresponding components:

$$(1) \qquad\qquad (f,g) = \int_a^b f(x)g(x)\,dx = (g,f).$$

The *norm* of a function f, the generalized concept of the length of a vector, is $(f,f)^{\frac{1}{2}}$; that is,

$$(2) \qquad\qquad \|f\| = \left\{ \int_a^b |f(x)|^2\,dx \right\}^{\frac{1}{2}}.$$

Two functions f and g are *orthogonal* on the interval if $(f,g) = 0$. This signifies nothing about actual perpendicularity, but instead that the product of the functions changes sign in such a way that

$$(3) \qquad\qquad \int_a^b f(x)g(x)\,dx = 0.$$

Corresponding to units of length that differ from one coordinate axis to another in vector analysis, we introduce a nonnegative *weight function p* and define the inner product with respect to p, of two functions, as

$$(4) \qquad\qquad (f,g) = \int_a^b p(x)f(x)g(x)\,dx = (g,f) \qquad\qquad [p(x) \geq 0].$$

The norm $\|f\|$ is then $(f,f)^{\frac{1}{2}}$; and f and g are orthogonal if $(f,g) = 0$.

Let a set of functions y_n $(n = 1, 2, \ldots)$ be orthogonal on the interval, with weight function p. It is convenient to normalize the set by dividing each function by its norm. Thus the set

$$(5) \qquad\qquad \phi_n(x) = \frac{y_n(x)}{\|y_n\|} \qquad\qquad (a \leq x \leq b, n = 1, 2, \ldots)$$

is orthogonal with unit norms, an *orthonormal set*. It corresponds to a set of mutually perpendicular unit reference vectors. A prescribed function f may possibly be represented as a generalized linear combination, an infinite series, of those orthonormal functions,

$$(6) \qquad\qquad f(x) = \sum_{n=1}^{\infty} c_n \phi_n(x) \qquad\qquad (a < x < b).$$

If so, and if it is true that $(\phi_m, f) = \sum_{n=1}^{\infty} c_n(\phi_m, \phi_n)$, we can determine the constants c_n. For $(\phi_m, \phi_n) = 0$ when $n \neq m$, and $\|\phi_m\| = 1$. Thus $(f, \phi_m) = c_m$; that is

$$(7) \qquad\qquad c_n = (f, \phi_n) = \int_a^b p(x)f(x)\phi_n(x)\,dx \qquad\qquad (n = 1, 2, \ldots).$$

Series (6) with coefficients (7) is the *generalized Fourier series* for f corresponding to the orthonormal set of functions ϕ_n. The coefficients c_n are called the *Fourier constants* for the function f. In case the orthogonal sets are generated by certain eigenvalue problems known as Sturm-Liouville problems, we shall establish useful conditions on f under which its generalized Fourier series converges to $f(x)$.

The set of functions $y_n(x) = \sin nx$ is a well-known example of orthogonal sets. The set is orthogonal with unit weight function on the interval $(0,\pi)$ and $\|y_n\|^2 = \pi/2$ for each n; thus the orthonormal set consists of the functions

$$(8) \qquad\qquad \phi_n(x) = \sqrt{\frac{2}{\pi}} \sin nx \qquad (0 < x < \pi, n = 1, 2, \ldots).$$

For this set series (6) becomes the *Fourier sine series*:

$$f(x) = \sqrt{\frac{2}{\pi}} \sum_{n=1}^{\infty} c_n \sin nx = \frac{2}{\pi} \sum_{n=1}^{\infty} \int_0^{\pi} f(\xi) \sin n\xi \, d\xi \, \sin nx,$$

which converges to $f(x)$ when $0 < x < \pi$ at each point where f is continuous, if f' is sectionally continuous on the interval.[1]

PROBLEMS

1. Verify Lagrange's identities (3) and (5), Sec. 94.

2. Use Lagrange's identity to show that $v(x)$ is an integrating factor for the differential equation

$$D_2 y = f(x), \qquad \text{where } D_2 y = \alpha_2(x)y''(x) + \alpha_1(x)y'(x) + \alpha_0(x)y(x),$$

if v is not identically zero and satisfies the homogeneous equation $D_2^* v = 0$, where D_2^* is the adjoint of D_2. Thus show that $D_2 y = f$ reduces to this equation of first order:

$$\alpha_2 v y' + [\alpha_1 v - (\alpha_2 v)']y = \int v(x)f(x)\,dx.$$

3. The *adjoint of the linear differential form*

$$D_n y = \alpha_0(x)y(x) + \alpha_1(x)y'(x) + \alpha_2(x)y''(x) + \cdots + \alpha_n(x)y^{(n)}(x)$$

of order n, is the form

$$D_n^* y = \alpha_0 y - (\alpha_1 y)' + (\alpha_2 y)'' - \cdots + (-1)^n(\alpha_n y)^{(n)}.$$

When $n = 3$, establish the *Lagrange identity*

$$v D_3 y - y D_3^* v = \frac{d}{dx} B_3(v,y).$$

[1] See the author's "Fourier Series and Boundary Value Problems," 2d ed., 1963; also, orthogonal sets are treated somewhat more fully in Chap. 3 of that book.

where B_3 is this bilinear form in v and y:

$$B_3 = \alpha_1 vy + [\alpha_2 vy' - (\alpha_2 v)'y] + [\alpha_3 vy'' - (\alpha_3 v)'y' + (\alpha_3 v)''y].$$

4.   Show that this special case of $D_n y$ in Prob. 3,

$$D_4 y = \alpha_0(x)y(x) + y^{(4)}(x),$$

is self-adjoint, and that it satisfies the Lagrange identity

$$vD_4 y - yD_4 v = \frac{d}{dx}(vy''' - v'y'' + v''y' - v'''y).$$

5.   Find Green's function for the elementary problem

$$y''(x) = F(x), \qquad y(0) = 0, \qquad y(1) = 0,$$

and solve for y in the form

$$y(x) = (x - 1)\int_0^x \xi F(\xi)\,d\xi + x\int_x^1 (\xi - 1)F(\xi)\,d\xi.$$

Ans. $G(x,\xi) = x(\xi - 1)$ when $x \le \xi$, $G(x,\xi) = G(\xi,x)$.

6.   Obtain Green's function (4), Sec. 82, for the problem

$$u''(x,s) - su(x,s) = -g(x), \qquad u(0,s) = u(1,s) = 0 \qquad (s > 0).$$

7.   Let F be a sectionally continuous function, with only one jump at a point x_0, for the sake of simplicity, on the interval $a < x < b$. Carry out the verification of the solution (Sec. 95)

$$y(x) = \int_a^b F(\xi)G(x,\xi)\,d\xi$$

of the boundary value problem

$$[r(x)y'(x)]' - Q(x)y(x) = F(x), \qquad y(a) = y(b) = 0 \qquad (x \ne x_0),$$

where G is Green's function for the problem. Note that the solution $y(x)$ is of class C' if $y'(x_0)$ is defined as $\int_a^b F(\xi)G_x(x_0,\xi)\,d\xi$.

8.   Let the constants α and β be real and let $w = u(x)$ and $w = v(x)$ be the solutions of the homogeneous self-adjoint equation $(rw')' - Qw = 0$ that satisfy these initial conditions:

$$u(a) = \sin \alpha, \qquad u'(a) = -\cos \alpha,$$

$$v(b) = \sin \beta, \qquad v'(b) = -\cos \beta.$$

Construct Green's function for this problem (Sec. 95):

$$[r(x)y'(x)]' - Q(x)y(x) = F(x) \qquad (a < x < b),$$

$$y(a)\cos \alpha + y'(a)\sin \alpha = 0, \qquad y(b)\cos \beta + y'(b)\sin \beta = 0,$$

in case $c \ne 0$, where

$$c = r(a)[v(a)\cos \alpha + v'(a)\sin \alpha]$$

$$= -r(b)[u(b)\cos \beta + u'(b)\sin \beta],$$

in the form

$$G(x,\xi) = \frac{u(x)v(\xi)}{c} \qquad\qquad \text{when } x \leqq \xi$$

$$= \frac{u(\xi)v(x)}{c} \qquad\qquad \text{when } \xi \leqq x.$$

9. When F is continuous $(a \leqq x \leqq b)$, verify the solution

$$y(x) = \int_a^b F(\xi)G(x,\xi)\,d\xi$$

of the boundary value problem in y for which Green's function G was constructed in Prob. 8.

10. When the constant k is neither zero nor an integer, use Green's function (Probs. 8 and 9) to solve this problem for y:

$$y''(x) + k^2 y(x) = F(x), \qquad y'(0) = 0, \qquad y'(\pi) = 0.$$

Ans. $y(x)k \sin k\pi = \displaystyle\int_0^x F(\xi) \cos k\xi \, d\xi \cos k(\pi - x) + \int_x^\pi F(\xi) \cos k(\pi - \xi) \, d\xi \cos kx.$

11. Write the differential equation in the problem

$$(x + 1)y''(x) + y'(x) = F(x), \qquad y(0) = 0, \qquad y'(1) = 0,$$

in self-adjoint form and use Green's function (Probs. 8 and 9) to solve the problem in the form

$$y(x) = -\int_0^x F(\xi) \log(\xi + 1)\,d\xi - \log(x + 1)\int_x^1 F(\xi)\,d\xi.$$

12. Interpret the symmetry property $G(x,x_0) = G(x_0,x)$ of Green's function in terms of displacements of the string considered in Sec. 95 and concentrated loads, at points x and x_0.

13. Suppose a function w of class C' is not identically zero and satisfies the homogeneous problem

$$[r(x)w'(x)]' - Q(x)w(x) = 0, \qquad w(a) = w(b) = 0.$$

Then if the problem

$$[r(x)y'(x)]' - Q(x)y(x) = F(x), \qquad y(a) = y(b) = 0,$$

has a solution of class C', use Lagrange's identity to show that F must be orthogonal to w on the interval (a,b):

$$\int_a^b F(x)w(x)\,dx = 0.$$

Illustrate that result using the problem

$$y''(x) + y(x) = -3 \sin 2x, \qquad y(0) = y(\pi) = 0,$$

which has the family of solutions $y = C \sin x + \sin 2x$.

14. Show that the set of functions

$$\frac{1}{\sqrt{c}}, \qquad \sqrt{\frac{2}{c}}\cos\frac{n\pi x}{c} \qquad\qquad (n = 1, 2, \ldots)$$

is orthonormal on the interval $(0,c)$ with unit weight function, and that the generalized Fourier series for a function f, with respect to the set, is the *Fourier cosine series*

$$\frac{1}{c}\int_0^c f(\xi)\,d\xi + \frac{2}{c}\sum_{n=1}^{\infty}\cos\frac{n\pi x}{c}\int_0^c f(\xi)\cos\frac{n\pi\xi}{c}\,d\xi \qquad (0 < x < c).$$

98 EIGENVALUE PROBLEMS

The homogeneous two-point boundary value problem

$$[r(x)y'(x,\lambda)]' - [q(x) + \lambda p(x)]y(x,\lambda) = 0 \qquad (a < x < b)$$
(1)
$$A_1 y(a,\lambda) + A_2 y'(a,\lambda) = 0, \qquad B_1 y(b,\lambda) + B_2 y'(b,\lambda) = 0,$$

involving a parameter λ in the manner indicated, is called a *Sturm-Liouville problem or system*.[1] The constants A_1, A_2, B_1, and B_2 are real and independent of λ. The functions p, q, r, and r' are real-valued and continuous, and $p(x) > 0$ and $r(x) > 0$, over the entire interval $a \leqq x \leqq b$.

Problem (1) has the trivial solution $y = 0$, for every value of λ. But for certain values λ_n of λ, called *eigenvalues* or *characteristic numbers*, the problem has a nontrivial solution

$$y(x,\lambda_n) = y_n(x),$$

an *eigenfunction* or *characteristic function*. Note that $y = y_m(x)$ and $y = y_n(x)$ are solutions of different problems obtained by writing λ_m and λ_n, respectively, for λ in problem (1). Also note that if $y_n(x)$ is an eigenfunction, then $Cy_n(x)$ is also one, where C is any constant other than zero.

The Sturm-Liouville problem is a particular type of eigenvalue problem. Such problems arise in the process of solving boundary value problems in partial differential equations by the classical method of separation of variables, a method we shall illustrate in this chapter.[2]

Since eigenfunctions satisfy the Sturm-Liouville differential equation, their derivatives of second order are to exist over the open interval (a,b). We assume the functions are of class C' (Sec. 96) on the interval $a \leqq x \leqq b$, and show later on that eigenfunctions of that class do exist. Also, we write the Sturm-Liouville equation in terms of our self-adjoint operator $\mathscr{D}_2$ in the form

$$\mathscr{D}_2 y - \lambda p y = 0, \qquad \text{where } \mathscr{D}_2 y = (ry')' - qy.$$

---

[1] The first extensive development of the theory of such systems was published by J. C. F. Sturm and J. Liouville in the first three volumes of *Journal de mathématique*, 1836–1838.

[2] See also the author's "Fourier Series and Boundary Value Problems," 2d ed., 1963.

Let y and z denote any two eigenfunctions of problem (1) corresponding to eigenvalues λ and μ, respectively. Then both functions satisfy the first boundary condition,

(2) $$A_1 y(a) + A_2 y'(a) = 0, \qquad A_1 z(a) + A_2 z'(a) = 0.$$

Since A_1 and A_2 are not both zero, the determinant of this pair of simultaneous linear equations in A_1 and A_2 must vanish; that is, $W(a) = 0$, where W is the Wronskian of the functions y and z:

(3) $$W(x) = \begin{vmatrix} y(x) & z(x) \\ y'(x) & z'(x) \end{vmatrix} = y(x)z'(x) - y'(x)z(x),$$

and similarly for the second boundary condition. Thus the functions y and z must be such that

(4) $$W(a) = 0, \qquad W(b) = 0.$$

Also, they satisfy the Sturm-Liouville equations

$$\mathscr{D}_2 y = \lambda p(x)y(x), \qquad \mathscr{D}_2 z = \mu p(x)z(x),$$

and from these and Lagrange's identity (5), Sec. 94, it follows that

(5) $$(\mu - \lambda)pyz = y\mathscr{D}_2 z - z\mathscr{D}_2 y$$

$$= \frac{d}{dx}[yz' - y'z)r] = \frac{d}{dx}[W(x)r(x)].$$

In view of conditions (4), therefore,

(6) $$(\mu - \lambda)\int_a^b pyz\, dx = W(b)r(b) - W(a)r(a) = 0.$$

In case $\lambda \neq \mu$, it follows that y and z are orthogonal on the interval (a,b), with weight function p:

(7) $$\int_a^b p(x)y(x)z(x)\, dx = 0 \qquad\qquad (\lambda \neq \mu).$$

If the eigenfunctions y and z correspond to the same eigenvalue, then $\mu = \lambda$ in Eq. (5), and so $(Wr)' = 0$ and Wr is constant; but the constant is zero because $W(a) = 0$. Since $r(x) > 0$, it follows that the Wronskian of y and z vanishes:

(8) $$W(x) = y(x)z'(x) - y'(x)z(x) \equiv 0 \qquad (a \leqq x \leqq b).$$

In that case we now prove y and z linearly dependent.

Consider the linear combination of y and z,

(9) $$X(x) = [A_2 z(a) - A_1 z'(a)]y(x) - [A_2 y(a) - A_1 y'(a)]z(x),$$

which also satisfies the homogeneous equation $\mathcal{D}_2 X = \lambda X$. The reader can verify that it satisfies the initial conditions $X(a) = 0$ and $X'(a) = 0$ because $W(a) = 0$. Therefore $X(x) \equiv 0$, the unique solution of that equation under those conditions. Moreover, the coefficient of $y(x)$ in formula (9) cannot vanish, in view of the second of boundary conditions (2) and because $z(a)$ and $z'(a)$ cannot both vanish, for the determinant of the pair of equations

$$A_2 z(a) - A_1 z'(a) = 0, \qquad A_1 z(a) + A_2 z'(a) = 0$$

in $z(a)$ and $z'(a)$ has the value $A_2{}^2 + A_1{}^2$. That value is positive for our real numbers A_1 and A_2.

Likewise, the coefficient of $z(x)$ in formula (9) is a nonzero constant. Since $X(x) \equiv 0$, it follows that there is a constant k such that

$$(10) \qquad\qquad z(x) = k y(x) \qquad\qquad (k \neq 0, a \leqq x \leqq b).$$

Finally, we show that all eigenvalues are real and have corresponding real-valued eigenfunctions. For if $y(x)$ is an eigenfunction corresponding to an eigenvalue λ, then by writing complex conjugates of all terms of the Sturm-Liouville problem (1), we can find that

$$\mathcal{D}_2 \bar{y} = \bar{\lambda} \bar{y}, \qquad A_1 \bar{y}(a) + A_2 \bar{y}'(a) = B_1 \bar{y}(b) + B_2 \bar{y}'(b) = 0.$$

Thus $\bar{y}$ is an eigenfunction corresponding to $\bar{\lambda}$ and, according to Eq. (6), it follows that

$$(\bar{\lambda} - \lambda) \int_a^b p(x) y(x) \bar{y}(x)\, dx = 0.$$

But y is not identically zero, and $y\bar{y} = |y|^2 \geqq 0$, while $p(x) > 0$, so the integral here has a positive value; therefore, $\bar{\lambda} - \lambda = 0$. Thus $\bar{\lambda} = \lambda$, so λ is real.

Now since y and $\bar{y}$ are eigenfunctions corresponding to the same real eigenvalue λ, then y and $\bar{y}$ are linearly dependent; that is, for some constant k, $\bar{y} = ky$. In case $k = -1$, then $\bar{y} = -y$ and $\overline{iy} = iy$, so iy is a real-valued eigenfunction. If $k \neq -1$, then $(1 + k)y$ is the eigenfunction $y + \bar{y}$, or $2 \operatorname{Re} y$, which is real-valued. That is, each eigenfunction can be made real-valued by multiplying it by an appropriate complex constant.

We have now established the following properties of characteristic numbers and functions.

Theorem 3 *Each eigenvalue λ_n of the Sturm-Liouville problem (1) is real. The eigenfunction y_n corresponding to λ_n is unique up to an arbitrary constant factor k other than zero, and k can be chosen so that the eigenfunction is real-valued. Eigenfunctions y_m and y_n corresponding to different eigenvalues λ_m and λ_n have the orthogonality property*

$$\int_a^b p(x) y_m(x) y_n(x)\, dx = 0 \qquad\qquad (\lambda_m \neq \lambda_n).$$

99   A REPRESENTATION THEOREM

In the sections to follow, the existence of an infinite sequence of eigenvalues λ_n of the Sturm-Liouville system will be established. We then show that the generalized Fourier series of orthonormal eigenfunctions, for each function f of a broad class, does converge to $f(x)$ on the interval (a,b).

Let y_n denote *real-valued* eigenfunctions of the system

$$(1) \qquad\qquad [r(x)y'(x)]' - [q(x) + \lambda p(x)]y(x) = 0 \qquad\qquad (a < x < b),$$

$$A_1 y(a) + A_2 y'(a) = 0, \qquad B_1 y(b) + B_2 y'(b) = 0,$$

in case eigenfunctions exist. Then the normalized eigenfunctions

$$(2) \qquad\qquad \phi_n(x) = \frac{y_n(x)}{\|y_n\|} \qquad \text{where } \|y_n\|^2 = \int_a^b p(x)[y_n(x)]^2\, dx$$

are orthonormal on the interval (a,b) with weight function p:

$$(3) \qquad\qquad \int_a^b p(x)\phi_m(x)\phi_n(x)\, dx = \begin{cases} 0 \text{ if } m \neq n \\ 1 \text{ if } m = n \end{cases} \qquad (m, n = 1, 2, \ldots).$$

The generalized Fourier series of those functions, for a function f on the interval, is

$$(4) \qquad\qquad \sum_{n=1}^{\infty} c_n \phi_n(x) \qquad \text{where } c_n = \int_a^b p(x)f(x)\phi_n(x)\, dx.$$

Our theorems on representing the inversion integral by a series of residues are useful in establishing the following representation theorem for the series (4).

Theorem 4   *Let p, q, r, r', and $(pr)''$ be continuous real-valued functions of x such that $p(x) > 0$ and $r(x) > 0$, when $a \leq x \leq b$, and let the real-valued constants A_1, A_2, B_1, and B_2 be independent of λ. Then the Sturm-Liouville problem (1) has an infinite sequence of eigenvalues $\lambda_n(n = 1, 2, \ldots)$, all real, and at most a finite number of them are positive. The corresponding real-valued orthonormal eigenfunctions ϕ_n are of class C' on the interval $a \leq x \leq b$; also $\phi_n(x)$ and, when $\lambda_n \neq 0$, $\lambda_n^{-\frac{1}{2}}\phi_n'(x)$ are bounded with respect to both x and λ_n. Each sectionally continuous function f with sectionally continuous derivatives f' and f'' is represented by its generalized Fourier series (4) at all points x $(a < x < b)$ where f is continuous.*[1]

The proof of the theorem is lengthy. We carry it out fully for the case $A_2 = B_2 = 0$ and indicate in the problems some variations needed to

---

[1] The theorem is true without the condition on f'', but in our proof it is convenient to assume f'' sectionally continuous.

prove the general case. The proof exhibits interesting devices for finding properties of solutions of differential equations when the equations are too general to solve.

100   THE REDUCED STURM-LIOUVILLE SYSTEM

Under the conditions stated in Theorem 4 on p, q, and r, the Sturm-Liouville problem can be reduced to a simpler form by substituting certain new variables t and z for x and y. In the Sturm-Liouville differential equation we first write $y(x) = v(x)X(x)$ and $x = x(t)$ or $t = t(x)$, where v and $t(x)$ are yet to be determined, then write $X(x) = z(t)$. If dots denote derivatives with respect to t, then the equation $(ry')' = (q + \lambda p)y$ takes the form

$$(1) \qquad \ddot{z} + \left(\frac{2rv' + r'v}{rv}\dot{x} - \frac{\ddot{x}}{\dot{x}} \right)\dot{z} = \left(Q + \lambda\frac{p\dot{x}^2}{r} \right)z,$$

where $Q(t)$ is independent of λ.

Equation (1) can now be simplified by determining $t(x)$ or $x(t)$ so that $p\dot{x}^2/r = C^2$, where C is a constant, then $v(x)$ so that the coefficient of $\dot{z}$ vanishes. Details are left to the problems. The results are

$$(2) \qquad t = \frac{1}{C}\int_a^x \left[\frac{p(\xi)}{r(\xi)} \right]^{\frac{1}{2}} d\xi, \qquad z(t) = (pr)^{\frac{1}{4}}y(x),$$

and, to reduce the interval $a < x < b$ to $0 < t < 1$, we write

$$(3) \qquad C = \int_a^b \left[\frac{p(\xi)}{r(\xi)} \right]^{\frac{1}{2}} d\xi.$$

Then the Sturm-Liouville equation takes the reduced form

$$(4) \qquad \ddot{z}(t) - [Q(t) + \mu]z(t) = 0 \qquad\qquad (0 < t < 1),$$

where $\mu = C^2\lambda$ and z is a function of t and μ.

Under the substitutions (2) the boundary conditions in the Sturm-Liouville system retain their linear homogeneous forms. Thus

$$(5) \qquad a_1 z(0) + a_2\dot{z}(0) = 0, \qquad b_1 z(1) + b_2\dot{z}(1) = 0,$$

where the a's and b's are real-valued constants related to the A's and B's. In particular, if $A_2 = B_2 = 0$ so that $y(a) = y(b) = 0$, then $z(0) = z(1) = 0$, and therefore $a_2 = b_2 = 0$.

The reduced Sturm-Liouville problem consisting of the simpler Sturm-Liouville equation (4) and conditions (5) transforms back into the original problem in y under the substitutions (2). Orthogonality of eigenfunctions $z_n(t)$ on the interval $(0,1)$ with unit weight function corresponds to orthogonality of $y_n(x)$ on (a,b) with weight function p. The reduced system can be used to prove Theorem 4.

101 A RELATED BOUNDARY VALUE PROBLEM

To shorten the analysis, we now consider the case where $a_2 = b_2 = 0$ in the boundary conditions of our reduced Sturm-Liouville problem. Except for changes in notation, that problem then reads

$$(1) \qquad X''(x) - [\lambda + q(x)]X(x) = 0, \qquad X(0) = X(1) = 0.$$

We introduce, to serve as a guide to the analysis, a boundary value problem in partial differential equations whose solution is associated with problem (1). Let $Y(x,t)$ denote transverse displacements in a string stretched between points $(0,0)$ and $(1,0)$ when the string is attached along its span to an elastic medium that may be nonuniform. If the string starts from rest with a prescribed initial displacement $Y = f(x)$, and if the unit of time is properly chosen, then

$$(2) \qquad \begin{aligned} Y_{tt}(x,t) &= Y_{xx}(x,t) - q(x)Y(x,t) \qquad (0 < x < 1, t > 0), \\ Y(0,t) &= Y(1,t) = Y_t(x,0) = 0, \qquad Y(x,0) = f(x). \end{aligned}$$

The differential equation and both *two-point* boundary conditions are homogeneous. Hence the method of separation of variables may apply to problem (2). First, to find all functions not identically zero of type $X(x)T(t)$ that satisfy *all homogeneous* conditions, we can write

$$XT'' = X''T - qXT, \qquad X(0) = X(1) = T'(0) = 0,$$

Consequently
$$\frac{T''(t)}{T(t)} = \frac{X''(x) - q(x)X(x)}{X(x)} = \lambda,$$

where the parameter λ is independent of x and t, because T''/T is independent of x and equal to a function that is independent of t. The function X therefore satisfies all conditions in problem (1). Thus λ must be an eigenvalue λ_n of that Sturm-Liouville problem and, except for an arbitrary constant factor, $X(x)$ must be the corresponding orthonormal eigenfunction $\phi_n(x)$. The function $T(t)$ then satisfies the system $T'' = \lambda_n T$, $T'(0) = 0$. The functions XT can therefore be written

$$(3) \qquad X(x)T(t) = c_n\phi_n(x)\cos t\sqrt{-\lambda_n} \qquad (n = 1, 2, \ldots),$$

where the constants c_n are arbitrary.

A sum of functions (3) also satisfies all homogeneous conditions in problem (2). (Note that the sum itself is not a product of a function of x alone by a function of t alone.) Formally, the series

$$(4) \qquad Y(x,t) = \sum_{n=1}^{\infty} c_n\phi_n(x)\cos r\sqrt{-\lambda_n}$$

satisfies all conditions in the problem, including the condition $Y(x,0) = f(x)$, if the numbers c_n are the Fourier constants of $f(x)$ corresponding to the orthonormal functions $\phi_n(x)$, so that

$$(5) \qquad\qquad f(x) = \sum_{n=1}^{\infty} c_n \phi_n(x) \qquad\qquad (0 < x < 1),$$

where

$$(6) \qquad\qquad c_n = \int_0^1 f(x)\phi_n(x)\, dx \qquad\qquad (n = 1, 2, \ldots).$$

Thus the *Sturm-Liouville expansion* (5) is needed to complete the formal solution of problem (2) by the method of separation of variables.

    If we now apply the Laplace transformation with respect to t to the boundary value problem (2) in $Y(x,t)$ and establish order properties of the transform $y(x,s)$, we may be able to show that $L_i^{-1}\{y\} = f(x)$ when $t = 0$, or that

$$L_i^{-1}\left\{ y(x,s) - \frac{f(x)}{s} \right\} = 0 \qquad\qquad \text{when } t = 0.$$

Then we can anticipate that the sum of residues of y may be just the series (5), and that careful analysis will show that the series converges to $f(x)$.

102 THE TRANSFORM $y(x,s)$

Formally, the Laplace transform of the function $Y(x,t)$ in problem (2), Sec. 101, satisfies the conditions

$$(1) \qquad\qquad \begin{aligned} y''(x,s) - [s^2 + q(x)]y(x,s) &= -sf(x), \\ y(0,s) = y(1,s) &= 0. \end{aligned}$$

    To construct Green's function for problem (1) here, we introduce, as in Sec. 96, the unique solutions $u(x,s)$ and $v(x,s)$ of these initial-value problems:

$$(2) \qquad u'' - (s^2 + q)u = 0, \qquad u(0,s) = 0, \qquad u'(0,s) = 1,$$

$$(3) \qquad v'' - (s^2 + q)v = 0, \qquad v(1,s) = 0, \qquad v'(1,s) = 1.$$

Note that u and v are actually functions of x and s^2, hence *even* functions of s; they are *entire* functions of s^2 (Sec. 94) and therefore of s. Then for all s such that $u(1,s) \neq 0$, Green's function can be written as

$$(4) \qquad\qquad G(x,\xi,s) = \begin{cases} \dfrac{u(x,s)v(\xi,s)}{u(1,s)} & (x \leq \xi) \\[3mm] \dfrac{u(\xi,s)v(x,s)}{u(1,s)} & (x \geq \xi). \end{cases}$$

Thus if f is sectionally continuous, the solution of problem (1) is

$$(5) \qquad y(x,s) = -s \int_0^1 f(\xi) G(x,\xi,s)\, d\xi = -s \frac{h(x,s)}{u(1,s)}.$$

where

$$(6) \qquad h(x,s) = v(x,s) \int_0^x f(\xi) u(\xi,s)\, d\xi + u(x,s) \int_x^1 f(\xi) v(\xi,s)\, d\xi.$$

For each fixed x, h is an entire function of s. This follows from the Cauchy-Riemann conditions, because the products uv are entire functions of s.

Consequently y is an analytic function of s except at the zeros of the entire function $u(1,s)$, which are necessarily isolated. But when $u(1,s) = 0$, we see from problem (2) that $u(x,s)$ is a solution of our Sturm-Liouville problem (1), Sec. 101, in which $\lambda = s^2$. If λ_n is the corresponding eigenvalue and $\sqrt{\lambda_n}$ denotes one of the two square roots of that real number, then $s = \pm s_n$ where $s_n = \sqrt{\lambda_n}$. Thus if ϕ_n is the normalized eigenfunction, $u(x, \pm s_n) = k\phi_n(x)$ for some nonzero constant k. Since $u'(0,s_n) = 1$, it follows that $\phi_n'(0) \neq 0$ and

$$(7) \qquad u(x, \pm s_n) = \frac{\phi_n(x)}{\phi_n'(0)} \qquad\qquad (s_n = \sqrt{\lambda_n}).$$

But $v(1, \pm s_n) = 0$ so when $u(1, \pm s_n) = 0$, then $v(x, \pm s_n) = Ku(x, \pm s_n)$, where $K \neq 0$ and, since $v'(1, \pm s_n) = 1$,

$$(8) \qquad v(x, \pm s_n) = \frac{\phi_n(x)}{\phi_n'(1)}.$$

The function ϕ_n, not identically zero, satisfies the conditions

$$(9) \qquad \phi_n''(x) - [s_n{}^2 + q(x)]\phi_n(x) = 0, \qquad \phi_n(0) = \phi_n(1) = 0 \qquad (s_n = \sqrt{\lambda_n}).$$

From Eq. (6) to (8) we find that

$$(10) \qquad \phi_n'(1)h(x, \pm s_n) = \frac{\phi_n(x)}{\phi_n'(0)} \int_0^1 f(\xi)\phi_n(\xi)\, d\xi$$

$$= c_n \frac{\phi_n(x)}{\phi_n'(0)},$$

where c_n is the Fourier constant for our function f. Also, note that

$$(11) \qquad \phi_n'(0) \int_0^1 u(x, \pm s_n)\phi_n(x)\, dx = \int_0^1 [\phi_n(x)]^2\, dx = 1.$$

Eliminating q between Eqs. (2) and (9), we find that

$$(s^2 - s_n{}^2)u\phi_n = u''\phi_n - u\phi_n'' = (u'\phi_n - u\phi_n')',$$

so, in view of the boundary conditions on u and ϕ_n,

(12) $$(s^2 - s_n{}^2) \int_0^1 u(x,s)\phi_n(x)\,dx = -\phi_n'(1)u(1,s).$$

Consequently formula (5) can be written

$$y(x,s) = \frac{\phi_n'(1)h(x,s)}{\displaystyle\int_0^1 u(x,s)\phi_n(x)\,dx} \cdot \frac{s}{s^2 - s_n{}^2}.$$

It follows that $\pm s_n$ are simple poles of y. If $s_n \neq 0$, the residue of y at $\pm s_n$ is, in view of formulas (10) and (11),

$$\frac{\phi_n'(1)h(x, \pm s_n)}{2\displaystyle\int_0^1 u(x, \pm s_n)\phi_n(x)\,dx} = \tfrac{1}{2}c_n\phi_n(x),$$

and the sum of those equal residues is $c_n\phi_n(x)$. Similarly, if $s_n = 0$, we find that the residue of y at s_n is $c_n\phi_n(x)$. Hence the sum of residues of the function y at all its poles $s = \pm s_n$ is formally the generalized Fourier series for f:

$$\sum_{n=1}^{\infty} c_n\phi_n(x).$$

We have yet to prove that an infinite sequence of eigenvalues exists and that the series of residues converges to $f(x)$.

103   EXISTENCE OF EIGENVALUES

Again, let $u(x,s)$ be the unique solution of the initial-value problem

(1) $$u'' - (s^2 + q)u = 0, \qquad u(0,s) = 0, \qquad u'(0,s) = 1.$$

Then the solution of this problem in $z(x,s)$:

(2) $$z'' - s^2 z = q(x)u(x,s), \qquad z(0,s) = 0, \qquad z'(0,s) = 1$$

can be written in the form

(3) $$z(x,s) = \frac{1}{s}\sinh sx + \frac{1}{s}\int_0^x q(\xi)u(\xi,s)\sinh s(x - \xi)\,d\xi$$

when $s \neq 0$, as is easily verified or derived (cf. Prob. 17, Sec. 104). The solution (3) is unique because the difference $w - z$ of two solutions w and z of problem (2) must satisfy a completely homogeneous initial-value problem, so that $w - z \equiv 0$. But problem (1) shows that $z = u(x,s)$ is the solution of problem (2). Therefore u satisfies the *integral equation*

$$(4) \qquad su(x,s) = \sinh sx + \int_0^x q(\xi)u(\xi,s) \sinh s(x - \xi)\, d\xi,$$

where s is any complex number.

Likewise, for all s the solution v of the initial value problem (3), Sec. 102, satisfies the integral equation

$$(5) \qquad sv(x,s) = \sinh s(x - 1) + \int_x^1 q(\xi)u(\xi,s) \sinh s(\xi - x)\, d\xi.$$

We write $\rho = |s|$ and $\sigma = |\text{Re } s|$. Then

$$e^{-\sigma x}|e^{sx} \pm e^{-sx}| \leq 1 + e^{-2\sigma x} \leq 2 \qquad \text{when } 0 \leq x \leq 1;$$

that is, $e^{-\sigma x}|\sinh sx| \leq 1$ and $e^{-\sigma x}|\cosh sx| \leq 1$ when $0 \leq x \leq 1$. For a fixed s let $M(s)$ denote the maximum value of the continuous function $\rho|u(x,s)|e^{-\sigma x}$ on the interval $0 \leq x \leq 1$. The function assumes its maximum value at some point $x = k$ on the interval. When we write $x = k$ in formula (4) and introduce the factor $e^{-\sigma k}$ there, we find that

$$M(s) \leq 1 + \int_0^k |q(\xi)u(\xi,s) \sinh s(k - \xi)|e^{-\sigma k}\, d\xi.$$

When the integrand here is written in the form

$$\rho^{-1}|q(\xi)|[\rho|u(\xi,s)|e^{-\sigma\xi}]e^{-\sigma(k-\xi)}|\sinh s(k - \xi)|,$$

we can see that

$$M(s) \leq 1 + \frac{q_0}{\rho} M(s), \qquad \text{where} \qquad q_0 = \int_0^1 |q(\xi)|\, d\xi.$$

Therefore

$$(6) \qquad M(s) \leq \left(1 - \frac{q_0}{\rho}\right)^{-1} < 2 \qquad \text{whenever } \rho > 2q_0.$$

A similar argument applies to v, using Eq. (5); thus

$$(7) \qquad \rho|u(x,s)|e^{-\sigma x} < 2, \qquad \rho|v(x,s)|e^{-\sigma(1-x)} < 2 \qquad \text{whenever } \rho > 2q_0.$$

From Eq. (4) we now find that

$$(8) \qquad su(x,s) = \sinh sx + s^{-1}e^{\sigma x}R_1(x,s),$$

$$(9) \qquad u'(x,s) = \cosh sx + s^{-1}e^{\sigma x}R_2(x,s),$$

where R_1 and R_2 are continuous functions whose absolute values are less than $2q_0$ when $\rho > 2q_0$. In particular,

$$su(1,s)e^{-\sigma} = e^{-\sigma}\sinh s + \frac{1}{s}R_1(1,s), \tag{10}$$

where
$$|R_1(1,s)| < 2q_0 \qquad \text{if} \qquad |s| > 2q_0.$$

We found that $u(1,s)$ cannot vanish unless s has real or pure imaginary values $\pm\sqrt{\lambda_n}$. If s is real and not zero, $s = \pm\sigma$, then $e^{-\sigma}\sinh s$ approaches $\pm\frac{1}{2}$ as $\sigma \to \infty$, and if follows from Eq. (10) that a number γ exists such that $u(1,s) \neq 0$ when $\sigma \geq \gamma$. All real zeros of the entire function $u(1,s)$, therefore, lie on a bounded interval of the real axis; hence they are finite in number.

When s is a pure imaginary number, $\sigma = 0$ and $s = i\eta$, so

$$\eta u(1,i\eta) = \sin\eta - \frac{1}{\eta}R_1(1,i\eta) \qquad (\eta \text{ real}).$$

Note that $\eta u(1,i\eta)$ is an odd function of η since $u(x,s)$ is an even function of s. Also $u(x,i\eta)$ is real-valued because it is the solution of the initial-value problem (1) in which the coefficients are all real, so that $\bar{u}$ satisfies the same problem and hence $\bar{u} = u$. Thus the function $\eta^{-1}R_1(1,i\eta)$ is real-valued and odd. It is continuous and vanishes as $\eta \to \infty$ so its graph intersects that of $\sin\eta$ at an infinite sequence of points $\eta = \pm\eta_j$ $(j = 1, 2, \ldots)$, where η_j is arbitrarily close to an integral multiple of π when η_j is sufficiently large.

An infinite sequence of eigenvalues

$$\lambda_1, \lambda_2, \ldots, \lambda_m, \lambda_{m+1}, \ldots \tag{11}$$

$$(\lambda_n \geq 0 \text{ if } n \leq m, \ \lambda_n < 0 \text{ if } n > m),$$

or simple poles $\pm s_n = \pm\sqrt{\lambda_n}$ of $y(x,s)$, therefore exists. When $n > m$, then $s_n = i\eta_n$, where η_n is real and, according to Eqs. (8) and (9),

$$u(x,s_n) = \frac{1}{\eta_n}\sin\eta_n x - \frac{1}{\eta_n{}^2}R_1(x,i\eta_n), \tag{12}$$

$$u'(x,s_n) = \cos\eta_n x - \frac{i}{\eta_n}R_2(x,i\eta_n). \tag{13}$$

With the aid of the last two equations we find that the norm $\|\eta_n u(x,s_n)\|$ tends to $1/\sqrt{2}$ as $n \to \infty$, and that a constant K, independent of n, exists such that the normalized eigenfunctions ϕ_n satisfy conditions

$$|\phi_n(x)| < K, \qquad |\lambda_n{}^{-\frac{1}{2}}\phi_n'(x)| < K \qquad (\lambda_n \neq 0). \tag{14}$$

Since u is of class C' in x, then ϕ_n is also of that class. Our principal results are stated as follows.

Theorem 5 *The reduced Sturm-Liouville problem* (1), *Sec.* 101, *has an infinite set* (11) *of eigenvalues* λ_n. *All eigenvalues are real, and no more than a finite number of them are positive. The value of* $\sqrt{-\lambda_n}$ *is arbitrarily close to an integral multiple of* π *when n is sufficiently large. The corresponding orthonormal eigenfunctions* ϕ_n *are of class C', and* $\phi_n(x)$ *and* $\phi'_n(x)$ *satisfy the boundedness conditions* (14).

104   THE GENERALIZED FOURIER SERIES

To prove that the series of residues of y converges to $f(x)$, we first find order properties of Green's function G and our function y.

Again we write $\sigma = |\mathrm{Re}\ s|$ and $\rho = |s|$. When $x \leq \xi$,

$$G(x,\xi,s) = \frac{u(x,s)v(\xi,s)}{u(1,s)} = \frac{[se^{-\sigma x}u(x,s)][se^{-\sigma(1-\xi)}v(\xi,s)]}{s^2 e^{-\sigma}u(1,s)e^{\sigma(\xi-x)}}.$$

In view of conditions (7) and Eq. (10), Sec. 103, then

(1) $$|G(x,\xi,s)| < \frac{4}{\rho|e^{-\sigma}\sinh s + s^{-1}R_1(1,s)|} \qquad (\rho > 2q_0).$$

This condition is valid also when $x \geq \xi$, owing to the symmetry of Green's function. The denominator vanishes only at points lying along the axis of imaginaries or at points on a segment $\sigma < \gamma$ of the real axis, since the zeros $\pm s_n$ of $u(1,s)$ are so distributed. In case all the zeros are pure imaginary numbers, we let γ denote any positive number.

Throughout either of the two half planes $\sigma \geqq \gamma$, the denominator in condition (1) is bounded away from zero, since

$$e^{-\sigma}|\sinh s| \geqq \tfrac{1}{2}(1 - e^{-2\gamma}) > 0$$

and $s^{-1}R_1(1,s) \to 0$ as $\rho \to \infty$. Therefore G is $\mathcal{O}(\rho^{-1})$ over those half planes. For points $s = \pm\sigma + i\eta$ on lines $\eta = \pm(N - \tfrac{1}{2})\pi$ $(N = 1, 2, \ldots)$,

$$e^{-\sigma}|\sinh s| = e^{-\sigma}\cosh\sigma = \tfrac{1}{2}(1 + e^{-2\sigma}) > \tfrac{1}{2}.$$

It follows from condition (1) that G is $\mathcal{O}(\rho^{-1})$ on those lines as $\rho \to \infty$. Hence for all points s in a right half plane or on the open rectangular paths C_N used in Theorem 10, Sec. 73, in which $\beta_N = (N - \tfrac{1}{2})\pi$, a constant M independent of x, ξ, and N exists such that

(2) $$|G(x,\xi,s)| < \frac{M}{\rho} \qquad \text{when } \mathrm{Re}\ s \geqq \gamma \text{ or when } s \text{ is on } C_N.$$

It follows that our function

(3) $$y(x,s) = -s\int_0^1 f(\xi)G(x,\xi,s)\,d\xi$$

is at least bounded over the half plane and on C_N.

Since for a fixed x, Green's function satisfies the homogeneous differential equation of which $u(\xi,s)$ and $v(\xi,s)$ are solutions, we can make the substitution

$$G(x,\xi,s) = \frac{1}{s^2}G_{\xi\xi}(x,\xi,s) - \frac{1}{s^2}q(\xi)G(x,\xi,s) \qquad (\xi \neq x)$$

into Eq. (3) to write

$$y(x,s) + \frac{1}{s}\int_0^1 f(\xi)G_{\xi\xi}(x,\xi,s)\,d\xi = \frac{1}{s}\int_0^1 f(\xi)q(\xi)G(x,\xi,s)\,d\xi.$$

The right-hand member here is $\mathcal{O}(s^{-2})$ on the half plane $\operatorname{Re} s \geq \gamma$ and on the paths C_N, in view of condition (2). The left-hand member of the equation therefore satisfies the same conditions, and so its inversion integral along the line $\operatorname{Re} s = \gamma$ has the value zero when $t = 0$; also (Sec. 73) its integral over C_N vanishes as $N \to \infty$. That is, if $\bar{C}_N$ denotes the *closed* rectangle with the side along $\operatorname{Re} s = \gamma$ added to C_N, then

(4) $$\int_{\bar{C}_N}\left[y(x,s) + \frac{1}{s}\int_0^1 f(\xi)G_{\xi\xi}(x,\xi,s)\,d\xi \right]ds \to 0 \qquad \text{as } N \to \infty.$$

Now let us assume f'', and therefore f' and f, sectionally continuous. It will suffice to let f have just one jump of an amount j_0 at a point x_0 $(0 < x_0 < 1)$. Let x be a fixed point of the interval at which f is continuous, say $0 < x < x_0$. The jump in G_ξ at the point $\xi = x$ is 1, as we can see from the expression for G in terms of u and v. Then

$$\int_0^1 f G_{\xi\xi}\,d\xi = \int_0^x f G_{\xi\xi}\,d\xi + \int_x^{x_0} f G_{\xi\xi}\,d\xi + \int_{x_0}^1 f G_{\xi\xi}\,d\xi$$

$$= f G_\xi|_0^x + f G_\xi|_x^{x_0} + f G_\xi|_{x_0}^1 - \int_0^1 f' G_\xi\,d\xi$$

$$= -f(x) - f(0)G_\xi(x,0,s) - j_0 G_\xi(x,x_0,s)$$

$$\quad + f(1)G_\xi(x,1,s) - \int_0^1 f' G_\xi\,d\xi,$$

if we write $f(0)$ for $f(+0)$ and $f(1)$ for $f(1-0)$. Note that $G_\xi(x,0,s)$ is $u'(0,s)v(x,s)/u(1,s)$ and $u'(0,s) = 1$, etc. Thus in terms of the function

(5) $$g(x,s) = f(0)\frac{v(x,s)}{su(1,s)} + j_0\frac{u(x,s)v'(x_0,s)}{su(1,s)} - f(1)\frac{u(x,s)}{su(1,s)}$$

the bracketed integrand in Eq. (4) is the sum

(6) $$y(x,s) - \frac{f(x)}{s} - g(x,s) - \frac{1}{s}\int_0^1 f'(\xi)G_\xi(x,\xi,s)\,d\xi.$$

The final member of the sum (6) is $\mathcal{O}(s^{-2})$ on $\bar{C}_N$ as we can see by integrating the sectionally continuous function $f'G_\xi$ by parts and recalling that G itself and therefore such terms as $j_iG(x,x_i,s)$ and $\int_0^1 f''G\,d\xi$ are $\mathcal{O}(s^{-1})$. Thus the integral around $\bar{C}_N$ of that member tends to zero as $N \to \infty$.

We find, corresponding to Eqs. (8) and (9), Sec. 103, that

(7) $$sv(x,s) = -\sinh s(1 - x) + s^{-1}e^{\sigma(1-x)}R_3(x,s),$$

(8) $$v'(x,s) = \cosh s(1 - x) + s^{-1}e^{\sigma(1-x)}R_4(x,s),$$

where R_3 and R_4 are bounded when $|s| > 2q_0$ and $0 \le x \le 1$.

We write the coefficient of j_0 in the sum (5) in the forms

$$\frac{1}{s^2u(1,s)}[\sinh sx + s^{-1}e^{\sigma x}R_1(x,s)][\cosh s(1 - x_0) + s^{-1}e^{\sigma(1-x_0)}R_4(x_0,s)]$$

$$= \frac{e^{-\sigma}\sinh sx\cosh s(1 - x_0)}{se^{-\sigma}\sinh s + R_1(1,s)} + g_1(x,x_0,s),$$

where it is readily verified that g_1 is $\mathcal{O}(s^{-2})$ on $\bar{C}_N$. Since $1/(a + b)$ can be written as $1/a - (b/a)/(a + b)$, that coefficient is

$$\frac{\sinh sx\cosh s(1 - x_0)}{s\sinh s} + \frac{R_1(1,s)}{s\sinh s}\frac{\sinh sx\cosh s(1 - x_0)}{se^{-\sigma}\sinh s + R_1(1,s)} + g_1(x,x_0,s)$$

$$= \frac{\sinh s(1 + x - x_0) - \sinh s(1 - x - x_0)}{2s\sinh s} + g_2(x,x_0,s) + g_1(x,x_0,s),$$

where g_2 is $\mathcal{O}(s^{-2})$ on $\bar{C}_N$. Note that $0 < 1 + x - x_0 < 1$ and $|1 - x - x_0| < 1$. Thus the two terms of weak order here are of the type

$$\psi(c,s) = \frac{\sinh cs}{s\sinh s} \qquad \text{where } -1 < c < 1.$$

Actually, ψ is the transform of a square-wave function which vanishes over an interval containing the origin $t = 0$ (Prob. 9, Sec. 104).

An examination of the integral of $|\psi|$ around $\bar{C}_N$ shows that $\int_{\bar{C}_N}\psi\,ds \to 0$ as $N \to \infty$ (Prob. 11 below). Similarly, the integrals around $\bar{C}_N$ of the coefficients of $f(0)$ and $f(1)$ in the sum (5) vanish as $N \to \infty$; the reduction in terms of the function ψ is simpler for those coefficients. Thus $\int_{\bar{C}_N}g\,ds \to 0$ as $N \to \infty$.

Consequently, condition (4) reduces to the equation

(9) $$\lim_{N \to \infty}\int_{\bar{C}_N}\left[y(x,s) - \frac{f(x)}{s}\right]ds = 0;$$

that is, the sum of the residues of the integrand at the poles inside C_N converges to zero as $N \to \infty$. But the sum of residues of y were found in Sec. 102 to be $\Sigma c_n \phi_n(x)$; therefore

$$(10) \qquad \sum_{n=1}^{\infty} c_n \phi_n(x) - f(x) = 0, \qquad \text{where } c_n = \int_0^1 f(x) \phi_n(x)\, dx.$$

There are no essential changes in the proof of formula (10) when $x_0 < x < 1$. Thus Theorem 4 is established for the reduced form of the Sturm-Liouville problem with boundary conditions $X(0) = X(1) = 0$. A similar procedure, but more involved, can be followed to prove the theorem when the more general boundary conditions of type (5), Sec. 100, are used.

Theorem 4 as stated in Sec. 99 then follows by transforming the reduced problem back to the original Sturm-Liouville problem with the substitutions (2), Sec. 100.

PROBLEMS

1. Prove that the eigenvalues λ_n of the Sturm-Liouville problem

$$y''(x) - (\lambda + h)y(x) = 0, \qquad 0 < x < c; \qquad y(0) = y(c) = 0,$$

where h is real and constant, consist of the numbers $-h - n^2\pi^2/c^2$ $(n = 1, 2, \ldots)$. Note that not more than a finite number of them are positive, and that $-h$ is not an eigenvalue. Show that the normalized eigenfunctions ϕ_n are $\sqrt{2/c} \sin(n\pi x/c)$ and that according to Theorem 4, if f'' is sectionally continuous $(0 < x < c)$, then $f(x)$ is represented at each point where f is continuous by its *Fourier sine series* on the interval $(0,c)$:

$$f(x) = \frac{2}{c} \sum_{n=1}^{\infty} \sin\frac{n\pi x}{c} \int_0^c f(\xi) \sin\frac{n\pi\xi}{c}\, d\xi \qquad (0 < x < c).$$

2. Write $f(x) = 1$ $(0 < x < c)$ in Prob. 1 to show that

$$\frac{4}{\pi} \sum_{n=1}^{\infty} \frac{1}{2n-1} \sin\frac{(2n-1)\pi x}{c} = 1 \qquad (0 < x < c).$$

3. Find all λ_n and $\phi_n(x)$ for the Sturm-Liouville problem

$$y''(x) - \lambda y(x) = 0, \qquad y'(0) = y'(c) = 0,$$

and obtain the *Fourier cosine series* expansion of $f(x)$,

$$f(x) = \frac{1}{c} \int_0^c f(\xi)\, d\xi + \frac{2}{c} \sum_{n=1}^{\infty} \cos\frac{n\pi x}{c} \int_0^c f(\xi) \cos\frac{n\pi\xi}{c}\, d\xi \qquad (0 < x < c),$$

as a special case of Theorem 4. Note that when $\lambda = 0$ the general solution of the Sturm–Liouville equation is $y = Ax + B$, where A and B are arbitrary constants.

4. Find all characteristic numbers and functions of the systems

(a) $y''(x) - \lambda y(x) = 0,$ \qquad $y(0) = y'(c) = 0,$

(b) $y''(x) - \lambda y(x) = 0,$ \qquad $y(0) = 0,$ \qquad $y(1) + hy'(1) = 0,$

where h is a positive constant. Note the orthogonality of the characteristic functions as given by Theorem 3.

Ans. $\lambda_n = -\alpha_n^2$, $y_n(x) = \sin \alpha_n x$, where in (a), $\alpha_n = (n - \tfrac{1}{2})\pi c^{-1}$, and in (b),
$$\alpha_n > 0 \text{ and } \tan \alpha_n = -h\alpha_n.$$

5. Use the substitutions (2), Sec. 100, to reduce the problem

$$(x^4 y')' - (\lambda - 2x^2)y = 0, \qquad y(1) = 0, \qquad y'(2) = 0,$$

to the form

$$z''(t) - \mu z(t) = 0, \qquad z(0) = 0, \qquad z(1) - z'(1) = 0.$$

Then use them to derive the following expressions for the eigenvalues and eigenfunctions of the problem in y:

$$\lambda_1 = 0, \qquad \lambda_n = -4\alpha_n^2 \qquad\qquad (n = 2, 3, \ldots)$$

where α_n are the positive roots of the equation $\tan \alpha = \alpha$;

$$y_1 = x^{-1} - x^{-2}, \qquad y_n = x^{-1} \sin[2\alpha_n(1 - x^{-1})] \qquad (n = 2, 3, \ldots).$$

6. For the fourth-order eigenvalue problem

$$\frac{d^4 y}{dx^4} - \lambda y = 0, \qquad y(0) = y'(0) = y(c) = y'(c) = 0,$$

prove that two eigenfunctions y and z and their corresponding eigenvalues λ and μ must satisfy the condition

$$(\lambda - \mu) \int_0^c y(x)z(x)\, dx = 0,$$

and hence that if $\lambda \neq \mu$, y is orthogonal to z on the interval $(0,c)$; also that all eigenvalues must be real.

7. In Sec. 100, complete the derivation of form (1) of the Sturm-Liouville equation and of the substitutions (2) that produce the reduced form (4).

8. Derive formulas (7) and (8), Sec. 104, for $sv(x,s)$ and $v'(x,s)$.

9. Let Ψ be this periodic square-wave function:

$$\Psi(c,t) = 0 \qquad\qquad \text{if } 0 < t < 1 - c \text{ or if } 1 + c < t < 2,$$
$$\Psi(c,t) = 1 \qquad\qquad \text{if } 1 - c < t < 1 + c,$$
$$\Psi(c, t + 2) = \Psi(t) \qquad \text{when } t > 0,$$

where $0 \leq c < 1$; also $\Psi(-c,t) = -\Psi(c,t)$. Show Ψ graphically as a function of t. Use Theorem 10, Sec. 23, to prove that its transform is the function $\psi(c,s)$ used in Sec. 104:

$$L\{\Psi(c,t)\} = \psi(c,s) = \frac{\sinh cs}{s \sinh s} \qquad (-1 < c < 1, \text{ Re } s > 0).$$

10. Apply Theorem 6, Sec. 68, to the square-wave-function $\Psi(c,t)$ in Prob. 9 to prove that $L_i^{-1}\{\psi\} = 0$ when $t = 0$.

11.   If we write $s = v + i\eta$, the sides of the closed rectangle $\bar{C}_N$ used in Sec. 104 lie on the lines $v = \gamma (\gamma > 0)$, $v = -\beta_N$, and $\eta = \pm \beta_N$, where $\beta_N = (N - \frac{1}{2})\pi$, $N = 1, 2, \ldots$. For the function

$$\psi(c,s) = \frac{\sinh cs}{s \sinh s} \qquad\qquad (-1 < c < 1)$$

establish this property, used in Sec. 104:

$$\lim_{N \to \infty} \int_{\bar{C}_N} \psi(c,s)\, ds = 0,$$

with the aid of the result found in Prob. 10.

12.   If the function f is *continuous* $(0 \leq x \leq 1)$ *and satisfies the boundary conditions in the reduced Sturm-Liouville problem* (1), Sec. 101, *so that* $f(0) = f(1) = 0$, *and if* f'' *is sectionally continuous, prove that condition* (9), Sec. 104, *is satisfied uniformly in* x; *hence that*

$$f(x) = \sum_{n=1}^{\infty} c_n \phi_n(x) \qquad\qquad (0 \leq x \leq 1)$$

and this generalized Fourier series for f *converges uniformly.*

13.   Write the reduced system (1), Sec. 101, in the form

$$X''(x) - q(x)X(x) = \lambda X(x), \qquad X(0) = X(1) = 0,$$

and show that, if zero is not an eigenvalue of the system, each eigenfunction X_n is a solution of the *homogeneous integral equation*

$$X_n(x) = \lambda_n \int_0^1 X_n(\xi) G(x,\xi,0)\, d\xi$$

where λ_n is the eigenvalue corresponding to X_n and G is Green's function (4), Sec. 102.

Notation used in Probs. 14 to 18: Let $N_\alpha[X]$ and $N_\beta[X]$ and their derivatives with respect to the real-valued parameters α and β denote these boundary values on the interval $0 \leq x \leq 1$:

$$N_\alpha[X] = X(0)\cos\alpha + X'(0)\sin\alpha, \qquad N_\beta[X] = X(1)\cos\beta + X'(1)\sin\beta,$$

$$N_\alpha'[X] = -X(0)\sin\alpha + X'(0)\cos\alpha, \quad N_\beta'[X] = -X(1)\sin\beta + X'(1)\cos\beta.$$

Also, $u(x,s)$ and $v(x,s)$ denote the solutions of these initial-value problems, so arranged that $N_\alpha[u] = N_\beta[v] = 0$:

(a) $u'' - (s^2 + q)u = 0$, $u(0,s) = \sin\alpha$, $u'(0,s) = -\cos\alpha$,

(b) $v'' - (s^2 + q)v = 0$, $v(1,s) = \sin\beta$, $v'(1,s) = -\cos\beta$,

and $y(x,s)$ is the solution of the problem

(c) $y'' - (s^2 + q)y = -sf(x)$, $N_\alpha[y] = N_\beta[y] = 0$.

The following problems display several steps in proving Theorem 4 *for the Sturm-Liouville problem*

(d) $X''(x) - [\lambda + q(x)]X(x) = 0$, $N_\alpha[X] = N_\beta[X] = 0$.

14. Use Green's function (Probs. 8 and 9, Sec. 97) to write the solution of the boundary value problem (c) above in the form

$$y(x,s) = \frac{s}{N_\beta[u]}\left[v(x,s)\int_0^x f(\xi)u(\xi,s)\,d\xi + u(x,s)\int_x^1 f(\xi)v(\xi,s)\,d\xi\right].$$

15. Let λ_n and ϕ_n be eigenvalues and normalized eigenfunctions of Prob. (d) above and write $s_n = \sqrt{\lambda_n}$. Note why $\pm s_n$ are zeros of $N_\beta[u]$ and singular points of the function y (Prob. 14). Recall that (Prob. 8, Sec. 97) $N_\beta[u] = -N_\alpha[v]$ and that (Sec. 98) eigenfunctions corresponding to the same eigenvalue are linearly dependent; then show that, corresponding to Eqs. (7) and (8), Sec. 102,

$$u(x,\pm s_n) = -\frac{\phi_n(x)}{N_\alpha'[\phi_n]}, \qquad v(x,\pm s_n) = -\frac{\phi_n(x)}{N_\beta'[\phi_n]}.$$

16. Verify that (cf. Sec. 102 and Prob. 15)

$$N_\beta[u]N_\beta'[\phi_n] = -(s^2 - s_n{}^2)\int_0^1 u(x,s)\phi_n(x)\,dx.$$

Then deduce that $\pm s_n$ are simple poles of $y(x,s)$ and that the sum of the residues of y at the two poles $\pm s_n$ is $c_n\phi_n(x)$ if $s_n \ne 0$, and if $s_n = 0$, the residue is $c_n\phi_n(x)$.

17. Verify or derive the solution of the initial-value problem

$$z'' - s^2 z = q(x)u(x,s), \qquad z(0,s) = \sin\alpha, \qquad z'(0,s) = -\cos\alpha,$$

where u is the solution of Prob. (a) above, in the form

$$z(x,s) = \sin\alpha\cosh sx - \frac{1}{s}\cos\alpha\sinh sx + \frac{1}{s}\int_0^x q(\xi)u(\xi,s)\sinh s(x-\xi)\,d\xi.$$

Hence show that u satisfies the integral equation

$$su(x,s) = s\sin\alpha\cosh sx - \cos\alpha\sinh sx + \int_0^x q(\xi)u(\xi,s)\sinh s(x-\xi)\,d\xi.$$

18. Prove that [cf. condition (6), Sec. 103, and see Prob. 17]

$$|su(x,s)|e^{-\sigma x} < 2(|s\sin\alpha| + |\cos\alpha| + |\cos\alpha|) \qquad (\sigma = |\mathrm{Re}\, s|, |s| > 2q_0)$$

105 STEADY TEMPERATURE IN A WALL

Let $U(x,y)$ be the steady-state temperatures in a semi-infinite wall bounded by the planes $x = 0$, $x = 1$, and $y = 0$ (Fig. 101), for which the face $x = 0$ is insulated and surface heat transfer takes place at the face $x = 1$, while the face $y = 0$ is kept at temperature $F(x)$, so that the boundary value problem becomes

(1) $$U_{xx}(x,y) + U_{yy}(x,y) = 0 \qquad (0 < x < 1, y > 0),$$
(2) $$U_x(0,y) = 0, \qquad U_x(1,y) = -hU(1,y) \qquad (y > 0),$$
(3) $$U(x,0) = F(x), \qquad \lim_{y\to\infty} U(x,y) = 0 \qquad (0 < x < 1).$$

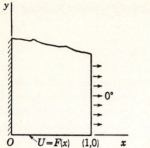

Fig. 101

The constant h is positive, and the function F is prescribed.

Although the variable y has a semi-infinite range, the problem is not well adapted to solution by using the Laplace transformation for reasons indicated in Sec. 52. The method of separation of variables does apply.

To find all functions of type $X(x)Y(y)$, other than zero itself, that satisfy all *homogeneous* conditions in the problem, we first note that, from condition (1),

$$X''(x)Y(y) + X(x)Y''(y) = 0.$$

Therefore
$$\frac{X''(x)}{X(x)} = -\frac{Y''(y)}{Y(y)} = \lambda,$$

where the parameter λ is independent of x and y (cf. Sec. 101). Then all the homogeneous conditions are satisfied if $X(x)$ is a solution of the Sturm-Liouville problem

(4) $X''(x) - \lambda X(x) = 0,$ $X'(0) = 0,$ $hX(1) + X'(1) = 0,$

and if Y satisfies the conditions

(5) $Y''(y) + \lambda Y(y) = 0,$ $\lim\limits_{y \to \infty} Y(y) = 0.$

When $\lambda = 0$, the general solution of the differential equation in problem (4) is $X(x) = Ax + B$, and this satisfies the two boundary conditions in the problem only if the constants A and B are both zero. Consequently zero is not an eigenvalue.

When $\lambda \neq 0$, the function that satisfies the first two of conditions (4) can be written

$$X(x) = C \cosh x\sqrt{\lambda}.$$

It satisfies the third condition if

(6) $h \cosh \sqrt{\lambda} + \sqrt{\lambda} \sinh \sqrt{\lambda} = 0.$

According to Theorem 3, all eigenvalues of problem (4) are real. If $\lambda > 0$, the left-hand member of (6) has a positive value. All eigenvalues of problem (4) are therefore negative.

We write $\lambda = -\alpha^2$. Then Eq. (6) becomes

$$(7) \qquad\qquad \tan \alpha = \frac{h}{\alpha},$$

an equation that has an infinite set of roots $\pm\alpha_n$ $(n = 1, 2, \ldots)$ that can be approximated graphically. In fact the number α_n is only slightly greater than $n\pi$ when n is large.

The complete set of eigenvalues of problem (4) is therefore

$$\lambda_n = -\alpha_n^2 \qquad\qquad (n = 1, 2, \ldots),$$

where the numbers α_n are the positive roots of Eq. (7), and the corresponding eigenfunctions are

$$X_n(x) = \cos \alpha_n x.$$

Let β_n denote the norm $\|X_n\|$. Then

$$\beta_n^2 = \int_0^1 \cos^2 \alpha_n x \, dx = \frac{1}{2}\left(1 + \frac{\sin \alpha_n \cos \alpha_n}{\alpha_n}\right)$$

and, since α_n is a root of Eq. (7), we find that

$$(8) \qquad\qquad \beta_n = 2^{-\frac{1}{2}}\left(1 + \frac{h}{h^2 + \alpha_n^2}\right)^{\frac{1}{2}}.$$

The orthonormal eigenfunctions of problem (4) are therefore

$$(9) \qquad\qquad \phi_n(x) = \frac{1}{\beta_n} \cos \alpha_n x \qquad\qquad (n = 1, 2, \ldots).$$

When $\lambda = -\alpha_n^2$, the solution of problem (5) is

$$Y(y) = c_n \exp(-\alpha_n y),$$

where the constant c_n is arbitrary, and therefore

$$X(x)Y(y) = c_n \frac{\cos \alpha_n x}{\beta_n} \exp(-\alpha_n y) \qquad\qquad (n = 1, 2, \ldots).$$

The sum of any number of these functions also satisfies all homogeneous conditions in problem (1) to (3); but unless $F(x)$ is a linear combination of the functions $\cos \alpha_n x$, no finite sum will satisfy the remaining condition

$$U(x,0) = F(x) \qquad\qquad (0 < x < 1).$$

The function represented by the infinite series

$$\sum_{n=1}^{\infty} \frac{c_n}{\beta_n} \cos \alpha_n x \exp(-\alpha_n y)$$

formally satisfies the homogeneous conditions. According to Theorem 4, this function reduces to $F(x)$ when $y = 0$, provided F'' is sectionally continuous, if the numbers c_n are the Fourier constants for F with respect to ϕ_n; that is, if

$$(10) \qquad c_n = \int_0^1 F(x)\phi_n(x)\,dx = \frac{1}{\beta_n}\int_0^1 F(x)\cos\alpha_n x\,dx.$$

Note that the series used to represent $F(x)$ here,

$$(11) \qquad F(x) = \sum_{n=1}^{\infty} \frac{c_n}{\beta_n}\cos\alpha_n x \qquad (0 < x < 1),$$

is a generalized Fourier series, not the Fourier cosine series for the interval $0 < x < 1$.

The formal solution of our problem is therefore

$$(12) \qquad U(x,y) = \sum_{n=1}^{\infty} \frac{c_n}{\beta_n}\exp(-\alpha_n y)\cos\alpha_n x,$$

where the constants β_n and c_n are given by formulas (8) and (10).

106   VERIFICATION OF THE SOLUTION

In the preceding temperature problem F'', and hence the prescribed function F itself, is sectionally continuous $(0 < x < 1)$. At each fixed point x where F is continuous, the Sturm-Liouville series (11), Sec. 105, therefore converges to $F(x)$:

$$(1) \qquad \sum_{n=1}^{\infty} \frac{c_n}{\beta_n}\cos\alpha_n x = F(x) \qquad (0 < x < 1).$$

We order the positive numbers α_n so that $\alpha_{n+1} > \alpha_n$. Then the functions $\exp(-\alpha_n y)$ in our formal solution (12), Sec. 105, are bounded, and nonincreasing with respect to n:

$$0 < \exp(-\alpha_n y) \leq 1, \qquad \exp(-\alpha_{n+1}y) \leq \exp(-\alpha_n y) \qquad (y \geq 0).$$

According to Abel's test (Prob. 6, Sec. 81), therefore, the series

$$(2) \qquad \sum_{n=1}^{\infty} \frac{c_n}{\beta_n}\cos\alpha_n x\exp(-\alpha_n y) \qquad (y \geq 0, 0 < x < 1)$$

converges uniformly with respect to y when $y \geq 0$. Since its terms are continuous functions of y, its sum U represents a continuous function of y when $y \geq 0$. Our formal solution therefore satisfies the condition

$$U(x,+0) = U(x,0) = F(x) \qquad (0 < x < 1).$$

From Eq. (8), Sec. 105, we see that $\beta_n > 1/\sqrt{2}$. Also, a fixed number M exists such that $|c_n|/\beta_n < M$. Whenever $y \geq y_0$ where $y_0 > 0$, the absolute values of the terms in series (2) are less than the constants $M \exp(-\alpha_n y_0)$; also the absolute values of the derivatives of those terms, of first and second order, with respect to x and y, are less than $M\alpha_n \exp(-\alpha_n y_0)$ and $M\alpha_n{}^2 \exp(-\alpha_n y_0)$, respectively. The series of those constant terms converges, according to the ratio test, since $\alpha_n - n\pi \to 0$ as $n \to \infty$. Hence series (2) and the series of derivatives of its terms converge uniformly with respect to x and y over each strip $0 \leq x \leq 1$, $y \geq y_0 > 0$. The terms are continuous functions, so the series converge to a continuous function $U(x,y)$ and its continuous partial derivatives.

In particular, U and U_x are continuous functions of x on the interval $0 \leq x \leq 1$, when $y > 0$. They satisfy the homogeneous boundary conditions

$$U_x(0,y) = 0, \qquad U_x(1,y) + hU(1,y) = 0$$

because the terms in series (2) satisfy those conditions. The terms also satisfy Laplace's equation (1), Sec. 105; hence the sum U satisfies that equation when $y > 0$ and $0 < x < 1$.

Finally, we note that the sum $U(x,y)$ satisfies the remaining condition $U \to 0$ and $y \to \infty$ because each term satisfies that condition and the remainder after N terms is arbitrarily small in absolute value, uniformly with respect to x and $y(y \geq y_0)$, when N is sufficiently large.

Thus formula (12), Sec. 105, is established as a solution of the boundary value problem in the temperature function U.

107 SINGULAR EIGENVALUE PROBLEMS[1]

If the differential equation in an eigenvalue problem has a singular point on the interval, or if the interval is unbounded, the eigenvalue problem is singular. A requirement of continuity or boundedness of eigenfunctions replaces a boundary condition at a singular point or at an infinitely distant end of the interval. The representation theory for such problems may parallel that used for the Sturm-Liouville system, but the singular aspects of the probems require variations in the development of the theory.[2]

Some singular problems have discrete eigenvalues like those in Sturm-Liouville problems. Others have *continuous* eigenvalues, consisting of all numbers on a bounded or unbounded segment of the axis of reals. When eigenvalues are continuous, the representation of an arbitrary function in terms of eigenfunctions takes the form of an integral with respect to λ, instead

---

[1] Further details, including proofs or references to proofs of representation theorems, and applications, are given in the author's "Fourier Series and Boundary Value Problems," chaps. 6, 8, and 9, 1963.

[2] See E. C. Titchmarsh, "Eigenfunction Expansions," 1946.

of a series, of eigenfunctions. Prominent examples of singular problems follow.

BESSEL'S EQUATION

An example of an eigenvalue problem with Bessel's differential equation of index zero, on an interval $(0,c)$, is

(1)
$$[xX'(x)]' - \lambda x X(x) = 0 \qquad (0 < x < 1),$$

$$X(0) = 0, \qquad X \text{ and } X' \text{ continous} \qquad (0 \le x \le c).$$

Here the differential equation is a Sturm-Liouville equation, but $r(x) = x$ and $r(0) = 0$. When we write it in the form

(2)
$$X''(x) + \frac{1}{x}X'(x) - \lambda X(x) = 0 \qquad (0 < x < c),$$

we see that one of the coefficients in that form is discontinuous at the end point $x = 0$ of the interval, so that end point is a singular point of the equation, and the eigenvalue problem is singular.

The solution of Eq. (2) that is continuous, together with its derivative, over the interval $0 \le x \le c$, is (Sec. 83) $X = J_0(x\sqrt{-\lambda})$, except for a constant factor. Thus $X(c) = 0$ if $J_0(c\sqrt{-\lambda}) = 0$. As in Sec. 83, let $\alpha_n (n = 1, 2, \ldots)$ denote the positive zeros of $J_0(\alpha)$. Then the eigenfunctions and eigenvalues of problem (1) are

(3)
$$X_n = J_0\left(\alpha_n\frac{x}{c}\right), \qquad \lambda_n = -\frac{\alpha_n{}^2}{c^2} \qquad [J_0(\alpha_n) = 0, n = 1, 2, \ldots].$$

The orthogonality of the eigenfunctions on the interval $(0,c)$ with weight function $p(x) = x$ can be established by the procedure used in the Sturm-Liouville theory. A representation theorem states that if f' is sectionally continuous on the interval, the generalized Fourier series of those eigenfunctions converges to $f(x)$ at each interior point where f is continuous. The representation can be written

(4)
$$f(x) = \frac{2}{c^2}\sum_{n=1}^{\infty}\frac{1}{[J_1(\alpha_n)]^2}J_0\left(\alpha_n\frac{x}{c}\right)\int_0^c \xi f(\xi)J_0\left(\alpha_n\frac{\xi}{c}\right)d\xi \qquad (0 < x < c).$$

There are similar results if the differential equation in the problem is Bessel's equation with a positive index v,

$$(xX')' - \left(\lambda x + \frac{v^2}{x}\right)X = 0 \qquad (0 < x < c),$$

or when the boundary condition at the end point $x = c$ reads

$$hX(c) + X'(c) = 0 \qquad (h \ge 0).$$

But if the interval does not contain the origin $x = 0$ as an end point or as an interior point, these eigenvalue problems in Bessel's equation, with a homogeneous boundary condition at each end of the interval, are non-singular special cases of the Sturm-Liouville problem.

LEGENDRE'S EQUATION

The eigenvalue problem

(5)
$$[(1 - x^2)X'(x)]' - \lambda X(x) = 0 \qquad (-1 < x < 1),$$

X and X' continuous on the closed interval $-1 \leq x \leq 1$,

is another example of a singular problem. Legendre's differential equation here has singular points at both end points $x = \pm 1$ of the interval. Its solution satisfies the continuity condition only if $\lambda = -n(n + 1)$, where $n = 0, 1, 2, \ldots$. Corresponding to those eigenvalues, the eigenfunctions are the *Legendre polynomials*

(6)
$$P_n(x) = \frac{1}{2^n} \sum_{j=0}^{m} \frac{(-1)^j}{j!} \frac{(2n - 2j)!}{(n - 2j)!(n - j)!} x^{n-2j} \qquad (0! = 1)$$

where $m = \frac{1}{2}n$ if n is even and $m = \frac{1}{2}(n - 1)$ if n is odd. They form an orthogonal set with weight function $p(x) = 1$ on the interval $(-1,1)$. The representation of a function f on that interval, such that f' is sectionally continuous, as a generalized Fourier series of the eigenfunctions, can be established rather easily by using properties of P_n. It can be written

(7)
$$f(x) = \sum_{n=0}^{\infty} (n + \tfrac{1}{2})P_n(x) \int_{-1}^{1} f(\xi)P_n(\xi)\, d\xi \qquad (-1 < x < 1).$$

FOURIER INTEGRALS

The singular eigenvalue problem

(8)
$$X'' - \lambda X = 0, \qquad x > 0; \qquad X'(0) = 0, \qquad |X(x)| < M$$
$$(0 < x < \infty),$$

where M is some constant, is one in which the interval is unbounded. The solution of the differential equation that satisfies the condition $X'(0) = 0$ is, except for a constant factor, $X = \cos x\sqrt{-\lambda}$, for any value of λ, including zero. That solution is bounded over the half line $x > 0$ if and only if λ is real and $\lambda \leq 0$. We write $\lambda = -\alpha^2$, where $\alpha \geq 0$. Then the eigenfunctions and eigenvalues are

(9)
$$X(x) = \cos \alpha x, \qquad \lambda = -\alpha^2 \qquad (\alpha \geq 0).$$

In this example the eigenvalues are continuous, $\lambda \leq 0$, rather than discrete. The eigenfunctions lack an orthogonality property, but a representation theorem follows from the Fourier integral formula (Sec. 68) when the function G there is even. If f' is sectionally continuous on each bounded interval of the half line $x > 0$ and if f is absolutely integrable over that half line, then

$$(10) \qquad f(x) = \frac{2}{\pi} \int_0^\infty \cos \alpha x \int_0^\infty f(\xi) \cos \alpha \xi \, d\xi \, d\alpha \qquad (x \geq 0).$$

Thus f is represented as an integral of the eigenfunctions $\cos \alpha x$, with respect to α, by the Fourier cosine integral formula

$$f(x) = \int_0^\infty A(\alpha) \cos \alpha x \, d\alpha, \qquad A(\alpha) = \frac{2}{\pi} \int_0^\infty f(\xi) \cos \alpha \xi \, d\xi \qquad (x \geq 0).$$

Further examples of singular eigenvalue problems will be cited in the problems below and in the following chapters.

PROBLEMS

Apply the method of separation of variables to the boundary value problems in partial differential equations that follow.

1. The steady-state temperature problem (cf. Sec. 105)

$$U_{xx}(x,y) + U_{yy}(x,y) = 0, \qquad |U(x,y)| < M, \qquad (0 < x < 1, y > 0)$$

$$U_x(0,y) = U_x(1,y) = 0, \qquad U(x,0) = F(x) \qquad (0 < x < 1),$$

where M is some constant. Show that $\lambda = 0$ is an eigenvalue of the associated Sturm-Liouville problem and obtain the formal solution in the form

$$U(x,y) = \int_0^1 F(\xi) \, d\xi + 2 \sum_{n=1}^\infty e^{-n\pi y} \cos n\pi x \int_0^1 F(\xi) \cos n\pi \xi \, d\xi.$$

Note that as $y \to \infty$, U tends to the mean value of F.

2. The problem in Sec. 90 on displacements $Y(x,t)$ in a string when $Y(x,0) = g(x)$, $Y_t(x,0) = Y(0,t) = Y(c,t) = 0$.

3. Prob. 1, Sec. 93, in the longitudinal displacements $Y(x,t)$ of a stretched bar, where $Y(x,0) = Ax$, $Y_t(x,0) = Y(0,t) = Y_x(1,t) = 0$.

4. Prob. 4, Sec. 93, for displacements $Y(x,t)$ in a string with initial velocity $Y_t(x,0) = g(x), 0 < x < c$.

5. (a) The problem in Sec. 82 on temperatures $U(x,t)$ in a bar when $U(x,0) = g(x)$, $0 < x < 1$, $U(0,t) = U(1,t) = 0$, and $k = 1$.

(b) When g' is sectionally continuous, verify the solution of the boundary value problem in part (a) using the series representation

$$U(x,t) = 2 \sum_{n=1}^\infty \exp(-n^2\pi^2 t) \sin n\pi x \int_0^1 g(\xi) \sin n\pi \xi \, d\xi.$$

6.   Give a physical interpretation of $U(x,y)$ in the problem

$$U_{xx} + U_{yy} = 0, \qquad U_x(0,y) = U(1,y) = 0, \qquad -KU_y(x,0) = F(x),$$

where U is bounded over the region $0 < x < 1, y > 0$, and $K > 0$. Write $m = (n - \tfrac{1}{2})\pi$ and derive the solution in the form

$$U(x,y) = \frac{2}{K} \sum_{n=1}^{\infty} \frac{1}{m} e^{-my} \cos mx \int_0^1 F(\xi) \cos m\xi \, d\xi.$$

7.   Let $U(x,y)$ be the steady-state temperatures in a thin plate in the shape of a semi-infinite strip, from which heat is transferred at the faces into a medium at temperature zero, so that

$$U_{xx} + U_{yy} - bU = 0 \qquad (0 < x < 1, y > 0, b > 0).$$

If U is bounded and satisfies the conditions

$$U(0,y) = 0, \qquad U_x(1,y) = -hU(1,y), \qquad U(x,0) = 1 \qquad (0 < x < 1).$$

and $h > 0$, derive the formula

$$U(x,y) = 2h \sum_{n=1}^{\infty} \frac{A_n}{\alpha_n} \exp\left[-y(b + \alpha_n^2)^{\frac{1}{2}}\right] \sin \alpha_n x \qquad (b > 0, h > 0),$$

where $A_n = (1 - \cos \alpha_n)/(h + \cos^2 \alpha_n)$ and $\alpha_1, \alpha_2,\ldots$ are the positive roots of the equation $\tan \alpha = -\alpha/h$.

8.   Let $U(x,y)$ be the steady-state temperatures in an infinite prism bounded by the planes $x = 0, y = 0, x = 1$, and $y = 1$. If $U = 1$ on the face $y = 1$ and if $U_x = -hU$ at $x = 1$, where $h > 0$, and $U = 0$ on the other two faces (Fig. 102), derive the formula

$$U(x,y) = 2h \sum_{n=1}^{\infty} \frac{A_n}{\alpha_n} \frac{\sinh \alpha_n y}{\sinh \alpha_n} \sin \alpha_n x,$$

where the numbers A_n and α_n are those described in Prob. 7.

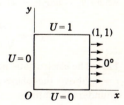

Fig. 102

9.   Write $m = (n - \tfrac{1}{2})\pi$ and obtain the solution of the problem

$$(t + 1)U_t(x,t) = U_{xx}(x,t) \qquad\qquad (0 < x < 1, t > 0),$$

$$U(0,t) = U_x(1,t) = 0, \qquad U(x,0) = 1 \qquad\qquad (0 < x < 1),$$

in the form

$$U(x,t) = 2 \sum_{n=1}^{\infty} \frac{1}{m}(t + 1)^{-m^2} \sin mx.$$

Verify that this function satisfies the condition $U(x,+0) = 1 (0 < x < 1)$.

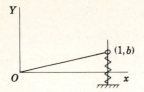

Fig. 103

10. The end $x = 1$ of a stretched string is elastically supported (Fig. 103) so that the transverse displacement $Y(x, t)$ satisfies the condition $Y_x(1,t) = -hY(1,t)$, where $h > 0$. Also, let Y satisfy the conditions

$$Y(0,t) = 0, \qquad Y(x,0) = bx, \qquad Y_t(x,0) = 0.$$

Show that the solution of the equation $Y_{tt}(x,t) = Y_{xx}(x,t)$ is then

$$Y(x,t) = 2bh(h + 1) \sum_{n=1}^{\infty} \frac{\sin \alpha_n \sin \alpha_n x}{\alpha_n^2(h + \cos^2 \alpha_n)} \cos \alpha_n t,$$

where $\alpha_1, \alpha_2, \ldots$ are the positive roots of the equation $\tan \alpha = -\alpha/h$.

11. The longitudinal displacements $Y(x,t)$ in a certain nonuniform bar satisfy the conditions

$$\frac{\partial}{\partial x}\left(e^x \frac{\partial Y}{\partial x}\right) = e^{-x}\frac{\partial^2 Y}{\partial t^2} \qquad (0 < x < 1, t > 0),$$

$$Y(0,t) = Y(1,t) = Y_t(x,0) = 0, \qquad Y(x,0) = F(x) \qquad (0 < x < 1).$$

Derive the formula

$$Y(x,t) = \sum_{n=1}^{\infty} c_n \phi_n(x) \cos \beta_n t,$$

where $\beta_n = n\pi e/(e - 1)$ and

$$\phi_n(x) = \sqrt{\frac{\beta_2}{\pi}} \sin [\beta_n(1 - e^{-x})], \qquad c_n = \int_0^1 F(x)e^{-x}\phi_n(x)\, dx.$$

12. The electrostatic potential $V(r,z)$ in the space bounded by the cylinders $r = a$ and $r = b$ above the plane $z = 0$ satisfies Laplace's equation $(rV_r)_r + rV_{zz} = 0$ and the boundary conditions

$$V(a,z) = V(b,z) = 0 \qquad\qquad (z > 0),$$

$$V(r,0) = F(r) \qquad\qquad (a < r < b);$$

also $V(r,z)$ is bounded. Derive the formula

$$V(r,z) = \sum_{n=1}^{\infty} A_n \exp{(-\alpha_n z)}\psi(\alpha_n, r),$$

where $\psi(\alpha,r) = J_0(\alpha r)Y_0(\alpha a) - J_0(\alpha a)Y_0(\alpha r)$. J_0 and Y_0 are Bessel functions of the first and second kind, linearly independent solutions of Bessel's equation. The numbers α_n

are the roots of the equation $\psi(a,b) = 0$, all real[1] $\alpha_m > 0$, and A_n are given by the formula

$$A_n \int_a^b r[\psi(\alpha_n,r)]^2 \, dr = \int_a^b rF(r)\psi(\alpha_n,r) \, dr.$$

13. The temperature $V(r,t)$ in a solid cylinder $r \leq 1$ of infinite length satisfies the conditions

$$V_t = V_{rr} + r^{-1}V_r \qquad (0 \leq r < 1, t > 0), \qquad V(r,0) = 1, \qquad V(1,t) = 0;$$

also V and V_r are continuous when $0 \leq r \leq 1$ and $t > 0$. Show that the method of separation of variables involves a singular eigenvalue problem of type (1), Sec. 107. With the aid of the integration formula

$$\int_0^c xJ_0(x) \, dx = cJ_1(c),$$

solve for V in the form

$$V(r,t) = 2 \sum_{n=1}^{\infty} \frac{J_0(\alpha_n r)}{\alpha_n J_1(\alpha_n)} \exp(-\alpha_n^2 t),$$

where the α_n's are the positive zeros of $J_0(\alpha)$.

14. The temperature $U(x,t)$ in a semi-infinite solid $x \geq 0$, whose initial temperature function is a step function, satisfies the conditions

$$U_t = kU_{xx}; \qquad U(x,0) = 1 - S_0(x - 1), x > 0; \qquad U_x(0,t) = 0,$$

and U is bounded. Show that the solution by separation of variables involves the singular eigenvalue problem (8), Sec. 107, and derive the formula

$$U(x,t) = \frac{2}{\pi} \int_0^{\infty} \frac{\sin \alpha}{\alpha} \cos \alpha x \exp(-\alpha^2 kt) \, d\alpha.$$

15. Prove that the eigenvalues and eigenfunctions of the singular problem

$$X'' - \lambda X = 0, 0 < x < \infty; \qquad X(0) = 0; \qquad |X(x)| \leq 1, x > 0;$$

are $\lambda = -\alpha^2 (\alpha > 0)$ and $X = \sin \alpha x$. If f' is sectionally continuous and absolutely integrable over the half line $x > 0$, note that, according to the Fourier integral formula (Sec. 68) for odd functions, f is represented in terms of those eigenfunctions by the *Fourier sine integral formula*

$$f(x) = \frac{2}{\pi} \int_0^{\infty} \sin \alpha x \int_0^{\infty} f(\xi) \sin \alpha \xi \, d\xi \qquad (x > 0).$$

16. Use the results found in Prob. 15 to solve the problem in Sec. 43 for the displacements $Y(x, t)$ of an initially displaced semi-infinite stretched string, namely,

$$Y_{tt} = a^2 Y_{xx}, \qquad x > 0, t > 0; \qquad Y(x,0) = \Phi(x), x > 0;$$

$$Y_t(x,0) = Y(0,t) = 0, \qquad |Y(x,t)| < M, \qquad |\Phi(x)| < M,$$

---

[1] See H. S. Carslaw and J. C. Jaeger, "Conduction of Heat in Solids," 2d ed., p. 206, 1959, and their tables, p. 493.

in the form

$$Y = \frac{1}{\pi} \int_0^\infty [\sin \alpha(x + at) + \sin \alpha(x - at)] \int_0^\infty \Phi(\xi) \sin \alpha \xi \, d\xi \, d\alpha.$$

Show how this reduces to the form (8), Sec. 43.

17. Prove that the eigenfunctions of the singular problem

$$X''(x) - \lambda X(x) = 0, \qquad |X(x)| \leq 1 \qquad\qquad (-\infty < x < \infty)$$

are $X = \cos \alpha(x - \xi)$, where $\lambda = -\alpha^2$ and $\alpha \geq 0$ and the parameter ξ has any real value. Note that according to the Fourier integral formula (2), Sec. 68,

$$f(x) = \frac{1}{\pi} \int_0^\infty \int_{-\infty}^\infty f(\xi) \cos \alpha(x - \xi) \, d\xi \, d\alpha \qquad\qquad (-\infty < x < \infty).$$

18. Note that the first few Legendre polynomials are

$$P_0(x) = 1, \qquad P_1(x) = x, \qquad P_2(x) = \tfrac{1}{2}(3x^2 - 1), \qquad P_3(x) = \tfrac{1}{2}(5x^3 - 3x).$$

Verify that they satisfy Legendre's equation in problem (5), Sec. 107, when $\lambda = -n(n + 1)$, $n = 0, 1, 2, 3$.

19. Legendre's polynomials have the property

$$P_n(-x) = (-1)^n P_n(x) \qquad\qquad (n = 0, 1, 2, \ldots),$$

so P_{2n} and P_{2n+1} are even and odd functions, respectively. Deduce that $X = P_{2n+1}(x)$ and $\lambda = -2(2n + 1)(n + 1)$ are eigenfunctions and eigenvalues of this singular problem on the interval (0,1):

$$[(1 - x^2)X']' - \lambda X = 0, 0 < x < 1; \qquad X(0) = 0,$$

where X and X' are continuous on the closed interval $0 \leq x \leq 1$.

20. If the condition $X(0) = 0$ is replaced by the condition $X'(0) = 0$ in Prob. 19, show that $\lambda = -2n(2n + 1)$ are eigenvalues and $X = P_{2n}(x)$ are eigenfunctions.

10

General Integral Transforms

108 LINEAR INTEGRAL TRANSFORMATIONS

Let (a,b) denote a bounded or unbounded interval of the x axis and let $K(x,\lambda)$ be a prescribed function of x on that interval and of a parameter λ. Then, as we pointed out in Sec. 1, a general linear integral transformation of functions F on the interval is

(1)
$$T\{F\} = \int_a^b F(x)K(x,\lambda)\,dx = f(\lambda).$$

The resulting function $f(\lambda)$ is the *transform* of F under that transformation with *kernel* K.

Of course the class of functions from which F and K are chosen must be restricted and a set Λ of values of the parameter λ must be specified so as to ensure not only the existence of the integral (1) but also the uniqueness of the inverse transform. That is, not more than one function F should correspond to a given transform $f(\lambda)$.

The class of functions, the *function space*, to which F belongs is to be a *linear space* in order that $AF(x) + BG(x)$ will belong to that space whenever F and G belong to it, for each pair of constants A and B. Then T is a *linear transformation*:

$$(2) \qquad T\{AF + BG\} = AT\{F\} + BT\{G\} = Af(\lambda) + Bg(\lambda).$$

If inverse transforms $T^{-1}\{f\}$ are unique, the linearity condition (2) can be written

$$(3) \qquad T^{-1}\{Af + Bg\} = AF + BG = AT^{-1}\{f\} + BT^{-1}\{g\};$$

that is, the inverse transformation is also linear.

When $a = 0$, $b = \infty$, and $K(x,\lambda) = e^{-\lambda x}$, the transformation T is the Laplace transformation $L\{F\} = f(\lambda)$. In that case we noted that the function space of F may consist of all real-valued functions of exponential order on the half line $x > 0$ which are sectionally continuous on each bounded interval, and the set Λ of values of λ may be all complex numbers in a right half plane. We restricted the function space somewhat further in order to ensure uniqueness of $L^{-1}\{f\}$. Also, we found a general formula, the inversion integral, for the inverse Laplace transformation.

We shall soon indicate how the transformation T may be specified so that it transforms a given linear differential form in F into an algebraic form in $f(\lambda)$, λ and prescribed boundary values of F at the ends of the interval (a,b). That gives the *basic operational property* of T for the reduction of boundary value problems in differential equations. As in the case of the Laplace transformation, where that operational property is $L\{F'\} = \lambda f(\lambda) - F(0)$, some further restrictions on the space of the functions F are needed to ensure the validity of the property.

The operational calculus of an integral transformation becomes more effective when its basic operational property and its inversion process are supplemented by other operational properties and tables of transforms.

A *convolution property* of T is one that gives a representation of $T^{-1}\{f(\lambda)g(\lambda)\}$ directly in terms of the functions F and G whose transforms are f and g. The representation is often simpler than one found by applying a formula for inverse transformations. In some cases the property is modified by replacing fg by a weighted product of transforms to give $T^{-1}\{\omega(\lambda)f(\lambda)g(\lambda)\}$, where the weight function ω is fixed for a given transformation. We have seen that the convolution is an important property of the Laplace transformation. Convolution properties are also prominent operational properties of various Fourier transformations. But for some of the less common integral transformations such properties are either unknown or so complicated that they are tedious to use.

109 KERNEL-PRODUCT AND CONVOLUTION PROPERTIES

For the Laplace transformation the kernel $K(x,\lambda) = e^{-\lambda x}$ has the simple product property

$$(1) \qquad K(x,\lambda)K(x_0,\lambda) = K(x + x_0, \lambda).$$

Then if $f(\lambda) = L\{F\}$ and $F(x) = 0$ when $x < 0$, it followed that L has the translation property (Sec. 11)

$$(2) \qquad f(\lambda)K(x_0,\lambda) = L\{F(x - x_0)\} \qquad\qquad (x_0 > 0).$$

This formula represents the product of the transform of F by the kernel as the transform of a function related to F. We call it a *kernel-product property* of the Laplace transformation. That property led us to the convolution property (Sec. 16)

$$(3) \qquad f(\lambda)g(\lambda) = L\left\{\int_0^x F(y)G(x - y)\,dy\right\} = L\{F * G\},$$

where $f(\lambda) = L\{F\}$ and $g(\lambda) = L\{G\}$.

Now, in the case of another transformation of the type

$$(4) \qquad T\{F\} = \int_a^b F(x)K(x,\lambda)\,dx = f(\lambda),$$

suppose we have a modified or weighted kernel-product property such that for some specified function $\omega(\lambda)$ and for each function F there is a function $P_F(x,x_0)$ for which

$$(5) \qquad f(\lambda)\omega(\lambda)K(x_0,\lambda) = T\{P_F(x,x_0)\} \qquad\qquad (a < x_0 < b).$$

Then a modified convolution property follows, formally.

For if $f(\lambda) = T\{F\}$ and $g(\lambda) = T\{G\}$, we can write

$$(6) \qquad f(\lambda)g(\lambda)\omega(\lambda) = \int_a^b g(\lambda)\omega(\lambda)K(y,\lambda)F(y)\,dy$$

$$= \int_a^b T\{P_G(x,y)\}F(y)\,dy$$

$$= \int_a^b \int_a^b P_G(x,y)K(x,\lambda)\,dx\,F(y)\,dy.$$

When the order of integration is reversed, the iterated integral becomes

$$\int_a^b K(x,\lambda)\int_a^b F(y)P_G(x,y)\,dy\,dx$$

which is the transform of the function

(7)
$$X_{FG}(x) = \int_a^b F(y)P_G(x,y)\,dy.$$

We call the function X_{FG} the *convolution* of F and G for the transformation T, with respect to the weight function ω. It is the inverse transform of the product $fg\omega$. That is,

(8)
$$f(\lambda)g(\lambda)\omega(\lambda) = T\{X_{FG}(x)\} = T\{X_{GF}(x)\},$$

where the last form $T\{X_{GF}\}$ follows by interchanging the functions F and G, and their transforms, in Eqs. (6).

Thus we have indicated how *a convolution property* (8) *follows from a kernel-product property* (5). The convolution is described by Eq. (7) in terms of the function P_F in property (5). An example is given in the following section. Other examples will be found in the problems and in the chapters that follow. A corresponding result for a generalized convolution property for iterated transforms of functions $H(x,y)$ will also be noted in the problems (Prob. 8, Sec. 110).

If we replace F in the transformation (4) by the delta symbol, we find, formally,

(9)
$$T\{\delta(x - x_0)\} = K(x_0,\lambda) \qquad \text{if } a < x_0 < b.$$

When F is replaced by $\delta(x - x_0)$, the convolution (7) becomes

(10)
$$X_{\delta G}(x) = P_G(x,x_0) \qquad [\delta = \delta(x - x_0), a < x_0 < b],$$

and the convolution property (8) reverts to our kernel-product property written in terms of g and G:

$$K(x_0,\lambda)g(\lambda)\omega(\lambda) = T\{P_G(x,x_0)\} \qquad (a < x_0 < b).$$

110   EXAMPLE

We illustrate the results in the preceding section by deriving a convolution formula for the *Fourier sine transformation* on the half line $x > 0$,

(1)
$$S_\lambda\{F\} = \int_0^\infty F(x)\sin\lambda x\,dx = f_s(\lambda) \qquad (\lambda > 0).$$

Other operational properties of S_λ will be developed in Sec. 116 and Chap. 13. We assume that our functions F are sectionally continuous on each bounded interval and absolutely integrable over the half line $x > 0$. It is convenient to introduce the *odd extension* F_1 of F defined over the full line $-\infty < x < \infty$:

(2) $F_1(x) = F(x)$ if $x > 0$, $F_1(-x) = -F_1(x)$ for all x.

We introduce also the continuous function

(3) $$H(x) = \int_{-\infty}^{x} F_1(\xi) \, d\xi = \int_{-\infty}^{0} F_1(\xi) \, d\xi + \int_{0}^{x} F_1(\xi) \, d\xi$$

$$(-\infty < x < \infty)$$

and note that it is *even*, $H(-x) = H(x)$ or $H(|x|) = H(x)$, and that it vanishes as $x \to \pm\infty$; also, $H'(x) = F_1(x)$.

To find a kernel-product property of S_λ, we write

$$\frac{2}{\lambda} f_s(\lambda) = \frac{1}{\lambda} \int_{-\infty}^{\infty} F_1(x) \sin \lambda x \, dx = \int_{-\infty}^{\infty} \frac{\sin \lambda x}{\lambda} H'(x) \, dx$$

$$= \left[\frac{\sin \lambda x}{\lambda} H(x) \right]_{-\infty}^{\infty} - \int_{-\infty}^{\infty} H(x) \cos \lambda x \, dx.$$

The bracketed term here vanishes. Thus when $x_0 > 0$,

$$\frac{2}{\lambda} f_s(\lambda) \sin \lambda x_0 = - \int_{-\infty}^{\infty} H(x) \cos \lambda x \sin \lambda x_0 \, dx$$

$$= \tfrac{1}{2} \int_{-\infty}^{\infty} H(x) [\sin \lambda(x - x_0) - \sin \lambda(x + x_0)] \, dx$$

$$= \tfrac{1}{2} \int_{-\infty}^{\infty} [H(y + x_0) - H(|y - x_0|)] \sin \lambda y \, dy.$$

The integrand of the final integral is an even function of y. Thus

$$\frac{2}{\lambda} f_s(\lambda) \sin \lambda x_0 = \int_{0}^{\infty} \sin \lambda y \int_{|y - x_0|}^{y + x_0} F_1(\xi) \, d\xi \, dy.$$

Since $F_1(\xi) = F(\xi)$ when $\xi > 0$, a *kernel-product property* for the transformation S_λ can be written

(4) $$f_s(\lambda) \frac{2}{\lambda} \sin \lambda x_0 = S_\lambda \left\{ \int_{|x - x_0|}^{x + x_0} F(\xi) \, d\xi \right\}$$ $$(x_0 > 0).$$

In terms of the notation used in formula (5), Sec. 109, then

(5) $$P_F(x, x_0) = \int_{|x - x_0|}^{x + x_0} F(\xi) \, d\xi, \qquad \omega(\lambda) = \frac{2}{\lambda}$$

when $K(x, \lambda) = \sin \lambda x$, $a = 0$, $b = \infty$, and T is S_λ.

The corresponding convolution of two functions F and G is

(6) $$X_{FG}(x) = \int_{0}^{\infty} F(y) P_G(x, y) \, dy = \int_{0}^{\infty} F(y) \int_{|x - y|}^{x + y} G(\xi) \, d\xi \, dy,$$

according to formula (7), Sec. 109. The *convolution property* reads

(7) $$f_s(\lambda) g_s(\lambda) \frac{2}{\lambda} = S_\lambda \left\{ \int_0^\infty F(y) \int_{|y-x|}^{y+x} G(\xi)\, d\xi\, dy \right\} \qquad (\lambda > 0),$$

where $f_s(\lambda) = S_\lambda\{F\}$ and $g_s(\lambda) = S_\lambda\{G\}$.

ILLUSTRATION OF FORMULA (7)

In the particular case

(8) $$F(x) = G(x) = e^{-ax} \qquad (a > 0),$$

an elementary integration shows that

(9) $$f_s(\lambda) = g_s(\lambda) = \frac{\lambda}{\lambda^2 + a^2} \qquad (\lambda > 0);$$

also, $$P_G(x,y) = \int_{|x-y|}^{x+y} e^{-a\xi}\, d\xi = \frac{1}{a}[e^{-a|x-y|} - e^{-a(x+y)}].$$

Then the convolution (6) becomes

$$X_{FG}(x) = \frac{1}{a} \int_0^x e^{-ay} e^{-a(x-y)}\, dy + \frac{1}{a} \int_x^\infty e^{-ay} e^{-a(y-x)}\, dy$$

$$- \frac{1}{a} \int_0^\infty e^{-ay} e^{-a(x+y)}\, dy = \frac{1}{a} x e^{-ax}.$$

Thus we have this useful Fourier sine transformation

(10) $$[f_s(\lambda)]^2 \frac{2}{\lambda} = \frac{2\lambda}{(\lambda^2 + a^2)^2} = S_\lambda \left\{ \frac{x}{a} e^{-ax} \right\} \qquad (a > 0).$$

PROBLEMS

1. Use the convolution property (7), Sec. 110, when

$$F(x) = e^{-ax}, \qquad G(x) = e^{-bx} \qquad (a > 0, b > 0, a \neq b)$$

to obtain this Fourier sine transformation:

$$\frac{(a^2 - b^2)\lambda}{(\lambda^2 + a^2)(\lambda^2 + b^2)} = S_\lambda\{e^{-bx} - e^{-ax}\} \qquad (a > 0, b > 0, a \neq b).$$

(This result can be found also by using partial fractions.) Evaluate the integral transformation on the right to verify the result.

2. Interchange the order of integration in formula (6), Sec. 110, to show that the convolution X_{FG} for the Fourier sine transformation S_λ is symmetric in F and G: $X_{FG}(x) = X_{GF}(x)$.

3. The *exponential Fourier transformation* of functions F on the entire x axis is (Chap 12)

$$E_\lambda\{F\} = \int_{-\infty}^{\infty} F(x)e^{-i\lambda x}\, dx = f_e(\lambda) \qquad (-\infty < \lambda < \infty).$$

We consider functions F that are sectionally continuous on each bounded interval, absolutely integrable from $-\infty$ to ∞ and bounded. Show that E_λ has the kernel-product property

$$f_e(\lambda)\exp(-i\lambda x_0) = E_\lambda\{F(x - x_0)\} \qquad (-\infty < x_0 < \infty)$$

and, formally, that the corresponding convolution is

$$X_{FG}(x) = \int_{-\infty}^{\infty} F(y)G(x - y)\, dy = X_{GF}(x),$$

so that if $g_e(\lambda) = E_\lambda\{G\}$, then

$$f_e(\lambda)g_e(\lambda) = E_\lambda\{X_{FG}\} = E_\lambda\left\{\int_{-\infty}^{\infty} F(y)G(x - y)\, dy\right\}.$$

4. Let $S_0(x)$ be the unit step function: $S_0(x) = 0$ when $x < 0$, $S_0(x) = 1$ when $x > 0$. Note that the exponential Fourier transforms of the two functions

$$F(x) = e^{-x}S_0(x), \qquad G(x) = e^{x}S_0(-x) \qquad (-\infty < x < \infty)$$

are $f_e(\lambda) = (1 + i\lambda)^{-1}$ and $g_e(\lambda) = (1 - i\lambda)^{-1}$. Use the convolution property of E_λ in Prob. 3 to derive this transformation:

$$\frac{1}{\lambda^2 + 1} = E_\lambda\{\tfrac{1}{2}e^{-|x|}\}.$$

Verify the result by evaluating the integral $E_\lambda\{e^{-|x|}\}$.

5. When $K(x,\lambda) = \cos \lambda x$, $a = 0$, and $b = \infty$, our transformation T becomes the *Fourier cosine transformation* C_λ on the half line (Chap. 13),

$$C_\lambda\{F\} = \int_0^{\infty} F(x)\cos \lambda x\, dx = f_c(\lambda) \qquad (\lambda > 0).$$

Let the functions F be sectionally continuous on each bounded interval and absolutely integrable over the half line. If F_2 is the even extension of F, so that $F_2(x) = F(|x|)$ when $-\infty < x < \infty$, then

$$f_c(\lambda)\cos \lambda x_0 = \tfrac{1}{2}\int_{-\infty}^{\infty} F_2(x)\cos \lambda x \cos \lambda x_0\, dx.$$

Show that a kernel-product property of C_λ is

$$f_c(\lambda)\cos \lambda x_0 = C_\lambda\{\tfrac{1}{2}[F(x + x_0) + F(|x - x_0|)]\} \qquad (x_0 > 0);$$

then that the corresponding convolution property is

$$f_c(\lambda)g_c(\lambda) = C_\lambda\left\{\tfrac{1}{2}\int_0^{\infty} F(y)[G(x + y) + G(|x - y|)]\, dy\right\};$$

that is, the convolution for C_λ when $\omega(\lambda) = 1$ is

$$X_{FG}(x) = \tfrac{1}{2} \int_0^\infty F(y)[G(x + y) + G(|x - y|)] \, dy.$$

6. Prove that the convolution for C_λ found in Prob. 5 is symmetric in F and G: $X_{FG}(x) = X_{GF}(x)$.

7. Let F be the step function $1 - S_0(x - k)$, where $k > 0$, and let G be e^{-ax} $(a > 0)$. If C_λ is the cosine transformation in Prob. 5, then

$$C_\lambda\{F\} = \frac{\sin k\lambda}{\lambda}, \qquad C_\lambda\{G\} = \frac{a}{\lambda^2 + a^2}.$$

Show that the convolution of F and G for C_λ is the function

$$X_{FG}(x) = a^{-1}(1 - e^{-ak} \cosh ax) \qquad \text{when } x \le k$$

$$= a^{-1} e^{-ax} \sinh ak \qquad \text{when } x \ge k,$$

so that we have the transformation

$$\frac{a \sin k\lambda}{\lambda(\lambda^2 + a^2)} = C_\lambda\{X_{FG}(x)\}.$$

8. *Generalized convolution*   For a function $H(x,y)$ of two variables, let $h(\lambda,y)$ denote the transform $T\{H\}$ with respect to x,

$$h(\lambda,y) = \int_a^b H(x,y)K(x,\lambda) \, dx \qquad (a < y < b),$$

and let $\hat{h}(\lambda)$ denote the transform of $h(\lambda,y)$ with respect to y when the same parameter λ is used,

$$\hat{h}(\lambda) = \int_a^b h(\lambda,y)K(y,\lambda) \, dy,$$

if H and T are such that this iterated transform exists.

In case T has a kernel-product property

$$h(\lambda,y)\omega(\lambda)K(y,\lambda) = \int_a^b Q_H(x,y)K(x,\lambda) \, dx,$$

show formally that it has the generalized convolution property

$$\hat{h}(\lambda)\omega(\lambda) = \int_a^b \left[\int_a^b Q_H(x,y) \, dy \right] K(x,\lambda) \, dx.$$

Note that this reduces to property (8), Sec. 109, in case $H(x,y) = F(x)G(y)$; then $Q_H = P_F(x,y)G(y)$. Examples are given in Probs. 9 and 10.

9. When the transformation in Prob. 8 is the exponential Fourier transformation E_λ described in Prob. 3 and $\omega(\lambda) = 1$, show that the generalized convolution property becomes

$$\hat{h}(\lambda) = E_\lambda\left\{ \int_{-\infty}^\infty H(x - y, y) \, dy \right\},$$

where
$$\hat{h}(\lambda) = \int_{-\infty}^{\infty} e^{-i\lambda y} \int_{-\infty}^{\infty} H(x,y)e^{-i\lambda x}\, dx\, dy.$$

10. If $H(x,y) = 1$ when $x > 0$ and $y > 0$ and $x + y < 1$, while $H(x,y) = 0$ for all other real x and y, show that the iterated exponential Fourier transform of H is

$$\hat{h}(\lambda) = \frac{(1 + i\lambda)e^{-i\lambda} - 1}{\lambda^2}$$

and verify that this agrees with the convolution property in Prob. 9.

111 STURM-LIOUVILLE TRANSFORMS

We are now ready to present a method of determining a linear integral transformation T that simplifies a given linear boundary value problem in a function F. The problem is to be such that in the differential equation the differential form in F with respect to some one of the independent variables x can be isolated and, in the boundary conditions, boundary values with respect to x can be isolated. The transformation T with respect to x is to be designed so as to replace that differential form by an algebraic form in the transform $T\{F\}$ and the prescribed boundary values on F at the ends of the interval over which the variable x ranges.

We begin with an important case that can be treated with the aid of the Sturm-Liouville theory in Chap. 9.

Let the differential form with respect to x be of order 2,

$$(1) \qquad D_2 F = A(x)F''(x) + B(x)F'(x) + C(x)F(x) \qquad (a < x < b),$$

and let the interval (a,b) be bounded. For the sake of simplicity we assume that all functions of x here, together with their derivatives of the first and second order, are continuous over the closed interval $a \le x \le b$, although those conditions can be relaxed. Also, we assume that $A(x) > 0$ over that closed interval. In applications to problems in partial differential equations, F will be a function of x and the remaining independent variables, while A, B, and C will be functions of x alone.

The boundary values with respect to x, to be prescribed as functions of the remaining variables, are

$$(2) \qquad N_\alpha F = F(a) \cos \alpha + F'(a) \sin \alpha,$$

$$(3) \qquad N_\beta F = F(b) \cos \beta + F'(b) \sin \beta,$$

where the constants α and β are real.

In order to design a transformation

$$T\{F\} = \int_a^b F(x)K(x,\lambda)\, dx = f(\lambda)$$

that transforms D_2F into an algebraic form in $f(\lambda)$, $N_\alpha F$, and $N_\beta F$, it is convenient to write D_2F in terms of a self-adjoint differential form

(4) $$\mathscr{D}_2 F = [r(x)F'(x)]' - q(x)F(x).$$

We do that by writing $D_2F = Ar^{-1}(rF'' + A^{-1}BrF') + CF$, where

(5) $$r(x) = \exp\left[\int_a^x \frac{B(\xi)}{A(\xi)} d\xi\right]$$

so that $r' = A^{-1}Br$, and introducing the functions

(6) $$p(x) = \frac{r(x)}{A(x)}, \qquad q(x) = -p(x)C(x).$$

Then

(7) $$D_2 F = \frac{1}{p(x)}\mathscr{D}_2 F = \frac{1}{p}[(rF')' - qF].$$

Note that $r(a) = 1$ here. Corresponding representations in terms of self-adjoint forms are not always possible for forms of order higher than 2. In such cases the original linear differential form may be used; then the adjoint of that form arises in the process of designing T.

We now write the kernel of our transformation T as

$$K(x,\lambda) = p(x)\Phi(x,\lambda)$$

where the weight function p is r/A, and the kernel Φ and the range of the parameter λ are yet to be determined. Then the transform of D_2F becomes $\int_a^b \Phi \mathscr{D}_2 F\, dx$ and, either by integration by parts or by using Lagrange's identity (5), Sec. 94, according to which

$$\Phi \mathscr{D}_2 F = F\mathscr{D}_2 \Phi + [(\Phi F' - \Phi'F)r]',$$

we can write

(8) $$T\{D_2F\} = \int_a^b F(x)\mathscr{D}_2\Phi(x,\lambda)\, dx + [(\Phi F' - \Phi'F)r]_a^b.$$

In order to express $F(a)$ and $F'(a)$ in terms of the prescribed boundary value $N_\alpha F$, we introduce the derivative of $N_\alpha F$ with respect to α,

$$N_\alpha' F = -F(a)\sin\alpha + F'(a)\cos\alpha,$$

and solve this equation simultaneously with Eq. (2) to get

$$F(a) = N_\alpha F\cos\alpha - N_\alpha' F\sin\alpha, \qquad F'(a) = N_\alpha' F\cos\alpha + N_\alpha F\sin\alpha.$$

At the lower limit the bracketed term in Eq. (8) can now be written

$$[\Phi(a,\lambda)F'(a) - \Phi'(a,\lambda)F(a)]r(a) = -N_\alpha F N_\alpha' \Phi + N_\alpha' F N_\alpha \Phi.$$

The value of this term is determined by $N_\alpha F$ and Φ if $N_\alpha \Phi = 0$.
Likewise at $x = b$ we find that

$$[(\Phi F' - \Phi'F)r]_{x=b} = -r(b)N_\beta F N'_\beta \Phi$$

if $N_\beta \Phi = 0$, where

$$N'_\beta \Phi = -\Phi(b,\lambda) \sin \beta + \Phi'(b,\lambda) \cos \beta.$$

The integral in Eq. (8) becomes $\lambda f(\lambda)$ if Φ satisfies the Sturm-Liouville equation $\mathscr{D}_2 \Phi = \lambda p \Phi$. Hence we have the desired transformation

$$(9) \qquad T\{D_2 F\} = \lambda f(\lambda) + N'_\alpha[\Phi]N_\alpha F - r(b)N'_\beta[\Phi]N_\beta F$$

provided that Φ is a nontrivial solution of the Sturm-Liouville problem

$$(10) \qquad (r\Phi')' - (q + \lambda p)\Phi = 0, \qquad N_\alpha \Phi = 0, \qquad N_\beta \Phi = 0.$$

Since problem (10) has nontrivial solutions when and only when λ is an eigenvalue λ_n, the kernel Φ must be an eigenfunction Φ_n corresponding to the eigenvalue λ_n; then

$$(11) \qquad T\{F\} = \int_a^b F(x)p(x)\Phi_n(x)\,dx = f(\lambda_n) \qquad (n = 1, 2, \dots).$$

We call this the *Sturm-Liouville transformation*. Its parameter λ ranges through the discrete real values $\lambda_1, \lambda_2, \dots$, the set of *all* the eigenvalues of the Sturm-Liouville problem (10). The transform $f(\lambda_n)$ is also *an infinite sequence of numbers*.

The *basic operational property* (9) of T now reads

$$(12) \qquad T\{D_2 F\} = \lambda_n f(\lambda_n) + N'_\alpha[\Phi_n]N_\alpha F - r(b)N'_\beta[\Phi_n]N_\beta F.$$

112   INVERSE TRANSFORMS

We select some convenient complete set of *real-valued* eigenfunctions Φ_n, not necessarily normalized, as the kernel in our Sturm–Liouville transformation (11) above. The generalized Fourier series for a function F on the interval (a,b) can be written

$$\sum_{n=1}^{\infty} \frac{\Phi_n(x)}{\|\Phi_n\|} \int_a^b F(\xi)p(\xi)\frac{\Phi_n(\xi)}{\|\Phi_n\|}\,d\xi = \sum_{n=1}^{\infty} \frac{\Phi_n(x)}{\|\Phi_n\|^2} f(\lambda_n),$$

where $\|\Phi_n\|^2 = \int_a^b p(\xi)[\Phi_n(\xi)]^2\,d\xi$.

When F is of *class C''*, that is, *when F, F', and F'' are continuous over the closed interval $a \leq x \leq b$*, that series converges to $F(x)$ over the open interval. Thus the series represents F in terms of its transform $f(\lambda_n)$.

A formula for the *inverse Sturm-Liouville transform* of $f(\lambda_n)$ is therefore

$$(1) \qquad F(x) = \sum_{n=1}^{\infty} f(\lambda_n)\|\Phi_n\|^{-2}\Phi_n(x) = T^{-1}\{f(\lambda_n)\} \qquad (a < x < b).$$

Uniqueness of inverse transforms among functions of class C'' follows from that inversion formula. For if two such functions F and H have identical transforms, then $T\{F - H\} = 0$, and formula (1) shows that $F(x) - H(x) \equiv 0$.

Alternate ways of finding inverse transforms are furnished by tables of transforms, by devices for representing $f(\lambda_n)$ in terms of known transforms, or by operational properties of T. The transform of any particular eigenfunction $\Phi_m(x)$, where m is some positive integer, can be written at once from the definition of T and the orthogonality of the eigenfunctions. Thus

$$(2) \qquad T\{\Phi_m\} = 0 \quad \text{if } n \neq m, \qquad T\{\Phi_m\} = \|\Phi_m\|^2 \quad \text{if } n = m;$$

that is, the transform $\phi_m(\lambda_n)$ is the sequence $0, 0, \ldots, \|\Phi_m\|^2, 0, 0, \ldots$.

We present now an operational property that gives $T^{-1}\{f(\lambda_n)/\lambda_n\}$ in terms of the function F and a solution of a differential equation associated with the Sturm-Liouville transformation T. We write $y(\lambda_n) = f(\lambda_n)/\lambda_n$ if $\lambda_n \neq 0$; then

$$(3) \qquad \lambda_n y(\lambda_n) = f(\lambda_n) \qquad\qquad (\lambda_n \neq 0).$$

Suppose first that T is such that no eigenvalue λ_n is zero. According to our basic operational property, Eq. (3) is true if $y(\lambda_n) = T\{Y\}$ and Y is a solution of class C'' of the boundary value problem

$$(4) \qquad D_2[Y(x)] = F(x), \qquad N_\alpha Y = N_\beta Y = 0.$$

The notation is that used in Sec. 111; thus $\mathscr{D}_2 Y = pF$ since

$$D_2 Y = p^{-1}\mathscr{D}_2 Y = \frac{(rY')' - qY}{p}.$$

The homogeneous problem corresponding to problem (4) is the Sturm-Liouville problem in which $\lambda = 0$; $\mathscr{D}_2 Y = 0$, $N_\alpha Y = N_\beta Y = 0$. Since zero is not an eigenvalue, that homogeneous problem has only the solution $Y(x) \equiv 0$. Therefore Green's function $G(x,\xi)$ for problem (4) exists, where

$$(rG_x)_x - qG = 0 \qquad (x \neq \xi), \qquad N_\alpha G = N_\beta G = 0, \qquad (a < \xi < b)$$

and G_x has the jump $1/r(\xi)$ at the point $x = \xi$. The unique solution of problem (4) is then

$$Y(x) = \int_a^b F(\xi)p(\xi)G(x,\xi)\,d\xi \qquad (a \leq x \leq b).$$

Thus *when zero is not an eigenvalue* of the Sturm-Liouville problem associated with T, then

$$(5) \qquad T^{-1}\left\{\frac{f(\lambda_n)}{\lambda_n}\right\} = \int_a^b F(\xi)p(\xi)G(x,\xi)\,d\xi ;$$

that is, $T\{Y\} = f(\lambda_n)/\lambda_n$ where Y *is the solution of problem* (4).

If one of the eigenvalues is zero, say $\lambda_1 = 0$, then $\lambda_n y(\lambda_n) = f(\lambda_n)$ when $n = 2, 3, \dots$; that is, if $N_\alpha Y = N_\beta Y = 0$ and Y is of class C'', then

$$T\{D_2[Y] - F\} = 0 \qquad \text{when } n = 2, 3, \dots .$$

According to transformation (2), then, for some constant k, Y satisfies the conditions

$$D_2[Y(x)] - F(x) = k\Phi_1(x), \qquad N_\alpha Y = N_\beta Y = 0.$$

Therefore $\lambda_n y(\lambda_n) - f(\lambda_n) = kT\{\Phi_1\}$ and, when $n = 1$, it follows that $-f(0) = k\|\Phi_1\|^2$.

Thus Y, *the inverse transform of* $f(\lambda_n)/\lambda_n$, $n \neq 1$, *with* $f(\lambda_1)/\lambda_1$ *undefined, is a solution of class* C'' *of the problem*

$$(6) \qquad D_2[Y(x)] = F(x) - f(0)\|\Phi_1\|^{-2}\Phi_1(x), \qquad N_\alpha Y = N_\beta Y = 0$$

$$(\lambda_1 = 0),$$

if that problem has such a solution. Recall, however, that Green's function does not exist for that problem, *nor does the problem have a unique solution,* since $A\Phi_1(x)$ is a solution of the homogeneous form of the problem when A is an arbitrary constant. Thus if $Y = Z(x)$ is one solution of problem (6), then $Y = Z(x) + A\Phi_1(x)$ is a family of solutions.

113 FURTHER PROPERTIES

Until special cases of Sturm-Liouville transformations are treated, cases in which the kernels are fairly elementary functions, only a few operational properties and transforms of particular functions can be presented.

In the general case, however, the basic operational property can be extended easily to iterates of the differential form D_2F. Let the function D_2F as well as F itself belong to the class $C''(a \leq x \leq b)$. Then

$$T\{D_2[D_2F]\} = \lambda_n T\{D_2F\} + N_\alpha[D_2F]N'_\alpha\Phi_n - r(b)N_\beta[D_2F]N'_\beta\Phi_n.$$

Let $D_2{}^2$ denote the iterated differential operator here. Then

$$(1) \qquad T\{D_2{}^2F\} = \lambda_n{}^2 f(\lambda_n) + \lambda_n N_\alpha[F]N'_\alpha\Phi_n - \lambda_n r(b)N_\beta[F]N'_\beta\Phi_n$$
$$+ N_\alpha[D_2F]N'_\alpha\Phi_n - r(b)N_\beta[D_2F]N'_\beta\Phi_n.$$

Thus T replaces the differential form $D_2{}^2F$ of fourth order by an algebraic form in the transform $f(\lambda_n)$ of F and the four boundary values $N_\alpha F$, $N_\alpha[D_2 F]$, $N_\beta F$, and $N_\beta[D_2 F]$.

Further iterations of D_2 can be treated in like manner.

Our transformation was developed in Sec. 111 by expressing the differential operator D_2 in terms of a self-adjoint operator $\mathscr{D}_2$, a procedure that can be followed for operators of the second order. To proceed directly without introducing the self-adjoint operator, we first write, as before,

$$T\{F\} = \int_a^b F(x)K(x,\lambda)\,dx = f(\lambda)$$

and $D_2 F = A(x)F''(x) + B(x)F'(x) + C(x)F(x)$. Then the transform

$$T\{D_2 F\} = \int_a^b K(x,\lambda)D_2[F(x)]\,dx$$

is to be written in terms of $f(\lambda)$ and the two boundary values $N_\alpha F$ and $N_\beta F$. In terms of the adjoint D_2^* of D_2, Lagrange's identity (3), Sec. 94, shows that

$$KD_2 F - FD_2^*K = \frac{d}{dx}[AKF' - (AK)'F + BKF]$$

$$= \frac{d}{dx}\left\{\left[\frac{AK}{r}F' - \left(\frac{AK}{r}\right)'F\right]r\right\}$$

where $r(x) = \exp\int_a^x [B(\xi)/A(\xi)]\,d\xi$, and $r' = rB/A$. Therefore when F and K are of class C'',

$$(2) \qquad T\{D_2 F\} = \int_a^b FD_2^*K\,dx + \left[AKF' - \left(\frac{AK}{r}\right)'rF\right]_a^b.$$

The remaining steps correspond to those used in Sec. 111, where we expressed the values of F and F' at the end points $x = a$ and $x = b$ in terms of $N_\alpha F$, $N_\alpha'F$, $N_\beta F$, and $N_\beta'F$. Details are left to the problems. We find that Eq. (2) can be written as the basic operational property

$$(3) \qquad T\{D_2 F\} = \lambda_n f(\lambda_n) + N_\alpha'\left[\frac{AK}{r}\right]N_\alpha F - r(b)N_\beta'\left[\frac{AK}{r}\right]N_\beta F$$

provided that $K = K_n(x)$ and $\lambda = \lambda_n$, where K_n and λ_n are the characteristic functions and numbers of the Sturm-Liouville problem

$$(4) \qquad D_2^*K - \lambda K = 0, \qquad N_\alpha\left[\frac{AK}{r}\right] = N_\beta\left[\frac{AK}{r}\right] = 0.$$

Note that

(5)
$$N_\alpha\left[\frac{AK}{r}\right] = \{AK\cos\alpha + [(AK)' - BK]\sin\alpha\}_{x=a},$$

$$N_\beta\left[\frac{AK}{r}\right] = [r(b)]^{-1}\{AK\cos\beta + [(AK)' - BK]\sin\beta\}_{x=b}.$$

Then the Sturm-Liouville transformation has the form

(6)
$$T\{F\} = \int_a^b F(x)K_n(x)\,dx = f(\lambda_n) \qquad (n = 1, 2, \dots).$$

Formulas (3) to (6) follow also from those in Sec. 111 when the latter are written in terms of the kernel K_n and the coefficients A, B, and C in D_2, where $K_n = p\Phi_n$.

The basic operational formula for $T\{D_2F\}$ must be modified in case F or F' has jumps. Suppose, for instance, that both functions are *continuous over the interval $a \leqq x \leqq b$ except at one interior point $x = c$ where they have jumps j_c and j_c':*

$$j_c = F(c + 0) - F(c - 0), \qquad j_c' = F'(c + 0) - F'(c - 0)$$

$$(a < c < b),$$

and that F'' is sectionally continuous. Then if we write

$$T\{D_2F\} = \int_a^c \Phi\mathscr{D}_2F\,dx + \int_c^b \Phi\mathscr{D}_2F\,dx$$

and use Lagrange's identity, we find that

(7)
$$T\{D_2F\} = \lambda_n f(\lambda_n) + N_\alpha'[\Phi_n]N_\alpha F - r(b)N_\beta'[\Phi_n]N_\beta F$$
$$+ r(c)[j_c\Phi_n'(c) - j_c'\Phi_n(c)],$$

where $T\{F\} = \int_a^b Fp\Phi_n\,dx$ and Φ_n and λ_n are the characteristic functions and numbers of the Sturm-Liouville problem (10), Sec. 111.

We cite one more property of Sturm-Liouville transforms $f(\lambda_n)$. When F is sectionally continuous, its Fourier constants

$$c_n = \|\Phi_n\|^{-1}\int_a^b F(x)p(x)\Phi_n(x)\,dx \qquad (n = 1, 2, \dots)$$

with respect to the orthonormal set of functions $\Phi_n/\|\Phi_n\|$ on the interval (a,b) tend to zero as $n \to \infty$. That is a consequence of Bessel's inequality,

$$c_1{}^2 + c_2{}^2 + \cdots + c_m{}^2 \leqq \|F\|^2 \qquad (m = 1, 2, \dots),$$

for orthonormal sets.[1] Thus the transform of each sectionally continuous function F must satisfy the condition

$$(8) \qquad \lim_{n \to \infty} \frac{f(\lambda_n)}{\|\Phi_n\|} = 0.$$

114 TRANSFORMS OF CERTAIN FUNCTIONS

As noted in Sec. 112, for a fixed positive integer m the Sturm-Liouville transform of the kernel Φ_m is zero when $n \neq m$, and $\|\Phi_m\|^2$ when $n = m$. Thus if $f(\lambda_n)$ is the transform of a function F and k is a constant,

$$(1) \qquad T\{F(x) + k\Phi_m(x)\} = \begin{cases} f(\lambda_n) & \text{when } n \neq m, \\ f(\lambda_m) + k\|\Phi_m\|^2 & \text{when } n = m. \end{cases}$$

Suppose T is such that *no eigenvalue λ_n is zero*. Then Green's function $G(x,c)$ for problem (4), Sec. 112, exists. It is continuous, but its derivative G_x has the jump $j'_c = [r(c)]^{-1}$ at the point $x = c$ $(a < c < x)$; also,

$$(2) \qquad D_2[G(x,c)] = 0 \qquad (x \neq c), \qquad N_\alpha G = N_\beta G = 0.$$

According to the modified operational property (7), Sec. 113, then

$$0 = \lambda_n T\{G(x,c)\} - r(c)[r(c)]^{-1}\Phi_n(c).$$

Thus the Sturm-Liouville transform of $G(x,c)$ is

$$(3) \qquad T\{G(x,c)\} = \frac{1}{\lambda_n}\Phi_n(c) \qquad\qquad (a < c < b).$$

The continuity of $G(x,c)$ and the sectional continuity of its first derivative are sufficient (Theorem 4, Sec. 99) for the validity of our inversion formula (1), Sec. 112. Therefore we have the representation

$$(4) \qquad G(x,c) = \sum_{n=1}^{\infty} \frac{1}{\lambda_n} \frac{\Phi_n(c)\Phi_n(x)}{\|\Phi_n\|^2} \qquad (a < x < b, a < c < b),$$

known as the *bilinear formula* for Green's function.

From transform (3) and property (5), Sec. 112, we find another Sturm-Liouville transform when no eigenvalue is zero:

$$(5) \qquad T\left\{ \int_a^b p(\xi)G(x,\xi)G(\xi,c)\,d\xi \right\} = \frac{1}{\lambda_n^2}\Phi_n(c) \qquad (a < c < b).$$

A further consequence of formula (3) is

$$(6) \qquad T\left\{ \frac{\partial}{\partial c}G(x,c) \right\} = \frac{1}{\lambda_n}\Phi'_n(c) \qquad (a < c < b).$$

[1] Churchill, R. V., "Fourier Series and Boundary Value Problems," 2d ed., p. 60, 1963.

The basic operational property can be used in various ways to obtain transforms of particular solutions of differential equations involving our differential operator D_2. For instance, when a function $H(x)$ of class C'' satisfies the differential equation $D_2H = 0$, it follows from property (12), Sec. 111, that its transform is

$$(7) \qquad h(\lambda_n) = \frac{1}{\lambda_n} \{ r(b)N'_\beta[\Phi_n]N_\beta H - N'_\alpha[\Phi_n]N_\alpha H \} \qquad (\lambda_n \neq 0).$$

If constant values, not both zero, are assigned to $N_\alpha H$ and $N_\beta H$ here, then H is the solution of the differential equation $D_2H = 0$ that satisfies those boundary conditions. We note that in case $\alpha = \beta = 0$, formula (7) becomes

$$(8) \qquad T\{H\} = \lambda_n^{-1}[r(b)\Phi'_n(b)H(b) - \Phi'_n(a)H(a)].$$

If $F(x) \equiv 1 \ (a \leq x \leq b)$, then $D_2F = C(x)$ and our formula for $T\{D_2F\}$ reduces to

$$(9) \qquad T\{C(x)\} = \lambda_n T\{1\} + N'_\alpha[\Phi_n] \cos \alpha - r(b)N'_\beta[\Phi_n] \cos \beta.$$

Thus when Φ_n, λ_n, and either of the two transforms $T\{C(x)\}$ or $T\{1\}$ are known, the other transform can be found. If $C(x) = C_0$, a constant other than one of the eigenvalues, then $T\{1\}$ can be found, since

$$(10) \qquad (C_0 - \lambda_n)T\{1\} = N'_\alpha[\Phi_n] \cos \alpha - r(b)N'_\beta[\Phi_n] \cos \beta.$$

115 EXAMPLE OF STURM-LIOUVILLE TRANSFORMATIONS

Let the differential form D_2F be the simple self-adjoint form F'' on the interval $(0,\pi)$, and let the prescribed boundary values be $F(0)$ and $F'(\pi)$. Then

$$(1) \qquad \mathscr{D}_2F = F''(x), \qquad 0 < x < \pi; \qquad N_\alpha F = F(0), \qquad N_\beta F = F'(\pi);$$

thus $\alpha = 0$, $\beta = \pi/2$, $r(x) = p(x) = 1$, and $q(x) = 0$. In this case the kernel Φ_n of the Sturm-Liouville transformation is the family of all eigenfunctions of the problem

$$(2) \qquad \Phi''(x) - \lambda\Phi(x) = 0, \qquad \Phi(0) = \Phi'(\pi) = 0.$$

We find that $\lambda = 0$ is not an eigenvalue, and that

$$(3) \qquad \Phi_n(x) = \sin (n - \tfrac{1}{2})x, \qquad \lambda_n = -(n - \tfrac{1}{2})^2 \quad (n = 1, 2, \dots).$$

Let S_m denote the transformation T here, where $m = n - \tfrac{1}{2}$. It is the *modified finite Fourier transformation*

$$(4) \qquad S_m\{F\} = \int_0^\pi F(x) \sin (n - \tfrac{1}{2})x \, dx = f_s(m) \qquad (n = 1, 2, \dots).$$

Since $N'_\alpha \Phi_n = \Phi'_n(0) = m$ and $N'_\beta \Phi_n = -\Phi_n(\pi) = (-1)^n$, the basic operational property of S_m for functions of class C'' is

$$(5) \qquad S_m\{F''\} = -m^2 f_s(m) + mF(0) - (-1)^n F'(\pi) \qquad (m = n - \tfrac{1}{2}).$$

We find that $\|\Phi_n\|^2 = \pi/2$, so the generalized Fourier series that furnishes an inversion formula for T_m is

$$(6) \qquad F(x) = \frac{2}{\pi} \sum_{n=1}^{\infty} f_s(m) \sin mx \qquad (m = n - \tfrac{1}{2}, 0 < x < \pi).$$

Transforms of a number of functions can be found by evaluating the integral $S_m\{F\}$ or by applying property (5) or formula (6). For instance,

$$(7) \qquad S_m\{1\} = \frac{1}{m}, \qquad S_m\{x\} = \frac{(-1)^{n-1}}{m^2} \qquad (m = n - \tfrac{1}{2}).$$

For other particular transforms see Probs. 5 to 8, Sec. 116, and Table 2, Sec. 126.

We can derive kernel-product and convolution properties for S_m, whose kernel is an odd function that is antiperiodic with the period 2π. Let F_3 be the extension of F which is odd and *antiperiodic* with period 2π; that is,

$$(8) \qquad F_3(x) = F(x), \qquad\qquad\qquad\qquad 0 < x < \pi;$$

$$F_3(x) = -F_3(-x) = -F_3(x + 2\pi), \qquad\qquad \text{all } x.$$

Then $F_3(x) \sin mx$, where $m = n - \tfrac{1}{2}$, is even and periodic with period 2π. We introduce the integral

$$(9) \qquad F_0(x) = \int_x^\pi F_3(\xi)\, d\xi = \int_0^\pi F_3(\xi)\, d\xi - \int_0^x F_3(\xi)\, d\xi,$$

which is even and antiperiodic with period 2π; also $F'_0(x) = -F_3(x)$ and $F_0(\pi) = F_0(-\pi) = 0$. Then

$$2S_m\{F\} = \int_{-\pi}^\pi F_3(x) \sin mx\, dx = -\int_{-\pi}^\pi F'_0(x) \sin mx\, dx.$$

An integration by parts here shows that

$$2S_m\{F\} = 2f_s(m) = m \int_{-\pi}^\pi F_0(x) \cos mx\, dx.$$

If x_0 is a constant $(0 < x_0 \leqq \pi)$, then

$$f_s(m) \frac{2}{m} \sin mx_0 = \frac{1}{2} \int_{-\pi}^\pi F_0(x)[\sin m(x + x_0) - \sin m(x - x_0)]\, dx$$

$$= \frac{1}{2} \int_{-\pi}^\pi [F_0(\xi - x_0) - F_0(\xi + x_0)] \sin m\xi\, d\xi$$

because the integrand is periodic with period 2π. Also, the integrand is an even function of ξ, so the last equation is the *kernel-product property*

(10) $\qquad f_s(m)\dfrac{2}{m}\sin mx_0 = S_m\{F_0(x - x_0) - F_0(x + x_0)\} \qquad (m = n - \tfrac{1}{2}).$

When $0 < x_0 \leqq \pi$ and $0 \leqq x \leqq \pi$, we find that

$$F_0(x - x_0) = \int_{|x-x_0|}^{\pi} F(\xi)\,d\xi,$$

(11)

$$F_0(x + x_0) = \begin{cases} \displaystyle\int_{x+x_0}^{\pi} F(\xi)\,d\xi & \text{if } x + x_0 \leqq \pi \\[2ex] \displaystyle -\int_{2\pi-x-x_0}^{\pi} F(\xi)\,d\xi & \text{if } x + x_0 \geqq \pi. \end{cases}$$

When $x_0 = \pi$, Eq. (10) becomes

(12) $\qquad f_s(m)\dfrac{(-1)^{n-1}}{m} = S_m\{F_0(\pi - x)\} = S_m\left\{\displaystyle\int_{\pi-x}^{\pi} F(\xi)\,d\xi\right\}.$

In terms of the notation used in Sec. 109, $\omega(\lambda) = 2/m$ and $P_F(x,x_0)$ is the function inside the braces in property (10). Hence the convolution of two functions $H(x)$ and $F(x)$ corresponding to our kernel-product property is

$$X_{HF}(x) = \int_0^{\pi} H(y)[F_0(x - y) - F_0(x + y)]\,dy,$$

which can be written

(13) $\qquad X_{HF}(x) = \displaystyle\int_0^{\pi} H(y)\left[\int_{|x-y|}^{\pi} F(\xi)\,d\xi - \int_{x+y}^{\pi} F_3(\xi)\,d\xi\right] dy.$

A *convolution property* for S_m is therefore

(14) $\qquad S_m\{X_{HF}(x)\} = \dfrac{2}{m}f_s(m)h_s(m) \qquad\qquad (m = n - \tfrac{1}{2}),$

where $h_s(m) = S_m\{H\}$. Similar derivations of convolution properties can be made for other Fourier transforms (cf. Sec. 110 and the chapters to follow).

Properties of S_m are summarized in Table 2, Sec. 126.

Our transformation S_m reduces boundary value problems in partial differential equations in which the differential form in the unknown function U with respect to a variable x is U_{xx}, and the values of U and U_x are prescribed at boundaries $x = 0$ and $x = \pi$, respectively.

This problem in $U(x,y)$ is an illustration:

$$a(y)U_{xx} + b(y)U + c(y)U_{yy} = Q(x,y) \qquad (0 < x < \pi, y > 0),$$

$$U(0,y) = 0, \qquad U_x(\pi,y) = p(y), \qquad U(x,0) = U_y(x,\infty) = 0,$$

assuming that the prescribed functions a, b, c, Q, and p are such that the problem has a solution.

Let $u_s(m,y)$ be $S_m\{U\}$ with respect to x. Then according to our formula for $S_m\{D_2U\}$, the transformed problem is, formally,

$$a(y)[-m^2u_s - (-1)^np(y)] + b(y)u_s + c(y)u_s'' = q_s(m,y) \qquad (y > 0)$$

$$u_s(m,0) = u_s'(m,\infty) = 0,$$

where $q_s(m,y) = S_m\{Q\}$ and $u_s' = du_s/dy$. If this problem in ordinary differential equations can be solved for $u_s(m,y)$, then $U(x,y)$ can be written with the aid of our inversion formula, or by using known transforms and our convolution property.

If a term $\partial^4U/\partial x^4$, an iteration of the form U_{xx}, is added to the above partial differential equation in U, and if the boundary conditions are augmented by prescribing $U_{xx}(0,y)$ and $U_{xxx}(\pi,y)$, the problem is again reduced by S_m since

$$S_m\left\{\frac{\partial^4U}{\partial x^4}\right\} = -m^2S_m\{U_{xx}\} + mU_{xx}(0,y) - (-1)^nU_{xxx}(\pi,y),$$

and $S_m\{U_{xx}\} = -m^2u_s(m,y) - (-1)^np(y)$. Note that neither the Laplace transformation nor separation of variables applies directly to these problems in U.

116   SINGULAR CASES

Our method of setting up an integral transformation that applies to some specified differential form and boundary values may require the kernel to satisfy a singular eigenvalue problem (Sec. 107). Singular problems arise if the interval of our variable x is unbounded, or if the differential equation in the eigenvalue problem has a singular point on the interval, usually at an end point of the interval. Conditions on the order of magnitude or regularity of functions at singular points or infinitely distant end points may replace prescribed boundary values at those points. We present two examples to illustrate the procedure.

Example 1   Let the interval be *unbounded*, $0 < x < \infty$, and let the differential form and the prescribed boundary value and order condition be these:

(1) $D_2F = F''(x), \quad x > 0; \qquad N_\alpha F = F(0), \quad F(\infty) = F'(\infty) = 0.$

Here D_2 is self-adjoint. The integral transformation

(2) $$T\{F\} = \int_0^\infty F(x)\Phi(x,\lambda)\,dx = f(\lambda)$$

requires additional conditions on F and Φ to ensure the existence of the improper integral.

We assume that F is of class C'' and absolutely integrable over the half line $x \geq 0$, and that $F(\infty) = F'(\infty) = 0$. Formula (8), Sec. 111, obtained either by using Lagrange's identity or by integrating by parts, now becomes

$$T\{F''(x)\} = \int_0^\infty F(x)\Phi''(x,\lambda)\,dx + [\Phi F' - \Phi' F]_0^\infty$$

$$= \lambda \int_0^\infty F(x)\Phi(x,\lambda)\,dx + \Phi'(0,\lambda)F(0)$$

if Φ satisfies the singular eigenvalue problem

(3) $\Phi'' = \lambda\Phi, \quad x > 0; \qquad \Phi(0,\lambda) = 0, \quad \Phi$ and Φ' bounded $\quad (x > 0)$.

If k is a constant, the solution of problem (3) is

$$\Phi(x,\lambda) = k \sinh x\sqrt{\lambda} \qquad\qquad (k \neq 0),$$

provided that λ is such that Φ and Φ' are bounded as $x \to \infty$. We write $\sqrt{\lambda} = \eta + i\mu$, where η and μ are real. Then $\sinh x\sqrt{\lambda}$ is bounded if and only if $\eta = 0$, so that $\lambda = -\mu^2$ and $\Phi = ik \sin \mu x$; Φ' is also bounded. Negative values of μ yield no new eigenfunctions. Thus $\lambda = -\mu^2$, $\mu > 0$, and if we write $ik = 1$, then

(4) $$\Phi(x,\lambda) = \sin \mu x, \qquad \lambda = -\mu^2 \qquad\qquad (\mu > 0).$$

Hence T is the *Fourier sine transformation* (Sec. 110)

(5) $$T\{F\} = S_\mu\{F\} = \int_0^\infty F(x) \sin \mu x\,dx = f_s(\mu) \qquad (\mu > 0)$$

with the basic operational property

(6) $$S_\mu\{F''\} = -\mu^2 f_s(\mu) + \mu F(0).$$

Its inverse transformation is given by the Fourier sine integral formula (Prob. 15, Sec. 107):

(7) $$F(x) = \frac{2}{\pi} \int_0^\infty f_s(\mu) \sin \mu x\,d\mu = \frac{2}{\pi}S_x\{f_s(\mu)\}.$$

For other properties of S_μ see Chap. 13 and Appendix D.

Example 2   As an illustration of a differential form with the end point $x = 0$ of the bounded interval $0 < x < b$ as a *singular point*, consider the form

$$(8) \qquad D_2F = F''(x) + \frac{1}{x}F'(x) \qquad (0 < x < b).$$

The form arises when the Laplacian operator is written in terms of cylindrical coordinates. Let the prescribed boundary value and regularity conditions be

$$(9) \qquad N_\beta F = F'(b), \qquad F \text{ of class } C''(0 \leqq x \leqq b).$$

In terms of the self-adjoint form $\mathscr{D}_2 F = (xF')'$ and the notation used in Sec. 111, we write

$$D_2 F = x^{-1}(xF')', \qquad p(x) = r(x) = x, \qquad \beta = \pi/2,$$

$$T\{F\} = \int_0^b F(x)x\Phi(x,\lambda)\,dx = f(\lambda).$$

Then formula (8), Sec. 111, for $T\{D_2F\}$ becomes

$$\int_0^b (xF')'\Phi\,dx = \int_0^b (x\Phi')'F\,dx + [x(\Phi F' - \Phi'F)]_0^b,$$

since $r(x) = x$. If Φ satisfies the singular eigenvalue problem

$$(10) \qquad (x\Phi')' = \lambda x\Phi, \qquad \Phi'(b,\lambda) = 0, \qquad \Phi \text{ in class } C'' \quad (0 \leqq x \leqq b),$$

then

$$T\{D_2F\} = \lambda f(\lambda) + b\Phi(b,\lambda)F'(b).$$

The differential equation $x\Phi'' + \Phi' - \lambda x\Phi = 0$ in problem (10) is Bessel's equation with index $n = 0$, whose solution of class C'' ($0 \leqq x \leqq b$) is $\Phi = kJ_0(x\sqrt{-\lambda})$, where k is a constant (Sec. 83). The condition $\Phi'(b,\lambda) = 0$ requires λ to be zeros of the function $J_1(b\sqrt{-\lambda})$. We write $b\sqrt{-\lambda} = \mu$. The odd function $J_1(\mu)$ has an infinite sequence of zeros $\mu_1 = 0, \pm\mu_2, \pm\mu_3, \ldots$, all real; its non-negative zeros μ_i yield all the eigenfunctions. Thus the eigenvalues are $\lambda = -\mu_i{}^2/b^2$, and

$$\Phi(x,\lambda) = J_0\left(\mu_i \frac{x}{b}\right) \qquad [J_1(\mu_i) = 0, \ \mu_i \geqq 0, \ i = 1, 2, \ldots].$$

Our transformation on the bounded interval $(0,b)$ is then the particular *finite Hankel transformation*

$$(11) \qquad T\{F\} = \int_0^b F(x)xJ_0\left(\mu_i \frac{x}{b}\right) dx = f_0(\mu_i),$$

where $J_1(\mu_i) = 0$, $\mu_i \geq 0$, and $\mu_i \to \infty$ as $i \to \infty$. It has the property

$$(12) \qquad T\left\{F'' + \frac{1}{x}F'\right\} = -b^{-2}\mu_i^2 f_0(\mu_i) + bJ_0(\mu_i)F'(b).$$

Recall that $\mu_1 = 0$ and $J_0(0) = 1$.

We shall treat Hankel transforms more fully in Chap. 14 where we shall note that Fourier-Bessel series furnish inversion formulas for the finite transformations.

PROBLEMS

1. When $D_2F = F''(x)$, $0 < x < b$, $N_\alpha F = F(0)$, and $N_\beta F = F(b)$, show that the Sturm-Liouville transformation becomes the *finite Fourier sine transformation* on the interval $(0,b)$,

$$T\{F\} = S_n\{F\} = \int_0^b F(x) \sin\frac{n\pi x}{b}\,dx = f_s(n) \qquad (n = 1, 2, \dots);$$

show also that if F of class C'' ($0 \leq x \leq b$), then

$$S_n\{F''\} = -\left(\frac{n\pi}{b}\right)^2 f_s(n) + \frac{n\pi}{b}[F(0) - (-1)^n F(b)],$$

and the inversion formula (1), Sec. 112, is the Fourier sine series representation of F on the interval $(0,b)$.

2. (a) If F is of class $C^{(4)}$ ($0 \leq x \leq b$), show that the transformation S_n in Prob. 1 has the property

$$S_n\{F^{(4)}(x)\} = \left(\frac{n\pi}{b}\right)^4 f_s(n) - \left(\frac{n\pi}{b}\right)^3 [F(0) - (-1)^n F(b)]$$

$$+ \frac{n\pi}{b}[F''(0) - (-1)^n F''(b)].$$

(b) If $F^{(4)}(x) = 0$, $F(0) = F(b) = F''(0) = 0$, and $F''(b) = 6b$, show that $F(x) = x(x^2 - b^2)$ and $f_s(n) = 6b^4(-1)^n/(n\pi)^3$.

3. When $D_2F = F''(x)$, $0 < x < b$, $N_\alpha F = F'(0)$, and $N_\beta F = F'(b)$, show that the Sturm-Liouville transformation becomes the *finite Fourier cosine transformation* on the interval $(0,b)$,

$$C_n\{F\} = \int_0^b F(x) \cos\frac{n\pi x}{b}\,dx = f_c(n) \qquad (n = 0, 1, 2, \dots);$$

show also that if F is class C'' ($0 \leq x \leq b$), then

$$C_n\{F''\} = -\left(\frac{n\pi}{b}\right)^2 f_c(n) - F'(0) + (-1)^n F'(b) \qquad (n = 0, 1, 2, \dots)$$

and that $C_n^{-1}\{f_c\}$ is given by the Fourier cosine series

$$F(x) = \frac{1}{b}f_c(0) + \frac{2}{b}\sum_{n=1}^{\infty} f_c(n)\cos\frac{n\pi x}{b} \qquad (0 < x < b).$$

4. Find the Sturm-Liouville transformation on the interval $(0,1)$ that resolves the differential form $F''(x)$ in terms of the transform of F and the boundary values $F'(0)$ and $hF(1) + F'(1)$, where h is a positive constant, and show that it has the operational property

$$T\{F''\} = -\mu_n{}^2 T\{F\} - F'(0) + [hF(1) + F'(1)] \cos \mu_n,$$

where μ_n $(n = 1, 2, \dots)$ are the positive roots of the equation $\tan \mu = h/\mu$ (cf. Sec. 105).

$$\text{Ans. } T\{F\} = \int_0^1 F(x) \cos \mu_n x \, dx \qquad (n = 1, 2, \dots).$$

5. Show that Green's function for the problem $Y''(x) = 0$, $Y(0) = Y'(\pi) = 0$ is the function

$$G(x,c) = -x \qquad (0 \leq x \leq c), \qquad G(x,c) = c \qquad (c \leq x \leq \pi);$$

then apply formulas (3) and (6), Sec. 114, when T is the modified finite Fourier transformation S_m defined in Sec. 115, where $m = n - \frac{1}{2}$, to obtain the inverse transforms

$$S_m{}^{-1}\left\{\frac{\sin mc}{m^2}\right\} = -G(x,c), \qquad S_m{}^{-1}\left\{\frac{\cos mc}{m}\right\} = S_0(x - c),$$

where $0 < c < \pi$ and S_0 is the unit step function.

6. Show that the transformation (Sec. 115)

$$S_m\{F\} = \int_0^\pi F(x) \sin mx \, dx = f_s(m) \qquad (m = n - \tfrac{1}{2})$$

has the property

$$S_m{}^{-1}\left\{\frac{f_s(m)}{m^2}\right\} = \int_0^x \int_r^\pi F(t) \, dt \, dr = \int_0^x tF(t) \, dt + x \int_x^\pi F(t) \, dt,$$

and that

$$S_m{}^{-1}\left\{\frac{1}{m^3}\right\} = \pi x - \frac{x^2}{2}, \qquad S_m{}^{-1}\left\{\frac{-6(-1)^n}{m^4}\right\} = 3\pi^2 x - x^3.$$

7. When $f_s(m) = e^{-my}$ where $m = n - \frac{1}{2}$ and $y > 0$ and independent of x, inversion formula (6), Sec. 115, for the modified finite Fourier transformation S_m can be written

$$F(x,y) = \frac{2}{\pi} \sum_{n=1}^\infty e^{-my} \sin mx = \frac{2}{\pi} \text{Im} \sum_{n=1}^\infty (\exp iz)^{n-\frac{1}{2}},$$

where $z = x + iy$ and $|\exp iz| < 1$. Sum the series here to show that

$$S_m{}^{-1}\{e^{-my}\} = \frac{1}{\pi} \text{Re}\left(\csc \frac{z}{2}\right) = \frac{1}{\pi} \frac{(\sin x/2)(\cosh y/2)}{\cosh^2 y/2 - \cos^2 x/2}$$

when $y > 0$. [Since $F(x,y)$ is the imaginary component of an analytic function of z, it is a harmonic function. See Prob. 2, Sec. 118, where e^{-my} arises. Also note that F satisfies conditions (8), Sec. 115, on F_3.]

8. For the transformation S_m in Sec. 115, where $m = n - \frac{1}{2}$, show that if $y > 0$ and independent of x and $z = x + iy$, then

$$S_m^{-1}\left\{\frac{e^{-my}}{m}\right\} = \frac{2}{\pi} \operatorname{Im} \sum_{n=1}^{\infty} \frac{(\exp iz)^m}{m}$$

$$= \frac{2}{\pi} \operatorname{Im}\left(\log\left|\frac{1 + \exp iz/2}{1 - \exp iz/2}\right|\right) = \frac{2}{\pi} \arctan\left(\frac{\sin x/2}{\sinh y/2}\right) \qquad (y > 0).$$

9. Complete the derivation of the basic operational property (3), Sec. 113, for the Sturm-Liouville transformation in the form $\int_a^b FK\,dx$.

10. Write $F(x) = e^{-kx}(k > 0)$ in the basic operational property (6), Sec. 116, of the Fourier sine transformation S_μ to obtain the transform

$$S_\mu\{e^{-kx}\} = \frac{\mu}{\mu^2 + k^2} \qquad (k > 0).$$

Differentiate with respect to k to show that

$$S_\mu\{xe^{-kx}\} = 2k\mu(\mu^2 + k^2)^{-2} \qquad (k > 0).$$

11. Derive the integral transformation on the half line $x \geq 0$ that reduces $F''(x)$ in terms of the boundary value $F'(0)$ when F is of class C'', absolutely integrable over the half line, and $F(\infty) = F'(\infty) = 0$. This is the *Fourier cosine transformation*

$$C_\mu\{F\} = \int_0^\infty F(x) \cos \mu x\,dx = f_c(\mu) \qquad (\mu \geq 0)$$

(Prob. 5, Sec. 110 and Chap 13). Show that

$$C_\mu\{F''(x)\} = -\mu^2 f_c(\mu) - F'(0).$$

12. Derive the integral transformation that reduces the form $F''(x) + x^{-1}F'(x)$ on the interval $(0,1)$ in terms of the boundary value $F(1)$ when F is of class C'' ($0 \leq x \leq 1$). This is the *finite Hankel transformation* (Chap. 14)

$$T\{F\} = \int_0^1 F(x)xJ_0(\mu_j x)\,dx = f(\mu_j) \qquad (j = 1, 2, \cdots),$$

where the real numbers μ_j are the positive roots of the equation $J_0(\mu) = 0$. Since $J_0'(\mu) = -J_1(\mu)$, show that

$$T\{F'' + x^{-1}F'\} = -\mu_j^2 f(\mu_j) + \mu_j J_1(\mu_j)F(1).$$

117 A PROBLEM IN STEADY TEMPERATURES

As an example of boundary value problems that are well adapted to solution by specific Sturm-Liouville transformations, we present the following one on steady-state temperatures in a semi-infinite slab in which heat is generated at a steady rate while surface heat transfer takes place at one face. Let the temperature function $U(x,y)$ satisfy the conditions

(1)
$$U_{xx}(x,y) + U_{yy}(x,y) + Q(x,y) = 0 \qquad (0 < x < 1, y > 0)$$

$$U(0,y) = P(y), \qquad U(1,y) + hU_x(1,y) = 0, \qquad U(x,0) = 0,$$

where h is a constant ($h \geq 0$), the functions P and Q are bounded, and we require U to be bounded.

Here the differential form with respect to x is U_{xx}, and the prescribed boundary values at $x = 0$ and $x = 1$ are $N_\alpha U = U(0,y)$ and $N_\beta U = U(1,y) \cos \beta + U_x(1,y) \sin \beta$ where $\tan \beta = h\ (0 \leq \beta < \pi/2)$. Hence the kernel $\Phi_n(x)$ and the parameter λ_n of the Sturm-Liouville transformation that applies to problem (1) are the characteristic functions and numbers of the problem

(2) $\Phi'' - \lambda\Phi = 0,$ $\Phi(0,\lambda) = 0,$ $\Phi(1,\lambda) + h\Phi'(1,\lambda) = 0.$

Thus we find that

(3) $\Phi_n(x) = \sin \mu_n x,$ $\lambda_n = -\mu_n^2,$ $\tan \mu_n = -h\mu_n$ $(\mu_n > 0),$

and we can see from the graphs of the functions $\tan \mu$ and $-h\mu$ that $\mu_n - (n - \frac{1}{2})\pi \to 0$ as $n \to \infty$ when $h > 0$; when $h = 0$, $\mu_n = n\pi$. Our transformation is

(4) $$T\{U\} = \int_0^1 U(x,y) \sin \mu_n x \, dx = u_n(y)   (n = 1, 2, \dots).$$

Now $N_\alpha U = P(y)$, $\alpha = 0$, and $N'_\alpha \Phi_n = \Phi'_n(0) = \mu_n$; also $N_\beta U = 0$, and $\lambda_n = -\mu_n^2$. Therefore

$$T\{U_{xx}\} = -\mu_n^2 u_n(y) + \mu_n P(y)$$

and if $q_n(y) = T\{Q(x,y)\}$, the transformed problem is

(5) $u_n''(y) - \mu_n^2 u_n(y) + \mu_n P(y) + q_n(y) = 0,$ $u_n(0) = 0,$

where u_n is bounded on the half line $y > 0$. With the aid of the equation $\tan \mu_n = -h\mu_n$, we find that

$$\|\Phi_n\|^2 = \tfrac{1}{2}(1 + h \cos^2 \mu_n).$$

After solving problem (5) for u_n, we have the formal result

(6) $$U(x,y) = 2 \sum_{n=1}^{\infty} \frac{u_n(y) \sin \mu_n x}{1 + h \cos^2 \mu_n}.$$

The form of the solution can be improved in special cases. When $P(y) = 0$ and $Q(x,y) \equiv Q_0$, a constant, then

$$q_n(y) = Q_0 T\{1\} = Q_0 \frac{1 - \cos \mu_n}{\mu_n}$$

and the solution of problem (5) becomes

(7) $$u_n(y) = Q_0 \frac{1 - \cos \mu_n}{\mu_n^3} - Q_0 \frac{1 - \cos \mu_n}{\mu_n^3} \exp\left(-\mu_n y\right).$$

The first fraction is $-T\{1\}/\lambda_n$. According to formula (4), Sec. 112, this is $-T\{Z\}$ where

$$Z''(x) = 1, \qquad Z(0) = 0, \qquad Z(1) + hZ'(1) = 0.$$

Hence we find that

$$Z(x) = -\frac{x}{2}\frac{1 + 2h}{1 + h} + \frac{1}{2}x^2 = -T^{-1}\left\{\frac{1 - \cos\mu_n}{\mu_n{}^3}\right\}$$

and the inverse of transform (7) can be written

$$(8) \quad U(x,y) = \frac{Q_0}{2}\left[\frac{1 + 2h}{1 + h}x - x^2 + 4\sum_{n=1}^{\infty}\frac{1 - \cos\mu_n}{1 + h\cos^2\mu_n}\frac{\exp(-\mu_n y)}{\mu_n{}^3}\sin\mu_n x\right].$$

Form (8) can be verified fully as the solution of problem (1) in this special case $P = 0$, $Q = Q_0$. Note that if only a finite number of terms of the infinite series is used in formula (8), the resulting function satisfies all conditions in problem (1) except the condition at the base $y = 0$. Also, we see that U has the property

$$\lim_{y \to \infty} U(x,y) = \frac{Q_0}{2}\left(\frac{1 + 2h}{1 + h}x - x^2\right).$$

The differential form U_{yy} and boundary conditions in problem (1) are such that the Fourier sine transformation with respect to y, on the half line $y > 0$, applies to the problem provided that the transforms of both the functions P and Q exist; but that is not the case when $Q(x,y) \equiv Q_0$. In case the second method does apply, we have an advantage of obtaining a different representation of the function U. Note that problem (1) is not adapted to a Laplace transformation. Since the partial differential equation is not homogeneous, the method of separation of variables does not apply.

For further properties of transformation (4) see Sec. 128.

118 OTHER BOUNDARY VALUE PROBLEMS

Other special and singular cases of Sturm-Liouville transformations will be applied to boundary value problems in the chapters to follow. Operational properties of some of them are adequate to represent solutions of problems in more than one form, sometimes in quite simple forms. But in each application of our process an eigenvalue problem must be solved to determine the kernel of the transformation; then the transformed problem must be solved to find the transform of the unknown function. Failure of the transformed problem to have a unique solution suggests that conditions are missing or incorrect in the boundary value problem.

To make some observations on types of problems to which the method applies, we consider here problems in two independent variables and Sturm-Liouville transformations. But when boundary conditions are properly modified, our observations apply to other cases including singular cases, differential forms of higher order, and problems in more than two independent variables. In contrast with the method of separation of variables *the method of integral transformations is not restricted to problems with homogeneous differential equations or with certain homogeneous boundary conditions.*

Let $V(x,y)$ denote the unknown function in our problems and let the differential form and prescribed boundary values on x be

$$D_2 V = A(x)V_{xx}(x,y) + B(x)V_x(x,y) + C(x)V(x,y) \qquad (a < x < b)$$

$$N_\alpha V = V(a,y)\cos \alpha + V_x(a,y)\sin \alpha,$$

$$N_\beta V = V(b,y)\cos \beta + V_x(b,y)\sin \beta.$$

Note that the coefficients A, B, and C are functions of x alone; α and β are real and constant. Then as before

$$T\{V\} = \int_a^b V(x,y)p(x)\Phi_n(x)\, dx = v_n(y) \qquad (n = 1, 2, \dots).$$

where Φ_n are the eigenfunctions of the Sturm-Liouville problem corresponding to the operators D_2, N_α, and N_β.

To indicate some types of boundary value problems in V to which transformation T applies, let $L_{i,y}$, $i = 1, 2, 3$, denote linear homogeneous differential operators with respect to y whose coefficients are functions of y alone. Those operators may be of any order; if the order is zero, the form $L_{i,y}V$ is simply a product $E(y)V(x,y)$.

Consider a linear boundary value problem consisting of a partial differential equation and boundary conditions

(1) $$L_{1,y}[V] + L_{2,y}[D_2 V] = F(x,y),$$

(2) $$N_\alpha V = G(y), \qquad N_\beta V = H(y),$$

and some boundary conditions with respect to y of type

(3) $$c_1 V + c_2\frac{\partial V}{\partial y} + \cdots + c_m\frac{\partial^m V}{\partial y^m} = P(x,c) \qquad \text{when } y = c,$$

where the c's are constants. When T is applied to Eq. (1) and (3) with interchanges of operations with respect to x and y, then, in view of conditions (2), the resulting differential equation is

(4) $$L_{1,y}[v_n(y)] + L_{2,y}[\lambda_n v_n(y)] = L_{2,y}[H(y)r(b)N_\beta'\Phi_n - G(y)N_\alpha'\Phi_n] + f_n(y),$$

where $f_n(y) = T\{F(x,y)\}$. If $p_n(c) = T\{P(x,c)\}$, the boundary conditions on v_n are of type

(5)
$$c_1 v_n + c_2 \frac{dv_n}{dy} + \cdots + c_m \frac{d^m v_n}{dy^m} = p_n(c) \qquad \text{when } y = c.$$

Then $V(x,y) = T^{-1}\{v_n(y)\}$. Unless tables or operational properties of T can be used to simplify the inverse transformation, we may write the formal solution as

(6)
$$V(x,y) = \sum_{n=1}^{\infty} \|\Phi_n\|^{-2} v_n(y) \Phi_n(x).$$

The transformation T may apply if the problem involves D_2 and some of its iterations. For example, the partial differential equation may have the form

(7)
$$L_{1,y}[V] + L_{2,y}[D_2 V] + L_{3,y}[D_2{}^2 V] = F(x,y)$$

if it is accompanied by boundary conditions at $x = a$ and $x = b$ that prescribe $N_\alpha V$, $N_\beta V$, $N_\alpha[D_2 V]$, and $N_\beta[D_2 V]$. Then even if Eq. (7) and those boundary conditions are homogeneous, the usual method of separation of variables may not apply.

A special case of Eq. (7) is the *biharmonic equation*

(8)
$$\frac{\partial^4 V}{\partial y^4} + 2\frac{\partial^4 V}{\partial y^2 \partial x^2} + \frac{\partial^4 V}{\partial x^4} = 0$$

satisfied by the static transverse displacements $V(x,y)$ in an unloaded elastic plate. Here $D_2 V = V_{xx}$, $L_{1,y} = \partial^4/\partial y^4$, $L_{2,y} = 2\partial^2/\partial y^2$, and $L_{3,y} = 1$. If displacements V and bending moments are given along the edges $x = a$ and $x = b$, then V and V_{xx} are prescribed on those edges.

PROBLEMS

1. Note why the transformation (Sec. 115)

$$S_m\{U(x,y)\} = \int_0^\pi U(x,y) \sin mx \, dx = u_n(y) \qquad (m = n - \tfrac{1}{2})$$

applies to this problem in steady temperatures:

$$U_{xx}(x,y) + U_{yy}(x,y) = 0, \ U \text{ bounded}, \qquad (0 < x < \pi, y > 0),$$

$$U(0,y) = U_x(\pi,y) = 0, \qquad U(x,0) = 1 \qquad (0 < x < \pi).$$

Show that $u_n(y) = m^{-1} e^{-my}$ and thus (Prob. 8, Sec. 116)

$$U(x,y) = \frac{2}{\pi} \arctan \frac{\sin x/2}{\sinh y/2} \qquad (0 < x < \pi, y > 0).$$

2. Use the transformation S_m (Sec. 115) with respect to x to find the solution of the problem

$$U_{xx}(x,y) + U_{yy}(x,y) = 0, \ U \text{ bounded}, \qquad (0 < x < \pi, y > 0),$$

$$U(0,y) = U_x(\pi,y) = 0, \qquad U_y(x,0) = -1,$$

in the form $U(x,y) = S_m^{-1}\{m^{-2}e^{-my}\}$; thus show that

$$U(x,y) = \frac{x}{\pi}\log\frac{\cosh y/2 + \cos x/2}{\cosh y/2 - \cos x/2} + \int_0^x tF(t,y)\,dt$$

where (Probs. 6 and 7, Sec. 116)

$$F(t,y) = \frac{1}{\pi}\operatorname{Re}\csc\frac{t+iy}{2} = \frac{1}{\pi}\frac{(\sin t/2)(\cosh y/2)}{\cosh^2 y/2 - \cos^2 t/2}.$$

3. Determine the Sturm-Liouville transformation T, with respect to x, that is adapted to the problem

$$(t+1)U_t(x,t) = U_{xx}(x,t) \qquad (0 < x < 1, t > 0),$$

$$U(x,0) = 0, \qquad U_x(0,t) = -1, \qquad U(1,t) = 0.$$

Find $T^{-1}\{1/\lambda_n\}$ by solving and transforming the elementary problem $Y''(x) = 0$, $Y'(0) = 1$, $Y(1) = 0$; then derive the formula

$$U(x,t) = 1 - x - 2\sum_{n=1}^{\infty}\frac{(t+1)^{-N}}{N}\cos\frac{(2n-1)\pi x}{2} \qquad \left[N = \frac{(2n-1)^2\pi^2}{4}\right].$$

4. Determine the integral transformation (a) with respect to y, (b) with respect to x, that applies to the problem

$$U_{xx}(x,y) + U_{yy}(x,y) + 2U_x(x,y) = 0 \qquad (0 < x < 1, 0 < y < 1),$$

$$U(0,y) = G(y), \qquad U(1,y) = 0, \qquad U_y(x,0) = 0, \qquad U_y(x,1) = H(x),$$

and write the transformed problem in each case.

Ans. (a) $u_c(x,n) = \int_0^1 U(x,y)\cos n\pi y\,dy \qquad (n = 0, 1, 2, \dots)$;
$u_c'' + 2u_c' - n^2\pi^2 u_c = (-1)^{n+1}H(x), u_c(0,n) = g_c(n), u_c(1,n) = 0.$
(b) $u(\lambda_n,y) = \int_0^1 U(x,y)e^x\sin n\pi x\,dx \qquad (\lambda_n = -1 - n^2\pi^2, n = 1, 2, \dots)$;
$u'' - (1 + n^2\pi^2)u = -n\pi G(y), u'(\lambda_n,0) = 0, u'(\lambda_n,1) = h(\lambda_n).$

5. The line $y = 0$ is a singularity of the equation

$$V_{xx}(x,y) + y^2 V_{yy}(x,y) + yV_y(x,y) + F(x) = 0.$$

If V is to be bounded over the rectangle $0 < x < \pi$, $0 < y < b$, solve that partial differential equation under the conditions

$$V_x(0,y) = V(\pi,y) = V(x,b) = 0.$$

Ans. $V(x,y) = \int_x^\pi\int_0^t F(r)\,dr\,dt - \frac{2}{\pi}\sum_{n=1}^\infty\frac{f(\lambda_n)}{m^2}\left(\frac{y}{b}\right)^m\cos mx,$

where $f(\lambda_n) = \int_0^\pi F(r)\cos mr\,dr$ and $m = n - \frac{1}{2}$.

6. Note why the Fourier sine transform (Sec. 110) on the half line $x > 0$ applies to this problem on transverse displacements of a stretched string with prescribed initial velocity (Prob. 5, Sec. 44):

$$Y_{tt}(x,t) = a^2 Y_{xx}(x,t) \qquad\qquad (x > 0, t > 0)$$

$$Y(x,0) = 0, \qquad Y_t(x,0) = G(x), \qquad Y(0,t) = 0.$$

Use property (4), Sec. 110, to show that

$$Y(x,t) = \frac{1}{2a} \int_{|x-at|}^{x+at} G(r)\, dr$$

and verify that solution of the problem.

Without solving the following problems for $V(x,y)$, determine procedures that are adapted to their solution.

7.
$$V_{xx}(x,y) + V_{yy}(x,y) - V(x,y) = 0, \qquad (x > , 0 < y < 1),$$

$$V(0,y) = 1, \qquad V_y(x,0) = 0, \qquad V_y(x,1) = -V(x,1) + e^{-x}.$$

Ans. Fourier sine transformation with respect to x (Sec. 110); the Sturm-Liouville transformation in Prob. 4, Sec. 116, with respect to y, where $h = 1$.

8.
$$V_{xx}(x,y) + x^{-1}V_x(x,y) - V_{yy}(x,y) = Q(x,y) \qquad (0 < x < 1, y > 0),$$

$$V(1,y) = 0, \qquad V(x,0) = 0, \qquad V_y(x,0) = 0.$$

Ans. The finite Hankel transformation in Prob. 12, Sec. 116; the Laplace transformation with respect to y.

11

Finite Fourier Transforms

119 FINITE FOURIER SINE TRANSFORMS

A useful special case of the Sturm-Liouville transformation is that in which
the differential form is the simple self-adjoint form $F''(x)$, and the prescribed
boundary values are the values of F at the ends of a bounded interval (Prob.
1, Sec. 116).

It is convenient to choose the origin and the unit of length so that the
interval is $(0,\pi)$. Then in terms of the notation used in Chap. 10,

$$D_2 F = F''(x), \qquad 0 < x < \pi; \qquad N_\alpha F = F(0), \qquad N_\beta F = F(\pi);$$

here $\alpha = \beta = 0$, $p = r = 1$, and $q = 0$. The kernel of the transformation T
consists of the eigenfunctions of the problem

$$\Phi'' - \lambda\Phi = 0, \qquad \Phi(0) = \Phi(\pi) = 0.$$

Thus $\lambda = -n^2 (n = 1, 2, \ldots)$, $\Phi = \sin nx$, and T becomes the *finite Fourier sine transformation*

(1)
$$S_n\{F\} = \int_0^\pi F(x) \sin nx \, dx = f_s(n) \qquad (n = 1, 2, \ldots).$$

The *basic operational property* (12), Sec. 111, becomes

(2)
$$S_n\{F''(x)\} = -n^2 f_s(n) + n[F(0) - (-1)^n F(\pi)]$$

when F is of class C'' ($0 \le x \le \pi$), or even if F and F' are continuous while F'' is sectionally continuous over the interval. It follows that

(3)
$$S_n\{F^{(4)}(x)\} = n^4 f_s(n) - n^3[F(0) - (-1)^n F(\pi)]$$
$$+ n[F''(0) - (-1)^n F''(\pi)]$$

if F is of class $C^{(4)}$. Formulas for transforms of higher-ordered iterates of the form F'' can be written easily.

Transforms of many elementary functions can be found by evaluating the integral $S_n\{F\}$. Some can be found readily from property (2); as examples, when $F(x) \equiv 1$, or when F is x, $x(\pi - x)$, or e^{cx}, that formula gives these transforms:

(4)
$$S_n\{1\} = \frac{1 - (-1)^n}{n}, \qquad S_n\{x\} = \frac{\pi}{n}(-1)^{n+1},$$

$$S_n\left\{\frac{x}{2}(\pi - x)\right\} = \frac{1 - (-1)^n}{n^3}, \qquad S_n\{e^{cx}\} = \frac{n - n(-1)^n e^{c\pi}}{n^2 + c^2}.$$

Other entries in Appendix B, Table B.1, can be derived by those methods. Some operational properties of S_n are included in that table of finite sine transforms.

Following the procedure used in Sec. 112, we write $f_s(n)/n^2 = y_s(n)$ where $y_s(n) = S_n\{Y(x)\}$. Then $n^2 y_s(n) = f_s(n)$, and this is true if Y satisfies the conditions

$$Y''(x) = -F(x), \qquad Y(0) = Y(\pi) = 0.$$

Thus we find that, when F is sectionally continuous,

(5)
$$S_n^{-1}\left\{\frac{f_s(n)}{n^2}\right\} = \frac{x}{\pi} \int_0^\pi (\pi - r)F(r) \, dr - \int_0^x (x - r)F(r) \, dr.$$

For S_n the series representation (1), Sec. 112, of the inverse Sturm-Liouville transformation becomes the Fourier sine series representation of $F(x)$:

(6)
$$S_n^{-1}\{f_s(n)\} = F(x) = \frac{2}{\pi} \sum_{n=1}^\infty f_s(n) \sin nx \qquad (0 < x < \pi),$$

since $\|\sin nx\|^2 = \pi/2$. This formula for the inverse sine transformation is valid over the interval if F and F' are sectionally continuous there and if, at each point x_0 where F is discontinuous, $F(x)$ is defined as the mean value of its one-sided limits,

$$(7) \qquad\qquad F(x_0) = \tfrac{1}{2}[F(x_0 + 0) + F(x_0 - 0)] \qquad\qquad (0 < x_0 < \pi).$$

For all real x the series (6) converges to this *odd periodic extension* F_1 of F, with period 2π:

$$(8) \quad\begin{aligned} F_1(x) &= F(x) \qquad \text{when} \qquad 0 < x < \pi, \qquad F_1(0) = F_1(\pi) = 0; \\ F_1(-x) &= -F_1(x), \qquad F_1(x + 2\pi) = F_1(x) \qquad\qquad (-\infty < x < \infty), \end{aligned}$$

because $\sin nx$ is odd and periodic with period 2π.

120 OTHER PROPERTIES OF S_n

Let F_1 be the odd periodic extension, with period 2π, of a function F that is sectionally continuous over the interval $(0,\pi)$, and let k be real and constant. Then

$$2f_s(n) \cos nk = \int_{-\pi}^{\pi} F_1(x) \sin nx \cos nk \, dx$$

since the integrand is an even function of x. The integral can be written in these forms:

$$\tfrac{1}{2} \int_{-\pi}^{\pi} F_1(x) \sin n(x + k) \, dx + \tfrac{1}{2} \int_{-\pi}^{\pi} F_1(x) \sin n(x - k) \, dx$$

$$= \tfrac{1}{2} \int_{-\pi}^{\pi} F_1(r - k) \sin nr \, dr + \tfrac{1}{2} \int_{-\pi}^{\pi} F_1(r + k) \sin nr \, dr$$

because the last two integrands are periodic functions of r with period 2π. The sum of those integrands is an even function, so

$$2f_s(n) \cos nk = \int_0^{\pi} [F_1(r - k) + F_1(r + k)] \sin nr \, dr;$$

that is, *if F is sectionally continuous and $S_n\{F\} = f_s(n)$, then*

$$(1) \qquad\qquad 2f_s(n) \cos nk = S_n\{F_1(x - k) + F_1(x + k)\}.$$

When $k = \pi$, then, since $F_1(x - \pi) = -F(\pi - x)$ when $0 < x < \pi$ and $F_1(x + \pi) = F_1(x - \pi)$, property (1) becomes

$$(2) \qquad\qquad f_s(n)(-1)^{n+1} = S_n\{F(\pi - x)\} \qquad\qquad (n = 1, 2, \ldots).$$

For example, since $S_n\{x\} = \pi n^{-1}(-1)^{n+1}$, it follows that

(3)
$$\frac{\pi}{n} = S_n\{\pi - x\}.$$

Kernel-product and convolution properties can be written for S_n. Again let F_1 be the odd periodic extension of F with period 2π. Then the function

(4)
$$F_0(x) = \int_0^x F_1(r)\, dr \qquad\qquad (-\infty < x < \infty)$$

is periodic with period 2π and even, $F_0(x) = F_0(|x|)$, and $F_0'(x) = F_1(x)$ wherever F_1 is continuous. Therefore

(5)
$$f_s(n) = \int_0^\pi F_0'(x) \sin nx\, dx = -n \int_0^\pi F_0(x) \cos nx\, dx.$$

If x_0 is a constant, then

$$\frac{2}{n} f_s(n) \sin nx_0 = \frac{1}{2} \int_{-\pi}^\pi F_0(x)[\sin n(x - x_0) - \sin n(x + x_0)]\, dx,$$

and this *kernel-product property* follows easily:

(6)
$$\frac{2}{n} f_s(n) \sin nx_0 = S_n\left\{ \int_{x-x_0}^{x+x_0} F_1(r)\, dr \right\}.$$

The corresponding convolution of two sectionally continuous functions F and H is (Sec. 109)

(7)
$$X_{FH}(x) = \int_0^\pi H(y) \int_{x-y}^{x+y} F_1(r)\, dr\, dy,$$

with the *convolution property, with weight function* $2/n$,

(8)
$$S_n\{X_{FH}(x)\} = \frac{2}{n} f_s(n) h_s(n)$$

where $h_s = S_n\{H\}$. The integral (7) can be written in terms of H and the function F (Prob. 13).

In case F is such that F_1 is *everywhere continuous* and F' is sectionally continuous, then $F_1(\pm\pi) = 0$ and

(9)
$$2f_s(n) = \int_{-\pi}^\pi F_1(x) \sin nx\, dx = \frac{1}{n} \int_{-\pi}^\pi F_1'(x) \cos nx\, dx.$$

Thus a *kernel-product property with weight function* $\omega = 2n$ can be obtained, namely,

(10)
$$2n f_s(n) \sin nx_0 = S_n\{F_1'(x - x_0) - F_1'(x + x_0)\}.$$

For S_n, property (8), Sec. 113, becomes

$$\lim_{n \to \infty} f_s(n) = 0$$

whenever F is sectionally continuous. Some further properties of S_n, and transforms of particular functions, are given in the problems.

PROBLEMS

1. Use properties of S_n to obtain these transforms:

(a) $S_n\{x^3\} = \pi(-1)^n\left(\dfrac{6}{n^3} - \dfrac{\pi^2}{n}\right);$

(b) $S_n\{\sin kx\} = (-1)^{n+1}\dfrac{n}{n^2 - k^2}\sin k\pi$ if $k \neq \pm 1, \pm 2, \ldots;$

(c) $S_n\{\cosh cx\} = \dfrac{n}{n^2 + c^2}[1 - (-1)^n \cosh c\pi].$

2. (a) If $F(x) = \sin mx$ where m is a positive integer, show that $f_s(n) = 0$ when $n \neq m$, and $f_s(m) = \pi/2$.

(b) Derive: $S_n^{-1}\left\{\dfrac{1}{n^3}(-1)^{n+1}\right\} = \dfrac{x}{6\pi}(\pi^2 - x^2).$

(c) Derive: $S_n^{-1}\left\{\dfrac{1}{n^3}\right\} = \dfrac{x}{6\pi}(x^2 - 3\pi x + 2\pi^2).$

3. If $F(\pi - x) = F(x)$, show that $f_s(n) = 0$ when n is even.

4. Let F and F' be continuous when $0 \leq x \leq \pi$ except that F' has a jump b at a point $x = c$ interior to the interval, and let F'' be sectionally continuous. Prove that

$$S_n\{F''\} = -n^2 f_s(n) + n[F(0) - (-1)^n F(\pi)] - b \sin nc.$$

5. Note the graph of the function $F(x) = |\sin 2x|$ and use the results found in Probs. 3 and 4 to show that

$$f_s(n) = 0 \quad \text{if } n = 2, 4, \ldots, \qquad f_s(n) = \frac{4 \sin n\pi/2}{4 - n^2} \quad \text{if } n = 1, 3, 5, \ldots.$$

6. Derive the inverse transform

$$S_n^{-1}\left\{\pi\frac{\sin nc}{n^2}\right\} = \begin{cases} (\pi - c)x & \text{if } 0 \leq x \leq c < \pi \\ (\pi - x)c & \text{if } c \leq x \leq \pi, \end{cases}$$

(a) with the aid of formula (3), Sec. 114, when T is S_n;

(b) with the aid of the kernel-product property (6), Sec. 120, when $f_s(n) = n^{-1}$.

7. If the real numbers $a_1, a_2, \ldots$ are such that the series $a_1 + a_2 + \cdots$ is absolutely convergent, prove that

$$S_n\left\{\frac{2}{\pi}\sum_{m=1}^{\infty} a_m \sin mx\right\} = a_n \qquad (n = 1, 2, \ldots).$$

8. When $y > 0$, the numbers $a_m = m^{-1}(-1)^{m+1}e^{-my}$, where $m = 1, 2, \ldots$, satisfy the conditions in Prob. 7. If $z = x + iy$, show that

$$\sum_{m=1}^{\infty} \frac{(-1)^{m+1}}{m} e^{-my} \sin mx = \operatorname{Im} \sum_{m=1}^{\infty} \frac{(-1)^{m+1}}{m} (e^{iz})^m$$

$$= \operatorname{Im} [\log (1 + e^{iz})],$$

and thus derive the inverse transform

$$S_n^{-1}\left\{\frac{(-1)^{n+1}}{n} e^{-ny}\right\} = \frac{2}{\pi} \arctan\left(\frac{\sin x}{e^y + \cos x}\right) \qquad \text{if } y > 0.$$

9. From the transform found in Prob. 9 note that

$$S_n^{-1}\left\{\frac{1}{n} e^{-ny}\right\} = \frac{2}{\pi} \arctan \frac{\sin x}{e^y - \cos x}$$

when $y > 0$; then show that

$$S_n^{-1}\left\{\frac{1 - (-1)^n}{n} e^{-ny}\right\} = \frac{2}{\pi} \arctan \frac{\sin x}{\sinh y} \qquad \text{if } y > 0.$$

10. When $-1 < b < 1$ and $-\pi/2 < \arctan t < \pi/2$, use transforms found in Probs. 8 and 9 to show that

$$\frac{b^n}{n} = S_n\left\{\frac{2}{\pi} \arctan \frac{b \sin x}{1 - b \cos x}\right\} \qquad (-1 < b < 1);$$

then differentiate with respect to b to get the transform

$$b^n = S_n\left\{\frac{2}{\pi} \frac{b \sin x}{1 + b^2 - 2b \cos x}\right\} \qquad (-1 < b < 1).$$

11. Given that, when $c \neq 0$, the function

$$Y(x) = \frac{\sin cx \sinh c(2\pi - x) - \sin c(2\pi - x) \sinh cx}{4c^2(\sin^2 c\pi + \sinh^2 c\pi)},$$

satisfies the conditions

$$\frac{d^4 Y}{dx^4} + 4c^4 Y = 0, \qquad Y(0) = Y(\pi) = Y''(\pi) = 0, \qquad Y''(0) = -1,$$

show that

$$S_n\{Y(x)\} = \frac{n}{n^4 + 4c^4}.$$

12. If Y_1 is the odd periodic extension of the function Y in Prob. 11, with period 2π, find the solution of the problem

$$Z^{(4)}(x) + 4c^4 Z(x) = 2F(x),$$

$$Z(0) = Z(\pi) = Z''(0) = Z''(\pi) = 0,$$

in the form

$$Z(x) = \int_0^\pi F(t) \int_{x-t}^{x+t} Y_1(r)\, dr\, dt.$$

13. Express the convolution integral (7), Sec. 120, in terms of the functions H and F by the formula

$$\int_0^\pi H(y) \int_{x-y}^{x+y} F_1(r)\, dr\, dy = \int_0^{\pi-x} H(y) \int_0^{x+y} F(r)\, dr\, dy$$

$$+ \int_{\pi-x}^\pi H(y) \int_0^{2\pi-x-y} F(r)\, dr\, dy - \int_0^x H(y) \int_0^{x-y} F(r)\, dr\, dy$$

$$- \int_x^\pi H(y) \int_0^{y-x} F(r)\, dr\, dy.$$

14. If F_1, the odd periodic extension of F with period 2π, is *everywhere continuous* and if F_1' and the function H are sectionally continuous, show that a convolution property for S_n corresponding to the kernel-product property (10), Sec. 120, is

$$S_n^{-1}\{2nf_s(n)h_s(n)\} = \frac{\partial}{\partial x} \int_0^\pi H(y)[F_1(x-y) - F_1(x+y)]\, dy.$$

15. Establish the following relation between the Fourier sine transformation on the interval $(0,b)$ introduced in Prob. 1, Sec. 116, and our transformation S_n on the interval $(0,\pi)$:

$$\int_0^b F(x) \sin \frac{n\pi x}{b}\, dx = S_n\left\{\frac{b}{\pi} F\left(\frac{b}{\pi}x\right)\right\}.$$

121   FINITE COSINE TRANSFORMS

The Sturm-Liouville transformation that resolves the differential form $F''(x)$ on the interval $(0,\pi)$ in terms of the transform of F and the boundary values $F'(0)$ and $F'(\pi)$ has a kernel which satisfies the eigenvalue problem

$$\Phi''(x) - \lambda\Phi(x) = 0, \qquad \Phi'(0) = \Phi'(\pi) = 0.$$

Thus $\lambda = -n^2$ and $\Phi(x) = \cos nx$, where $n = 0, 1, 2, \ldots$. In this case $\Phi(x) = 1$ is an eigenfunction corresponding to the eigenvalue $\lambda = 0$. The transformation is the *finite Fourier cosine transformation*

(1) $$C_n\{F(x)\} = \int_0^\pi F(x) \cos nx\, dx = f_c(n) \qquad (n = 0, 1, 2, \ldots)$$

with the operational property

(2) $$C_n\{F''(x)\} = -n^2 f_c(n) - F'(0) + (-1)^n F'(\pi) \qquad (n = 0, 1, 2, \ldots)$$

when F and F' are continuous, and F'' is sectionally continuous, over the interval $0 \leq x \leq \pi$.

When F is of class $C^{(4)}$, it follows that

(3) $\qquad C_n\{F^{(4)}(x)\} = n^4 f_c(n) + n^2[F'(0) - (-1)^n F'(\pi)] - F'''(0) + (-1)^n F'''(\pi).$

The formula for $C_n\{F^{(6)}\}$ can be written by replacing F by F'' in formula (3), and so on for transforms of the higher even-ordered derivatives. The boundary values that arise are those of the odd-ordered derivatives.

Since $\|\cos nx\|^2 = \pi/2$ when $n = 1, 2, \ldots$, then if $m = 1, 2, \ldots$,

(4) $\qquad C_n\{\cos mx\} = 0 \qquad$ if $n \neq m$, $C_m\{\cos mx\} = \tfrac{1}{2}\pi$;

(5) $\qquad C_n\{1\} \qquad = 0 \qquad$ if $n = 1, 2, \ldots$, $C_0\{1\} = \pi.$

If A is a constant and $f_c(n) = C_n\{F\}$, then

(6) $\qquad C_n\{F(x) + A\} = f_c(n) \qquad\qquad$ when $n = 1, 2, \ldots$

$\qquad\qquad\qquad\qquad\quad = f_c(0) + \pi A \qquad\qquad$ when $n = 0.$

Also, $f_c(n) \to 0$ as $n \to \infty$ when F is sectionally continuous. Note that $f_c(0)/\pi$ is the mean value of F on the interval $(0,\pi)$. Thus $C_0\{x\}/\pi = \pi/2$, while property (2) shows that

$$C_n\{x\} = -[1 - (-1)^n]n^{-2} \qquad \text{when } n = 1, 2, \ldots.$$

When F is sectionally continuous, a function Y whose transform $y_c(n)$ is $f_c(n)/n^2$ when $n = 1, 2, \ldots$, can be found by writing $n^2 y_c(n) = f_c(n)$. That equation is satisfied, according to properties (2) and (6), if

$$Y''(x) = -F(x) + B, \quad Y'(0) = Y'(\pi) = 0,$$

where B is to be determined by the conditions on Y'. We find that

$$Y(x) = \int_0^x \int_t^\pi F(r)\, dr\, dt + \frac{1}{2\pi} f_c(0)(x - \pi)^2 + C.$$

The constant C is determined by the value assigned to $y_c(0)$. We can simplify the iterated integral to write

(7) $\qquad \dfrac{f_c(n)}{n^2} = C_n\left\{ \displaystyle\int_x^\pi (x - r)F(r)\, dr + \frac{1}{2\pi} f_c(0)(x - \pi)^2 + A \right\}$

when $n = 1, 2, \ldots$, where $A = C + \displaystyle\int_0^\pi r F(r)\, dr.$

For C_n the series representation of the inverse Sturm-Liouville transform is the Fourier cosine series

(8) $\qquad C_n^{-1}\{f_c(n)\} = \dfrac{1}{\pi} f_c(0) + \dfrac{2}{\pi} \displaystyle\sum_{n=1}^{\infty} f_c(n) \cos nx = F(x)$

when $0 < x < \pi$, valid if F and F' are sectionally continuous and if F is defined as its mean value at each point of discontinuity. For all real x the series represents the *even periodic extension* F_2 of F with period 2π:

(9) $F_2(x) = F(x)$ when $0 < x < \pi$,

$$= F_2(|x|) = F_2(x + 2\pi) \text{ for all } x.$$

This *kernel-product property* for C_n can be written by using the even periodic extension (9) of F:

(10) $2f_c(n) \cos nx_0 = C_n\{F_2(x - x_0) + F_2(x + x_0)\}$ $(n = 0, 1, 2, \ldots)$.

If $0 \leqq x_0 \leqq \pi$, then $F_2(x - x_0) = F(|x - x_0|)$ here. It follows that

(11) $f_c(n)(-1)^n = C_n\{F(\pi - x)\}$ $(n = 0, 1, 2, \ldots)$.

The convolution property corresponding to formula (10) is

(12) $C_n^{-1}\{2f_c(n)h_c(n)\} = \displaystyle\int_0^\pi [F_2(x - y) + F_2(x + y)]H(y)\,dy,$

where F and H are sectionally continuous. The integral here will be written in terms of H and the function F in Sec. 123.

122   TABLES OF FINITE FOURIER TRANSFORMS

Tables of $C_n\{F\}$ and $S_n\{F\}$ will be found in Appendix B. Transforms of particular functions are found by evaluating the integrals $C_n\{F\}$ or $S_n\{F\}$, by using partial fractions or other decompositions of transforms, by differentiating or integrating with respect to a parameter, by using operational properties of the transformations, or by summing the series that represents the inverse transform. Relations between sine and cosine transforms, given in the following section, also aid in extending the tables.

An example of the use of the Fourier cosine series will be given here, corresponding to Probs. 7 and 8, Sec. 120. Suppose

(1) $f_c(n,y) = \dfrac{1}{n}e^{-ny}$ when $n = 1, 2, \ldots,$ and $f_c(0,y) = 0$,

where $y > 0$. Then the series with terms $f_c(n,y)$ is absolutely convergent and $f_c(n, y) = C_n\{F(x,y)\}$ when

$$F(x,y) = \frac{2}{\pi} \sum_{n=1}^{\infty} \frac{1}{n}e^{-ny} \cos nx = \frac{2}{\pi}\text{Re} \sum_{n=1}^{\infty} \frac{1}{n}(e^{iz})^n$$

where $z = x + iy$ and $y > 0$. It follows that

$$F(x,y) = -\frac{2}{\pi}\text{Re}\,[\log(1 - e^{iz})] = -\frac{1}{\pi}\log(1 + e^{-2y} - 2e^{-y}\cos x)$$

$$= -\frac{1}{\pi}\log[2e^{-y}(\cosh y - \cos x)].$$

Thus we find that

(2) $$\pi F(x,y) = y - \log 2 - \log (\cosh y - \cos x) \qquad (y > 0).$$

The method used to derive formula (2) is not sound if $y = 0$. But in that case the function F is still defined:

(3) $$F(x,0) = -\frac{1}{\pi} \log [2(1 - \cos x)] = -\frac{2}{\pi} \log \left(2 \sin \frac{x}{2}\right),$$

and our results suggest that

(4) $$C_n \left\{ -\frac{2}{\pi} \log \left(2 \sin \frac{x}{2}\right) \right\} = \begin{cases} 0 & \text{if } n = 0 \\ \dfrac{1}{n} & \text{if } n = 1, 2, \ldots . \end{cases}$$

The improper integral $C_n\{F(x,0)\}$ can be evaluated with the aid of the theory of residues[1] to prove that formula (4) is correct.

123   JOINT PROPERTIES OF C_n AND S_n

Some operational properties of our two finite Fourier transformations have simple forms when stated in terms of both transformations.

If F is continuous and F' is sectionally continuous, integration by parts shows that, when $n = 0, 1, 2, \ldots$,

(1) $$S_n\{F'(x)\} = -nC_n\{F(x)\},$$

(2) $$C_n\{F'(x)\} = nS_n\{F(x)\} - F(0) + (-1)^n F(\pi).$$

Alternate forms of those properties are

(3) $$S_n\{H(x)\} = -nC_n \left\{ \int_0^x H(r)\, dr \right\},$$

(4) $$C_n \left\{ H(x) - \frac{1}{\pi} h_c(0) \right\} = nS_n \left\{ \int_0^x H(r)\, dr - \frac{x}{\pi} h_c(0) \right\},$$

when H is sectionally continuous.

Let F_1 and F_2 be the odd and even periodic extensions, with period 2π, of a function F that is sectionally continuous on the interval $(0, \pi)$, and let k be real and constant. Then by the method used to derive property (1), Sec. 120, we find that

(5) $$2f_s(n) \sin nk = C_n\{F_1(x + k) - F_1(x - k)\},$$

(6) $$2f_c(n) \sin nk = S_n\{F_2(x - k) - F_2(x + k)\}.$$

---

[1] Cf. sec. 73 of the author's "Complex Variables and Applications," 2d ed., 1960.

When P and Q are sectionally continuous on the interval $(-\pi, \pi)$ and periodic with period 2π, the function

$$(7) \qquad P * Q(x) = \int_{-\pi}^{\pi} P(x - r)Q(r)\, dr$$

is periodic with period 2π and continuous. Let us call it the *faltung integral of P and Q on the interval* $(-\pi, \pi)$. We find that

$$(8) \qquad P * Q(x) = Q * P(x),$$

and that the faltung is an even function if P and Q are both even or both odd; it is odd if either P or Q is even and the other odd.

Now let F_2 and H_2 be the even periodic extensions, with period 2π, of two functions F and H that are sectionally continuous on the interval $(0, \pi)$. By writing the faltung of F_2 and H_2 on $(-\pi, \pi)$ as a sum of integrals and introducing new variables of integration, we find that

$$(9) \qquad F_2 * H_2(x) = \int_0^{\pi} [F_2(x - r) + F_2(x + r)]H(r)\, dr = \sum_{i=1}^{4} I_i(x),$$

when $0 < x < \pi$, where I_i are these integrals involving F and H:

$$(10) \qquad
\begin{aligned}
I_1 &= \int_0^x F(r)H(x - r)\, dr, & I_2 &= \int_x^{\pi} F(r)H(r - x)\, dr, \\
I_3 &= \int_0^{\pi - x} F(r)H(x + r)\, dr, & I_4 &= \int_{\pi - x}^{\pi} F(r)H(2\pi - x - r)\, dr.
\end{aligned}$$

In view of Eq. (9) the *convolution property* (12), Sec. 121, takes the form

$$(11) \qquad 2f_c(n)h_c(n) = C_n\{F_2 * H_2(x)\}.$$

Similarly, if F_1 and H_1 are the odd periodic extensions of F and H, with period 2π, it turns out that

$$(12) \qquad F_1 * H_1(x) = \int_0^{\pi} [F_1(x - r) - F_1(x + r)]H(r)\, dr$$

$$= I_1(x) - I_2(x) - I_3(x) + I_4(x),$$

$$(13) \qquad F_1 * H_2(x) = \int_0^{\pi} [F_1(x - r) + F_1(x + r)]H(r)\, dr$$

$$= \int_0^{\pi} F(r)[H_2(x - r) - H_2(x + r)]\, dr$$

$$= I_1(x) + I_2(x) - I_3(x) - I_4(x).$$

Then from the product properties (5) and (6) we find these *joint convolution properties*:

(14) $$-2f_s(n)h_s(n) = C_n\{F_1 * H_1(x)\},$$

(15) $$2f_s(n)h_c(n) = S_n\{F_1 * H_2(x)\}.$$

PROBLEMS

1. Find these transforms, with the aid of properties of C_n:

(a) $C_n\{x\} = \dfrac{(-1)^n - 1}{n^2}$ if $n = 1, 2, \ldots, C_0(x) = \dfrac{\pi^2}{2}$;

(b) $C_n\{(\pi - x)^2\} = \dfrac{2\pi}{n^2}$ if $n = 1, 2, \ldots, C_0\{(\pi - x)^2\} = \dfrac{\pi^3}{3}$;

(c) $C_n\{\cos kx\} = k \sin k \dfrac{(-1)^{n+1}}{n^2 - k^2}$ if $k \neq 0, \pm 1, \pm 2, \ldots$;

(d) $C_n\{\cosh c(\pi - x)\} = \dfrac{c \sinh c\pi}{n^2 + c^2}$ if $c \neq 0$.

2. With the aid of transform (b), Prob. 1, show that if

$$y_c(n) = \frac{1}{n^4} \quad \text{when } n = 1, 2, \ldots, \text{ and } y_c(0) = 0,$$

then

$$C_n^{-1}\{y_c(n)\} = \frac{\pi^3}{45} - \frac{\pi}{6}x^2 + \frac{1}{6}x^3 - \frac{1}{24\pi}x^4.$$

3. If $a \neq 0$ and $b \neq 0$, derive the transforms

(a) $\dfrac{(a^2 - b^2)(-1)^n}{(n^2 + a^2)(n^2 + b^2)} = C_n\left\{\dfrac{\cosh bx}{b \sinh b\pi} - \dfrac{\cosh ax}{a \sinh a\pi}\right\}$ $(a^2 \neq b^2)$;

(b) $2a^3 \sinh^2 a\pi C_n^{-1}\left\{\dfrac{(-1)^n}{(n^2 + a^2)^2}\right\}$

       $= (a\pi \cosh a\pi + \sinh a\pi) \cosh ax - ax \sinh a\pi \sinh ax.$

4. We noted in Prob. 9, Sec. 120, that when $y > 0$,

$$\frac{\pi}{n}e^{-ny} = S_n\left\{2 \arctan \frac{\sin x}{e^y - \cos x}\right\} \quad (n = 1, 2, \ldots).$$

Use formula (2), Sec. 103, and its special case $C_0\{F'(x)\} = F(\pi) - F(0)$ to prove that, when $n = 0, 1, 2, \ldots,$ and $y > 0$,

(a) $\pi e^{-ny} = C_n\left\{1 + \dfrac{\cos x - e^{-y}}{\cosh y - \cos x}\right\} = C_n\left\{\dfrac{\sinh y}{\cosh y - \cos x}\right\}.$

Then deduce that

(b) $b^n = C_n\left\{\dfrac{1}{\pi}\dfrac{1 - b^2}{1 + b^2 - 2b \cos x}\right\}$ if $-1 < b < 1$ and $b \neq 0$.

5. (a) From $S_n^{-1}\{n^{-1}e^{-ny}\}$ given in Prob. 4, deduce that

$$\frac{1}{n^2}e^{-ny} = C_n\left\{-\frac{2}{\pi}\int_0^x \arctan\frac{\sin r}{e^y - \cos r}\,dr\right\} \qquad \text{if } n = 1, 2, \ldots, \text{ and } y > 0.$$

(b) Differentiate with respect to y in formula (a), Prob. 4, to show that

$$ne^{-ny} = C_n\left\{\frac{1}{\pi}\frac{\cos x \cosh y - 1}{(\cos x - \cosh y)^2}\right\} \qquad \text{when } y > 0.$$

6. Use C_n to find the solution of the problem

$$Y''(x) - c^2 Y(x) = -F(x), \qquad Y'(0) = Y'(\pi) = 0, \qquad (c \neq 0)$$

in the form $2c \sinh c\pi Y(x) = F_2 * H_2(x)$ where $H(x) = \cosh c(\pi - x)$, and reduce the solution to the form

$$c \sinh c\pi Y(x) = \cosh cx \int_x^\pi F(r) \cosh c(\pi - r)\,dr$$

$$+ \cosh c(\pi - x) \int_0^x F(r) \cosh cr\,dr.$$

7. If F and F' are continuous when $0 \leq x \leq \pi$ except at an interior point $x = c$ of that interval, where $F(x)$ and $F'(x)$ have jumps b and b', respectively, and if F'' is sectionally continuous, prove that

$$C_n\{F''\} = -n^2 f_c(n) - F'(0) + (-1)^n F'(\pi) - bn \sin nc - b' \cos nc.$$

8. When p is not necessarily an integer, let us write

$$f_c(p) = \int_0^\pi F(x) \cos px\,dx, \qquad f_s(p) = \int_0^\pi F(x) \sin px\,dx.$$

When k is a constant, show that

(a) $2S_n\{F(x) \cos kx\} = f_s(n + k) + f_s(n - k)$;

(b) $2C_n\{F(x) \sin kx\} = f_s(n + k) - f_s(n - k)$;

(c) $2C_n\{F(x) \cos kx\} = f_c(n - k) + f_c(n + k)$;

(d) $2S_n\{F(x) \sin kx\} = f_c(n - k) - f_c(n + k)$.

124 POTENTIAL IN A SLOT

Let $V(x,y)$ denote the electrostatic potential in a space bounded by the planes $x = 0$, $x = \pi$, and $y = 0$, in which there is a uniform distribution of space charge of density $h/(4\pi)$. Then the function V satisfies Poisson's equation

(1) $$V_{xx}(x,y) + V_{yy}(x,y) = -h \qquad (0 < x < \pi, y > 0).$$

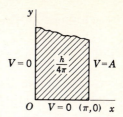

Fig. 104

Let the planes $x = 0$ and $y = 0$ be kept at potential zero and the plane $x = \pi$ at another fixed potential $V = A$ (Fig. 104). Then

(2) $$V(0,y) = 0, \qquad V(\pi,y) = A \qquad\qquad (y > 0),$$

$$V(x, 0) = 0 \qquad\qquad (0 < x < \pi),$$
(3)
$$|V(x,y)| < M$$

throughout the region, where M is some constant.

The determination of $V(x,y)$ is a type of problem that arises in the subject of electronics. The function V here can also be interpreted as steady-state temperatures in a semi-infinite slab containing a uniform source of heat with strength proportional to h.

Our boundary value problem [(1) to (3)] is not adapted to either the method of separation of variables or solution by Laplace transforms. The finite Fourier sine transformation with respect to x does apply, because the differential form in that variable is V_{xx} and the prescribed boundary values on the interval $0 < x < \pi$ are $V(0,y)$ and $V(\pi,y)$.

We write $v_s(n,y) = S_n\{V(x,y)\}$ and transform the members of Eq. (1), using conditions (2), to find formally that

$$-n^2 v_s(n,y) - An(-1)^n + \frac{d^2}{dy^2} v_s(n,y) = -h S_n\{1\}.$$

We also write conditions (3) in terms of transforms. Thus the problem in $v_s(n,y)$ becomes

$$\frac{d^2 v_s}{dy^2} - n^2 v_s = An(-1)^n - h S_n\{1\} \qquad (n = 1, 2, \ldots),$$

$$v_s(n,0) = 0, \qquad |v_s(n,y)| < M\pi.$$

The solution of that problem is

(4) $$v_s(n,y) = \frac{h S_n\{1\} - An(-1)^n}{n^2}(1 - e^{-ny})$$

$$= \left[h \frac{1 - (-1)^n}{n^3} + A \frac{(-1)^{n+1}}{n} \right](1 - e^{-ny}).$$

The formula for the potential can therefore be written

$$(5) \qquad V(x,y) = \frac{2}{\pi} \sum_{n=1}^{\infty} v_s(n,y) \sin nx.$$

In addition to that form of the solution in terms of an infinite series we can derive a closed form of the solution. Referring to Appendix B, Table B.1, we find that

$$S_n^{-1}\left\{\frac{1-(-1)^n}{n^3}\right\} = \frac{x}{2}(\pi - x), \qquad S_n^{-1}\left\{\frac{(-1)^{n+1}}{n}\right\} = \frac{x}{\pi},$$

$$S_n^{-1}\left\{(-1)^{n+1}\frac{e^{-ny}}{n}\right\} = \frac{2}{\pi}\arctan\frac{\sin x}{e^y + \cos x},$$

and

$$S_n^{-1}\left\{\frac{1-(-1)^n}{n}e^{-ny}\right\} = \alpha(x,y),$$

where

$$(6) \qquad \alpha(x,y) = \frac{2}{\pi}\arctan\frac{\sin x}{\sinh y} \qquad\qquad (y \geqq 0).$$

From our formula for $S_n^{-1}\{f_s(n)/n^2\}$ it follows that

$$S_n^{-1}\left\{\frac{1-(-1)^n}{n^3}e^{-ny}\right\} = U(x,y),$$

where

$$(7) \qquad U(x,y) = \frac{x}{\pi}\int_0^{\pi}(\pi - r)\alpha(r,y)\,dr - \int_0^x (x - r)\alpha(r,y)\,dr.$$

Our potential function is therefore

$$(8) \qquad V(x,y) = h\left[\frac{x}{2}(\pi - x) - U(x,y)\right] + \frac{A}{\pi}\left(x - 2\arctan\frac{\sin x}{e^y + \cos x}\right),$$

where $U(x,y)$ is given by Eq. (7) in terms of the function (6). Both inverse tangent functions involved here are known solutions of Laplace's equation; consequently it is not difficult to verify that the function (8) satisfies all conditions in our boundary value problem [(1) to (3)].

125 SUCCESSIVE TRANSFORMATIONS

A semi-infinite slab of finite thickness is initially at temperature zero, and its base is kept at that temperature. One of its parallel faces is insulated and the other is subjected to a prescribed constant and uniform inward flux of heat. Units are chosen so that the boundary value problem in the temperature function $U(x,y,t)$ becomes

$$U_t = U_{xx} + U_{yy} \qquad (0 < x < \pi,\, y > 0,\, t > 0),$$

(1)
$$U(x,y,0) = 0, \qquad U_x(0,y,t) = -1,$$

$$U_x(\pi,y,t) = 0, \qquad U(x,0,t) = 0;$$

also $|U| < Mt$ throughout the slab, for some constant M.

The presence of the differential form U_{xx} together with prescribed values of U_x at $x = 0$ and $x = \pi$ indicates that the finite Fourier cosine transformation, with respect to x, can be used here. The form U_t and a prescribed value of U at $t = 0$ indicate the Laplace transformation with respect to t. Let us begin with the cosine transformation.

The transform with respect to x,

(2)
$$W(n,y,t) = C_n\{U(x,y,t)\} \qquad (n = 0, 1, 2, \ldots),$$

satisfies this problem in partial differential equations:

(3)
$$W_t = -n^2 W + 1 + W_{yy}, \qquad W(n,y,0) = W(n,0,t) = 0,$$

where $|W(n,y,t)| < M\pi t$. The Laplace transform $w(n,y,s)$ of W, with respect to t, therefore satisfies a corresponding boundedness condition and the conditions

(4)
$$(s + n^2)w - \frac{d^2 w}{dy^2} = \frac{1}{s}, \qquad w(n,0,s) = 0.$$

The solution of this problem can be written as

(5)
$$w(n,y,s) = \frac{1}{s(s + n^2)} - \frac{1}{s}\frac{\exp(-y\sqrt{s + n^2})}{s + n^2}.$$

With the aid of the inverse transform of $s^{-1}\exp(-y\sqrt{s})$ and the operational property on replacing s by $s + n^2$, we find that

$$L^{-1}\left\{ \frac{1}{s}\frac{\exp(-y\sqrt{s + n^2})}{s + n^2} \right\} = E_n(y,t),$$

where $y \geq 0$ and

(6)
$$E_n(y,t) = \int_0^t \exp(-n^2\tau)\,\mathrm{erfc}\,\frac{y}{2\sqrt{\tau}}\,d\tau \qquad (n = 0, 1, 2, \ldots),$$

The inverse Laplace transform of the function (5) is therefore

$$W(n,y,t) = \frac{1}{n^2} - \frac{\exp(-n^2 t)}{n^2} - E_n(y,t) \qquad (n \neq 0,\, y \geq 0),$$

(7)
$$W(0,y,t) = t - E_0(y,t) \qquad (y \geq 0).$$

The inverse cosine transform of $f_c(n) = n^{-2}$, $f_c(0) = 0$, is $\frac{1}{2}(x - \pi)^2/\pi - \pi/6$. Thus $C_n^{-1}\{W\}$ can be written

$$(8) \qquad U(x,y,t) = \frac{(x - \pi)^2}{2} - \frac{\pi}{6} + \frac{t}{\pi} - \frac{E_0(y,t)}{\pi}$$

$$-\frac{2}{\pi} \sum_{n=1}^{\infty} \left[\frac{\exp(-n^2 t)}{n^2} + E_n(y,t) \right] \cos nx.$$

An integration by parts in formula (6) shows that $E_n(y,t)$ is of the order of n^{-4} for large n, when $t > 0$ and $y > 0$. The series in our solution (8) can be differentiated termwise, and the solution satisfies all conditions in the boundary value problem (1).

PROBLEMS

1. A steady-state temperature function V satisfies the conditions

$$V_{xx}(x,y) + V_{yy}(x,y) = 0 \qquad\qquad (0 < x < \pi, y > 0),$$

$$V(0,y) = 0, \qquad V(\pi,y) = A \qquad\qquad (y > 0),$$

$$V(x,0) = B \qquad\qquad (0 < x < \pi);$$

also $V(x,y)$ is bounded. Derive the formula

$$V(x,y) = \frac{A}{\pi}x - \frac{2A}{\pi} \arctan \frac{\sin x}{e^y + \cos x} + \frac{2B}{\pi} \arctan \frac{\sin x}{\sinh y}.$$

2. A bounded harmonic function U satisfies the conditions

$$U_{xx}(x,y) + U_{yy}(x,y) = 0 \qquad\qquad (0 < x < \pi, y > 0),$$

$$U(0,y) = U(\pi,y) = 0, \qquad U(x,+0) = F(x) \qquad \text{when } 0 < x < \pi.$$

Show formally that $u_s(n,y) = f_s(n)e^{-ny} = 2f_s(n)q_c(n,y)$, where

$$C_n^{-1}\{q_c(n,y)\} = \frac{1}{2\pi} \frac{\sinh y}{\cosh y - \cos x} \qquad\qquad \text{when } y > 0,$$

an even periodic function of x with period 2π. Use our joint convolution property to represent U in this form, when $y > 0$:

$$U(x,y) = \frac{1}{2\pi} \int_0^\pi F(r) \left[\frac{\sinh y}{\cosh y - \cos(r - x)} - \frac{\sinh y}{\cosh y - \cos(r + x)} \right] dr.$$

3. When $F(x) = A \sin x$ in Prob. 2, note the values of $u_s(n,y)$ to show that $U(x,y) = Ae^{-y} \sin x$. Verify that solution.

4. Solve this problem for V when $0 \leq x \leq \pi$ and $y > 0$:

$$V_{xx}(x,y) + V_{yy}(x,y) = G(x) \qquad\qquad (0 < x < \pi, y > 0),$$

$$V(0,y) = V(\pi,y) = V(x,+0) = 0,$$

where V is bounded. Obtain these formulas for V:

$$V(x,y) = F(x) + \frac{2}{\pi} \sum_{n=1}^{\infty} \frac{1}{n^2} g_s(n) e^{-ny} \sin nx$$

$$= F(x) - U(x,y),$$

where U is the solution of Prob. 2 and

$$F(x) = \int_0^x (x - r)G(r)\, dr - \frac{x}{\pi} \int_0^\pi (\pi - r)G(r)\, dr.$$

5. Let $V(x,y)$ satisfy the conditions (Fig. 105)

$$V_{xx} + V_{yy} = 0 \qquad\qquad (0 < x < \pi, 0 < y < y_0),$$

$$V(0,y) = 0, \qquad V(\pi,y) = 1, \qquad V_y(x,0) = V(x,y_0) = 0.$$

Derive the formula

$$V(x,y) = \frac{x}{\pi} + \frac{2}{\pi} \sum_{n=1}^{\infty} \frac{(-1)^n}{n} \frac{\cosh ny}{\cosh ny_0} \sin nx.$$

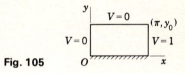

Fig. 105

6. The steady-state temperature $U(x,y)$ in a semi-infinite slab with insulated faces satisfies the conditions

$$U_{xx} + U_{yy} = 0 \quad \text{and} \quad |U(x,y)| < M \quad \text{when} \quad 0 < x < \pi, y > 0,$$

$$U_x(0,y) = U_x(\pi,y) = 0, U(x,+0) = F(x).$$

(*a*) Obtain this formula for $U(x,y)$:

$$U = \frac{1}{2\pi} \int_0^\pi F(r) \left[\frac{\sinh y}{\cosh y - \cos(x - r)} + \frac{\sinh y}{\cosh y - \cos(x + r)} \right] dr.$$

(*b*) When $F(x) = A$, note the values of $u_c(n,y)$ and deduce that $U(x,y) = A$.

(*c*) When $F(x) = A \cos x$, note the values of $u_c(n,y)$; thus show and then verify that

$$U(x,y) = Ae^{-y} \cos x.$$

7. The faces $x = 0$ and $x = \pi$ of a semi-infinite slab $0 \le x \le \pi, y \ge 0$, are insulated and the base $y = 0$ is kept at temperature $V = 0$. A steady source of heat $KQ(x)$ *with mean value zero*, $\int_0^\pi Q(x)\, dx = 0$, is distributed throughout the slab; thus $V_{xx} + V_{yy} + Q = 0$, where $V(x,y)$ is the steady-state temperature. If V is bounded, derive the formulas

$$V(x,y) = F(x) - \frac{2}{\pi} \sum_{n=1}^{\infty} \frac{1}{n^2} q_c(n) e^{-ny} \cos nx$$

$$= F(x) - U(x,y),$$

where $F(x) = \int_x^\pi (x - r)Q(r)\,dr$ and U is the temperature function in Prob. 6 corresponding to this function F.

8. If V is a bounded harmonic function in the semicircular region $r < a, 0 < \theta < \pi$, whose values are prescribed as $F(\theta)$ on the semicircle and as zero on the bounding diameter (Fig. 106), then

$$r^2 V_{rr}(r,\theta) + r V_r(r,\theta) + V_{\theta\theta}(r,\theta) = 0, \qquad |V(r,\theta)| < M \qquad (r < a, 0 < \theta < \pi),$$

$$V(r,0) = V(r,\pi) = 0, \qquad V(a - 0, \theta) = F(\theta).$$

Derive this *Poisson integral formula*[1] for V:

$$V(r,\theta) = \frac{1}{2\pi}\int_0^\pi F(\alpha)[P(r, \alpha - \theta) - P(r, \alpha + \theta)]\,d\alpha,$$

when $r < a$ and $0 \le \theta \le \pi$, where P is *Poisson's kernel*,

$$P(r,\theta) = \frac{a^2 - r^2}{a^2 + r^2 - 2ar\cos\theta}.$$

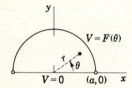

$V = F(\theta)$

$V = 0$ $\qquad (a,0) \qquad x$ **Fig. 106**

9. In Prob. 8, if the condition on the bounding diameter of the semicircular region is replaced by the condition that the normal derivative of V vanishes there,

$$V_\theta(r,0) = V_\theta(r,\pi) = 0,$$

derive this *Poisson integral formula* for V:

$$V(r,\theta) = \frac{1}{2\pi}\int_0^\pi F(\alpha)[P(r, \alpha - \theta) + P(r, \alpha + \theta)]\,d\alpha \qquad (r < a, 0 \le \theta \le \pi).$$

10. The transverse displacement $Y(x,t)$ in a stretched string with a steady transverse force distributed along it satisfies the conditions

$$Y_{tt}(x,t) = a^2 Y_{xx}(x,t) + F(x) \qquad \text{when } 0 < x < \pi \text{ and } t > 0,$$

$$Y(0,t) = Y(\pi,t) = Y(x,0) = Y_t(x,0) = 0.$$

Find the solution, and verify it, in the form

$$Y(x,t) = \frac{1}{2a^2}[2G(x) - G_1(x + at) - G_1(x - at)]$$

where G_1 is the odd periodic extension, with period 2π, of the function

$$G(x) = \frac{x}{\pi}\int_0^\pi (\pi - r)F(r)\,dr - \int_0^x (x - r)F(r)\,dr.$$

[1] Cf. sec. 101 of the author's "Complex Variables and Applications," 2d ed., 1960.

11. The ends of a stretched string are fixed at the origin and the point $(\pi,0)$. The string is initially at rest along the horizontal x axis, then it drops under its own weight. Thus the vertical displacements $Y(x,t)$ satisfy the equation $Y_{tt} = a^2 Y_{xx} + g$, where g is the acceleration of gravity. If $Q(x)$ denotes the odd periodic extension of the function $\frac{1}{2}x(\pi - x)$, where $0 < x < \pi$, with period 2π, derive the formula

$$Y(x,t) = \frac{g}{2a^2}[2Q(x) - Q(x - at) - Q(x + at)].$$

12. If a steady transverse load is distributed along a beam, the transverse displacements $Y(x,t)$ satisfy an equation

$$\frac{\partial^2 Y}{\partial t^2} = -a^2\frac{\partial^4 Y}{\partial x^4} + F(x).$$

If the ends $x = 0$ and $x = \pi$ are hinged so that Y and Y_{xx} vanish there, and if the initial displacement and velocity are zero, derive the formula

$$Y(x,t) = \frac{1}{a^2}G(x) - \frac{2}{\pi a^2}\sum_{n=1}^{\infty}\frac{f_s(n)}{n^4}\cos n^2 at \sin nx,$$

where $G^{(4)}(x) = F(x)$ and $G(x) = G''(x) = 0$ at $x = 0$ and at $x = \pi$.

13. Note that for the elementary problem

$$\Theta_t(x,t) = \Theta_{xx}(x,t) \qquad\qquad (0 < x < \pi, t > 0),$$

$$\Theta(x,0) = 1, \qquad \Theta(0,t) = \Theta(\pi,t) = 0,$$

the temperature function can be written

$$\Theta(x,t) = \frac{2}{\pi}\sum_{n=1}^{\infty}\theta_s(n,t)\sin nx,$$

where

$$\theta_s(n,t) = S_n\{1\}\exp(-n^2 t).$$

Use Fourier transforms to show that the problem

$$U_t(x,t) = f(t)U_{xx}(x,t) + g(t) \qquad\qquad (0 < x < \pi, t > 0),$$

$$U(x,0) = U(0,t) = U(\pi,t) = 0,$$

a problem not adapted to the Laplace transformation, can be solved formally in terms of Θ in the form

$$U(x,t) = \int_0^t g(\tau)\Theta[x, F(\tau,t)]\, d\tau,$$

where

$$F(\tau,t) = \int_\tau^t f(r)\, dr.$$

14. Obtain the solution of the diffusion problem

$$U_t(x,t) = U_{xx}(x,t) - h(t)U(x,t) + A \qquad\qquad (0 < x < \pi, t > 0),$$

$$U(0,t) = U(\pi,t) = U(x,0) = 0,$$

in the form of a series. Also derive the form

$$U(x,t) = \frac{A}{H(t)} \int_0^t H(\tau)\Theta(x, t - \tau)\, d\tau,$$

where $\Theta(x,t)$ is the function described in Prob. 13, and

$$H(t) = \exp\left[\int_0^t h(r)\, dr\right].$$

15. Derive the formula for the temperatures $U(x,y,t)$ when

$$U_t = U_{xx} + U_{yy} + A \qquad (x > 0, 0 < y < \pi, t > 0),$$

$$U(x,y,0) = U(0,y,t) = U(x,0,t) = U(x,\pi,t) = 0,$$

and $U(x,y,t)$ is bounded, in the form

$$U(x,y,t) = A \int_0^t \Theta(y,\tau)\, \text{erf}\, \frac{x}{2\sqrt{\tau}}\, d\tau,$$

where Θ is the function described in Prob. 13.

126 A MODIFIED SINE TRANSFORMATION

In Sec. 115 we presented the modified finite Fourier sine transformation

(1) $$S_m\{F(x)\} = \int_0^\pi F(x) \sin mx\, dx = f_s(m)$$

where $m = n - \frac{1}{2}$ and $n = 1, 2, \ldots$, as an example of Sturm-Liouville transformations. It resolves $F''(x)$ on the interval $(0,\pi)$ in terms of $f_s(m)$ and the boundary values $F(0)$ and $F'(\pi)$, when F is of class C'':

(2) $$S_m\{F''(x)\} = -m^2 S_m\{F(x)\} + mF(0) - (-1)^n F'(\pi).$$

We introduced the odd *antiperiodic* extension F_3 of F with period 2π; that is,

(3) $$F_3(x) = F(x) \qquad \text{when } 0 < x < \pi,$$

$$F_3(-x) = -F_3(x), \qquad F_3(x + 2\pi) = -F_3(x) \qquad \text{for all } x.$$

In terms of F_3 the properties of S_m represented by Eqs. (10), (13), and (14), Sec. 115, can be written

(4) $$\frac{2}{m} f_s(m) \sin mc = S_m\left\{\int_{x-c}^{x+c} F_3(r)\, dr\right\},$$

(5) $$\frac{2}{m} f_s(m)h_s(m) = S_m\left\{\int_0^\pi H(y) \int_{x-y}^{x+y} F_3(r)\, dr\, dy\right\}.$$

Table 2 summarizes properties of S_m and lists transforms of some functions.

Table 2   Modified sine transforms[1] $f_s(n - \tfrac{1}{2})$

$f_s(m)$ $(m = n - \tfrac{1}{2}, n = 1, 2, \ldots)$	$F(x)$ $(0 < x < \pi)$
1 $\displaystyle\int_0^\pi F(x) \sin mx \, dx = S_m\{F\}$	$F(x)$
2 $f_s(m)$ $(m = n - \tfrac{1}{2})$	$\displaystyle\frac{2}{\pi} \sum_{n=1}^\infty f_s(m) \sin mx$
3 $-m^2 f_s(m) + mF(0) - (-1)^n F'(\pi)$	$F''(x)$
4 $\dfrac{1}{m} f_s(m)(-1)^{n+1}$	$\displaystyle\int_{\pi-x}^\pi F(r)\, dr$
5 $\dfrac{2}{m} f_s(m) \sin mc$	$\displaystyle\int_{x-c}^{x+c} F_3(r)\, dr$
6 $\dfrac{1}{m^2} f_s(m)$	$\displaystyle\int_0^x rF(r)\, dr + x \int_x^\pi F(r)\, dr$
7 $\dfrac{2}{m} f_s(m) h_s(m)$	$\displaystyle\int_0^\pi H(y) \int_{x-y}^{x+y} F_3(r)\, dr\, dy$
8 $\dfrac{1}{m}$	1
9 $\dfrac{\cos mc}{m}$ $(0 < c < \pi)$	$\begin{cases} 0 & \text{if } 0 < x < c \\ 1 & \text{if } c < x < \pi \end{cases}$
10 $\dfrac{1}{m^2}(-1)^{n+1}$	x
11 $\dfrac{\sin mc}{m^2}$ $(0 < c < \pi)$	$\begin{cases} x & \text{if } 0 \leq x \leq c \\ c & \text{if } c \leq x \leq \pi \end{cases}$
12 $\dfrac{2}{m^3}$	$2\pi x - x^2$
13 $\dfrac{6}{m^4}(-1)^n$	$x^3 - 3\pi^2 x$
14 $\dfrac{c}{m^2 + c^2}(-1)^{n+1}$	$\dfrac{\sinh cx}{\cosh c\pi}$
15 $\dfrac{m - ce^{c\pi}(-1)^n}{m^2 + c^2}$	e^{cx}
16 $\dfrac{k(-1)^{n+1}}{m^2 - k^2}$ $(k \neq \pm\tfrac{1}{2}, \pm\tfrac{3}{2}, \ldots)$	$\dfrac{\sin kx}{\cos k\pi}$
17 e^{-my} $(y > 0)$	$\dfrac{1}{\pi} \dfrac{(\sin x/2)(\cosh y/2)}{\cosh^2 y/2 - \cos^2 x/2}$
18 $\dfrac{1}{m} e^{-my}$ $(y > 0)$	$\dfrac{2}{\pi} \arctan \dfrac{\sin x/2}{\sinh y/2}$

[1] For details see Secs. 115 and 126 and the problems following Sec. 116.

Note that the modified cosine transformation C_m that corresponds to S_m is not essentially different from S_m because

$$(6) \qquad C_m\{F(x)\} = \int_0^\pi F(x) \cos mx \, dx = (-1)^{n+1} S_m\{F(\pi - x)\}.$$

Thus C_m transforms $F''(x)$ in terms of the boundary values $F'(0)$ and $F(\pi)$. That variation of boundary values is taken care of by S_m itself when we replace the coordinate x by $\pi - x$.

127   GENERALIZED COSINE TRANSFORMS

The transformation

$$(1) \qquad C_h\{F(x)\} = \int_0^1 F(x) \cos q_n x \, dx = f_c(q_n).$$

where q_n $(n = 1, 2, \ldots)$ are the positive roots of the equation

$$(2) \qquad \tan q_n = \frac{h}{q_n} \qquad \text{and} \qquad h > 0,$$

is the special case (Prob. 4, Sec. 116) of the Sturm-Liouville transformation on the interval $(0,1)$ that has the property

$$(3) \qquad C_h\{F''(x)\} = -q_n^2 f_c(q_n) - F'(0) + [hF(1) + F'(1)] \cos q_n.$$

The boundary value $hF(1) + F'(1)$ arises in applications involving elastic supports or surface heat transfer.

Since $q_n \sin q_n = h \cos q_n$, we find that

$$\|\cos q_n x\|^2 = \int_0^1 \cos^2 q_n x \, dx = \frac{h + \sin^2 q_n}{2h}.$$

Thus the inversion formula becomes

$$(4) \qquad F(x) = 2h \sum_{n=1}^\infty f_c(q_n) \frac{\cos q_n x}{h + \sin^2 q_n} \qquad (0 < x < 1).$$

From Eq. (2) we can see graphically that $q_n - n\pi \to 0$ as $n \to \infty$.

Let F_2 denote the even periodic extension of F with period 2:

$$(5) \quad F_2(x) = F(|x|) \quad \text{if } -1 < x < 1, \qquad F_2(x + 2) = F_2(x) \quad \text{for all } x.$$

When $0 \le y \le 1$, the formulas

$$2f_c(q_n) \cos q_n y = C_h\{F_2(x + y) + F_2(x - y)\}$$

$$- 2 \sin q_n \int_0^y F(x + 1 - y) \sin q_n x \, dx,$$

$$(6) \quad \sin q_n \int_0^y F(x) \sin q_n x \, dx = h \int_{1-y}^1 e^{-hx} \int_{1-y}^x e^{hr} F(1 - r) \, dr \cos q_n x \, dx$$

can be derived or verified by making several elementary manipulations. From them we can find this *kernel-product property*:

(7) $\qquad 2f_c(q_n)\cos q_n y = C_h\{F_2(x + y) + F_2(x - y) - 2he^{hz}R(z)\}$

where $z = 2 - x - y$ and

$$R(z) = \int_z^1 e^{-hr}F(r)\,dr \quad \text{if } z < 1, \qquad R(z) = 0 \quad \text{if } z > 1.$$

The corresponding *convolution property* for C_h is

(8) $\qquad\qquad\qquad 2f_c(q_n)g_c(q_n) = C_h\{X_{FG}(x)\}$

in terms of this convolution of F and G:

(9) $\qquad X_{FG}(x) = \int_0^1 G(y)[F_2(x + y) + F_2(x - y)]\,dy$

$$- 2hG(1 - x) * F(1 - x) * e^{-hx}.$$

Here the asterisk denotes the convolution integral used with Laplace transforms (Sec. 16); thus

(10) $\quad G(1 - x) * F(1 - x) * e^{-hx} = e^{-hx}\int_0^x e^{ht}G(1 - t)\int_0^{x-t} e^{hr}F(1 - r)\,dr\,dt.$

Unfortunately, our convolution is not a simple one.[1]

When $y = 1$, formula (7) becomes

(11) $\qquad f_c(q_n)\cos q_n = C_h\left\{F(1 - x) - he^{-hx}\int_0^x e^{hr}F(1 - r)\,dr\right\}.$

From property (3) we find in the usual way that

(12) $\qquad \dfrac{f_c(q_n)}{q_n^2} = C_h\left\{\int_0^1 \left(1 + \dfrac{1}{h} - r\right)F(r)\,dr + \int_0^x (r - x)F(r)\,dr\right\}.$

These transforms can be found in various ways:

(13) $\qquad C_h\{1\} = \dfrac{\sin q_n}{q_n} = h\dfrac{\cos q_n}{q_n^2}; \qquad C_h\left\{1 + \dfrac{1}{h} - x\right\} = \dfrac{1}{q_n^2};$

(14) $\qquad C_h\{e^{-hx}\} = \dfrac{h}{q_n^2 + h^2}; \qquad C_h\{\cos \pi x\} = \dfrac{h\cos q_n}{q_n^2 - \pi^2}.$

---

[1] Another derivation of the property can be found in the author's paper, Generalized Finite Fourier Cosine Transforms, *Mich. Math. Jour.*, vol. 3, pp. 85–94, 1955–56.

A RELATED TRANSFORMATION

Since the above kernel $\cos q_n x$ is an even function of x, the transformation

$$(15) \qquad \bar{C}_h\{F(x)\} = \int_{-1}^{1} F(x) \cos q_n x \, dx = \bar{f}_c(q_n)$$

of functions defined on the interval $(-1,1)$, where q_n $(n = 1, 2, \ldots)$ are again the positive roots of Eq. (2), is expressed in terms of C_h by the equation

$$(16) \qquad \bar{C}_h\{F(x)\} = C_h\{F(x)\} + C_h\{F(-x)\} = C_h\{F(x) + F(-x)\}.$$

It follows from property (3) that, when F is of class C'' on the interval $-1 \leq x \leq 1$, $\bar{C}_h$ resolves $F''(x)$ in terms of the boundary values $hF(-1) - F'(-1)$ and $hF(1) + F'(1)$:

$$(17) \quad \bar{C}_h\{F''(x)\} = -q_n^2 \bar{f}_c(q_n) + [hF(-1) - F'(-1) + hF(1) + F'(1)] \cos q_n.$$

Since $h > 0$, the kernel of $\bar{C}_h$ is the set of all eigenfunctions of the problem

$$\Phi''(x) = \lambda \Phi(x), \qquad h\Phi(-1) - \Phi'(-1) = h\Phi(1) + \Phi'(1) = 0,$$

corresponding to the eigenvalues $\lambda = -q_n^2$. The inversion formula for $\bar{C}_h$ is

$$(18) \qquad F(x) = h \sum_{n=1}^{\infty} \bar{f}_c(q_n) \frac{\cos q_n x}{h + \sin^2 q_n} \qquad (-1 < x < 1).$$

128 A GENERALIZED SINE TRANSFORM[1]

The transformation

$$(1) \qquad S_k\{F(x)\} = \int_{0}^{1} F(x) \sin p_n x \, dx = f_s(p_n),$$

where p_n $(n = 1, 2, \ldots)$ are the positive roots of the equation

$$(2) \qquad \tan p_n = -\frac{p_n}{k} \qquad \text{and} \qquad k > 0,$$

was used in Sec. 117 to solve a problem involving surface heat transfer. It has the property

$$(3) \qquad S_k\{F''(x)\} = -p_n^2 f_s(p_n) + F(0)p_n + [kF(1) + F'(1)] \sin p_n,$$

and the inversion formula

$$(4) \qquad F(x) = 2k \sum_{n=1}^{\infty} f_s(p_n) \frac{\sin p_n x}{k + \cos^2 p_n} \qquad (0 < x < 1).$$

---

[1] The transformations in Secs. 126 to 128 are among the generalized finite Fourier transformations treated by Ida Roettinger (Kaplan) in *Quart. Appl. Math.*, vol. 5, pp. 298–319, 1947, and in *Jour. Math. Physics*, vol. 27, pp. 232–239, 1948.

The transformation has the further properties

(5) $$S_k^{-1}\left\{\frac{f_s(p_n)}{p_n^2}\right\} = \left(1 - \frac{kx}{1+k}\right)\int_0^1 rF(r)\,dr + \int_x^1 (x-r)F(r)\,dr,$$

(6) $$\frac{1}{p_n}S_k\{F(x)\} = \int_0^1 \cos p_n x \int_x^1 F(r)\,dr\,dx,$$

(7) $$S_k\left\{F(1-x) - ke^{-kx}\int_0^x e^{kr}F(1-r)\,dr\right\} = \sin p_n \int_0^1 F(x)\cos p_n x\,dx.$$

Let F_1 be the odd periodic extension of F with period 2:

(8)
$$F_1(x) = F(x) \qquad \text{if } 0 < x < 1,$$
$$F_1(x) = -F_1(-x) = F_1(x+2) \qquad \text{for all } x.$$

When $0 \leqq y \leqq 1$, we can obtain the integration formula

(9) $$\frac{2}{p_n}f_s(p_n)\sin p_n y = S_k\left\{\int_{x-y}^{x+y} F_1(r)\,dr\right\}$$

$$+ 2\sin p_n \int_0^y \cos p_n x \int_{x+1-y}^1 F(r)\,dr\,dx,$$

from which we find the *kernel-product property*

(10) $$\frac{2}{p_n}f_s(p_n)\sin p_n y = S_k\left\{\int_{x-y}^{x+y} F_1(r)\,dr + 2e^{kz}R(z)\right\}$$

where $z = 2 - x - y$ and

$$R(z) = \int_z^1 e^{-kr}F(r)\,dr \quad \text{if } z < 1, \qquad R(z) = 0 \quad \text{if } z > 1.$$

The corresponding *convolution property* is

(11) $$\frac{2}{p_n}f_s(p_n)g_s(p_n) = S_k\{X_{FG}(x)\}$$

when $g_s(p_n) = S_k\{G(x)\}$ and

(12) $$X_{FG}(x) = \int_0^1 G(y)\int_{x-y}^{x+y} F_1(r)\,dr\,dy + 2G(1-x) * F(1-x) * e^{-kx},$$

where the final term, the iterated convolution used with Laplace transforms, is given by formula (10), Sec. 127.

We note some particular transforms.

(13) $$S_k\{1\} = \frac{1 - \cos p_n}{p_n}; \qquad S_k\{x\} = \frac{1 + k}{p_n^2} \sin p_n;$$

(14) $$S_k\{e^{-kx}\} = \frac{p_n}{k^2 + p_n^2}; \qquad S_k\{\sin \pi x\} = \frac{\pi \sin p_n}{\pi^2 - p_n^2}.$$

PROBLEMS

1. Use the transformations S_m (Sec. 126) to solve for $Y(x)$ when

$$\frac{d^4 Y}{dx^4} - Y - 2x, \qquad Y(0) = Y''(0) = Y'(\pi) = Y'''(\pi) = 0.$$

Ans. $Y(x) = \operatorname{sech} \pi \sinh x - 2x - \sin x.$

2. Let the steady-state temperatures $V(x,y)$ in a semi-infinite wall $0 \le x \le \pi, y \ge 0$, satisfy the conditions

$$V_{xx}(x,y) + V_{yy}(x,y) = 0, \qquad |V(x,y)| < M \qquad (0 < x < \pi, y > 0),$$

$$V(0,y) = 0, \qquad V_x(\pi,y) = 1, \qquad V(x,+0) = 0,$$

where M is some constant. Find the solution of that boundary value problem in the forms

$$V(x,y) = x + \frac{2}{\pi} \sum_{n=1}^{\infty} \frac{(-1)^n}{m^2} e^{-my} \sin mx \qquad (m = n - \tfrac{1}{2})$$

$$= x - \frac{2}{\pi} \int_{\pi - x}^{x} \arctan\left(\frac{\sin r/2}{\sinh y/2}\right) dr.$$

3. A harmonic function $V(r,\theta)$ in the semicircular region $r < 1, 0 < \theta < \pi$, satisfies these conditions:

$$r^2 V_{rr}(r,\theta) + r V_r(r,\theta) + V_{\theta\theta}(r,\theta) = 0 \qquad (r < 1, 0 < \theta < \pi),$$

$$V(r,0) = 1, \qquad V_\theta(r,\pi) = 0, \qquad V(1,\theta) = 0;$$

also, V is bounded over the region. Derive the formula

$$V(r,\theta) = 1 - \frac{2}{\pi} \arctan \frac{2\sqrt{r} \sin \theta/2}{1 - r}.$$

4. If $V(r,\theta)$ satisfies the conditions stated in Prob. 3 in the part of the upper half plane exterior to the semicircle, the unbounded region $r > 1, 0 < \theta < \pi$, show that

$$V(r,\theta) = 1 - \frac{2}{\pi} \arctan \frac{2\sqrt{r} \sin (\theta/2)}{r - 1} \qquad (r > 1, 0 < \theta < \pi).$$

5. For the transformation C_h in Sec. 127 show that, if $0 \leq c \leq 1$,

(a) $C_h^{-1}\left\{\dfrac{\cos q_n c}{q_n^{\,2}}\right\} = \begin{cases} 1 + \dfrac{1}{h} - c & \text{if } 0 \leq x \leq c, \\[2mm] 1 + \dfrac{1}{h} - x & \text{if } c \leq x \leq 1; \end{cases}$

(b) $C_h^{-1}\left\{\dfrac{\sin q_n c}{q_n}\right\} = \begin{cases} 1 & \text{if } 0 < x < c, \\ 0 & \text{if } c < x < 1. \end{cases}$

6. The longitudinal displacements $Y(x,t)$ in an elastic bar with an elastic support at one end and constant pressure applied at the other end satisfy the conditions

$$Y_{tt}(x,t) = a^2 Y_{xx}(x,t) \qquad (0 < x < 1,\, t > 0),$$

$$Y_x(0,t) = -B, \qquad Y_x(1,t) = -hY(1,t), \qquad Y(x,0) = Y_t(x,0) = 0,$$

where $h > 0$. Derive the formula

$$Y(x,t) = B\left[1 + \frac{1}{h} - x - 2h \sum_{n=1}^{\infty} \frac{\cos(q_n at)\cos(q_n x)}{q_n^{\,2}(h + \sin^2 q_n)}\right],$$

where $q_1, q_2, \ldots$ are the positive roots of the equation $\tan q_n = h/q_n$. Also, use transform (a) in Prob. 5 to show that, when $at \leq 1$,

$$Y(x,t) = B(at - x) \quad \text{if } x \leq at, \qquad Y(x,t) = 0 \quad \text{if } x \geq at.$$

7. Use the transformation C_h in Sec. 127 to solve the following problem for steady-state temperatures in a semi-infinite wall with heat transfer at one face into a medium at uniform temperature A:

$$V_{xx}(x,y) + V_{yy}(x,y) = 0 \qquad (0 < x < 1,\, y > 0),$$

$$V_x(0,y) = 0, \qquad -V_x(1,y) = h[V(1,y) - A] \qquad (y > 0),$$

$$V(x,0) = 0, \qquad V_y(x,\infty) = 0 \qquad (0 < x < 1),$$

where $h > 0$. Obtain the solution in the form

$$V(x,y) = A\left[1 - 2h^2 \sum_{n=1}^{\infty} \frac{\cos q_n \cos q_n x}{h + \sin^2 q_n} \frac{\exp(-q_n y)}{q_n^{\,2}}\right],$$

where $\tan q_n = h/q_n$ and $q_n > 0$.

8. Let the temperatures $U(x,t)$ in a slab with surface heat transfer at one face, and with heat generated within the slab, satisfy the conditions

$$U_t(x,t) = U_{xx}(x,t) + F(x) \qquad (0 < x < 1,\, t > 0),$$

$$U(0,t) = 0, \qquad -U_x(1,t) = bU(1,t), \qquad U(x,0) = 0,$$

where $b > 0$. Derive the formula

$$U(x,t) = \left(1 - \frac{bx}{1+b}\right)\int_0^1 rF(r)\,dr + \int_x^1 (x - r)F(r)\,dr$$

$$- 2b \sum_{n=1}^{\infty} \frac{f_s(p_n)}{p_n^{\,2}} \frac{\sin p_n x}{b + \cos^2 p_n}\exp(-p_n^{\,2}t),$$

where $p_1, p_2, \ldots$ are the positive roots of the equation $\tan p_n = -p_n/b$ and

$$f_s(p_n) = \int_0^1 F(x) \sin p_n x \, dx.$$

129   FINITE EXPONENTIAL TRANSFORMS $E_n\{F\}$

A linear integral transformation of functions $F(x)$ on an interval $(-\pi,\pi)$ that resolves $F'(x)$ in terms of the *cyclic boundary value* $F(\pi) - F(-\pi)$ is often useful in solving problems in which x is an angle, such as the polar coordinate θ, or in finding periodic solutions of problems in differential equations. For a transformation

$$T\{F(x)\} = \int_{-\pi}^{\pi} F(x)\Phi(x\,\lambda) \, dx = f(\lambda)$$

we find that, when F and Φ are of class C' $(-\pi \leqq x \leqq \pi)$,

$$T\{F'(x)\} = F\Phi]_{-\pi}^{\pi} - \int_{-\pi}^{\pi} F\Phi' \, dx = \lambda f(\lambda) + \Phi(\pi,\lambda)[F(\pi) - F(-\pi)]$$

if Φ and λ satisfy the homogeneous conditions

(1) $\qquad\qquad \Phi'(x,\lambda) = -\lambda\Phi(x,\lambda), \qquad \Phi(\pi,\lambda) = \Phi(-\pi,\lambda).$

This last equation is a periodic boundary condition on Φ. The nontrivial solutions of problem (1) are $\Phi(\lambda,x) = e^{-inx}$ and $\lambda = in$ where $n = 0, \pm 1, \pm 2, \ldots$. Thus T is the *finite Fourier exponential transformation*

(2) $\qquad\quad E_n\{F(x)\} = \int_{-\pi}^{\pi} F(x)e^{-inx} \, dx = f_e(n) \qquad (n = 0, \pm 1, \pm 2, \ldots),$

with the *basic operational property*

(3) $\qquad\qquad E_n\{F'(x)\} = inE_n\{F(x)\} + (-1)^n[F(\pi) - F(-\pi)].$

When F is of class C^n on the interval $-\pi \leqq x \leqq \pi$, or even if F and F' are continuous there and F'' is sectionally continuous, we find, by replacing F and F' in formula (3), that

(4) $\qquad\quad E_n\{F''(x)\} = -n^2 f_e(n) + in(-1)^n[F(\pi) - F(-\pi)]$

$$+(-1)^n[F'(\pi) - F'(-\pi)].$$

The formula can be extended easily to transforms of derivatives of higher order in terms of $f_e(n)$ and cyclic boundary values of F and its derivatives.

The set of functions e^{-inx} $(n = 0, \pm 1, \pm 2, \ldots)$ is orthogonal in the Hermitian sense on the interval $(-\pi,\pi)$; that is, when $m = 0, \pm 1, \pm 2, \ldots$,

(5) $\qquad\quad \int_{-\pi}^{\pi} \exp(-inx)\overline{\exp(-imx)} \, dx = \int_{-\pi}^{\pi} e^{i(m-n)x} \, dx = 0 \quad$ if $m \neq n$.

That set is the basis for the exponential form of the Fourier series on that interval.[1] In terms of $f_e(n)$ that series representation of $F(x)$ can be written as

$$(6) \qquad F(x) = \frac{1}{2\pi} \lim_{m \to \infty} \sum_{n=-m}^{n=m} f_e(n)e^{inx} \qquad (-\pi < x < \pi),$$

valid if F and F' are sectionally continuous over the interval and F is defined as its mean value from the right and left at each point of discontinuity. Note that the sum in Eq. (4) includes the term corresponding to $n = 0$. The equation is a formula for $E_n^{-1}\{f_e(n)\}$.

In case the infinite series of terms $|f_e(n)|$, where n runs from $-\infty$ to ∞, converges, then

$$(7) \qquad F(x) = E_n^{-1}\{f_e(n)\} = \frac{1}{2\pi} \sum_{n=-\infty}^{\infty} f_e(n)e^{inx} \qquad (-\pi < x < \pi);$$

that is, the sum of the series here exists, and so it is the same as the principal value used in formula (6). In fact, whenever the series of terms $|f_e(n)|$ converges, then series (7) converges uniformly to a continuous sum $F(x)$ whose transform is $f_e(n)$.

In terms of Fourier sine and cosine transforms

$$E_n\{F(x)\} = \int_0^{\pi} [F(-x) + F(x)] \cos nx \, dx$$

$$+ i \int_0^{\pi} [F(-x) - F(x)] \sin nx \, dx$$

where $n = 0, \pm 1, \pm 2, \ldots$. In particular, when F is an *even* sectionally continuous function on the interval $(-\pi, \pi)$, we have a useful formula for obtaining exponential transforms from cosine transforms:

$$(8) \qquad E_n\{F(x)\} = 2C_{|n|}\{F(x)\} \qquad \text{if } F(-x) = F(x) \qquad (n = 0, \pm 1, \pm 2, \ldots).$$

When $m = \pm 1, \pm 2, \ldots$, we find easily that

$$(9) \qquad f_e(n - m) = E_n\{e^{imx}F(x)\} \qquad (n = 0, \pm 1, \pm 2, \ldots).$$

From formula (5) and the linearity of E_n it follows that

$$(10) \qquad E_n\{1\} = 0 \qquad \text{if } n = \pm 1, \pm 2, \ldots, \qquad \text{but} \qquad E_0\{1\} = 2\pi;$$

$$(11) \qquad E_n\{F(x) + A\} = f_e(n) \qquad \text{if } n \neq 0,$$

$$E_0\{F(x) + A) = f_e(0) + 2\pi A.$$

[1] Churchill, R. V., "Fourier Series and Boundary Value Problems," 2d ed., sec. 43, 1963.

From property (3) we find that, when F is sectionally continuous,

(12) $$\frac{1}{in} f_e(n) = E_n\left\{\int_0^x F(r)\,dr - \frac{x}{2\pi} f_e(0) + A\right\} \qquad \text{if } n \neq 0.$$

130   OTHER PROPERTIES OF E_n

Let F_π denote the periodic extension of F with period 2π,

(1) $F_\pi(x) = F(x)$ if $-\pi < x < \pi$, $F_\pi(x + 2\pi) = F_\pi(x)$ for all x.

In case F_π is of class C' on every interval, or if F_π is everywhere continuous and F' is sectionally continuous over the interval $(-\pi,\pi)$, then according to property (3), Sec. 129,

(2) $$E_n\{F'_\pi(x)\} = inE_n\{F(x)\} = inf_e(n) \qquad (n = 0, \pm 1, \pm 2, \ldots).$$

If F_π has a continuous derivative of order m everywhere, then

(3) $$E_n\{F_\pi^{(m)}(x)\} = (in)^m f_e(n) \qquad (m = 1, 2, \ldots).$$

This *kernel-product property* for $E_n\{F\}$ is easily found:

(4) $$e^{-iny} f_e(n) = \int_{-\pi+y}^{\pi+y} F(x - y)e^{-inx}\,dx = E_n\{F_\pi(x - y)\}$$

for all real y. When $y = \pi$, it becomes

(5) $$(-1)^n f_e(n) = E_n\{F_\pi(x - \pi)\} = \int_0^{2\pi} F(x - \pi)e^{-inx}\,dx.$$

Note that $F_\pi(x - \pi) = F(x + \pi)$ if $-\pi < x < 0$, and $F_\pi(x - \pi) = F(x - \pi)$ if $0 < x < \pi$. The corresponding *convolution property* is

(6) $$f_e(n)g_e(n) = E_n\{X_{FG}(x)\}$$

where $g_e(n) = E_n\{G(x)\}$ and

(7) $$X_{FG}(x) = \int_{-\pi}^{\pi} G(y)F_\pi(x - y)\,dy.$$

In terms of G and F directly, that convolution has the form

(8) $$X_{FG}(x) = \int_{-\pi}^{\pi+x} G(y)F(x - y)\,dy + \int_{\pi+x}^{\pi} G(y)F(x - y + 2\pi)\,dy$$

when $-\pi < x < 0$; but when $0 < x < \pi$,

$$X_{FG}(x) = \int_{-\pi}^{-\pi+x} G(y)F(x - y - 2\pi)\,dy + \int_{-\pi+x}^{\pi} G(y)F(x - y)\,dy.$$

In form (7) the convolution has the properties
$$X_{FG}(x) = X_{GF}(x) = X_{FG}(x + 2\pi).\qquad (9)$$

Table 3 lists transforms of several functions, transforms that can be derived by using our properties of E_n including the relation (8), Sec. 129, between E_n and $C_{|n|}$ when F is even. Some of the properties of E_n are included in the table.

Table 3   Finite exponential transforms $E_n\{F\}$

	$f_e(n) \qquad (n = 0, \pm 1, \pm 2, \ldots)$	$F(x) \qquad (-\pi < x < \pi)$
1	$\displaystyle\int_{-\pi}^{\pi} F(x)e^{-inx}\,dx = E_n\{F(x)\}$	$F(x)$
2	$f_e(n) \quad (n = 0, \pm 1, \pm 2, \ldots)$	$\dfrac{1}{2\pi}\lim\limits_{m\to\infty}\sum\limits_{n=-m}^{m} f_e(n)e^{inx}$
3	$inf_e(n) + (-1)^n[F(\pi) - F(-\pi)]$	$F'(x)$
4	$f_e(n - m) \quad (m = \pm 1, \pm 2, \ldots)$	$e^{imx}F(x)$
5	$e^{-iny}f_e(n)$	$F_\pi(x - y)$ (Sec. 130)
6	$f_e(n)g_e(n)$	$\displaystyle\int_{-\pi}^{\pi} G(y)F_\pi(x - y)\,dy$
7	$0 \quad\text{if}\quad n \neq 0, f_e(0) = 2\pi$	1
8	$0 \quad\text{if}\quad n \neq \pm 1, f_e(\pm 1) = \pi$	$\cos x$
9	$0 \quad\text{if}\quad n \neq \pm 1, f_e(1) = -\pi i, f_e(-1) = \pi i$	$\sin x$
10	$2\pi i\dfrac{(-1)^n}{n} \quad\text{if}\quad n \neq 0, f_e(0) = 0$	x
11	$\dfrac{2\pi i}{n} \quad\text{if}\quad n \neq 0, f_e(0) = 0$	$\begin{cases} x + \pi & \text{if}\ -\pi < x < 0 \\ x - \pi & \text{if}\ 0 < x < \pi \end{cases}$
12	$\dfrac{(-1)^n}{c - in} \quad (c^2 \neq 0, -1, -2, \ldots)$	$\dfrac{e^{cx}}{2\sinh c\pi}$
13	$\dfrac{2\sinh c\pi}{c - in} \quad (c^2 \neq 0, -1, -2, \ldots)$	$\begin{cases} e^{c\pi}e^{cx} & \text{if}\ -\pi < x < 0 \\ e^{-c\pi}e^{cx} & \text{if}\ 0 < x < \pi \end{cases}$
14	$4\pi\dfrac{(-1)^n}{n^2} \quad\text{if}\quad n \neq 0, f_e(0) = \tfrac{2}{3}\pi^3$	x^2
15	$\dfrac{(-1)^n}{n^2 + c^2} \quad (c^2 \neq 0, -1, -2, \ldots)$	$\dfrac{\cosh cx}{2c\sinh c\pi}$
16	$\dfrac{(-1)^{n+1}}{n^2 - c^2} \quad (c \neq 0, \pm 1, \pm 2, \ldots)$	$\dfrac{\cos cx}{2c\sin c\pi}$
17	$\dfrac{(-1)^{n+1}}{n^2 - 1} \quad\text{if}\quad n \neq \pm 1, f_e(\pm 1) = 0$	$\dfrac{1}{4\pi}(2x\sin x + \cos x)$
18	$b^{|n|} \quad (-1 < b < 1, b \neq 0)$	$\dfrac{1}{2\pi}\dfrac{1 - b^2}{1 + b^2 - 2b\cos x}$

Applications of E_n are illustrated in the set of problems that follows. Finally, we note that by writing $x = (2\pi/c)t - \pi$,

$$(10) \qquad E_n\{F(x)\} = (-1)^n \int_0^c e^{-in\omega t} G(t)\, dt$$

where $\omega = 2\pi/c$ and $G(t) = \omega F(\omega t - \pi)$. The integral in Eq. (10) is the exponential transformation of $G(t)$ on the interval $0 < t < c$. Intervals other than $(-\pi,\pi)$ have been used as a basis for finite exponential transformations.[1]

PROBLEMS

1. If $F(x)$ is sectionally continuous over the interval $(-\pi,\pi)$ and $f_e(0) = 0$, prove that, when $n = \pm 1, \pm 2, \ldots$,

$$\frac{f_e(n)}{n^2} = E_n\left\{\int_{-\pi}^x (r - x)F(r)\, dr - \frac{x}{2\pi}\int_{-\pi}^\pi rF(r)\, dr\right\}.$$

2. Use E_n with respect to t to find the periodic solution $Y(t)$, with period 2π, of the equation

$$Y'(t) - cY(t) + P(t) = 0,$$

where $P(t + 2\pi) = P(t)$ for all t, and $c^2 \neq 0, -1, -2, \ldots$.

$$Ans. \quad Y(t)(1 - e^{-2c\pi}) = e^{ct}\int_t^{t+2\pi} P(r)e^{-cr}\, dr.$$

3. If the external force acting on the mass m shown in Fig. 15 (Sec. 29) is periodic, the unit of time can be chosen so that the displacement $X(t)$ of m satisfies an equation

$$mX''(t) = -kX(t) + P(t) \qquad where \qquad P(t + 2\pi) = P(t) \text{ and } k > 0.$$

Write $\omega_0 = \sqrt{k/m}$. When $\omega_0 \neq 1, 2, \ldots$, and P is sectionally continuous over its period, obtain these formulas for the periodic displacements with period 2π:

$$X(t) = \frac{1}{2m\pi}\sum_{n=-\infty}^{\infty}\frac{p_e(n)}{\omega_0{}^2 - n^2}e^{int} = \frac{\int_{-\pi}^\pi P(t + \pi - \tau)\cos\omega_0\tau\, d\tau}{2m\omega_0\sin\omega_0\pi}$$

$$= (2m\omega_0\sin\omega_0\pi)^{-1}\int_t^{t+2\pi} P(r)\cos\omega_0(t + \pi - r)\, dr.$$

4. Let $P(x)$ be a prescribed periodic function with period 2π such that P'' is everywhere continuous. Use E_n to find the periodic function $Y(x)$ of period 2π that satisfies the integral equation

$$\int_{-\pi}^\pi Y(x - r)\cosh cr\, dr = P(x) \qquad\qquad where\ c > 0.$$

$$Ans. \quad 2cY(x)\sinh c\pi = c^2 P(x - \pi) - P''(x - \pi).$$

[1] Kaplan, W., "Operational Methods for Linear Systems," chap. 4, 1962; Doetsch, G., Integration von Differentialgleichungen vermittels der endlichen Fourier-Transformation, *Mathematische Annalen*, vol. 112, pp. 52–68, 1935.

5. If $P(t)$ and $Y(x,t)$ are both periodic in t with period 2π, derive the solution of the problem

$$Y_{tt}(x,t) = a^2 Y_{xx}(x,t), \qquad Y(0,t) = P(t), \qquad Y_x(\pi a,t) = 0$$

in the form

$$Y(x,t) = \frac{1}{2}\left[P\left(t + \frac{x}{a}\right) + P\left(t - \frac{x}{a}\right)\right].$$

6. In Prob. 6, Sec. 85, let the unit of time be chosen so that the temperature $V(x,t)$ in the semi-infinite solid $x \geq 0$ satisfies the conditions

$$V_t(x,t) = kV_{xx}(x,t), \qquad V(0,t) = A \sin t, \qquad V(\infty,t) = 0;$$

also $V(x, t + 2\pi) = V(x,t)$. Show that the exponential transform $v_e(x,n)$ vanishes except when $n = \pm 1$, and that

$$V(x,t) = A \exp\left(-\frac{x}{\sqrt{2k}}\right) \sin\left(t - \frac{x}{\sqrt{2k}}\right).$$

7. Heat is generated at a steady rate uniformly along the radius $\theta = \pi$ of a thin, circular plate $r < 1$, $-\pi < \theta \leq \pi$ (Fig. 107) whose faces are insulated. If the edge $r = 1$ is kept at temperature zero, the steady temperatures $U(r,\theta)$ in the plate are such that

$$r^2 U_{rr}(r,\theta) + rU_r(r,\theta) + U_{\theta\theta}(r,\theta) = 0 \qquad (r < 1, -\pi < \theta < \pi),$$

$$\frac{1}{r}[U_\theta(r,\pi) - U_\theta(r, -\pi + 0)] = A, \qquad U(r,\pi) = U(r, -\pi + 0), \qquad U(1,\theta) = 0.$$

Also, U is bounded over the region $r < 1$, $-\pi < \theta < \pi$, and U and its partial derivatives of the first order are continuous there. Show that, with respect to θ,

$$E_n\{U(r,\theta)\} = A\frac{(-1)^n}{n^2 - 1}(r - r^{|n|}) \qquad \text{if } n \neq \pm 1,$$

$$= \frac{A}{2}r \log r \qquad\qquad\quad \text{if } n = \pm 1.$$

Thus derive the formula

$$U(r,\theta) = \frac{A}{2\pi}\left[1 - r\theta \sin\theta - \left(\frac{r}{2} - r \log r\right)\cos\theta - 2\sum_{n=2}^{\infty} \frac{(-1)^n}{n^2 - 1}r^n \cos n\theta\right].$$

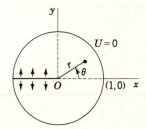

Fig. 107

8. If V is a bounded harmonic function interior to the circle $r = a$ that assumes values $F(\theta)$ on the circle, then

$$r^2 V_{rr}(r,\theta) + r V_r(r,\theta) + V_{\theta\theta}(r,\theta) = 0 \qquad (r < a, \ -\pi \leqq \theta \leqq \pi),$$

$$V(a - 0, \theta) = F(\theta) \qquad \text{when } -\pi < \theta < \pi;$$

also, V and its partial derivatives of first and second order are continuous everywhere inside the circle and thus periodic in θ with period 2π. Show that, with respect to θ,

$$E_n\{V(r,\theta)\} = \left(\frac{r}{a}\right)^{|n|} E_n\{F(\theta)\}$$

when F is sectionally continuous. Thus derive *Poisson's integral formula* for the harmonic function in the disk (cf. Probs. 8 and 9, Sec. 125),

$$V(r,\theta) = \frac{1}{2\pi} \int_{-\pi}^{\pi} \frac{(a^2 - r^2)F(\alpha)}{a^2 + r^2 - 2ar \cos(\alpha - \theta)} \, d\alpha \qquad (r < a, \ -\pi \leqq \theta \leqq \pi).$$

12

Exponential Fourier Transforms

The exponential Fourier transformation is a prominent singular linear integral transformation of functions $F(x)$ on the entire x axis. Conditions of validity on its operational properties are severe. For most properties we use the requirement that F be absolutely integrable from $x = -\infty$ to $x = \infty$ in order to ensure convergence or uniform convergence of improper integrals that arise. Proofs and statements of some properties of the transformation can be simplified by using Lebesgue integrals instead of our Riemann integrals and requiring F to belong to certain classes of Lebesgue integrable functions;[1] but this calls for familiarity with the theory of Lebesgue integrals.

131   THE TRANSFORMATION $E_\alpha\{F\}$

Let F denote a real- or complex-valued function of a real variable x such that $F(x)$ is defined over the entire x axis. A linear integral transformation of such

[1] See, for instance, books by Wiener and Titchmarsh on the theory of Fourier integrals, listed in the Bibliography.

functions is

$$T\{F\} = \int_{-\infty}^{\infty} F(x)K(x,\lambda)\,dx = f(\lambda).$$

Let us determine the kernel K and the domain of the parameter λ so that T transforms derivatives F' into products $\lambda f(\lambda)$.

For the purpose of designing the transformation so that it has that property, we shall assume that, for each fixed value of λ, the kernel K has a continuous derivative K' with respect to x and that K and K' are bounded for all x. We assume that the functions F are everywhere continuous with sectionally continuous derivatives $F'(x)$ on each bounded interval (a,b), that $F(\pm\infty) = 0$, and that F is absolutely integrable from $-\infty$ to ∞. Then

$$\int_a^b F'(x)K(x,\lambda)\,dx = F(x)K(x,\lambda)]_a^b - \int_a^b F(x)K'(x,\lambda)\,dx,$$

and the right-hand member here has a limit as $a \to -\infty$ and $b \to \infty$, so the integral on the left has the same limit, and it follows that

$$T\{F'\} = \int_{-\infty}^{\infty} F'(x)K(x,\lambda)\,dx = -\int_{-\infty}^{\infty} F(x)K'(x,\lambda)\,dx.$$

Thus $T\{F'\} = \lambda f(\lambda)$ if K is a solution other than zero of the homogeneous equation

$$K'(x,\lambda) = -\lambda K(x,\lambda)$$

and if K satisfies our requirements of continuity and boundedness. We may select the solution $K(x,\lambda) = e^{-\lambda x}$. This function is bounded for all real x only if $\lambda = i\alpha$ where α is any real number; then K' is also bounded with respect to x. Thus T becomes the *exponential Fourier transformation*

$$(1) \qquad E_\alpha\{F(x)\} = \int_{-\infty}^{\infty} F(x)e^{-i\alpha x}\,dx = f_e(\alpha) \qquad (-\infty < \alpha < \infty),$$

with the operational property

$$(2) \qquad E_\alpha\{F'(x)\} = i\alpha f_e(\alpha).$$

If F' is everywhere continuous and F'' is sectionally continuous, and if $F(\pm\infty) = F'(\pm\infty) = 0$ and $\int_{-\infty}^{\infty}|F(x)|\,dx$ exists, then

$$(3) \qquad \int_a^b F''(x)e^{-i\alpha x}\,dx = [e^{-i\alpha x}(F' + i\alpha F)]_a^b - \alpha^2 \int_a^b F(x)e^{-i\alpha x}\,dx$$

and, when $a \to -\infty$ and $b \to \infty$, we find that

$$(4) \qquad E_\alpha\{F''(x)\} = -\alpha^2 f_e(\alpha) \qquad (-\infty < \alpha < \infty).$$

Likewise, the operational property (2) can be extended as follows:

Theorem 1 *If $F(x)$ is absolutely integrable from $-\infty$ to ∞, if its derivative $F^{(m-1)}$ is everywhere continuous while $F^{(m)}$ is sectionally continuous on each bounded interval, and if*

$$F(\pm\infty) = F'(\pm\infty) = \cdots = F^{(m-1)}(\pm\infty) = 0,$$

then

(5) $$E_\alpha\{F^{(m)}(x)\} = (i\alpha)^m f_e(\alpha) \qquad (m = 1, 2, \ldots).$$

In case F has a jump discontinuity at a point $x = c$ but otherwise satisfies our conditions for the validity of property (2), we find that

(6) $$E_\alpha\{F'(x)\} = i\alpha f_e(\alpha) - e^{-i\alpha c}[F(c + 0) - F(c - 0)].$$

The literature on exponential Fourier transforms contains minor variations in the choice of the kernel and in notation. The kernels $e^{i\alpha x}$ and $(2\pi)^{-\frac{1}{2}} e^{i\alpha x}$, for instance, are sometimes used.

132 THE INVERSE TRANSFORMATION

In Sec. 68 we stated conditions under which a function is represented by its Fourier integral formula

(1) $$F(x) = \frac{1}{2\pi} \lim_{\beta \to \infty} \int_{-\beta}^{\beta} e^{i\alpha x} \int_{-\infty}^{\infty} F(\tau) e^{-i\alpha\tau} \, d\tau \, d\alpha \qquad (-\infty < x < \infty).$$

Since the inner integral here is $f_e(\alpha)$, the formula represents the inverse of the exponential Fourier transformation:

(2) $$F(x) = \frac{1}{2\pi} \lim_{\beta \to \infty} \int_{-\beta}^{\beta} f_e(\alpha) e^{i\alpha x} \, d\alpha = E_\alpha^{-1}\{f_e(\alpha)\}.$$

Sufficient conditions for its validity can be stated as follows, even if F is a complex-valued function (Probs. 9 and 10, Sec. 133).

Theorem 2 *Let $F(x)$ be absolutely integrable from $-\infty$ to ∞ and let F' be sectionally continuous over each bounded interval. Also let $F(x)$ be defined at each point c of discontinuity as $\frac{1}{2}[F(c + 0) + F(c - 0)]$. Then its transform $f_e(\alpha)$ exists, and the inversion formula (2) is valid for all real x.*

In case $f_e(\alpha)$ is also absolutely integrable from $-\infty$ to ∞, or in case the integral $\int_{-\infty}^{\infty} f_e(\alpha) e^{i\alpha x} \, d\alpha$ exists, formula (2) can be written

(3) $$F(x) = \frac{1}{2\pi} \int_{-\infty}^{\infty} f_e(\alpha) e^{i\alpha x} \, d\alpha = \frac{1}{2\pi} E_{-x}\{f_e(\alpha)\} \qquad (-\infty < x < \infty).$$

By replacing α by $-\alpha$ in the integral here, we can write

$$(4) \qquad\qquad F(x) = \frac{1}{2\pi} E_x\{f_e(-\alpha)\}.$$

Thus under appropriate conditions on F and f_e the inverse of the operator E_α can be written in terms of E_α itself.

That the principal value of the improper integral $E_{-x}\{f_e(\alpha)\}$ in the inversion formula (2) may be needed is illustrated by the following example. Consider the function F such that

$$F(x) = 0 \quad \text{if} \quad x < 0, \qquad F(x) = e^{-x} \quad \text{if} \quad x > 0, \qquad F(0) = \tfrac{1}{2},$$

which satisfies the conditions in Theorem 2. Here $F'(x) = -F(x)$ when $x \neq 0$ and, according to formula (6), Sec. 131,

$$E_\alpha\{F'\} = -f_e(\alpha) = i\alpha f_e(\alpha) - 1;$$

thus f_e is not absolutely integrable from $-\infty$ to ∞ since

$$f_e(\alpha) = \frac{1}{1 + i\alpha} = \frac{1 - i\alpha}{\alpha^2 + 1} \qquad \text{and} \qquad |f_e(\alpha)| = \frac{1}{\sqrt{1 + \alpha^2}}.$$

When $x = 0$, the integral $E_{-x}\{f_e(\alpha)\}$ does not exist because $(\alpha^2 + 1)^{-1}$ is integrable from $-\infty$ to ∞ while $\alpha(\alpha^2 + 1)^{-1}$ is not. But

$$\frac{-i}{2\pi} \int_{-\beta}^{\beta} \frac{\alpha}{\alpha^2 + 1}\, d\alpha = 0 \qquad \text{and} \qquad \lim_{\beta \to \infty} \frac{-i}{2\pi} \int_{-\beta}^{\beta} \frac{\alpha}{\alpha^2 + 1}\, d\alpha = 0;$$

therefore

$$\lim_{\beta \to \infty} \frac{1}{2\pi} \int_{-\beta}^{\beta} f_e(\alpha)\, d\alpha = \frac{1}{2\pi} \int_{-\infty}^{\infty} \frac{d\alpha}{\alpha^2 + 1} = \frac{1}{2} = F(0).$$

No two functions F and G that satisfy the conditions on F in Theorem 2 can have identical transforms. For if F and G are absolutely integrable from $-\infty$ to ∞, then $F - G$ is also, since $|F - G| \leq |F| + |G|$. Thus we see that the function $F - G$ satisfies all those conditions. But $E_\alpha\{F - G\} = 0$ if F and G have identical transforms and, according to the inversion formula (2), $F(x) - G(x) = 0$ for all x. In that class of functions, therefore, there is a one-to-one correspondence between functions and their transforms.

Theorem 2 gives conditions on F that are sufficient for the validity of inversion formula (2). We now establish sufficient conditions on the function $f_e(\alpha)$ instead, conditions that also establish f_e as a transform and as an inverse transform; but one of our conditions is apt to be difficult to apply.

Theorem 3   *Let a function $f(\alpha)$ have a sectionally continuous derivative f' over each bounded interval, and let f be absolutely integrable from $-\infty$*

to ∞. *Then its transform* $E_x\{f\}$ *exists. We write*

(5) $\qquad 2\pi\Phi(x) = E_x\{f(\alpha)\} = \int_{-\infty}^{\infty} f(\tau)e^{-i\tau x}d\tau \qquad (-\infty < x < \infty).$

If in addition $\Phi(-x)$ *has a transform for all real* α, *then*

(6) $\qquad\qquad\qquad f(\alpha) = E_\alpha\{\Phi(-x)\}$

and, for all real x, *the inversion formula*

(7) $\qquad \Phi(-x) = \dfrac{1}{2\pi}\int_{-\infty}^{\infty} f(\alpha)e^{i\alpha x}\,d\alpha = \dfrac{1}{2\pi}E_{-x}\{f(\alpha)\}$

is valid. Also, from Eq. (5) *it follows that*

(8) $\qquad\qquad\qquad E_\alpha\{f(x)\} = 2\pi\Phi(\alpha).$

Under the conditions on f the integral (5) converges uniformly to a function $2\pi\Phi(x)$ that is continuous for all real x. Since the integral $E_\alpha\{\Phi(-x)\}$ exists, by hypothesis, we can write it as $E_{-\alpha}\{\Phi(x)\}$, or in the forms

$$\int_{-\infty}^{\infty} \Phi(x)e^{i\alpha x}\,dx = \frac{1}{2\pi}\int_{-\infty}^{\infty} e^{i\alpha x}\int_{-\infty}^{\infty} f(\tau)e^{-i\tau x}\,d\tau\,dx.$$

But the right-hand member here represents $f(\alpha)$ for all real α, according to the Fourier integral formula (1), since f satisfies the conditions of validity of that formula. Thus

$$f(\alpha) = \int_{-\infty}^{\infty} \Phi(-x)e^{-i\alpha x}\,dx = E_\alpha\{\Phi(-x)\}.$$

Equation (5) shows that $\Phi(-x)$ is represented by the inversion formula (7).

To illustrate Theorem 3, let us find the inverse transform of

$$f(\alpha) = \exp{(-|\alpha|)},$$

a function that is everywhere continuous, absolutely integrable from $-\infty$ to ∞, for which f' is continuous except for a jump at the point $\alpha = 0$. For that function

$$2\pi\Phi(x) = \int_{-\infty}^{0} e^{\tau}e^{-i\tau x}\,d\tau + \int_{0}^{\infty} e^{-\tau}e^{-i\tau x}\,d\tau.$$

The integrals here can be written as simple Laplace transforms ($s = 1$), and we find that

$$\Phi(x) = \frac{1}{\pi}\frac{1}{1 + x^2} = \Phi(-x).$$

Since Φ is absolutely integrable over the x axis, $E_\alpha\{\Phi\}$ exists and

$$(9) \qquad \exp(-|\alpha|) = E_\alpha\left\{\frac{1}{\pi}\frac{1}{1+x^2}\right\}.$$

Also, according to formula (8)

$$(10) \qquad E_\alpha\{\exp(-|x|)\} = \frac{2}{1+\alpha^2}.$$

133   OTHER PROPERTIES OF E_α

We now point out several properties of the exponential Fourier transformation that follow readily from properties of integrals noted in Secs. 14, 15, and 56. It is convenient to assume throughout this section that $F(x)$ *is sectionally continuous over each bounded interval on the x axis.*

Suppose that, for all real α, *the integral*

$$E_\alpha\{F(x)\} = \int_{-\infty}^{\infty} F(x)e^{-i\alpha x}\,dx = f_e(\alpha)$$

exists. Then if c is real and constant, we find that

$$(1) \qquad E_\alpha\{F(x-c)\} = e^{-ic\alpha}f_e(\alpha),$$

$$(2) \qquad E_\alpha\{F(-x)\} = f(-\alpha),$$

$$(3) \qquad E_\alpha\{F(cx)\} = \frac{1}{|c|}f_e\left(\frac{\alpha}{c}\right) \qquad\qquad \text{if } c \neq 0,$$

$$(4) \qquad E_\alpha\{e^{icx}F(x)\} = f_e(\alpha - c).$$

Also the complex conjugates of f_e and F are related:

$$(5) \qquad \overline{f_e(-\alpha)} = E_\alpha\{\overline{F(x)}\}.$$

If F is an even or an odd function, its transform f_e is even or odd, respectively, and f_e can be written in terms of Fourier cosine or sine transforms (Chap. 13):

$$(6) \qquad f_e(\alpha) = 2\int_0^{\infty} F(x)\cos\alpha x\,dx = f_e(-\alpha) \quad \text{if } F(-x) = F(x),$$

$$(7) \qquad f_e(\alpha) = -2i\int_0^{\infty} F(x)\sin\alpha x\,dx = -f_e(-\alpha) \quad \text{if } F(-x) = -F(x).$$

Property (1) is the *kernel-product property* of E_α that we noted in Prob. 3, Sec. 110.

Suppose now that F *is absolutely integrable* from $-\infty$ to ∞. Then the integral $E_\alpha\{F\}$ *converges uniformly* with respect to α, and we find that

$$(8) \qquad f_e(\alpha) \text{ is continuous for all real } \alpha.$$

Also, f_e is bounded; in fact, the Riemann-Lebesgue theorem noted in Prob. 8, Sec. 65, shows that

$$\lim_{\alpha \to \pm \infty} f_e(\alpha) = 0.$$

In case $|xF(x)|$ is also integrable from $-\infty$ to ∞, then

(10) $$f'_e(\alpha) = E_\alpha\{-ixF(x)\}$$

since the member on the right represents a uniformly convergent integral (Sec. 15).

Example If $F(x) = \exp(-x^2)$, then

$$f_e(\alpha) = \int_{-\infty}^{\infty} \exp\left[-(x^2 + i\alpha x)\right] dx$$

$$= \exp\left(-\frac{\alpha^2}{4}\right) \int_{-\infty}^{\infty} \exp\left[-\left(x + i\frac{\alpha}{2}\right)^2\right] dx$$

$$= \exp\left(-\frac{\alpha^2}{4}\right) \int_{-\infty}^{\infty} \exp\left(-r^2\right) dr = \sqrt{\pi} \exp\left(-\frac{\alpha^2}{4}\right),$$

since the integral of $\exp(-z^2)$ over each horizontal line $z = x + i\alpha/2$ in the complex plane is equal to the integral over the real axis. Thus

(11) $$E_\alpha\{\exp(-x^2)\} = \sqrt{\pi} \exp\left(-\frac{\alpha^2}{4}\right).$$

Then from properties (3) and (10) we find these transforms:

(12)
$$E_\alpha\{\exp(-c^2 x^2)\} = \frac{\sqrt{\pi}}{|c|} \exp\left(-\frac{\alpha^2}{4c^2}\right) \qquad (c \text{ real}, c \neq 0),$$

$$E_\alpha\{x \exp(-x^2)\} = -i\frac{\sqrt{\pi}}{2} \alpha \exp\left(-\frac{\alpha^2}{4}\right).$$

PROBLEMS

1. If $F(x) = 1$ when $-b < x < b$ and $F(x) = 0$ elsewhere, show that

$$f_e(\alpha) = \frac{2}{\alpha} \sin b\alpha \quad \text{if } \alpha \neq 0, \quad \text{and} \quad f_e(0) = 2b.$$

2. In terms of our unit step function S_0, where $S_0(x) = 0$ if $x < 0$ and $S_0(x) = 1$ if $x > 0$, we found in Sec. 132 that

$$E_\alpha\{e^{-x}S_0(x)\} = \frac{1}{1 + i\alpha} = \frac{1 - i\alpha}{1 + \alpha^2}.$$

Use properties of E_α in Sec. 133 to deduce that

(a) $E_\alpha\{e^{-cx}S_0(x)\} = \dfrac{1}{c + i\alpha}$ if $c > 0$;

(b) $E_\alpha\{xe^{-x}S_0(x)\} = \dfrac{1}{(1 + i\alpha)^2}$;

(c) $E_\alpha\{e^xS_0(-x)\} = \dfrac{1}{1 - i\alpha}$.

3.   Use transform (10), Sec. 132, of $\exp(-|x|)$ to prove that

(a) $E_\alpha\{\exp(-c|x|)\} = \dfrac{2c}{c^2 + \alpha^2}$ if $c > 0$;

(b) $E_\alpha\{\exp(-|x - c|)\} = \dfrac{2}{1 + \alpha^2}\, e^{-ic\alpha}$ if c is real;

(c) $E_\alpha\{x \exp(-|x|)\} = \dfrac{-4i\alpha}{(1 + \alpha^2)^2}$.

4.   From transform (9), Sec. 132, of $(1 + x^2)^{-1}$ deduce that

(a) $E_\alpha\left\{\dfrac{a}{a^2 + x^2}\right\} = \pi \exp(-a|\alpha|)$ if $a > 0$;

(b) $E_\alpha\left\{\dfrac{a}{(x - c)^2 + a^2}\right\} = \pi \exp(-a|\alpha| - ic\alpha)$ if $a > 0$ and c is real.

5.   Apply Theorem 3 to the function $f(\alpha) = \exp(-\alpha^2)$ to establish the transform

(a) $2\sqrt{\pi} \exp(-\alpha^2) = E_\alpha\left\{\exp\left(-\dfrac{x^2}{4}\right)\right\}$;

then deduce that

(b) $\sqrt{\pi} \exp\left(-\dfrac{\alpha^2}{4}\right) = E_\alpha\{\exp(-x^2)\}$.

6. (a) Show formally that

$$E_\alpha\{xF'(x)\} = -\frac{d}{d\alpha}[\alpha f_e(\alpha)].$$

(b) Apply E_α formally to find a solution Y, for which $E_\alpha\{Y\}$ exists, of the differential equation

$$2Y''(x) + xY'(x) + Y(x) = 0.$$

Verify that solution.   *Ans.* $Y(x) = A \exp(-x^2/4)$.

7.   If $a > 0$ and

$F(x) = \cos^2 ax$ when $a|x| \leq \dfrac{\pi}{2}$,   and $F(x) = 0$ when $a|x| \geq \dfrac{\pi}{2}$, show that $F''(x) = 2a^2[1 - 2F(x)]$ when $a|x| < \pi/2$ and hence that

$$E_\alpha\{F(x)\} = \frac{4a^2}{\alpha(4a^2 - \alpha^2)} \sin \frac{\pi\alpha}{2a}.$$

8.   When $c > 0$ and $F(x) = 0$ if $|x| > c$ while

$$F(x) = -1 \quad \text{if } -c < x < 0, \quad \text{and} \quad F(x) = 1 \quad \text{if } 0 < x < c,$$

show that

$$E_\alpha\{F(x)\} = \frac{2}{i\alpha}(1 - \cos c\alpha).$$

9.   Write $F(x) = U(x) + iV(x)$ where U and V are real-valued functions of x $(-\infty < x < \infty)$. Let F be sectionally continuous on each bounded interval of the x axis (Sec. 56). If F is absolutely integrable from $-\infty$ to ∞, point out why U and V also have that property.

10.   Theorem 2, Sec. 132, gives conditions under which a function F is represented by the Fourier integral formula, conditions that were stated in Sec. 68 when F is real-valued. With the aid of the observation in Prob. 9, show why Theorem 2 applies when $F(x) = U(x) + iV(x)$ where U and V are real-valued functions.

134   THE CONVOLUTION INTEGRAL FOR E_α

The kernel-product property (1), Sec. 133, for each fixed real t is

$$e^{-it\alpha}f_e(\alpha) = E_\alpha\{F(x - t)\}.$$

As noted in Prob. 3, Sec. 110, the corresponding convolution of two functions F and G with transforms $f_e(\alpha)$ and $g_e(\alpha)$ is

$$(1) \qquad X_{FG}(x) = \int_{-\infty}^{\infty} F(t)G(x - t)\, dt = F * G(x) \qquad (-\infty < x < \infty).$$

Note that the convolution or faltung operation here, $F * G(x)$, is not the one used with Laplace transforms since the integral has limits $-\infty$ and ∞ rather than 0 and x. Both convolutions are the same in case $F(x) = G(x) = 0$ whenever $x < 0$.

Before establishing conditions on F and G under which $E_\alpha\{X_{FG}\} = f_e(\alpha)g_e(\alpha)$, we need some properties of the operation $F * G$ and information on the nature of the function $X_{FG}(x)$.

We assume in this section that our functions $F(x)$ and $G(x)$ are sectionally continuous on each bounded interval of the x axis, and that they are bounded and absolutely integrable from $-\infty$ to ∞.

Since a constant M exists such that $|G(x - t)| < M$, then $|F(t)G(x - t)| < M|F(t)|$. But $M|F(t)|$ is independent of x and integrable from $-\infty$ to ∞.

Hence the convolution integral (1) converges *absolutely and uniformly for all real x.* Moreover,

$$\int_a^b F(t)G(x - t) \, dt = \int_{x-b}^{x-a} G(y)F(x - y) \, dy$$

and since the first integral tends to $X_{FG}(x)$ as $a \to -\infty$ and $b \to \infty$, then the second one has the same limit. It follows that the operation $F * G$ is commutative:

(2)
$$X_{FG}(x) = \int_{-\infty}^{\infty} G(y)F(x - y) \, dy = X_{GF}(x).$$

If F_1 and F_2 satisfy our conditions on F, it is clear that

(3)
$$[F_1(x) + F_2(x)] * G(x) = F_1 * G(x) + F_2 * G(x).$$

To establish the continuity of $X_{FG}(x)$, first assume that $G(x)$ is continuous on an interval $|x| \leq 2c$. Then $G(x - t)$ is continuous in x and t together when $-c \leq x \leq c$ and $-c \leq t \leq c$. But $F(t)$ is continuous over the interval $|t| \leq c$ except possibly for a finite number of jumps at points t_i. Thus the integral

(4)
$$I_c(x) = \int_{-c}^{c} F(t)G(x - t) \, dt$$

can be written as a sum of integrals from $-c$ to t_1, t_1 to t_2, etc., whose integrands are continuous in x and t. Consequently I_c is continuous when $-c \leq x \leq c$.

Next let G be the unit step function $S_0(x - k)$ where $-2c < k < 2c$. Then $G(x - t) = S_0(x - t - k) = 1$ if $t < x - k$, and it vanishes if $t > x - k$. If $-c \leq t \leq c$, then $t \leq x - k$ when $c \leq x - k$, and so

$$I_c(x) = \int_{-c}^{c} F(t)S_0(x - t - k) \, dt = \int_{-c}^{c} F(t) \, dt \qquad \text{if } x \geq k + c.$$

Since $S_0(x - t - k) = 0$ when $t > -c > x - k$, then

$$I_c(x) = 0 \qquad\qquad\qquad \text{if } x \leq k - c.$$

Thus $I_c(x)$ has constant values on each of the half lines $x \geq k + c$ and $x \leq k - c$. On the remaining interval of the x axis, $k - c < x < k + c$, we note that both t and $x - k$ lie on the interval $(-c,c)$, and thus

$$I_c(x) = \int_{-c}^{c} F(t)S_0(x - k - t) \, dt = \int_{-c}^{x-k} F(t) \, dt \qquad (k - c \leq x \leq k + c).$$

This continuous function of x assumes the above constant values at the end points $k - c$ and $k + c$ of the interval. Therefore I_c is continuous for all x when $G(x) = S_0(x - k)$.

Finally, let G be sectionally continuous over the interval $(-2c,2c)$, with jumps $j_1, j_2, \ldots, j_n$ at points $k_1, k_2, \ldots, k_n$ $(k_1 < k_2 < \cdots < k_n)$ Then it is clear graphically that the function

$$G_n(x) = G(x) - j_1 S_0(x - k_1) - \cdots - j_n S_0(x - k_n)$$

is continuous on the interval when properly defined at the points $x = k_i$. Thus G is a linear combination of the continuous function G_n and unit step functions. It follows that the function I_c represented by the integral (4) is continuous when $-c \leqq x \leqq c$ if F and G are sectionally continuous over each bounded interval.

The continuity of the convolution integral (1) for all x now follows from the uniform convergence of that improper integral. We write

$$(5) \qquad X_{FG}(x) = \int_{-c}^{c} F(t)G(x - t)\, dt + R_c(x)$$

and the absolute value of the remainder $R_c(x)$ can be made arbitrarily small, uniformly for all x, by making c sufficiently large. Let x be fixed and its increments Δx be such that $|\Delta x| < b$, where b is some positive constant, and select c such that $c > |x| + b$. Then

$$|X_{FG}(x + \Delta x) - X_{FG}(x)| \leqq |I_c(x + \Delta x) - I_c(x)| + |R_c(x + \Delta x)| + |R_c(x)|$$

and the right-hand member can be made arbitrarily small by first taking c large, then $|\Delta x|$ small because I_c is a continuous function.

We have now established these properties of $F * G(x)$:

Theorem 4   *Let two functions F and G be bounded and absolutely integrable from $-\infty$ to ∞, and sectionally continuous over each bounded interval. Then their convolution integral*

$$F * G(x) = \int_{-\infty}^{\infty} F(t)G(x - t)\, dt \qquad (-\infty < x < \infty)$$

converges absolutely and uniformly to a function $X_{FG}(x)$ that is bounded and continuous for all real x. Also, $F * G(x)$ has the commutative and distributive properties (2) and (3).

135   CONVOLUTION THEOREM

We now prove that the product of two transforms is the transform of the convolution of the original functions.

Theorem 5   *The conditions on functions F and G stated in Theorem 4 are sufficient not only for the existence of the transforms $f_e(\alpha)$ and $g_e(\alpha)$, but* *also for the existence of the transform of the convolution*

(1)
$$X_{FG}(x) = \int_{-\infty}^{\infty} F(t)G(x-t)\,dt \quad (-\infty < x < \infty),$$

and the transform of that convolution is the product $f_e(\alpha)g_e(\alpha)$:

(2)
$$E_\alpha\{X_{FG}(x)\} = f_e(\alpha)g_e(\alpha) \quad (-\infty < \alpha < \infty).$$

Also, $X_{FG}(x)$ *is absolutely integrable from* $-\infty$ *to* ∞.

The existence and even the continuity of the transforms f_e and g_e were noted earlier (Sec. 133).

Since the functions $|F|$ and $|G|$ satisfy the conditions on F and G in Theorem 4, the convolution $|F| * |G(x)|$ is a continuous function

(3)
$$Q(x) = \int_{-\infty}^{\infty} |F(t)|\,|G(x-t)|\,dt \quad (-\infty < x < \infty),$$

and the integral here converges uniformly with respect to x. With an inversion of order of integration (Sec. 15), we can write

$$\int_0^b Q(x)\,dx = \int_{-\infty}^{\infty} |F(t)| \int_0^b |G(x-t)|\,dx\,dt.$$

That integral is bounded for all positive numbers b, since

$$\int_0^b Q(x)\,dx = \int_{-\infty}^{\infty} |F(t)| \int_{-t}^{b-t} |G(y)|\,dy\,dt$$

$$\leqq \int_{-\infty}^{\infty} |F(t)|\,dt \int_{-\infty}^{\infty} |G(y)|\,dy.$$

It is a monotone nondecreasing function of b because $Q(x) \geqq 0$. Therefore it it has a limit as $b \to \infty$. Likewise

$$\lim_{a \to -\infty} \int_a^0 Q(x)\,dx, \quad \text{or} \quad \int_{-\infty}^0 Q(x)\,dx,$$

exists, and hence the improper integral

(4)
$$\int_{-\infty}^{\infty} Q(x)\,dx = \int_{-\infty}^{\infty} \int_{-\infty}^{\infty} |F(t)|\,|G(x-t)|\,dt\,dx$$

converges. Since $|e^{-i\alpha x}X_{FG}(x)| \leqq Q(x)$ for all α, it follows that the Fourier integral

(5)
$$E_\alpha\{X_{FG}(x)\} = \int_{-\infty}^{\infty} e^{-i\alpha x}X_{FG}(x)\,dx$$

converges absolutely and uniformly with respect to α. In particular, X_{FG} is absolutely integrable.

To prove that the transform (5) is $f_e(\alpha)g_e(\alpha)$, we may use the kernel-product property

$$(6) \qquad e^{-i\alpha t}g_e(\alpha) = E_\alpha\{G(x - t)\} \qquad (-\infty < t < \infty),$$

which is satisfied for each fixed t under our conditions on F and G. Then

$$f_e(\alpha)g_e(\alpha) = \int_{-\infty}^{\infty} F(t)e^{-i\alpha t}g_e(\alpha)\, dt$$

$$= \int_{-\infty}^{\infty} F(t)E_\alpha\{G(x - t)\}\, dt.$$

The integral here is the limit as $c \to \infty$ of the integral

$$(7) \qquad \int_{-c}^{c} \int_{-\infty}^{\infty} F(t)G(x - t)e^{-i\alpha x}\, dx\, dt.$$

Here the integral with respect to x converges uniformly in t when $|t| < c$ because F is bounded, $|F(t)| < M$, and the remainder from $x = b$ to ∞ satisfies the conditions

$$\left| \int_{b}^{\infty} F(t)G(x - t)e^{-i\alpha x}\, dx \right| = \left| \int_{b-t}^{\infty} F(t)e^{-i\alpha y}G(y)\, dy \right|$$

$$\leq M \int_{b-c}^{\infty} |G(y)|\, dy.$$

The last member is independent of t and vanishes as $b \to \infty$; similarly for the remainder from $-\infty$ to $x = a$ as $a \to -\infty$.

It follows (Sec. 15) that the order of integration can be interchanged to write the iterated integral (7) as

$$\int_{-\infty}^{\infty} q(c,x)\, dx$$

where q is the function

$$(8) \qquad q(c,x) = e^{-i\alpha x} \int_{-c}^{c} F(t)G(x - t)\, dt.$$

The integral here, and hence q, is a continuous function of x (Sec. 134). We have now shown that

$$(9) \qquad f_e(\alpha)g_e(\alpha) = \lim_{c \to \infty} \int_{-\infty}^{\infty} q(c,x)\, dx.$$

We note that $q(c,x) \to e^{-i\alpha x}X_{FG}(x)$ as $c \to \infty$, and that limit is uniform with respect to x because the improper integral X_{FG} converges uniformly for all x.

Since

$$|q(c,x)| \leqq \int_{-c}^{c} |F(t)G(x - t)|\, dt \leqq \int_{-\infty}^{\infty} |F(t)G(x - t)|\, dt = Q(x)$$

and Q is integrable from $-\infty$ to ∞, then the integral in Eq. (9) converges uniformly with respect to c $(c > 0)$. Our earlier observation [Eq. (5)] that $E_\alpha\{X_{FG}\}$ exists completes the facts that will enable us to interchange the limit and the integral in formula (9).

To prove that

$$\lim_{c \to \infty} \int_{-\infty}^{\infty} q(c,x)\, dx = \int_{-\infty}^{\infty} e^{-i\alpha x} X_{FG}(x)\, dx,$$

we write the difference of the two integrals here in the form

$$(10) \qquad \int_{a}^{b} [e^{-i\alpha x} X_{FG}(x) - q(c,x)]\, dx + \int_{-\infty}^{a} e^{-i\alpha x} X_{FG}(x)\, dx$$

$$- \int_{-\infty}^{a} q(c,x)\, dx + \int_{b}^{\infty} e^{-i\alpha x} X_{FG}(x)\, dx - \int_{b}^{\infty} q(c,x)\, dx.$$

The improper integrals here converge absolutely; those involving q converge uniformly with respect to c. Thus their absolute values can be made small, independently of c, by taking a large and negative and b large and positive. Then by taking c large and positive, the absolute value of the first integral can be made small because the integrand tends to zero uniformly in x as $c \to \infty$. This completes the proof of Theorem 5 since it follows that

$$f_e(\alpha)g_e(\alpha) = \int_{-\infty}^{\infty} e^{-i\alpha x} X_{FG}(x)\, dx = E_\alpha\{X_{FG}(x)\}.$$

The convolution property here was illustrated in Prob. 4, Sec. 110. In Probs. 9 and 10, Sec. 110, we noted and illustrated a generalized convolution property for E_α, one that represents the iterated transform $\hat{h}(\alpha)$ of a function $H(x,y)$ of two variables as the transform of the function

$$\int_{-\infty}^{\infty} H(x - t, t)\, dt.$$

136   TABLES OF TRANSFORMS

Appendix C consists of a table that summarizes properties of E_α and lists exponential Fourier transforms of several functions.

Tables listed in the Bibliography under the names Erdélyi (vol. 1) and Oberhettinger give exponential Fourier transforms of some other functions.

Also, chap. 5 of W. Kaplan's "Operational Methods for Linear Systems," 1962, presents short tables of transforms of functions and generalized functions and a few additional properties of E_α.

Tables of Laplace transforms furnish a source of exponential Fourier transforms. In correspondence with the Laplace transform of a particular function, we can list a certain exponential Fourier transform. For if $F(t)$ is defined for all real t and has a Laplace transform $L\{F(t)\} = f(s)$ when $s = \gamma + i\alpha$, then

$$f(\gamma + i\alpha) = \int_0^\infty e^{-(\gamma + i\alpha)t} F(t)\, dt$$

$$= \int_{-\infty}^\infty e^{-i\alpha t} S_0(t) e^{-\gamma t} F(t)\, dt.$$

Thus, in terms of the Laplace transform $f(s)$ of $F(t)$,

$$E_\alpha\{S_0(x) e^{-\gamma x} F(x)\} = f(\gamma + i\alpha).$$

137 BOUNDARY VALUE PROBLEMS

Problems in differential equations that are adapted to solution by means of the exponential Fourier transformation are singular in that the variable, called x here, with respect to which the transformation is made, ranges from $-\infty$ to ∞. But they are singular *boundary value* problems with rather severe boundary conditions since the formula for the transform of a derivative with respect to x assumes that the function itself vanishes as $x \to \pm\infty$. That condition is more severe than the condition of exponential order as $t \to \infty$ used in initial-value problems treated by the Laplace transformation. Conditions of integrability with respect to x from $-\infty$ to ∞ further restrict the variety of problems to which E_α applies.

Of course, formal procedures may yield verifiable solutions of some problems in which the implied conditions are not satisfied. But the variety of problems that can be solved by using E_α is small compared with the varieties that are adapted to solution by finite Fourier or Laplace transformations.

Here and in the set of problems to follow we give some examples of problems that can be solved readily by using E_α.

Example 1   Let $F(x)$ be such that F' is continuous and F itself is absolutely integrable, from $-\infty$ to ∞, and such that $F(\pm\infty) = 0$. To find a solution of the boundary value problem

(1)
$$W_x(x,y) + W_y(x,y) + 2yW(x,y) = F(x) \qquad (-\infty < x < \infty, y > 0),$$
$$W(x,0) = 0, \qquad W(\pm\infty,y) = 0,$$

we let $w_e(\alpha,y)$ denote the exponential Fourier transform of W with respect to x and write $f_e(\alpha) = E_\alpha\{F(x)\}$. Then, formally, $E_\alpha\{W_x\} = i\alpha w_e$, and the transform of problem (1) becomes

$$(i\alpha + 2y)w_e(\alpha,y) + \frac{d}{dy}w_e(\alpha,y) = f_e(\alpha) \qquad (y > 0),$$

(2)

$$w_e(\alpha,0) = 0.$$

The solution of problem (2) in linear ordinary differential equations of first order is

(3) $$w_e(\alpha,y) = \exp(-y^2) \int_0^y f_e(\alpha)e^{i\alpha(t-y)} \exp(t^2)\, dt.$$

According to property (1), Sec. 133, the inverse transform of $f_e(\alpha)e^{i\alpha(t-y)}$ is $F(x + t - y)$. By formally inverting the order of the operators E_α^{-1} and integration with respect to t, we can write

(4) $$W(x\ y) = \exp(-y^2) \int_0^y F(x - y + t) \exp(t^2)\, dt.$$

It is not difficult to verify that the function (4) satisfies all conditions in problem (1).

Example 2 To obtain a formula for the bounded harmonic function $V(x,y)$ in the half plane $y > 0$ that assumes values $F(x)$ on the boundary $y = 0$, we solve this problem:

$$V_{xx}(x,y) + V_{yy}(x,y) = 0 \qquad (-\infty < x < \infty, y > 0),$$

(5)

$$V(x, +0) = F(x), \qquad |V(x, y)| < M,$$

where M is some constant and, over the entire x axis, the function F is bounded and continuous except possibly for a finite number of finite jumps.

To solve the problem using the transformation E_α with respect to x, we shall assume some auxiliary conditions on F, conditions associated with our method of solution but not necessarily required in order that the resulting formula for $V(x,y)$ will be valid. We assume that $F(x)$ is such that its transform $f_e(\alpha)$ exists and that the problem will have a solution $V(x,y)$ whose transform $v_e(\alpha,y)$ with respect to x exists and is bounded when $y > 0$ for each fixed α. Further conditions on V are needed to write $E_\alpha\{V_{xx}\} = -\alpha^2 v_e$ and to justify steps in transforming problem (5) into the problem

$$-\alpha^2 v_e(\alpha,y) + \frac{d^2}{dy^2}v_e(\alpha,y) = 0 \qquad (y > 0),$$

(6)

$$v_e(\alpha,0) = f_e(\alpha), \qquad |v_e(\alpha,y)| < N(\alpha),$$

where N is independent of y.

The solution of problem (6) is

(7) $$v_e(\alpha,y) = f_e(\alpha) \exp(-y|\alpha|) \qquad (y \geq 0).$$

We found earlier that

$$\exp(-y|\alpha|) = E_\alpha \left\{ \frac{1}{\pi} \frac{y}{x^2 + y^2} \right\} \qquad \text{when } y > 0.$$

Thus our convolution property indicates that

(8) $$V(x,y) = \frac{1}{\pi} \int_{-\infty}^{\infty} \frac{yF(t)}{(x-t)^2 + y^2} \, dt \qquad (y > 0).$$

That formal solution of problem (5) is known as the *Poisson integral formula for the half plane* $y > 0$, or the *Schwarz integral formula*. It can be fully verified as a solution under the conditions on F first stated with problem (5), without the auxiliary conditions.[1]

If $F(x)$ is a constant F_0, for instance, it satisfies the conditions stated in problem (5), but $f_e(\alpha)$ does not exist unless $F_0 = 0$. In that case formula (8) reduces to

(9) $$V(x,y) = \frac{F_0}{\pi} \arctan\frac{t-x}{y} \Bigg]_{-\infty}^{\infty} = \frac{F_0}{\pi}\left(\frac{\pi}{2} + \frac{\pi}{2}\right) = F_0,$$

which is clearly a solution of problem (5).

In case $F(x) = S_0(x)$, formula (8) becomes

(10) $$V(x,y) = \frac{1}{\pi} \int_0^{\infty} \frac{y \, dt}{(t-x)^2 + y^2} = \frac{1}{\pi}\left(\frac{\pi}{2} + \arctan\frac{x}{y}\right).$$

This function is bounded and harmonic in the half plane $y > 0$; it vanishes as $y \to +0$ if $x < 0$, and $V(x,y) \to 1$ as $y \to +0$ if $x > 0$.

PROBLEMS

1. Let $F(x)$ be sectionally continuous over an interval $-c < x < c$, while $F(x) = 0$ when $x < -c$ and when $x > c$. Use the exponential Fourier transformation to find a solution of the differential equation

$$Y''(x) - Y(x) + 2F(x) = 0 \qquad (-\infty < x < \infty)$$

such that Y' is everywhere continuous and $Y(\pm\infty) = Y'(\pm\infty) = 0$, in the form

$$Y(x) = e^{-x} \int_{-c}^{x} e^t F(t) \, dt + e^x \int_x^c e^{-t} F(t) \, dt.$$

Verify that solution. Note that the first integral here vanishes when $x < -c$ and the second one vanishes when $x > c$, since $F(t) = 0$ when $|t| > c$.

[1] For a full verification see the author's "Complex Variables and Applications," 2d ed., sec. 108, 1960.

2. Use E_α and the inversion formula to solve the problem

$$V_{xx}(x,y) + V_{yy}(x,y) + x \exp(-x^2) = 0 \qquad (-\infty < x < \infty, y > 0),$$

$$V(x,0) = 0, \qquad |V(x,y)| < My \qquad\qquad \text{when } y > 0,$$

where M is some constant, in the form

$$V(x\ y) = \frac{1}{2\sqrt{\pi}} \int_0^\infty (1 - e^{-\alpha y})\frac{\sin \alpha x}{\alpha} \exp\left(-\frac{\alpha^2}{4}\right) d\alpha.$$

3. Let $h(y)$ be continuous when $y \geq 0$, and let $F(x)$ have a continuous derivative F' for all x and vanish as $x \to \pm\infty$. Derive the solution of the problem

$$W_x(x,y) + W_y(x,y) + h(y)W(x,y) = 0 \qquad (-\infty < x < \infty, y > 0),$$

$$W(x,0) = F(x), \qquad W(\pm\infty,y) = 0,$$

in the form

$$W(x,y) = F(x - y) \exp\left[-\int_0^y h(t)\, dt\right].$$

Verify that solution.

4. The transverse displacements $Y(x,t)$ in a string stretched along the entire x axis satisfy the conditions

$$Y_{tt}(x,t) = a^2 Y_{xx}(x,t) \qquad (-\infty < x < \infty, t > 0),$$

$$Y(x,0) = F(x), \qquad Y_t(x,0) = 0;$$

also $F(x) \to 0$ and $Y(x,t) \to 0$ as $x \to \pm\infty$. Transform with respect to x to derive the formula

$$Y(x,t) = \tfrac{1}{2}[F(x - at) + F(x + at)].$$

5. Transform with respect to x to solve this problem formally for temperatures $U(x,t)$:

$$U_t(x,t) = kU_{xx}(x,t) \qquad (-\infty < x < \infty, t > 0),$$

$$U(x,0) = F(x).$$

$$Ans. \quad U(x,t) = \frac{1}{2\sqrt{\pi kt}} \int_{-\infty}^\infty F(\xi) \exp\left[-\frac{(x-\xi)^2}{4kt}\right] d\xi$$

$$= \frac{1}{\sqrt{\pi}} \int_{-\infty}^\infty F(x + 2r\sqrt{kt}) \exp(-r^2)\, dr.$$

13

Fourier Transforms on the Half Line

Fourier sine and cosine transforms of functions defined over the half line $x > 0$ were introduced in Chap. 10 as examples of general integral transforms. They transform the differential form F'' in terms of the transform of F itself and the initial values $F(0)$ and $F'(0)$, respectively. Operational properties of the two transformations are noted here in greater detail and summarized in Appendix D; some applications are given. Also, we present a modified transformation that resolves F'' in terms of a boundary value $F'(0) - hF(0)$.

138 FOURIER SINE TRANSFORMS $f_s(\alpha)$

In Sec. 116 we derived a linear integral transformation that transforms $F''(x)$ in terms of the boundary value $F(0)$ and the transform of F when F is defined over the positive x axis. It is the *Fourier sine transformation*

$$(1) \qquad S_\alpha\{F(x)\} = \int_0^\infty F(x) \sin \alpha x \, dx = f_s(\alpha) \qquad (\alpha > 0),$$

whose basic operational property is

(2) $$S_\alpha\{F''(x)\} = -\alpha^2 f_s(\alpha) + \alpha F(0).$$

The inverse transformation $S_\alpha^{-1}\{f_s\}$ can be expressed in terms of the transformation itself:

(3) $$F(x) = \frac{2}{\pi} \int_0^\infty f_s(\alpha) \sin \alpha x \, d\alpha = \frac{2}{\pi} S_x\{f_s(\alpha)\} \qquad (x > 0).$$

The transform $f_s(\alpha)$ exists as a continuous function of the parameter α if F is sectionally continuous over each bounded interval $0 < x < c$ and absolutely integrable from 0 to ∞. Our conditions are sufficient, but not necessary for the existence of $f_s(\alpha)$; for example, $S_\alpha\{1/x\} = \pi/2$ when $\alpha > 0$. Under those conditions it follows from the Riemann-Lebesgue theorem (Prob. 8, Sec. 65) that

(4) $$\lim_{\alpha \to \infty} f_s(\alpha) = 0.$$

The operational property (2) is valid if F'' is sectionally continuous over each bounded interval while F and F' are continuous when $x \geqq 0$ and vanish as $x \to \infty$, and if F itself is absolutely integrable from 0 to ∞. The condition that F be absolutely integrable can be replaced by the condition that $f_s(\alpha)$ exists for all positive α because integration by parts shows that, whenever $c > 0$,

(5) $$\int_0^c F''(x) \sin \alpha x \, dx = -\alpha^2 \int_0^c F(x) \sin \alpha x \, dx$$

$$+ \left[F'(x) \sin \alpha x - \alpha F(x) \cos \alpha x \right]_0^c.$$

As $c \to \infty$, the bracketed term tends to $\alpha F(0)$; thus the integral on the left has the limit $S_\alpha\{F''\}$ when the integral on the right has a limit.

By replacing F by F'' in Eq. (5) and integrating by parts, we find that

(6) $$S_\alpha\{F^{(4)}(x)\} = \alpha^4 f_s(\alpha) - \alpha^3 F(0) + \alpha F''(0),$$

under these conditions: F and its derivatives up to F''' are continuous when $x \geqq 0$ and vanish as $x \to \infty$, $F^{(4)}$ is sectionally continuous on each bounded interval, and $f_s(\alpha)$ exists; similarly for transforms of further iterates of the operator d^2/dx^2.

Sufficient conditions for the inversion formula (3), the Fourier sine integral formula, are that F be absolutely integrable from 0 to ∞ and that F' be sectionally continuous on each bounded interval. Then the formula represents $F(x)$ where F is continuous, and the mean value of $F(x + 0)$ and $F(x - 0)$ where F has a jump.

The sine transformation is a special case of the exponential Fourier transformation. For if F_1 denotes the odd extension of F to the entire x axis, that is,

$$F_1(x) = F(x) \quad \text{if } x > 0, \qquad F_1(-x) = -F_1(x) \qquad \text{for all } x,$$

then as noted in Sec. 133,

$$E_\alpha\{F_1(x)\} = -2i \int_0^\infty F_1(x) \sin \alpha x \, dx.$$

Therefore

(7) $$S_\alpha\{F(x)\} = \frac{i}{2} E_\alpha\{F_1(x)\} = E_\alpha\left\{\frac{ix}{2|x|}F(|x|)\right\}.$$

Some properties of S_α follow from properties of E_α. In particular, according to Theorem 3, Sec. 132, the inversion formula (3) is valid if $f_s'(\alpha)$ is sectionally continuous on each bounded interval, if f_s is absolutely integrable from 0 to ∞, and if $S_x\{f_s(\alpha)\}$ exists.

Property (2) is modified in case F and F' have jumps b and b', respectively, at a point $x = c$:

(8) $$S_\alpha\{F''(x)\} = -\alpha^2 f_s(\alpha) + \alpha F(0) + \alpha b \cos \alpha c - b' \sin \alpha c \qquad (c > 0).$$

From property (2) and the inversion formula we find that

$$S_\alpha\{e^{-x}\} = \frac{\alpha}{1+\alpha^2}, \qquad S_\alpha\left\{\frac{x}{1+x^2}\right\} = \frac{\pi}{2}e^{-\alpha}.$$

From property (8) we find that the function

$$G(x) = c - x \quad \text{if } 0 \leq x \leq c, \qquad G(x) = 0 \qquad \text{if } x \geq c,$$

has this transform:

$$S_\alpha\{G(x)\} = \alpha^{-2}(\alpha c - \sin \alpha c).$$

Under our conditions for the existence of $f_s(\alpha)$ we find that

(9) $$f_s(\alpha k) = \frac{1}{k} S_\alpha\left\{F\left(\frac{x}{k}\right)\right\} \qquad \text{if } k > 0.$$

If in addition $x^2 F(x)$ is absolutely integrable from 0 to ∞, we can differentiate integral (1) with respect to α to see that

(10) $$f_s''(\alpha) = S_\alpha\{-x^2 F(x)\}.$$

It is convenient to introduce the cosine transformation on the half line before taking up further properties of S_α.

139 FOURIER COSINE TRANSFORMS $f_c(\alpha)$

In Prob. 11, Sec. 116, we noted that the integral transformation that resolves F'' on the half line $x > 0$ in terms of the transform of F and the boundary value $F'(0)$ is the *Fourier cosine transformation*

$$(1) \qquad C_\alpha\{F(x)\} = \int_0^\infty F(x) \cos \alpha x \, dx = f_c(\alpha) \qquad (\alpha > 0).$$

Its basic operational property is

$$(2) \qquad C_\alpha\{F''(x)\} = -\alpha^2 f_c(\alpha) - F'(0).$$

The Fourier cosine integral formula furnishes an inverse transformation C_α^{-1} in terms of $C_x\{f_c(\alpha)\}$:

$$(3) \qquad F(x) = \frac{2}{\pi} \int_0^\infty f_c(\alpha) \cos \alpha x \, d\alpha = \frac{2}{\pi} C_x\{f_c(\alpha)\} \qquad (x > 0).$$

Sufficient conditions for the existence of $f_c(\alpha)$, or for the validity of the inversion formula (3) or the operational properties of C_α, are those used for the corresponding formulas for S_α in the preceding section.

We find, corresponding to property (6), Sec. 138, that

$$(4) \qquad C_\alpha\{F^{(4)}(x)\} = \alpha^4 f_c(\alpha) + \alpha^2 F'(0) - F'''(0).$$

Property (2) can be extended easily to formulas for the transform of higher derivatives of F of even order. In case F and F' are continuous except for jumps b and b', respectively, at a point $x = c$, we find that property (2) is modified as follows:

$$(5) \qquad C_\alpha\{F''(x)\} = -\alpha^2 f_c(\alpha) - F'(0) - \alpha b \sin \alpha c - b' \cos \alpha c \qquad (c > 0).$$

We find that, corresponding to other properties of S_α,

$$(6) \qquad f_c(\alpha k) = \frac{1}{k} C_\alpha\left\{F\left(\frac{x}{k}\right)\right\} \qquad \text{if } k > 0,$$

$$(7) \qquad f_c''(\alpha) = C_\alpha\{-x^2 F(x)\},$$

$$(8) \qquad \lim_{\alpha \to \infty} f_c(\alpha) = 0.$$

The relation between C_α and the exponential Fourier transformation (Sec. 133) can be written

$$(9) \qquad C_\alpha\{F(x)\} = \tfrac{1}{2} E_\alpha\{F_2(x)\} = \tfrac{1}{2} E_\alpha\{F(|x|)\},$$

where F_2 is the even extension of F over the entire x axis.

Conditions of validity of the kernel-product property

$$(10) \qquad 2f_c(\alpha) \cos \alpha k = C_\alpha\{F(x + k) + F(|x - k|)\} \qquad (k > 0)$$

were noted in Prob. 5, Sec. 110. The corresponding convolution property for C_α will be established in Sec. 141.

140   FURTHER PROPERTIES OF S_α AND C_α

Some useful operational properties of the sine and cosine transformations involve the two transformations jointly.

   If F is continuous when $x \geq 0$ and $F(\infty) = 0$, if F' is sectionally continuous on each bounded interval, and if *either F or F'* is absolutely integrable from 0 to ∞, then upon integrating by parts between limits 0 and c and letting c tend to infinity, we find that

(1) $$S_\alpha\{F'(x)\} = -\alpha C_\alpha\{F(x)\},$$

(2) $$C_\alpha\{F'(x)\} = \alpha S_\alpha\{F(x)\} - F(0).$$

If G is sectionally continuous and absolutely integrable from 0 to ∞, then the function

$$F(x) = \int_x^\infty G(r)\, dr \qquad\qquad (x \geq 0)$$

satisfies our conditions since $F'(x) = -G(x)$ and $F(\infty) = 0$. Thus property (1) can be written

(3) $$S_\alpha\{G(x)\} = \alpha C_\alpha\left\{\int_x^\infty G(r)\, dr\right\}.$$

   By differentiating the integrals $S_\alpha\{F\}$ and $C_\alpha\{F\}$ with respect to α, we find that

(4) $$f_s'(\alpha) = C_\alpha\{xF(x)\},$$

(5) $$f_c'(\alpha) = S_\alpha\{-xF(x)\},$$

assuming the sectional continuity of F and the absolute integrability of $xF(x)$ from 0 to ∞.

   Let F_1 and F_2 be the odd and even extensions of F:

$$F_1(x) = \frac{x}{|x|}F(|x|), \qquad F_2(x) = F(|x|), \qquad \text{when } -\infty < x < \infty.$$

Assuming the sectional continuity and absolute integrability of F, we find these properties corresponding to the kernel-product property (10) of the preceding section:

$$2f_s(\alpha)\sin \alpha k = C_\alpha\{F_1(x + k) - F_1(x - k)\},$$

(6) $$2f_s(\alpha)\cos \alpha k = S_\alpha\{F_1(x + k) + F_1(x - k)\},$$

$$2f_c(\alpha)\sin \alpha k = S_\alpha\{F_2(x - k) - F_2(x + k)\},$$

where k is a real-valued constant.

The existence of the transforms f_s and f_c for all positive α, and the natural agreement that $f_s(-\alpha) = -f_s(\alpha)$ and $f_c(-\alpha) = f_c(\alpha)$, enables us to write the relations

(7)
$$f_s(\alpha + k) + f_s(\alpha - k) = S_\alpha\{2F(x) \cos kx\},$$
$$f_s(\alpha + k) - f_s(\alpha - k) = C_\alpha\{2F(x) \sin kx\},$$

(8)
$$f_c(\alpha + k) + f_c(\alpha - k) = C_\alpha\{2F(x) \cos kx\},$$
$$f_c(\alpha - k) - f_c(\alpha + k) = S_\alpha\{2F(x) \sin kx\}.$$

141   CONVOLUTION PROPERTIES

Let F and G be two bounded functions that are sectionally continuous on each interval $0 < x < c$ and absolutely integrable from 0 to ∞. Then

$$2f_c(\alpha) = E_\alpha\{F_2(x)\}, \qquad 2g_c(\alpha) = E_\alpha\{G_2(x)\}$$

where $F_2(x) = F(|x|)$ and $G_2(x) = G(|x|)$ when $-\infty < x < \infty$. According to Theorem 5, Sec. 135, the product

$$4f_c(\alpha)g_c(\alpha) = E_\alpha\{F_2(x)\}E_\alpha\{G_2(x)\}$$

is the exponential Fourier transform of the convolution

$$\int_{-\infty}^{\infty} F_2(t)G_2(x - t)\, dt.$$

That integral is easily written in the form

$$\int_0^{\infty} F(r)[G_2(x + r) + G_2(x - r)]\, dr$$

which is an even function of x, so its exponential transform is twice its cosine transform.

It follows that C_α has the convolution property

(1)
$$2f_c(\alpha)g_c(\alpha) = C_\alpha\left\{\int_0^{\infty} F(r)[G(x + r) + G(|x - r|)]\, dr\right\}.$$

Under the same conditions on F and G, and in the same manner, we can establish these joint properties of sine and cosine transforms:

(2)
$$2f_s(\alpha)g_s(\alpha) = C_\alpha\left\{\int_0^{\infty} F(r)[G(x + r) - G_1(x - r)]\, dr\right\},$$

(3)
$$2f_s(\alpha)g_c(\alpha) = S_\alpha\left\{\int_0^\infty F(r)[G(|x - r|) - G(x + r)]\, dr\right\}$$

$$= S_\alpha\left\{\int_0^\infty G(t)[F(x + t) + F_1(x - t)]\, dt\right\},$$

where F_1 and G_1 are the odd extensions of F and G.

To establish a convolution property for the sine transformation alone, we impose an additional condition on the function G, namely, that its integral

$$H(x) = \int_x^\infty G(t)\, dt \qquad\qquad (x \geqq 0)$$

be absolutely integrable from $x = 0$ to $x = \infty$. Now H is continuous when $x \geqq 0$, and $H(\infty) = 0$. According to property (3), Sec. 140, then

$$g_s(\alpha) = \alpha C_\alpha\{H(x)\} = \alpha h_c(\alpha)$$

and it follows from formula (3) that

$$2f_s(\alpha)g_s(\alpha) = 2\alpha f_s(\alpha)h_c(\alpha)$$

$$= \alpha S_\alpha\left\{\int_0^\infty F(r)[H(|x - r|) - H(x + r)]\, dr\right\}.$$

Thus we have the convolution property that was obtained formally in Sec. 110:

(4)
$$\frac{2}{\alpha} f_s(\alpha)g_s(\alpha) = S_\alpha\left\{\int_0^\infty F(r)\int_{|x-r|}^{x+r} G(t)\, dt\, dr\right\}.$$

142 TABLES OF SINE AND COSINE TRANSFORMS

Appendix D consists of tables that summarize properties of Fourier sine and cosine transformations on the half line and list transforms of particular functions.

Some transforms can be found by evaluating integrals with the aid of Laplace transforms (Chap. 3). The results shown in Probs. 14 and 15, Sec. 35, for example, can be written as

(1)
$$S_\alpha\left\{\frac{1}{c + x}\right\} = \left(\frac{\pi}{2} - \operatorname{Si}\alpha c\right)\cos\alpha c + \operatorname{Ci}\alpha c \sin\alpha c,$$

(2)
$$C_\alpha\left\{\frac{1}{c + x}\right\} = \left(\frac{\pi}{2} - \operatorname{Si}\alpha c\right)\sin\alpha c - \operatorname{Ci}\alpha c \cos\alpha c,$$

where $c > 0$. The evaluation of improper integrals by using contour integrals and residue theory, illustrated in sec. 72 of the author's "Complex Variables and Applications," 2d ed., 1960, is another method of deriving some of the transforms listed in Appendix D. Other transforms can be found by direct integration, by using operational properties of S_α and C_α, or by differentiating or integrating a known transform with respect to a parameter.

Extensive tables of sine and cosine transforms will be found in vol. 1 of the "Tables of Integral Transforms" edited by A. Erdélyi. Those tables, listed in the Bibliography, give transforms of several hundred special functions. The book by Ditkin and Prudnikov, also listed in the Bibliography, includes extensive tables of those transforms.

143   STEADY TEMPERATURES IN A QUADRANT

Let $V(x,y)$ denote steady temperatures in a solid that fills the space $x > 0$, $y > 0$, $-\infty < z < \infty$ (Fig. 108). If the face $x = 0$ is insulated, if $V(x,0) = F(x)$, and if a steady source of heat varying only with x is distributed throughout the solid, then

$$(1) \qquad\qquad V_{xx}(x,y) + V_{yy}(x,y) + H(x) = 0 \qquad (x > 0, y > 0),$$

$$(2) \qquad\qquad V_x(0,y) = 0, \qquad V(x,0) = F(x);$$

also, V is to satisfy boundness conditions as $x \to \infty$ or $y \to \infty$.

To help keep temperatures steady, we assume that the total rate of generation of heat throughout the solid is zero, that is,

$$(3) \qquad\qquad \int_0^\infty H(x)\, dx = 0.$$

Let us assume also that whenever $x \geq 0$, the integral

$$(4) \qquad\qquad P(x) = \int_x^\infty \int_0^r H(t)\, dt\, dr$$

exists and that P has a cosine transform. Since $P''(x) = -H(x)$ and $P'(0) = 0$, the cosine transforms of P and H satisfy the relation

$$(5) \qquad\qquad \alpha^2 p_c(\alpha) = h_c(\alpha).$$

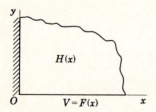

Fig. 108

Our problem in V is adapted to the Fourier cosine transformation with respect to x on the half line $x > 0$, in view of the differential form V_{xx} in Eq. (1) and the prescribed boundary value $V_x(0,y)$. Let $v_c(\alpha,y)$ denote the transform $C_\alpha\{V\}$ where the transformation is made with respect to x, and write $f_c(\alpha) = C_\alpha\{F(x)\}$. Then, formally,

$$(6) \qquad -\alpha^2 v_c(\alpha,y) + \frac{d^2}{dy^2} v_c(\alpha,y) + h_c(\alpha) = 0, \qquad v_c(\alpha,0) = f_c(\alpha),$$

where $y > 0$. The bounded solution of problem (6) is

$$v_c(\alpha,y) = \frac{1}{\alpha^2} h_c(\alpha) + \left[f_c(\alpha) - \frac{1}{\alpha^2} h_c(\alpha) \right] e^{-\alpha y}$$

or, in view of Eq. (5),

$$(7) \qquad v_c(\alpha,y) = p_c(\alpha) + [f_c(\alpha) - p_c(\alpha)]e^{-\alpha y} \qquad\qquad (\alpha > 0,\ y > 0).$$

From Appendix D, Table D.2, we see that

$$e^{-\alpha y} = C_\alpha \left\{ \frac{2}{\pi} \frac{y}{x^2 + y^2} \right\}$$

and, according to the convolution property (1), Sec. 141, it follows from Eq. (7) that

$$(8) \quad V(x,y) = P(x) + \frac{y}{\pi} \int_0^\infty [F(r) - P(r)] \left[\frac{1}{(x + r)^2 + y^2} + \frac{1}{(x - r)^2 + y^2} \right] dr.$$

We note some special cases. First, if $F(x) = P(x)$, then V is a function of x alone:

$$(9) \qquad\qquad V(x,y) = P(x) = \int_x^\infty \int_0^r H(t)\, dt\, dr,$$

a result that is easily verified as a solution.

In case no source is present, then $H(x) = 0$, and

$$(10) \qquad V(x,y) = \frac{1}{\pi} \int_0^\infty F(r) \left[\frac{y}{(x + r)^2 + y^2} + \frac{y}{(x - r)^2 + y^2} \right] dr$$

when $x \geq 0$ and $y > 0$. That is an integral formula for the harmonic function V in the quadrant such that $V_x(0,y) = 0$ and $V(x,+0) = F(x)$. In this case where $H(x) = 0$, the problem can be solved also by using the sine transformation with respect to y.

When $H(x) = 0$ and $F(x) = 1$, we find from formula (10) that $V(x,y) = 1$, and this is clearly a solution of the problem even though $f_c(x)$ does not exist.

In case $H(x) = 0$, and $F(x) = 1$ when $0 < x < c$ while $F(x) = 0$ when $x > c$, formula (10) gives the verifiable solution

$$V(x,y) = \frac{1}{\pi}\left(\arctan\frac{c + x}{y} + \arctan\frac{c - x}{y}\right).$$

144   DEFLECTIONS IN AN ELASTIC PLATE

As noted in Sec. 118, the static deflections or transverse displacements $Z(x,y)$ in a thin elastic plate with no transverse load satisfy the *biharmonic equation*

$$\nabla^2(\nabla^2 Z) = \nabla^4 Z = 0$$

where $\nabla^2 Z = Z_{xx} + Z_{yy}$. The bending moment about an edge $x = c$ is proportional to $Z_{xx}(c,y)$, so if that edge is simply supported, then $Z(c,y) = 0$ and $Z_{xx}(c,y) = 0$, and similarly for the other edges.

Let the plate have the form of a semi-infinite strip $x \geq 0$, $0 \leq y \leq \pi$. If there is no load and the edges $y = 0$ and $y = \pi$ are simply supported, while the edge $x = 0$ has a deflection $b \sin y$ but no bending moment is applied about the axis $x = 0$, then deflections $Z(x,y)$ satisfy the conditions

(1)
$$\frac{\partial^4 Z}{\partial x^4} + 2\frac{\partial^4 Z}{\partial x^2 \partial y^2} + \frac{\partial^4 Z}{\partial y^4} = 0 \qquad (x > 0, 0 < y < \pi),$$

(2)
$$Z(0,y) = b \sin y, \qquad Z_{xx}(0,y) = 0,$$

(3)
$$Z(x,0) = Z_{yy}(x,0) = Z(x,\pi) = Z_{yy}(x,\pi) = 0;$$

also Z and Z_x should tend to zero as $x \to \infty$.

The differential operators with respect to x in this problem are $\partial^2/\partial x^2$ and its iterate $\partial^4/\partial x^4$, on the half line $x > 0$, and the prescribed values at the boundary $x = 0$ are the values of Z and $\partial^2 Z/\partial x^2$. Thus the sine transformation with respect to x may apply. We write

$$z_s(\alpha,y) = S_\alpha\{Z(x,y)\};$$

then, formally,

$$S_\alpha\left\{\frac{\partial^4 Z}{\partial x^4}\right\} = \alpha^4 z_s(\alpha,y) - b\alpha^3 \sin y,$$

$$S_\alpha\left\{\frac{\partial^4 Z}{\partial x^2 \partial y^2}\right\} = \frac{\partial^2}{\partial y^2}S_\alpha\left\{\frac{\partial^2 Z}{\partial x^2}\right\} = \frac{\partial^2}{\partial y^2}[-\alpha^2 z_s(\alpha,y) + \alpha b \sin y]$$

$$= -\alpha^2\frac{\partial^2 z_s}{\partial y^2} - \alpha b \sin y.$$

Then our boundary value problem in Z transforms into the following problem in $z_s(\alpha,y)$:

(4)

$$\frac{d^4 z_s}{dy^4} - 2\alpha^2 \frac{d^2 z_s}{dy^2} + \alpha^4 z_s = b(2\alpha + \alpha^3) \sin y \qquad (0 < y < \pi, \, \alpha > 0),$$

$$z_s = \frac{d^2 z_s}{dy^2} = 0 \qquad \text{when } y = 0 \text{ and when } y = \pi.$$

We find that $A \sin y$ is a particular solution of the differential equation here if $(1 + \alpha^2)^2 A = b(2\alpha + \alpha^3)$, and that particular solution satisfies the conditions at the edges $y = 0$ and $y = \pi$. Thus

(5) $$z_s(\alpha,y) = b\frac{\alpha(2 + \alpha^2)}{(1 + \alpha^2)^2} \sin y = b\left[\frac{\alpha}{1 + \alpha^2} + \frac{\alpha}{(1 + \alpha^2)^2}\right] \sin y.$$

Referring to Appendix D, Table D.1, we see that

(6) $$Z(x,y) = b(1 + \tfrac{1}{2}x)e^{-x} \sin y.$$

That function does satisfy conditions (1), (2), and (3) and it vanishes, together with its derivatives, as $x \to \infty$.

Such problems in Z are also adapted to the finite Fourier transformation S_n with respect to y, or to successive transformations S_α with respect to x and S_n with respect to y.

PROBLEMS

1. Derive properties (6), Sec. 140.

2. For the function

$$G(x) = c - x \qquad \text{when } 0 \leq x \leq c, \qquad G(x) = 0 \qquad \text{when } x \geq c,$$

prove that $C_\alpha\{G(x)\} = (1 - \cos \alpha c)\alpha^{-2}$.

3. The static transverse displacements $Y(x)$ in a string stretched upon an elastic support along the positive x axis, with its end $x = 0$ free to slide along the Y axis and with a uniform force over the span $0 < x < b$ in the direction of the negative Y axis, satisfies the conditions

$$Y''(x) - h^2 Y(x) - F(x) = 0 \qquad \qquad (x > 0),$$

$$Y'(0) = 0, \qquad Y(\infty) = 0,$$

where $F(x) = A$ when $0 < x < b$ and $F(x) = 0$ when $x > b$, A, b, and h being positive constants. Also, Y and Y' are to be continuous when $x \geq 0$. Show formally that

$$C_\alpha\{Y(x)\} = -\frac{A}{h^2}\left(\frac{\sin \alpha b}{\alpha} - \frac{\alpha}{h^2 + \alpha^2} \sin \alpha b\right)$$

and hence that

$$-h^2 Y(x) = \begin{cases} A(1 - e^{-bh} \cosh hx) & \text{when } 0 \leq x \leq b, \\ Ae^{-hx} \sinh bh & \text{when } x \geq b. \end{cases}$$

Verify that solution.

4. Use a transformation with respect to x in the boundary value problem

$$Y_{tt}(x,t) = Y_{xx}(x,t) \qquad\qquad (t > 0, x > 0),$$

$$Y_x(0,t) = -1, \qquad Y(\infty,t) = 0, \qquad Y(x,0) = Y_t(x,0) = 0,$$

to derive the solution

$$Y(x,t) = \begin{cases} t - x & \text{when } 0 \leq x \leq t, \\ 0 & \text{when } x \geq t. \end{cases}$$

5. Use a transformation with respect to x to find the solution of the boundary value problem

$$Y_{tt}(x,t) = a^2 Y_{xx}(x,t) \qquad\qquad (t > 0, x > 0),$$

$$Y(x,0) = \Phi(x), \qquad Y_t(x,0) = 0, \qquad Y(0,t) = 0, \qquad Y(\infty,t) = 0,$$

in the form (8), Sec. 43. Also, use superposition of the solutions of this problem and Prob. 6, Sec. 118, to write the solution when the initial conditions are

$$Y(x,0) = \Phi(x), \qquad Y_t(x,0) = G(x) \qquad\qquad (x > 0).$$

6. Find the solution of the problem

$$Y_{tt}(x,t) = Y_{xx}(x,t) + 2F(x) \qquad\qquad (t > 0, x > 0),$$

$$Y(x,0) = Y_t(x,0) = Y(0,t) = 0,$$

in terms of the odd extension F_1 or F, in the form

$$Y(x,t) = \int_0^t \int_{x-r}^{x+r} F_1(\xi) \, d\xi \, dr.$$

7. (a) Transform Laplace's equation $V_{xx} + V_{yy} = 0$ with respect to x to derive the integral formula

$$V(x,y) = \frac{1}{\pi} \int_0^\infty F(r) \left[\frac{y}{(r-x)^2 + y^2} - \frac{y}{(r+x)^2 + y^2} \right] dr$$

for the harmonic function in the quadrant $x \geq 0$, $y > 0$, such that $V(0,y) = 0$ and $V(x,+0) = F(x)$.

(b) In case $F(x) = 1$ whenever $x > 0$, then, even though $f_s(\alpha)$ does not exist, show that the formula in part (a) gives the valid solution

$$V(x,y) = \frac{2}{\pi} \arctan \frac{x}{y} = 1 - \frac{2}{\pi} \operatorname{Im} [\operatorname{Log}(x + iy)].$$

(c) Verify the solution given by the integral formula in part (a) when

$$F(x) = 1 \quad \text{if} \quad 0 < x < 1, \quad \text{and} \quad F(x) = 0 \quad \text{if} \quad x > 1.$$

8. The steady temperatures $W(r,x)$ in a semi-infinite cylinder satisfy the conditions

$$W_{rr}(r,x) + \frac{1}{r} W_r(r,x) + W_{xx}(r,x) = 0 \qquad (0 < r < 1, x > 0),$$

$$W(1,x) = 0, \qquad W(r,+0) = 1,$$

and W is to be continuous when $0 \leq r \leq 1$ and $x > 0$. Use a transformation with respect to x to derive the formula

$$W(r,x) = \frac{2}{\pi} \int_0^\infty \left[1 - \frac{J_0(i\alpha r)}{J_0(i\alpha)} \right] \frac{\sin \alpha x}{\alpha} \, d\alpha = 1 - \frac{2}{\pi} \int_0^\infty \frac{J_0(i\alpha r)}{J_0(i\alpha)} \frac{\sin \alpha x}{\alpha} \, d\alpha.$$

9. Temperatures in a semi-infinite solid satisfy the conditions

$$U_t(x,t) = U_{xx}(x,t) + H(x,t) \qquad (t > 0, x > 0),$$

$$U(x,0) = F(x), \qquad U(0,t) = G(t).$$

Derive formally the solution

$$U(x,t) = \frac{2}{\pi} \int_0^\infty u_s(\alpha,t) \sin \alpha x \, d\alpha$$

where

$$u_s(\alpha,t) \exp(\alpha^2 t) = f_s(\alpha) + \int_0^t [\alpha G(\tau) + h_s(\alpha,t)] \exp(\alpha^2 \tau) \, d\tau,$$

f_s and h_s being sine transforms of F and H with respect to x.

10. (a) When $G(t) = H(x,t) = 0$ in Prob. 9, obtain the formula for $U(x,t)$ in the form given in Prob. 10, Sec. 47.

(b) When $F(x) = H(x,t) = 0$ in Prob. 9, obtain the formula for $U(x,t)$ in the form derived in Sec. 47.

11. The static deflections $Z(x,y)$ in a thin elastic plate in the form of a quadrant satisfy the conditions

$$\frac{\partial^4 Z}{\partial x^4} + 2\frac{\partial^4 Z}{\partial x^2 \partial y^2} + \frac{\partial^4 Z}{\partial y^4} = 0 \qquad (x > 0, y > 0),$$

$$Z(0,y) = Z_{xx}(0,y) = 0 \qquad (y > 0),$$

$$Z(x,0) = \frac{bx}{1 + x^2}, \qquad Z_{yy}(x,0) = 0 \qquad (x > 0);$$

also Z and its derivatives are to vanish as $x \to \infty$ and $y \to \infty$. Derive the formulas

$$Z(x,y) = \frac{b}{2} \int_0^\infty (2 + \alpha y) \exp[-(1 + y)\alpha] \sin \alpha x \, d\alpha$$

$$= \frac{bx}{x^2 + (1 + y)^2} + bxy\frac{1 + y}{[x^2 + (1 + y)^2]^2}.$$

The first form is helpful in verifying the solution.

12. In Prob. 11 let the conditions on the edge $y = 0$ be replaced by the conditions

$$Z(x,0) = 0, \qquad Z_{yy}(x,0) = \frac{bx}{(1 + x^2)^2} \qquad\qquad (x > 0),$$

then obtain the solution in the forms

$$Z(x,y) = -\frac{by}{4} \int_0^\infty \exp\left[-(1 + y)\alpha\right] \sin \alpha x \, d\alpha$$

$$= -\frac{b}{4} \frac{xy}{x^2 + (1 + y)^2}.$$

145 A MODIFIED FOURIER TRANSFORMATION T_α

We follow the procedure used in Chap. 10 to design a transformation

$$T\{F(x)\} = \int_0^\infty F(x)\Phi(x,\lambda) \, dx = f(\lambda)$$

of functions F on the half line $x > 0$ that resolves F'' in terms of $f(\lambda)$ and the boundary value $F'(0) - hF(0)$, where the constant h is positive.

Suppose that F is of class C'' and absolutely integrable over the half line $x \geq 0$ and that $F(\infty) = F'(\infty) = 0$. If for each fixed λ the kernel Φ and its derivative Φ' with respect to x are bounded and if Φ is of class C'', we can integrate by parts from 0 to c and let c tend to infinity to write

$$T\{F''\} = \int_0^\infty F\Phi'' \, dx + [F'\Phi - F\Phi']_0^\infty$$

$$= \int_0^\infty F(x)\Phi''(x,\lambda) \, dx - \begin{vmatrix} F'(0) - hF(0) & F(0) \\ \Phi'(0,\lambda) - h\Phi(0,\lambda) & \Phi(0,\lambda) \end{vmatrix}.$$

Thus if Φ and λ satisfy the singular eigenvalue problem

(1) $\Phi''(x,\lambda) = \lambda\Phi(x,\lambda)$ when $x > 0$, $\Phi'(0,\lambda) - h\Phi(0,\lambda) = 0$,

where Φ and Φ' are bounded, then

$$T\{F''(x)\} = \lambda f(\lambda) - \Phi(0,\lambda)[F'(0) - hF(0)].$$

A solution of that eigenvalue problem is $\lambda = -\alpha^2$, where α is real and positive, and $\Phi(x,\lambda) = \Phi_\alpha(x)$ where

(2) $\Phi_\alpha(x) = \alpha \cos \alpha x + h \sin \alpha x$ $(\alpha > 0)$.

Our transformation T is therefore the *modified Fourier transformation* T_α:

(3) $T_\alpha\{F(x)\} = \int_0^\infty F(x)\Phi_\alpha(x) \, dx = f_T(\alpha)$ $(\alpha > 0)$,

with the operational property

(4) $$T_\alpha\{F''(x)\} = -\alpha^2 f_T(\alpha) - \alpha[F'(0) - hF(0)].$$

From formulas (2) and (3) we see that T_α can be expressed in terms of Fourier cosine and sine transformations:

(5) $$T_\alpha\{F(x)\} = \alpha C_\alpha\{F(x)\} + hS_\alpha\{F(x)\};$$

that is, $$f_T(\alpha) = \alpha f_c(\alpha) + hf_s(\alpha).$$

Thus particular transforms and some properties of T_α can be obtained from tables and properties of C_α and S_α; in fact, property (4) can be verified by writing the transformations there in terms of sine and cosine transformations.

If we write

(6) $$\theta(\alpha) = \arctan \frac{\alpha}{h} \qquad \text{where } 0 < \theta(\alpha) < \frac{\pi}{2},$$

then $\Phi_\alpha = \sqrt{h^2 + \alpha^2} \sin(\alpha x + \theta)$, and T_α takes the form

(7) $$T_\alpha\{F(x)\} = \sqrt{h^2 + \alpha^2} \int_0^\infty F(x) \sin[\alpha x + \theta(\alpha)] \, dx.$$

In Sec. 140 we noted conditions on F under which

$$\alpha C_\alpha\{F(x)\} = -S_\alpha\{F'(x)\}.$$

Under those conditions form (5) for T_α can be written

(8) $$T_\alpha\{F(x)\} = f_T(\alpha) = S_\alpha\{hF(x) - F'(x)\}.$$

Formally then we can invert the sine transformation to write

(9) $$hF(x) - F'(x) = \frac{2}{\pi} \int_0^\infty f_T(\alpha) \sin \alpha x \, d\alpha \qquad (x > 0).$$

An inverse of the linear differential operation

(10) $$hF(x) - F'(x) = P(x)$$

is found by solving the differential equation here for F. The particular inverse that is bounded when P is bounded can be written

(11) $$F(x) = e^{hx} \int_x^\infty e^{-hr} P(r) \, dr = \int_0^\infty e^{-ht} P(x + t) \, dt.$$

When $P(x)$ is the right-hand member of Eq. (9), then

$$F(x) = \frac{2}{\pi} \int_0^\infty e^{-ht} \int_0^\infty f_T(\alpha) \sin \alpha(x + t) \, d\alpha \, dt.$$

We invert the order of integration here and use the known Laplace transform of $\sin \alpha(x + t)$ with respect to t, with parameter h, to write this formula for the inverse transformation $T_\alpha^{-1}\{f_T(\alpha)\}$:

$$(12) \qquad F(x) = \frac{2}{\pi} \int_0^\infty f_T(\alpha) \frac{\alpha \cos \alpha x + h \sin \alpha x}{h^2 + \alpha^2} \, d\alpha$$

$$= \frac{2}{\pi} \int_0^\infty f_T(\alpha) \Phi_\alpha(x) \frac{d\alpha}{h^2 + \alpha^2} \qquad (x > 0).$$

The inversion formula (12) is valid when F satisfies our conditions (Sec. 138) under which $F(x)$ is represented by its Fourier sine integral formula.[1]

Note that if $h = 0$, formula (12) reduces to the inversion formula for the Fourier cosine transformation.

146   CONVOLUTION FOR T_α

The following development of kernel-product and convolution formulas for our modified Fourier transformation is formal. The formulas are more complicated than those for the sine and cosine transformations, so they are apt to be troublesome to apply. We illustrate in the following section how our expression (8), Sec. 145, for T_α in terms of S_α may be used to circumvent our convolution formula for T_α.

The expression just referred to can be written

$$(1) \qquad f_T(\alpha) = S_\alpha\{P(x)\} \quad \text{if} \quad P(x) = hF(x) - F'(x),$$

where $f_T(\alpha) = T_\alpha\{F(x)\}$. Now $P(x) = -Q'(x)$ if

$$(2) \qquad Q(x) = F(x) + h \int_x^\infty F(r) \, dr$$

and, since $S_\alpha\{-Q'\} = \alpha C_\alpha\{Q\}$, we have this alternate representation of $T_\alpha\{F\}$:

$$(3) \qquad f_T(\alpha) = \alpha q_c(\alpha) = \alpha C_\alpha\{Q(x)\}.$$

To obtain a kernel-product formula, we write

$$\frac{2}{\alpha} f_T(\alpha) \Phi_\alpha(k) = 2\alpha q_c(\alpha) \cos \alpha k + 2h q_c(\alpha) \sin \alpha k$$

---

[1] The proof, with the aid of Fourier sine and cosine integral formulas, is not difficult even for a somewhat more general integral formula. See R. V. Churchill, Generalized Fourier Integral Formulas, *Mich. Math. Jour.*, vol. 2, pp. 133–139, 1953–54.

where $k > 0$, and use the kernel-product properties (Appendix D)

$$2q_c(\alpha) \cos \alpha k = C_\alpha\{Q(|x - k|) + Q(x + k)\},$$
$$2q_c(\alpha) \sin \alpha k = S_\alpha\{Q(|x - k|) - Q(x + k)\},$$

to write

(4) $$\frac{2}{\alpha} f_T(\alpha)\Phi_\alpha(k) = T_\alpha\{Q(|x - k|)\} + \alpha C_\alpha\{Q(x + k)\} - hS_\alpha\{Q(x + k)\}.$$

In formulas (1) and the definition of f_T we can replace h by $-h$ and F by Q. Thus

$$\alpha q_c(\alpha) - h q_s(\alpha) = -S_\alpha\{hQ(x) + Q'(x)\}.$$

If we replace h by $-h$ in the definition (2) of Q and write

$$R(x) = F(x) - h \int_x^\infty F(r)\, dr,$$

we find that

$$hQ(x) + Q'(x) = F'(x) + h^2 \int_x^\infty F(r)\, dr = -hR(x) + R'(x).$$

Then, in view of Eqs. (1),

$$-S_\alpha\{hQ(x) + Q'(x)\} = T_\alpha\{R(x)\},$$

and the right-hand member of Eq. (4) becomes

$$T_\alpha\{Q(|x - k|) + R(x + k)\} = T_\alpha\{P_F(x,k)\}$$

where

$$P_F(x,k) = F(|x - k|) + h \int_{|x-k|}^\infty F(r)\, dr + F(x + k) - h \int_{x+k}^\infty F(r)\, dr.$$

Our kernel-product formula is therefore

(5) $$\frac{2}{\alpha} f_T(\alpha)\Phi_\alpha(k) = T_\alpha\{P_F(x,k)\} \qquad\qquad (k > 0),$$

where

(6) $$P_F(x,k) = F(|x - k|) + F(x + k) + h \int_{|x-k|}^{x+k} F(t)\, dt.$$

The corresponding convolution (Sec. 109) for T_α is

(7) $$X_{FG}(x) = \int_0^\infty G(r)P_F(x,r)\, dr$$

$$= \int_0^\infty G(r)\left[F(|x - r|) + F(x + r) + h \int_{|x-r|}^{x+r} F(t)\, dt \right] dr,$$

and if $g_T(\alpha) = T_\alpha\{G(x)\}$, the convolution formula is

$$(8) \qquad \frac{2}{\alpha} f_T(\alpha) g_T(\alpha) = T_\alpha\{X_{FG}(x)\}.$$

147 SURFACE HEAT TRANSFER

Let $V(x,y)$ denote steady temperatures in a quadrant $x \geq 0, y \geq 0$, $-\infty < z < \infty$, whose two faces are subject to the linear law of surface heat transfer. If the medium outside the face $x = 0$ is at temperature zero and that outside the face $y = 0$ is at a temperature that varies with x, then

$$\begin{aligned} V_{xx}(x,y) + V_{yy}(x,y) &= 0 & (x > 0, y > 0), \\ V_x(0,y) &= hV(0,y), \qquad V_y(x,0) = kV(x,0) - F(x); \end{aligned}$$
(1)

also, V and F are to satisfy appropriate conditions of regularity. The constants h and k are positive.

Our transformation T_α with respect to x applies here in view of the differential form V_{xx}, on the half line $x > 0$ and a prescribed value of $V_x - hV$ at the boundary $x = 0$. If $v_T(\alpha,y) = T_\alpha\{V(x,y)\}$ and $f_T(\alpha) = T_\alpha\{F(x)\}$, then problem (1) transforms into the problem

$$-\alpha^2 v_T(\alpha,y) + \frac{d^2}{dy^2} v_T(\alpha,y) = 0 \qquad (y > 0),$$
(2)
$$\frac{dv_T}{dy} = kv_y - f_T(\alpha) \qquad \text{when } y = 0.$$

The bounded solution of this problem can be written as

$$(3) \qquad v_T(\alpha,y) = f_T(\alpha)\frac{e^{-\alpha y}}{k + \alpha} = f_T(\alpha)e^{ky}\int_y^\infty e^{-(k+\alpha)\tau}\, d\tau.$$

According to our inversion formula (12), Sec. 145, then

$$(4) \qquad V(x,y) = \frac{2}{\pi}\int_0^\infty f_T(\alpha)e^{-\alpha y}\frac{\alpha\cos\alpha x + h\sin\alpha x}{(k + \alpha)(h^2 + a^2)}\, d\alpha.$$

That form of the solution of problem (1) can be verified.

To present another form of the solution, let us express the transforms in terms of sine transforms. We write

$$(5) \qquad U(x,y) = hV(x,y) - V_x(x,y).$$

Then according to formula (1), Sec. 146,

$$(6) \qquad v_T(\alpha,y) = S_\alpha\{U(x,y)\} = u_s(\alpha,y).$$

Similarly,

(7) $\qquad\qquad f_T(\alpha) = p_s(\alpha) \qquad$ if $P(x) = hF(x) - F'(x).$

We use formula (11), Sec. 145, to recover V and F from U and P:

(8)
$$V(x,y) = \int_0^\infty e^{-h\xi} U(x + \xi, y)\, d\xi,$$

$$F(x) = \int_0^\infty e^{-h\xi} P(x + \xi)\, d\xi$$

The second representation of v_T in Equation (3) can now be written in the form

(9) $\qquad\qquad u_s(\alpha,y) = e^{ky} \int_y^\infty e^{-\alpha\tau} p_s(\alpha) e^{-k\tau}\, d\tau.$

Since

$$e^{-\alpha\tau} = \frac{2}{\pi} C_\alpha \left\{ \frac{\tau}{\tau^2 + x^2} \right\},$$

it follows from the joint convolution property (3), Sec. 141, that

$$e^{-\alpha\tau} p_s(\alpha) = \frac{1}{\pi} S_\alpha \left\{ \int_0^\infty P(r) \left[\frac{\tau}{\tau^2 + (x - r)^2} - \frac{\tau}{\tau^2 + (x + r)^2} \right] dr \right\}.$$

Formally then, the inverse of the sine transform (9) is

(10) $\quad U(x,y) = \dfrac{e^{ky}}{\pi} \displaystyle\int_y^\infty e^{-k\tau} \int_0^\infty P(r) \left[\dfrac{\tau}{\tau^2 + (x - r)^2} - \dfrac{\tau}{\tau^2 + (x + r)^2} \right] dr\, d\tau.$

Formula (8) then gives a representation of $V(x,y)$.

The solution represented by Eqs. (8) and (10) can be simplified and verified in case $F(x) = 1$, in which case $P(x) = h$, even though the transforms $f_T(\alpha)$ and $p_s(\alpha)$ do not exist.

14

Hankel Transforms

148 INTRODUCTION

Transforms produced by linear integral transformations whose kernels are Bessel functions are called Hankel transforms.[1] They are sometimes referred to as Bessel transforms. There is a great variety of those transformations, first because of the variety of Bessel functions, solutions of Bessel's differential equation

$$(1) \qquad x^2 y''(x) + xy'(x) + (\mu^2 x^2 - \nu^2)y(x) = 0$$

containing two parameters μ and ν. Furthermore, as in the case of Fourier transforms, the transformation may be made over either a bounded interval and involve various boundary conditions, or over a half line.

[1] Friedrich W. Bessel (1784–1846), German astronomer and mathematician; Hermann Hankel (1839–1873), German mathematician.

The origin is a singular point of Bessel's differential equation. If $v \geqq 0$, the solution of the equation that is continuous everywhere, including the origin, is

$$(2) \qquad y(x) = CJ_v(\mu x)$$

where C is any constant and J_v is Bessel's function of the first kind with the series representation

$$(3) \qquad J_v(t) = \sum_{j=0}^{\infty} \frac{(-1)^j}{j!\,\Gamma(v+j+1)} \left(\frac{t}{2}\right)^{v+2j} .$$

In the more common applications the parameter v is a nonnegative integer. Then Eq. (3) becomes

$$(4) \qquad J_n(t) = \left(\frac{t}{2}\right)^n \sum_{j=0}^{\infty} \frac{(-1)^j}{j!\,(n+j)!} \left(\frac{t}{2}\right)^{2j} \qquad (n = 0, 1, 2, \ldots).$$

We restrict our attention to that case and first treat finite transformations, then transformations on the half line $x > 0$.[1]

Useful operational properties of Hankel transformations are much fewer than those for Fourier and Laplace transformations. But again, applications to the solution of linear boundary value problems do not depend on the homogeniety of differential equations or boundary conditions.

149 FINITE HANKEL TRANSFORMATIONS

We consider functions on a bounded interval extending to the right of the origin and take our unit of length as the length of the interval, and this differential form in such functions:

$$(1) \qquad D_2 F = F''(x) + \frac{1}{x}F'(x) - \frac{n^2}{x^2}F(x) \qquad (0 < x < 1),$$

where n is zero or a positive integer. That form arises when the Laplacian operator ∇^2 is written in terms of cylindrical coordinates; in that case our variable x here is the radial coordinate usually called r or ρ (cf. Sec. 83). In terms of a self-adjoint form we write

$$(2) \qquad D_2 F = \frac{1}{x}\left\{[xF'(x)]' - \frac{n^2}{x}F(x)\right\} \qquad (0 < x < 1, n = 0, 1, 2, \ldots),$$

so in the notation used in Sec. 111, $p(x) = r(x) = x$ and $q(x) = -n^2/x$.

---

[1] As noted in Sec. 83, basic properties and representations of J_n are developed in the author's "Fourier Series and Boundary Value Problems." For extensive treatments of Bessel functions see G. N. Watson, "A Treatise on the Theory of Bessel Functions," 2d ed., 1944, or A. Gray, G. B. Mathews, and T. M. MacRobert, "A Treatise on Bessel Functions and Their Applications to Physics," 2d ed., 1952. For a shorter account see F. Bowman, "Introduction to Bessel Functions," 1958.

Let us first find the integral transformation

(3) $$T\{F\} = \int_0^1 F(x)x\Phi(x,\lambda)\,dx = f(\lambda),$$

with weight function $p(x) = x$, that transforms D_2F in terms of λ, $f(\lambda)$, and *the boundary value* $F(1)$. According to Eq. (8), Sec. 111, we can write

$$T\{D_2F\} = \int_0^1 F(x)\left[(x\Phi')' - \frac{n^2}{x}\Phi\right]dx + [x(\Phi F' - \Phi'F)]_0^1.$$

Thus if Φ and λ are such that

(4) $$(x\Phi')' - \frac{n^2}{x}\Phi = \lambda x\Phi(x,\lambda) \qquad \text{and} \qquad \Phi(1,\lambda) = 0,$$

then

(5) $$T\{D_2F\} = \lambda f(\lambda) - \Phi'(1,\lambda)F(1).$$

We assume, say, that Φ and F are of class C'' on the interval $0 \le x \le 1$.

It is convenient to write $\lambda = -\mu^2$. The differential equation (4) is Bessel's equation with index n,

(6) $$x^2\Phi'' + x\Phi' + (\mu^2x^2 - n^2)\Phi = 0.$$

Its solution of class C'' $(0 \le x \le 1)$ is $\Phi = J_n(\mu x)$, and the condition $\Phi(1,\lambda) = 0$ is satisfied if μ is such that $J_n(\mu) = 0$. For each fixed n $(n = 0, 1, 2,\ldots)$, the equation $J_n(\mu) = 0$ has an infinite sequence of roots, all real, and only the positive roots need be considered because they correspond to the full set of linearly independent eigenfunctions of the singular eigenvalue problem (4). Let μ_j, or $\mu_j(n)$, denote those roots:

(7) $$J_n(\mu_j) = 0 \qquad \text{where } j = 1, 2,\ldots, \mu_j > 0, \text{ and } \mu_j \text{ depends on } n.$$

Our transformation T then becomes the *finite Hankel transformation* on the interval $(0,1)$,

(8) $$H_{nj}\{F(x)\} = \int_0^1 F(x)xJ_n(\mu_jx)\,dx = f_H(\mu_j) \qquad (j = 1, 2,\ldots)$$

with the operational property

$$H_{nj}\{D_2F\} = -\mu_j^2 f_H(\mu_j) - \mu_j J_n'(\mu_j)F(1).$$

Here J_n' denotes the derivative of J_n with respect to its argument, so $dJ_n(\mu_jx)/dx = \mu_j J_n'(\mu_jx)$. Note that for each fixed n the transform $f_H(\mu_j)$ is a sequence of real numbers.

In view of the known formula

(9) $$tJ_n'(t) = nJ_n(t) - tJ_{n+1}(t)$$

and the fact that $J_n(\mu_j) = 0$, it follows that $J_n'(\mu_j) = -J_{n+1}(\mu_j)$. Thus the *basic operational property* of H_{nj} can be written

(10) $$H_{nj}\left\{F''(x) + \frac{1}{x}F'(x) - \frac{n^2}{x^2}F(x)\right\} = -\mu_j^2 f_H(\mu_j) + \mu_j J_{n+1}(\mu_j)F(1),$$

when F is of class C'' $(0 \le x \le 1)$ and $\mu_1, \mu_2, \ldots$ are the positive zeros of $J_n(\mu)$. Numerical values of the first four zeros of $J_0(\mu)$ were given in Sec. 83.

When $n = 0$, $D_2[1] = 0$ and $D_2[x^2 - 1] = 4$. Thus these particular transforms follow from formula (10):

(11) $$H_{0j}\{1\} = \frac{1}{\mu_j}J_1(\mu_j),$$

(12) $$H_{0j}\{1 - x^2\} = \frac{4}{\mu_j^2}H_{0j}\{1\} = \frac{4}{\mu_j^3}J_1(\mu_j),$$

where $J_0(\mu_j) = 0$ and $\mu_j > 0$.

Also from property (10) we find that

(13) $$H_{nj}\{x^n\} = \frac{1}{\mu_j}J_{n+1}(\mu_j),$$

(14) $$H_{nj}\{J_n(cx)\} = J_n(c)\frac{\mu_j}{\mu_j^2 - c^2}J_{n+1}(\mu_j) \qquad (c > 0),$$

where μ_j are the positive zeros of $J_n(\mu)$ when n is some nonnegative integer, and c is not one of those zeros.

150 INVERSION OF H_{nj}

For a fixed n the set of functions $J_n(\mu_j x)$, where $j = 1, 2, \ldots$, is orthogonal on the interval $(0,1)$ with weight function x, and norms of the functions are known. We can express that orthogonality in this form:

(1) $$H_{nj}\{J_n(\mu_i x)\} = \begin{cases} 0 & \text{if } j \ne i \qquad (i = 1, 2, \ldots) \\ \frac{1}{2}[J_{n+1}(\mu_i)]^2 & \text{if } j = i, \end{cases}$$

where $\mu_1, \mu_2, \ldots$ are the positive zeros of $J_n(\mu)$.

For a function F on the interval the generalized Fourier series in terms of the functions of that set is the *Fourier-Bessel series*

$$2 \sum_{j=1}^{\infty} \frac{J_n(\mu_j x)}{[J_{n+1}(\mu_j)]^2} \int_0^1 F(\xi)\xi J_n(\mu_j \xi)\, d\xi.$$

That series converges to $F(x)$ on the interval $0 < x < 1$ if F' is sectionally continuous there and if F is defined as the mean value of its limits from the right and left at each point where F is discontinuous. Then we can write

$$(2) \qquad F(x) = 2 \sum_{j=1}^{\infty} [J_{n+1}(\mu_j)]^{-2} f_H(\mu_j) J_n(\mu_j x) \qquad (0 < x < 1),$$

where f_H is our Hankel transform of F corresponding to the zeros μ_j of $J_n(\mu)$. Equation (2) is an *inversion formula* for the transformation H_{nj}.

Kernel-product and convolution properties are not known for finite Hankel transformations. Some general properties of Sturm-Liouville transforms noted in Chap. 10 can be extended to this singular case. The basic formula for $H_{nj}\{D_2 F\}$ when F or F' has a jump on the interval $0 < x < 1$, for instance, can be written by the method used to obtain formula (7), Sec. 113 (see Prob. 3, Sec. 152, for the case $n = 0$). The following is another such property.

If $f_H(\mu_j) = H_{nj}\{F(x)\}$, then $f_H(\mu_j)/\mu_j^2$ is the transform $y_H(\mu_j)$ of a function $Y(x)$ such that $D_2 Y = -F(x)$ and $Y(1) = 0$, according to our basic operational property for H_{nj}. Thus

$$(3) \qquad x^2 Y''(x) + x Y'(x) - n^2 Y(x) = -x^2 F(x), \qquad Y(1) = 0,$$

and Y and Y' are to be continuous and Y'' sectionally continuous over the interval $0 \le x \le 1$. A particular solution of the homogeneous form of the differential equation here is x^n. If we write $Y = x^n Z$, problem (3) is reduced to one of first order in Z'. In that way we find this formula for $Y(x)$, the inverse transform of $f_H(\mu_j)/\mu_j^2$:

$$(4) \qquad H_{nj}^{-1}\left\{\frac{f_H(\mu_j)}{\mu_j^2}\right\} = \int_x^1 \left(\frac{x}{r}\right)^n \int_0^r \left(\frac{t}{r}\right)^{n+1} F(t)\, dt\, dr$$

where $J_n(\mu_j) = 0$, $\mu_j > 0$, $n = 0, 1, 2, \ldots$, a result that is valid if F is sectionally continuous on the interval $(0,1)$.

151   MODIFIED FINITE TRANSFORMATIONS H_{nh}

If our differential form $D_2 F$ is to be reduced by the transformation on the interval $(0,1)$ in terms of *a boundary value* $hF(1) + F'(1)$, where $h \ge 0$, we denote the transformation by H_{nh}. The special case $n = h = 0$ was treated briefly in Sec. 116.

If $\Phi(x,\lambda)$ is the kernel, with weight function x, then as in Sec. 149 we can write

$$H_{nh}\{D_2 F\} = \int_0^1 F(x)\left[(x\Phi')' - \frac{n^2}{x}\Phi\right] dx + \Phi(1,\lambda)F'(1) - \Phi'(1,\lambda)F(1).$$

We now require that

(1) $\qquad (x\Phi')' - \dfrac{n^2}{x}\Phi = \lambda x\Phi,$ and $\qquad h\Phi + \Phi' = 0$ when $x = 1,$

so that

(2) $\qquad H_{nh}\{D_2 F\} = \lambda H_{nh}\{F\} + \Phi(1,\lambda)[hF(1) + F'(1)].$

If we write $\lambda = -\beta^2$, a solution of the singular eigenvalue problem (1) on the interval $(0,1)$ is $\Phi = J_n(\beta_i x)$ where $\beta_1, \beta_2, \ldots$ are the roots of the equation

(3) $\qquad hJ_n(\beta) + \beta J_n'(\beta) = 0$

where n is some fixed nonnegative integer.

For each pair of the numbers h and n the roots of Eq. (3) are all real. A full set of eigenfunctions of the singular problem (1) arises from the non-negative roots, roots that make up an infinite sequence $\beta_1, \beta_2, \ldots$ where $0 \leqq \beta_1 < \beta_2 < \cdots$. The sequence depends on the values of h and n. In case $n = h = 0$, we find that $\beta_1 = 0$, then the first eigenfunction is the constant $J_0(0) = 1$; otherwise $\beta_1 > 0$.

Our transformation is then the *modified finite Hankel transformation*

(4) $\qquad H_{nh}\{F(x)\} = \displaystyle\int_0^1 F(x)xJ_n(\beta_i x)\,dx = f_H(\beta_i) \qquad (i = 1, 2, \ldots),$

where β_i are the nonnegative zeros of the function $hJ_n(\beta) + \beta J_n'(\beta)$. The transform $f_H(\beta_i)$ is a sequence of numbers, a sequence that depends on the values given to n and h. The *basic operational property* (2) now takes the form

(5) $\qquad H_{nh}\left\{F''(x) + \dfrac{1}{x}F'(x) - \dfrac{n^2}{x^2}F(x)\right\} = -\beta_i^2 f_H(\beta_i)$

$$+ J_n(\beta_i)[hF(1) + F'(1)],$$

valid if F is of class C'' on the interval $0 \leqq x \leqq 1$.

Again, for each fixed pair of constants n and h, where $n = 0, 1, 2, \ldots$ and $h \geqq 0$, the set of functions $J_n(\beta_1 x), J_n(\beta_2 x), \ldots$, is orthogonal on the interval $(0,1)$, with weight function x. Using known formulas for norms of $J_n(\beta_i x)$, we can write

(6) $\qquad H_{nh}\{J_n(\beta_j x)\} = \begin{cases} 0 & \text{if } i \neq j, \qquad\qquad \text{where } j = 1, 2, \ldots, \\[2mm] \dfrac{\beta_j^2 + h^2 - n^2}{2\beta_j^2}[J_n(\beta_j)]^2 & \text{if } i = j, \end{cases}$

except that in case $n = h = 0$, then $\beta_1 = 0$ and $J_0(0) = 1$ so

(7) $\qquad H_{00}\{1\} = \displaystyle\int_0^1 xJ_0(\beta_i x)\,dx = \tfrac{1}{2}$ when $i = 1.$

The Fourier-Bessel series, the generalized Fourier series of those orthogonal functions corresponding to a function F, converges to $F(x)$ under the conditions stated for the representation (2), Sec. 150, and serves as the inversion formula

$$(8) \qquad F(x) = 2 \sum_{i=1}^{\infty} \frac{\beta_i^2}{\beta_i^2 + h^2 - n^2} \frac{f_H(\beta_i)}{[J_n(\beta_i)]^2} J_n(\beta_i x) \qquad (0 < x < 1);$$

but if $n = h = 0$, then $\beta_1 = 0$, and the first term in the summation is the constant $f_H(0) = \int_0^1 t F(t)\, dt$.

Some further properties of finite Hankel transforms and transforms of particular functions are included in the problems following Sec. 152.

The development of Hankel transforms has only minor modifications when the integer n is replaced by an arbitrary positive number v.

152   A BOUNDARY VALUE PROBLEM

In a half cylinder of semi-infinite length with internal heat generation, let steady temperatures $U(r,\phi,z)$ satisfy the conditions

$$(1) \qquad U_{rr} + \frac{1}{r} U_r + \frac{1}{r^2} U_{\phi\phi} + U_{zz} + Q(r) = 0$$

$$(0 < r < 1,\, 0 < \phi < \pi,\, z > 0),$$

$$(2) \qquad U(1,\phi,z) = 1, \qquad U(r,0,z) = U(r,\pi,z) = U(r,\phi,0) = 0.$$

Equation (1) is Poisson's equation $\nabla^2 U = -Q(r)$ in cylindrical coordinates r, ϕ, z (Fig. 109). Unless its nonhomogeneous term $Q(r)$ vanishes, that term has no Fourier sine transform on the half line $z > 0$.

Let $u_s(r,n,z)$ denote the finite Fourier sine transform of U with respect to ϕ on the interval $(0,\pi)$. Then, formally

$$(3) \qquad \frac{\partial^2 u_s}{\partial r^2} + \frac{1}{r} \frac{\partial u_s}{\partial r} - \frac{n^2}{r^2} u_s + \frac{\partial^2 u_s}{\partial z^2} + Q(r) S_n\{1\} = 0$$

$$(0 < r < 1,\, z > 0),$$

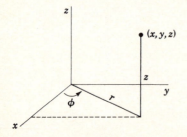

Fig. 109

(4) $$u_s(1,n,z) = S_n\{1\}, \qquad u_s(r,n,0) = 0,$$

where $S_n\{1\} = [1 - (-1)^n]n^{-1}$ and $n = 1, 2, \ldots$.

The differential form with respect to r in Eq. (3) and the boundary condition at the cylindrical surface $r = 1$ are adapted to the finite Hankel transformation H_{nj} with respect to r. We write

$$u_{sH}(\mu_j,n,z) = H_{nj}\{u_s(r,n,z)\} \qquad (j = 1, 2, \ldots),$$

where μ_j are the positive zeros of $J_n(\mu)$. Then according to property (10), Sec. 149,

$$-\mu_j{}^2 u_{sH} + \mu_j J_{n+1}(\mu_j)S_n\{1\} + \frac{d^2}{dz^2}u_{sH} + q_H(\mu_j)S_n\{1\} = 0$$

when $z > 0$, and $u_{sH}(\mu_j,n,0) = 0$, where $q_H(\mu_j)$ is the transform $H_{nj}\{Q(r)\}$. The bounded solution of this problem is

(5) $$u_{sH}(\mu_j,n,z) = S_n\{1\}\frac{q_H(\mu_j) + \mu_j J_{n+1}(\mu_j)}{\mu_j{}^2}[1 - \exp(-\mu_j z)].$$

The inverse transformations are represented by series:

$$u_s(r,n,z) = 2 \sum_{j=1}^{\infty} \frac{u_{sH}(\mu_j,n,z)}{[J_{n+1}(\mu_j)]^2}J_n(\mu_j r);$$

then

(6) $$U(r,\phi,z) = \frac{4}{\pi} \sum_{n=1}^{\infty} \sum_{j=1}^{\infty} \frac{u_{sH}(\mu_j,n,z)}{[J_{n+1}(\mu_j)]^2}J_n(\mu_j r)\sin n\phi$$

where u_{sH} is given by formula (5) and μ_j are the positive zeros of $J_n(\mu)$, which *depend on the value of n.*

In case the condition $U(1,\phi,z) = 1$ is replaced by

$$U_r(1,\phi,z) = -hU(1,\phi,z) + P(\phi,z),$$

a condition of heat transfer at the cylindrical surface, the modified finite Hankel transformation, treated in Sec. 151, can be made with respect to the variable r.

The more common applications of Hankel transforms involve the special case $n = 0$ and the Bessel function J_0.

PROBLEMS

1. Derive the finite Hankel transforms (13) and (14), Sec. 149, of the functions x^n and $J_n(cx)$.

2. Obtain the finite Hankel transforms

(a) $$H_{0j}\{x^2\} = \frac{\mu_j{}^2 - 4}{\mu_j{}^3}J_1(\mu_j),$$

(b) $H_{0j}\{1 - x^4\} = \dfrac{16}{\mu_j^2} H_{0j}\{x^2\}$,

where $J_0(\mu_j) = 0$ and $\mu_j > 0$ $(j = 1, 2, \ldots)$.

3. If F and F' are continuous on the interval $0 \leq x \leq 1$ except possibly for finite jumps at an interior point $x = c$, and if F'' is sectionally continuous on the interval, show that

$$H_{0j}\left\{F''(x) + \frac{1}{x}F'(x)\right\} = -\mu_j^2 f_H(\mu_j) + \mu_j J_1(\mu_j)F(1)$$

$$- c\mu_j J_1(\mu_j c)[F(c + 0) - F(c - 0)] - cJ_0(\mu_j c)[F'(c + 0) - F'(c - 0)]$$

where μ_j are the positive zeros of $J_0(\mu)$ and

$$f_H(\mu_j) = H_{0j}\{F\} = \int_0^1 F(x)xJ_0(\mu_j x)\,dx \qquad (j = 1, 2, \ldots).$$

4. Let $G(x,c)$ be Green's function (Sec. 95) for the differential form D_2 when $n = 0$, such that

$$xG_{xx}(x,c) + G_x(x,c) = 0 \qquad (0 < x < 1, x \neq c),$$

$$G(1,c) = 0, \qquad G_x(c + 0, c) - G_x(c - 0, c) = \frac{1}{c}$$

and G is continuous when $0 \leq x \leq 1$, where $0 < c < 1$. Show that

$$G(x,c) = \log c \qquad \text{when } 0 \leq x \leq c,$$

$$= \log x \qquad \text{when } c \leq x \leq 1;$$

then use the operational property found in Prob. 3 to obtain the transform

$$H_{0j}\{G(x,c)\} = -\frac{1}{\mu_j^2} J_0(\mu_j c) \qquad (j = 1, 2, \ldots),$$

where $J_0(\mu_j) = 0$ and $\mu_j > 0$. Compare formula (3), Sec. 114.

5. If n and h are not both zero and $\beta_1, \beta_2, \ldots$ are the positive zeros of $hJ_n(\beta) + \beta J_n'(\beta)$, show that

$$\int_0^1 x^{n+1} J_n(\beta_i x)\,dx = H_{nh}\{x^n\} = \frac{h + n}{\beta_i^2} J_n(\beta_i) \qquad (n = 0, 1, 2, \ldots).$$

6. When $n = 0$ and $h > 0$ and β_i are the positive zeros of the function $hJ_0(\beta) + \beta J_0'(\beta)$, or $hJ_0(\beta) - \beta J_1(\beta)$, and if

$$f_H(\beta_i) = H_{0h}\{F\} = \int_0^1 F(x)xJ_0(\beta_i x)\,dx \qquad (i = 1, 2, \ldots),$$

derive the formula

$$\frac{1}{\beta_i^2} f_H(\beta_i) = H_{0h}\left\{\int_x^1 \int_0^y \frac{t}{y}F(t)\,dt\,dy + \frac{1}{h}\int_0^1 tF(t)\,dt\right\}.$$

Also, from the transform $H_{0h}\{1\}$ found in Prob. 5, show that

$$\frac{h}{\beta_i^4} J_0(\beta_i) = H_{0h}\left\{\frac{1}{2h} + \frac{1-x^2}{4}\right\}.$$

7. When $n = h = 0$ in the transformation H_{nh}, note that

$$H_{00}\left\{F''(x) + \frac{1}{x}F'(x)\right\} = -\beta_i^2 H_{00}\{F\} + F'(1)J_0(\beta_i) \quad \text{if } i = 2, 3, \ldots,$$

$$= F'(1) \quad \text{if } i = 1, \qquad \text{since } \beta_1 = 0,$$

where $\beta_1, \beta_2, \ldots$ are the nonnegative zeros of $J_1(\beta)$. Show that

$$H_{00}\{1\} = 0 \quad \text{if } i = 2, 3, \ldots, \qquad H_{00}\{1\} = \tfrac{1}{2} \quad \text{if } i = 1;$$

$$H_{00}\{x^2 + b\} = \begin{cases} \dfrac{2}{\beta_i^2} J_0(\beta_i) & \text{if } i = 2, 3 \ldots, \\[2mm] \dfrac{1}{4} + \dfrac{b}{2} & \text{if } i = 1. \end{cases}$$

8. If $c > 0$ and $J_0(c) \neq 0$, show that the function Y of class C'' $(0 \leq x \leq 1)$ that satisfies the conditions

$$Y''(x) + \frac{1}{x}Y'(x) + c^2 Y(x) = x^2 - 1 \qquad Y(1) = 0,$$

has the transform

$$H_{0j}\{Y(x)\} = \frac{4}{c^4} J_1(\mu_j)\left(\frac{\mu_j}{\mu_j^2 - c^2} - \frac{1}{\mu_j} - \frac{c^2}{\mu_j^3}\right)$$

where $J_0(\mu_j) = 0$ and $\mu_j > 0$, hence (Sec. 149) that

$$Y(x) = \frac{4}{c^4}\left[\frac{J_0(cx)}{J_0(c)} - 1\right] + \frac{x^2 - 1}{c^2}.$$

9. Solve the problem

$$Y''(x) + \frac{1}{x}Y'(x) - c^2 Y(x) + F(x) = 0, \qquad 0 < x < 1; \qquad Y(1) = 0,$$

in the form

$$Y(x) = 2\sum_{j=1}^{\infty} \frac{f_H(\mu_j)}{\mu_j^2 + c^2} \frac{J_0(\mu_j x)}{[J_1(\mu_j)]^2}$$

where $J_0(\mu_j) = 0$, $\mu_j > 0$, and

$$f_H(\mu_j) = \int_0^1 F(t)t J_0(\mu_j t)\, dt \qquad\qquad (j = 1, 2, \ldots).$$

In case $F(x) = J_0(\mu_1 x)$, show that the solution becomes

$$Y(x) = \frac{J_0(\mu_1 x)}{\mu_1^2 + c^2}.$$

10. Use the Hankel transformation H_{0j} with respect to r to derive, in only a few steps, the formula (9), Sec. 83, for temperatures $U(r,t)$ in a circular cylinder of infinite length, when

$$U_t(r,t) = U_{rr}(r,t) + \frac{1}{r}U_r(r,t) \qquad (0 \le r < 1, t > 0),$$

$$U(r,0) = 0, \qquad U(1,t) = 1.$$

11. Use results found in Prob. 7 to show that the appropriate Hankel transform of $U(r,t)$, where

$$U_t(r,t) = U_{rr}(r,t) + \frac{1}{r}U_r(r,t) \qquad (0 \le r < 1, t > 0),$$

$$U(r,0) = 0, \qquad U_r(1,t) = A,$$

has the value

$$\frac{A}{2}\left[\frac{2}{\beta_i^2}J_0(\beta_i) - \frac{2}{\beta_i^2}J_0(\beta_i)\exp(-\beta_i^2 t) \right] \qquad \text{if } i = 2, 3, \ldots,$$

$$\frac{A}{2}\left[\frac{1}{4} + \frac{1}{2}\left(4t + \frac{1}{2}\right) \right] \qquad \text{if } i = 1,$$

and thus obtain the formula for $U(r,t)$ found in Prob. 8, Sec. 83.

12. In Prob. 8, Sec. 144, an integral representation was found for the temperature distribution $W(r,x)$ in a semi-infinite circular cylinder whose axis is the positive x axis, when

$$\nabla^2 W = W_{rr} + \frac{1}{r}W_r + W_{xx} = 0 \qquad (0 < r < 1, x > 0),$$

$$W(1,x) = 0 \quad \text{if } x > 0, \qquad W(r,0) = 1 \qquad \text{if } r < 1;$$

also, W is to be continuous and bounded when $0 \le r \le 1$ and $x > 0$. Use a Hankel transformation to derive this series representation of W:

$$W(r,x) = 2 \sum_{j=1}^{\infty} \frac{1}{\mu_j} \frac{J_0(\mu_j r)}{J_1(\mu_j)} \exp(-\mu_j x)$$

when $0 \le r < 1$ and $x > 0$, where $J_0(\mu_j) = 0$ and $\mu_j > 0$.

13. Let transverse displacements $Z(r,t)$ in a membrane stretched over a circular frame satisfy the conditions

$$Z_{tt} = Z_{rr} + \frac{1}{r}Z_r - 2kZ_t \qquad (0 \le r < 1, t > 0),$$

$$Z(1,t) = Z(r,0) = 0, \qquad Z_t(r,0) = v_0.$$

If the coefficient of damping $2k$ is such that $0 < k < \mu_1$ where $\mu_1, \mu_2, \ldots$ are the positive zeros of $J_0(\mu)$ in order of increasing magnitude, derive the formula

$$Z(r,t) = 2v_0 e^{-kt} \sum_{j=1}^{\infty} \frac{J_0(\mu_j r)}{\mu_j J_1(\mu_j)} \frac{\sin[t(\mu_j^2 - k^2)^{\frac{1}{2}}]}{(\mu_j^2 - k^2)^{\frac{1}{2}}}.$$

14. Let μ_j denote the positive zeros of $J_1(\mu)$ and write

$$H_{1j}\{F(r)\} = \int_0^1 F(r)rJ_1(\mu_j r)\,dr = f_H(\mu_j) \qquad (j = 1, 2, \ldots).$$

Derive the formula

$$\frac{1}{\mu_j^2}f_H(\mu_j) = H_{1j}\left\{\int_r^1 \frac{r}{\rho}\int_0^\rho \left(\frac{t}{\rho}\right)^2 F(t)\,dt\,d\rho\right\},$$

and since $H_{1j}\{r\} = \mu_j^{-1}J_2(\mu_j)$, according to transform (13), Sec. 149, show that

$$\frac{1}{\mu_j^3}J_2(\mu_j) = H_{1j}\left\{\frac{r - r^3}{8}\right\}.$$

15. If $y = r \sin \phi$ and $\nabla^2 V + Ay = 0$ inside a half cylinder $0 < r < 1, 0 < \phi < \pi$, $z > 0$, and if $V = 0$ on the boundaries, then $V(r,\phi,z)$ satisfies the conditions

$$V_{rr} + \frac{1}{r}V_r + \frac{1}{r^2}V_{\phi\phi} + V_{zz} + Ar\sin\phi = 0 \qquad (0 < r < 1,\ 0 < \phi < \pi,\ z > 0),$$

$$V(1,\phi,z) = V(r,0,z) = V(r,\pi,z) = V(r,\phi,0) = 0.$$

Reduce the problem to one in $W(r,z)$ by writing

$$V(r,\phi,z) = A \sin \phi\, W(r,z)$$

and, with the aid of the transform found in Prob. 14, derive the solution

$$W(r,z) = \frac{r - r^3}{8} - 2 \sum_{j=1}^\infty \frac{J_1(\mu_j r)}{\mu_j^3 J_2(\mu_j)} \exp(-\mu_j z)$$

where $J_1(\mu_j) = 0$ and $\mu_j > 0$.

153 NONSINGULAR HANKEL TRANSFORMATIONS

If our differential form

$$(1) \qquad D_2 F = F''(x) + \frac{1}{x}F'(x) - \frac{n^2}{x^2}F(x),$$

where n is a fixed nonnegative integer, applies to functions F on a *bounded interval* $0 < a \leq x \leq b$, one that does not have the origin as an end point or an interior point, then D_2 has no singular point on the interval. The linear integral transformation that resolves $D_2 F$ in terms of the transform of F itself and prescribed values of F or F', or of a linear combination of F and F', at the points $x = a$ and $x = b$ is a *nonsingular special case of the Sturm-Liouville transformation* introduced in Chap. 10.

The kernel $\Phi(x, -\mu^2)$ satisfies Bessel's equation

$$(2) \qquad (x\Phi')' - \frac{n^2}{x}\Phi = -\mu^2 x\Phi \qquad (a < x < b)$$

and homogeneous boundary conditions at the end points of the interval. The weight function $p(x)$ in the integral transformation is x itself. To satisfy both end conditions, the general solution of Eq. (2),

$$(3) \qquad \Phi = AJ_n(\mu x) + BY_n(\mu x)$$

is needed, where Y_n is Bessel's function of the second kind.[1] The ratio of the constants A and B, and the discrete values $\mu_1, \mu_2, \ldots$ of μ, are determined by the homogeneous boundary conditions on Φ.

In Prob. 12, Sec. 107, we used a generalized Fourier series to solve the following problem for the electrostatic potential $V(r,z)$ in the charge-free space between cylindrical surfaces $r = a$ and $r = b$, above the plane $z = 0$.

$$V_{rr}(r,z) + \frac{1}{r}V_r(r,z) + V_{zz}(r,z) = 0 \qquad (0 < a < r < b, z > 0),$$

$$(4)$$

$$V(a,z) = V(b,z) = 0, \qquad V(r,0) = F(r).$$

The solution found there can be obtained by using the nonsingular Hankel transformation

$$(5) \qquad H_0\{F(r)\} = \int_a^b F(r)r\psi(\alpha_i,r)\,dr \qquad (i = 1, 2, \ldots),$$

where

$$(6) \qquad \psi(\alpha,r) = J_0(\alpha r)Y_0(\alpha a) - J_0(\alpha a)Y_0(\alpha r)$$

and α_i are the positive zeros of the function $\psi(\alpha,b)$. But the transformation H_0 applies also to problems in which the differential equation and all boundary conditions in problem (4) are made nonhomogeneous.

154   HANKEL TRANSFORMATIONS $H_{n\alpha}$ ON THE HALF LINE $(x > 0)$

The best known of the various Hankel transformations are singular transformations of the type

$$T\{F\} = \int_0^\infty F(x)x\Phi(x,\lambda)\,dx = f(\lambda);$$

that is, they apply to functions $F(x)$ defined over the positive half of the x axis. The transformation T corresponding to a fixed integer n ($n = 0, 1, 2, \ldots$) replaces our differential form

$$D_2F = F''(x) + \frac{1}{x}F'(x) - \frac{n^2}{x^2}F(x) = \frac{1}{x}\left[(xF')' - \frac{n^2}{x}F\right]$$

---

[1] The function $Y_n(x)$ is of the order of x^{-n} as $x \to 0$ if $n = 1, 2, \ldots$, while $Y_0(x)$ is of the order of $\log x$ as $x \to 0$. A summary of representations and properties of Bessel functions, including Y_n, with references, is given by A. Erdélyi et al., "Higher Transcendental Functions," vol. 2, chap. 7, 1953.

on the half line $x > 0$ by $\lambda f(\lambda)$. That operational property is the principal one for applications of such transforms to problems in differential equations.

To construct T so that it has that property, we write

$$(1) \qquad T\{D_2 F\} = \int_0^\infty \left[(xF')' - \frac{n^2}{x} F \right] \Phi \, dx$$

$$= \int_0^\infty \left[(x\Phi')' - \frac{n^2}{x} \Phi \right] F \, dx + [x(\Phi F' - \Phi' F)]_0^\infty.$$

We assume, say, that F and Φ are of class C'' on the half line $x \geq 0$ and that the functions F and Φ are such that the last bracketed term vanishes as $x \to \infty$ and the second improper integral exists. Then we require that

$$(2) \qquad (x\Phi')' - \frac{n^2}{x} \Phi = \lambda x \Phi \qquad\qquad (x > 0)$$

so that $T\{D_2 F\} = \lambda T\{F\}$. Equation (2) is Bessel's equation. We write $\lambda = -\alpha^2$; then if α is real, the bounded solution of that equation that is of class C'' when $x \geq 0$ is $\Phi = A J_n(\alpha x)$. All linearly independent solutions are represented if $A = 1$ and $\alpha \geq 0$.

Thus T becomes the *Hankel transformation*

$$(3) \qquad H_{n\alpha}\{F(x)\} = \int_0^\infty F(x) x J_n(\alpha x) \, dx = f_H(\alpha) \qquad (\alpha \geq 0),$$

where n has one of the values $0, 1, 2, \ldots$. Its *basic operational property* is

$$(4) \qquad H_{n\alpha}\left\{ F''(x) + \frac{1}{x} F'(x) - \frac{n^2}{x^2} F(x) \right\} = -\alpha^2 f_H(\alpha).$$

Here the Hankel transform $f_H(\alpha)$ is a function of α that depends on the choice of n. The notation $f_{Hn}(\alpha)$ is more descriptive but rather cumbersome.

An asymptotic representation[1] of $J_n(t)$ for large values of t shows that $J_n(t)$ is $\mathcal{O}(t^{-\frac{1}{2}})$ as $t \to \infty$; that is, $\sqrt{t}|J_n(t)|$ is bounded. From the formula

$$(5) \qquad t J_n'(t) = n J_n(t) - t J_{n+1}(t) \qquad\qquad (n = 0, 1, 2, \ldots)$$

it follows that $J_n'(t)$ has that same order as $t \to \infty$. Now in Eq. (1), $\Phi = J_n(\alpha x)$ and $d\Phi/dx = \alpha J_n'(\alpha x)$ where J_n' is the derivative of J_n with respect to the argument of J_n; also, $(x\Phi')' - n^2 x^{-1}\Phi = -\alpha^2 x J_n(\alpha x)$. Then if $F(x)$ is $\mathcal{O}(x^{-k})$ as $x \to \infty$, where $k > \frac{3}{2}$, the integrand of the second improper integral in that equation is $\mathcal{O}(x^{-k+\frac{1}{2}})$ and, since $k - \frac{1}{2} > 1$, that integral is absolutely convergent. If in addition F' is $\mathcal{O}(x^{-k+1})$, the final term in Eq. (1) vanishes; thus the right-hand member of the equation exists, and so $T\{D_2 F\}$ exists and is equal to $-\alpha^2 H_{n\alpha}\{F\}$.

---

[1] Erdélyi, A., *et al.*, "Higher Transcendental Functions," vol. 2, p. 85.

Operational property (4) is therefore valid if F is class C'' when $x \geqq 0$, if F is $\mathcal{O}(x^{-k})$, and if F' is $\mathcal{O}(x^{-k+1})$ as $x \to \infty$, where $k > \frac{3}{2}$.

The Hankel transform (3) of F exists if F is sectionally continuous on each bounded interval of the positive x axis and if F is $\mathcal{O}(x^{-k})$ as $x \to \infty$, where $k > \frac{3}{2}$.

The Fourier-Bessel integral representation of $F(x)$ corresponding to the Fourier sine or cosine integral formula, furnishes an *inverse transformation*

$$(6) \qquad F(x) = \int_0^\infty f_H(\alpha) \alpha J_n(\alpha x)\, d\alpha = H_{nx}\{f_H(\alpha)\} \qquad (x > 0).$$

It is valid if $F(x)$ is $\mathcal{O}(x^{-k})$ as $x \to \infty$, where $k > \frac{3}{2}$, if F' is sectionally continuous over each bounded interval, and if $F(x)$ is defined as the mean value of its limits from the right and left at each of its points of discontinuity.

Our conditions under which the above formulas are valid are sufficient. They can be relaxed. The transform (3) of $F(x)$ may exist, for instance, when F is unbounded at the origin but such that $x^{n+1}F(x)$ is $\mathcal{O}(x^c)$ as $x \to 0$, where $c > -1$. No major changes are involved when the integer n in this section is replaced by v where $v > -\frac{1}{2}$.

155 FURTHER PROPERTIES OF $H_{n\alpha}$

If $f_H(\alpha) = H_{n\alpha}\{F(x)\}$, we find that

$$(1) \qquad f_H(\alpha k) = H_{n\alpha}\left\{ \frac{1}{k^2} F\!\left(\frac{x}{k}\right) \right\} \qquad\qquad \text{when } k > 0.$$

A property on division of Hankel transforms $f_H(\alpha)$ by α^2 follows from our basic operational property (4), Sec. 154, by solving the differential equation

$$Y''(x) + \frac{1}{x} Y'(x) - \frac{n^2}{x^2} Y(x) = -F(x) \qquad (x > 0)$$

for $Y(x)$, since $-\alpha^2 y_H(\alpha) = -f_H(\alpha)$. The formula is

$$(2) \qquad \frac{1}{\alpha^2} f_H(\alpha) = H_{n\alpha}\left\{ \int_x^A \left(\frac{x}{y}\right)^n \int_B^y \left(\frac{t}{y}\right)^{n+1} F(t)\, dt\, dy \right\}$$

where n is a nonnegative integer and A and B are nonnegative constants or infinity, to be chosen if possible, so that the transform on the right exists.

If F satisfies our conditions under which the operational property (4), Sec. 154, is valid except for finite jumps $j_c = F(c+0) - F(c-0)$ and $j_c' = F'(c+0) - F'(c-0)$ in F and F' at a point $x = c$, the property takes the modified form

$$(3) \qquad H_{n\alpha}\left\{F'' + \frac{1}{x}F' - \frac{n^2}{x^2}F\right\} = -\alpha^2 f_H(\alpha) + \alpha c J_n'(\alpha c) j_c - c J_n(\alpha c) j_c'.$$

A kernel-product property for $H_{n\alpha}$ arises from the formula[1]

$$(4) \qquad J_n(x)J_n(y) = \frac{x^n y^n}{2^n \sqrt{\pi}\Gamma(n + \frac{1}{2})} \int_0^\pi \frac{J_n[R(x,y,\theta)]}{[R(x,y,\theta)]^n} \sin^{2n}\theta \, d\theta$$

where $R(x,y,\theta)$ is the length of the side of a triangle opposite the angle θ included between two sides of length x and y; that is

$$(5) \qquad R(x,y,\theta) = (x^2 + y^2 - 2xy \cos \theta)^{\frac{1}{2}}.$$

If we use formula (4), the product $J_n(\alpha y)H_{n\alpha}\{F(x)\}$ can be written as an iterated integral with respect to x and θ. It is an integral over the upper half plane if x and θ are interpreted as polar coordinates. When that integral is written in terms of R and ϕ, polar coordinates about the point $(y,0)$ on the axis $\theta = 0$, we obtain the kernel-product property

$$(6) \qquad \frac{1}{\alpha^n} y J_n(\alpha y) f_H(\alpha) = H_{n\alpha}\{P_F(x,y)\}$$

where $f_H(\alpha) = H_{n\alpha}\{F(x)\}$ and

$$(7) \qquad P_F(x,y) = \frac{x^n y^{n+1}}{\sqrt{\pi}2^n \Gamma(n + \frac{1}{2})} \int_0^\pi \frac{F[R(x,y,\phi)]}{[R(x,y,\phi)]^n} \sin^{2n}\phi \, d\phi.$$

The corresponding convolution $X_{FG}(x)$ of two functions F and G such that

$$(8) \qquad \frac{1}{\alpha^n} f_H(\alpha)g_H(\alpha) = H_{n\alpha}\{X_{FG}(x)\}$$

is then formally, as indicated in Sec. 109,

$$(9) \qquad X_{FG}(x) = \int_0^\infty F(y)P_G(x,y) \, dy.$$

Formulas (4) to (9) are presented here to display their complexity. In case $n = 0$, the results are included in Table 4, but even in that case the kernel-product and convolution properties are difficult to use. The integer n can be replaced by ν, where $\nu \geq 0$, in those formulas.[2]

---

[1] Watson, G. N., *op. cit.*, p. 367.

[2] Derivations and conditions of validity of those kernel-product and convolution properties are included in J. S. Klein's doctoral dissertation, "Some Results in the Theory of Hankel Transforms," University of Michigan, 1958. Earlier, J. Delsarte, *J. Math. Pures Appl.* (9), vol. 17, pp. 213–231, 1936, and *Acta Math.*, vol. 69, pp. 259–317, 1938, gave convolution properties associated with Hankel transforms.

156   TABLES OF TRANSFORMS $H_{n\alpha}\{F\}$

Table 4 summarizes properties of the transformation $H_{n\alpha}$ on the half line $x > 0$ and lists transforms of some particular functions, for the important *special case n = 0*. Entry 5, giving a possible inverse transform of $f_H(\alpha)/\alpha^2$, is the special case of formula (2), Sec. 155, when $n = 0$, $A = \infty$, and $B = \infty$. Derivations of some of the transforms listed in the table are indicated in problems following Sec. 157.

Table 4   Hankel transforms ($n = 0$)

	$f_H(\alpha)$ $(\alpha > 0, n = 0)$	$F(x)$ $(x > 0)$
1	$\displaystyle\int_0^\infty F(x)xJ_0(\alpha x)\,dx = H_{0\alpha}\{F\}$	$F(x)$
2	$f_H(\alpha)$	$\displaystyle\int_0^\infty f_H(\alpha)\alpha J_0(\alpha x)\,d\alpha = H_{0x}\{f_H\}$
3	$-\alpha^2 f_H(\alpha)$	$F''(x) + \dfrac{1}{x}F'(x)$
4	$f_H(k\alpha)$ $(k > 0)$	$\dfrac{1}{k^2}F\!\left(\dfrac{x}{k}\right)$
5	$\dfrac{1}{\alpha^2}f_H(\alpha)$	$-\displaystyle\int_x^\infty \int_y^\infty \dfrac{t}{y}F(t)\,dt\,dy$
6	$\pi J_0(\alpha k)f_H(\alpha)$ $(k > 0)$	$\displaystyle\int_0^\pi F[(x^2 + k^2 - 2kx\cos\theta)^{\frac{1}{2}}]\,d\theta$
7	$\pi f_H(\alpha)g_H(\alpha)$	$\displaystyle\int_0^\infty yG(y)\int_0^\pi F[(x^2 + y^2 - 2xy\cos\theta)^{\frac{1}{2}}]\,d\theta\,dy$
8	$\dfrac{1}{\alpha}$	$\dfrac{1}{x}$
9	$\dfrac{1}{\sqrt{\alpha^2 + c^2}}$ $(c > 0)$	$\dfrac{1}{x}e^{-cx}$
10	$c(\alpha^2 + c^2)^{-3/2}$ $(c > 0)$	e^{-cx}
11	$e^{-c\alpha}$ $(c > 0)$	$c(x^2 + c^2)^{-3/2}$
12	$\dfrac{1}{\alpha}e^{-c\alpha}$ $(c > 0)$	$\dfrac{1}{\sqrt{x^2 + c^2}}$
13	$\dfrac{1}{\alpha^2}e^{-c\alpha}$ $(c > 0)$	$\dfrac{1}{2}\log\dfrac{\sqrt{x^2 + c^2} - c}{\sqrt{x^2 + c^2} + c}$
14	$\dfrac{c}{\alpha}J_1(c\alpha)$ $(c > 0)$	$\begin{cases} 1 & \text{if } 0 < x < c \\ 0 & \text{if } x > c \end{cases}$
15	$\dfrac{1}{\alpha}J_0(2\sqrt{c\alpha})$ $(c > 0)$	$\dfrac{1}{x}J_0\!\left(\dfrac{c}{x}\right)$

An extensive table of Hankel transforms on the half line $x > 0$ is given by A. Erdélyi *et al.*, "Tables of Integral Transforms," vol. 2, 1954. In those tables the transformation is written, except for variations in the letters used, as

$$\int_0^\infty G(x)\sqrt{\alpha x}J_\nu(\alpha x)\,dx = g(\alpha;\nu).$$

Thus in terms of the notation we use here,

$$F(x) = \frac{G(x)}{\sqrt{x}} \quad \text{and} \quad f_H(\alpha) = \frac{g(\alpha;\nu)}{\sqrt{\alpha}}.$$

Those tables list several properties of transforms $g(\alpha;\nu)$ in which different values of the index ν occur in the same property. Such properties follow from relations such as Eq. (5), Sec. 154, between J_n', J_n, and J_{n+1}, or the recurrence relation

$$2\nu J_\nu(\alpha x) = \alpha x[J_{\nu-1}(\alpha x) + J_{\nu+1}(\alpha x)].$$

The tables list transforms of particular functions in cases $\nu = 0$ and $\nu = 1$, and for the general index ν. Many of the functions and transforms involve higher transcendental functions.

Another table will be found in the book by Ditkin and Prudnikov; other treatments and applications of Hankel transforms are included in books by I. N. Sneddon ("Fourier Transforms") and C. J. Tranter, all listed in the Bibliography.

157   AXIALLY SYMMETRIC HEAT SOURCE

The following is an example of a boundary value problem for which a Hankel transformation $H_{0\alpha}$ with respect to r is the natural method of solution. Find a formula for steady temperatures $V(r,z)$ in a semi-infinite solid $z \geq 0$ throughout which heat is generated at a steady rate that varies only with the radial coordinate r, when the face $z = 0$ is kept at temperature zero. Then

$$V_{rr}(r,z) + \frac{1}{r}V_r(r,z) + V_{zz}(r,z) + F(r) = 0$$

(1) $$(r > 0, 0 \leq \phi < 2\pi, z > 0),$$

$$V(r,0) = 0,$$

where r, ϕ, and z are cylindrical coordinates (Fig. 109); also F and V are to satisfy order properties as $r \to \infty$ and $z \to \infty$ such that a solution exists.

Let $v_H(\alpha,z)$ denote the Hankel transform of V with respect to r, for which $n = 0$:

(2)
$$v_H(\alpha,z) = H_{0\alpha}\{V(r,z)\} = \int_0^\infty V(r,z)rJ_0(\alpha r)\,dr \qquad (\alpha > 0),$$

and write $f_H(\alpha) = H_{0\alpha}\{F(r)\}$. Then the transformed problem is, formally,

$$-\alpha^2 v_H(\alpha,z) + \frac{d^2}{dz^2}v_H(\alpha,z) + f_H(\alpha) = 0 \qquad (\alpha > 0, z > 0),$$

(3)
$$v_H(\alpha,0) = 0.$$

The bounded solution of that problem is

(4)
$$v_H(\alpha,z) = f_H(\alpha)\frac{1 - e^{-\alpha z}}{\alpha^2}.$$

Thus a formal solution of problem (1) is

(5)
$$V(r,z) = \int_0^\infty f_H(\alpha)\frac{1 - e^{-\alpha z}}{\alpha}J_0(\alpha r)\,d\alpha.$$

Another form of the solution can be written by using Table 4, first to represent the inverse transform of $\alpha^{-2}f_H(\alpha)$ as an iterated integral in terms of F, then by using the convolution integral (entry 7) to write an inverse transform of either the product $[\alpha^{-2}f_H(\alpha)]e^{-z\alpha}$ or $f_H(\alpha)(\alpha^{-2}e^{-z\alpha})$. But those forms are complicated, even when F is specified as the simple step function $F(r) = A$ when $0 < r < c$, $F(r) = 0$ when $r > c$, representing the case in which heat is generated uniformly throughout a cylindrical core of the solid.

In the special case

(6)
$$F(r) = \frac{k}{(r^2 + k^2)^{3/2}} \qquad \text{where } k > 0,$$

then $f_H(\alpha) = e^{-k\alpha}$, and formula (5) can be verified as a solution of problem (1). But another form of the solution can be written since

(7)
$$v_H(\alpha,z) = \frac{1}{\alpha^2}e^{-k\alpha} - \frac{1}{\alpha^2}\exp\left[-(z + k)\alpha\right].$$

From Table 4 it follows that

(8)
$$2V(r,z) = \log\frac{W(r,0) - 1}{W(r,0) + 1} - \log\frac{W(r,z) - 1}{W(r,z) + 1}$$

where

(9)
$$W(r,z) = \left[\frac{r^2}{(z + k)^2} + 1\right]^{1/2}$$

Formula (8) can be verified as a solution of problem (1) when F is the function (6). According to that formula, the temperatures on the z axis tend to infinity as $z \to \infty$, since

(10) $$V(+0,z) = \log \frac{z + k}{k}.$$

PROBLEMS

1. Use the known Laplace transformation

$$L\{J_0(\alpha t)\} = (s^2 + \alpha^2)^{-1/2} \qquad\qquad (s > 0, \ \alpha > 0)$$

to establish the transform $H_{0\alpha}\{x^{-1}e^{-cx}\}$ listed in Table 4. Also, differentiate with respect to the parameter c to show that, if $c > 0$ and $m = 1, 2, \ldots,$

$$H_{0\alpha}\{x^{m-1}e^{-cx}\} = (-1)^m \frac{\partial^m}{\partial c^m}(\alpha^2 + c^2)^{-1/2}$$

2. Note that our conditions of validity of the inversion formula (6), Sec. 154, are satisfied by the function $F(x) = e^{-cx}$ when $c > 0$. Thus

$$H_{0x}\{H_{0\alpha}[e^{-cx}]\} = e^{-cx} \qquad\qquad \text{when } x > 0.$$

From the transform $H_{0\alpha}\{e^{-cx}\}$ established in Prob. 1, deduce transform $H_{0\alpha}\{c(x^2 + c^2)^{-3/2}\}$ listed in Table 4.

3. Use transformations 5 and 14 in Table 4 to indicate that, when $c > 0$,

$$\frac{4c}{\alpha^3}J_1(c\alpha) = H_{0\alpha}\{G(x)\}$$

where $$G(x) = \begin{cases} c^2 - x^2 + 2c^2 \log \dfrac{x}{c} & \text{if } \ 0 < x \leqq c, \\ 0 & \text{if } \ x \geqq c. \end{cases}$$

4. Let $V(r,z)$ be bounded in the half space $z > 0$, $r \geqq 0$, $0 \leqq \phi < 2\pi$, where r,z, and ϕ are cylindrical coordinates, and satisfy the conditions

$$\nabla^2 V(r,z) = V_{rr} + \frac{1}{r}V_r + V_{zz} = 0 \qquad\qquad (r > 0, \ z > 0),$$

$$V(r,+0) = \frac{1}{(r^2 + c^2)^{1/2}},$$

where $c > 0$. Derive and verify the solution

$$V(r,z) = [r^2 + (z + c)^2]^{-1/2}.$$

5. If $V(r,\phi,z)$ satisfies Poisson's equation

$$V_{rr} + \frac{1}{r}V_r + \frac{1}{r^2}V_{\phi\phi} + V_{zz} + F(r) \sin \phi = 0$$

in the region $r > 0, 0 < \phi < \pi, z > 0$, and if $V = 0$ on the boundaries $\phi = 0, \phi = \pi$, and $z = 0$, derive formally the formula

$$V(r,\phi,z) = \sin \phi \int_0^\infty \frac{1 - e^{-\alpha z}}{\alpha} J_1(\alpha z) \int_0^\infty F(\rho) J_1(\alpha \rho) \, d\rho \, d\alpha.$$

6. Let $V(r,z)$ denote steady temperatures in a slab $r \geq 0, 0 \leq \phi < 2\pi, 0 \leq z \leq 1$, where r, ϕ, and z are cylindrical coordinates. If the face $z = 0$ is kept at temperature $V = 0$ and the face $z = 1$ is insulated except that heat is supplied through a circular region, such that

$$V_z(r,1) = \begin{cases} 1 & \text{if } 0 < r < c, \\ 0 & \text{if } r > c, \end{cases}$$

then $\nabla^2 V = 0$ when $r \geq 0$ and $0 < z < 1$, and $V(r,0) = 0$. Derive the formula

$$V(r,z) = c \int_0^\infty \frac{\sinh \alpha z}{\cosh z} \frac{J_1(\alpha c)}{\alpha} J_0(\alpha r) \, d\alpha.$$

15
Legendre and Other Integral Transforms

We have presented the operational calculus of various integral transformations, with emphasis on properties that are useful in solving problems in differential equations. We gave special attention to the Laplace transformation because a great variety of prominent problems, especially in partial differential equations, can be solved or simplified by using Laplace transforms. A different variety, not so extensive, can be solved by using various Fourier transforms, and still more special types of problems are adapted to Hankel transformations. Of course those and other transformations may be useful outside the field of differential equations, including applications to theory of functions, integral equations, or difference equations.

Unless a transformation has a substantial variety of applications, a development of its operational properties here is hardly justified. Procedures given in Chap. 10 may be used to design transformations needed for particular problems. Legendre transforms, treated below, have some good applications, especially to problems in potential theory that involve spherical polar coordinates.

158   THE LEGENDRE TRANSFORMATION T_n ON THE INTERVAL $(-1,1)$

Let r, ϕ, and θ denote the *spherical coordinates* of points (X,Y,Z) such that (Fig. 110)

$$X^2 + Y^2 + Z^2 = r^2, \qquad \frac{Y}{X} = \tan\phi, \qquad Z = r\cos\theta$$

$$(0 \leqq \theta \leqq \pi).$$

In terms of those coordinates the Laplacian of a function $V(r,\phi,\theta)$ takes the form

(1) $$\nabla^2 V = \frac{1}{r^2}\left[(r^2 V_r)_r + \frac{1}{\sin^2\theta}V_{\phi\phi} + \frac{1}{\sin\theta}(V_\theta \sin\theta)_\theta \right].$$

In case V is independent of ϕ, that is, $V = V(r,\theta)$ so the values of V have axial symmetry about the Z axis, the differential form with respect to θ that appears in $\nabla^2 V$ is

$$\frac{1}{\sin\theta}(V_\theta \sin\theta)_\theta = \frac{\partial}{\sin\theta\,\partial\theta}\left[(1-\cos^2\theta)\frac{\partial V(r,\theta)}{\sin\theta\,\partial\theta} \right].$$

If we write $x = \cos\theta$, the form becomes

(2) $$\mathscr{D}_2 V = \frac{\partial}{\partial x}\left[(1-x^2)\frac{\partial V}{\partial x} \right] \qquad (-1 < x < 1).$$

That self-adjoint form $\mathscr{D}_2 V$ is singular at the points $x = \pm 1$. We can construct a transformation of functions F on the interval $(-1,1)$,

$$T\{F(x)\} = \int_{-1}^{1} F(x)\Phi(x,\lambda)\,dx = f(\lambda),$$

such that $T\{\mathscr{D}_2 F\} = \lambda f(\lambda)$ by first writing

$$T\{\mathscr{D}_2 F\} = \int_{-1}^{1} \Phi \mathscr{D}_2 F\,dx = \int_{-1}^{1} \Phi(x,\lambda)[(1-x^2)F']'\,dx$$

$$= \int_{-1}^{1} F\mathscr{D}_2\Phi\,dx + [(1-x^2)(\Phi F' - \Phi'F)]_{-1}^{1},$$

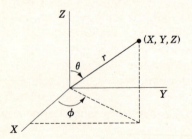

Fig. 110

then requiring that $\mathscr{D}_2\Phi = \lambda\Phi$ and that Φ and F be of class C'' on the interval $-1 \leqq x \leqq 1$. Thus

(3) $$[(1 - x^2)\Phi'(x,\lambda)]' = \lambda\Phi(x,\lambda) \qquad \text{when } -1 < x < 1$$

and Φ is to be of class C'' on the closed interval $-1 \leqq x \leqq 1$.

That singular eigenvalue problem in *Legendre's differential equation* (3) has as its set of eigenvalues

(4) $$\lambda = -n(n + 1) \qquad\qquad (n = 0, 1, 2, \ldots).$$

The corresponding eigenfunctions are the *Legendre polynomials* $P_n(x)$ of degree n:

(5) $$\Phi(x,\lambda) = P_n(x) = \frac{1}{2^n} \sum_{j=0}^{m} \frac{(-1)^j}{j!} \frac{(2n - 2j)!}{(n - 2j)!(n - j)!} x^{n-2j}$$

where $m = \frac{1}{2}n$ if n is even or zero and $m = \frac{1}{2}(n - 1)$ if n is odd.[1]

Thus $T\{F\}$ is the *Legendre integral transformation*

(6) $$T_n\{F(x)\} = \int_{-1}^{1} F(x)P_n(x)\,dx = f_L(n) \qquad (n = 0, 1, 2, \ldots)$$

on the interval $(-1,1)$, with the *operational property*

(7) $$T_n\{[(1 - x^2)F'(x)]'\} = -n(n + 1)f_L(n).$$

According to Eq. (5), the first few Legendre polynomials are

$$P_0(x) = 1, \qquad P_1(x) = x, \qquad P_2(x) = \tfrac{1}{2}(3x^2 - 1),$$

$$P_3(x) = \tfrac{1}{2}(5x^3 - 3x), \qquad P_4(x) = \tfrac{1}{8}(35x^4 - 30x^3 + 3).$$

We note that P_n is an even or odd function according as n is even or odd:

$$P_n(-x) = (-1)^n P_n(x) \qquad\qquad (n = 0, 1, 2, \ldots).$$

In case F and F' are continuous on the interval $-1 \leqq x \leqq 1$ except at an interior point $x = c$ where they have jumps j_c and j'_c, respectively, while F'' is sectionally continuous, property (7) is replaced by

(8) $$T_n\{\mathscr{D}_2 F\} = -n(n + 1)f_L(n) + (1 - c^2)[P'_n(c)j_c - P_n(c)j'_c].$$

The set of functions P_n is orthogonal on the interval $(-1,1)$. That orthogonality and the norms of P_n are expressed by the transform of P_m, where m has one of the values $0, 1, 2, \ldots$:

(9) $$T_n\{P_m(x)\} = \begin{cases} 0 & \text{if } n \neq m \\ (m + \tfrac{1}{2})^{-1} & \text{if } n = m. \end{cases}$$

[1] For basic properties of Legendre polynomials see the author's "Fourier Series and Boundary Value Problems," 2d ed., chap. 9, 1963.

The corresponding generalized Fourier series representation of a function F on the interval is *Legendre's series*

$$(10) \qquad F(x) = \sum_{n=0}^{\infty} (n + \tfrac{1}{2}) f_L(n) P_n(x) \qquad (-1 < x < 1).$$

That *inversion formula* for T_n is valid if F' is sectionally continuous over the interval and if F is defined as the mean value of its limits from the left and right at each point where F is discontinuous.

When F is a polynomial of degree m, it is a linear combination of $P_0, P_1, \ldots,$ and P_m, represented by series (10) in which $f_L(n) = 0$ when $n > m$.

Under the substitution $x = \cos \theta$, our transformation

$$T_n\{F(\cos \theta)\} = \int_0^\pi F(\cos \theta) P_n(\cos \theta) \sin \theta \, d\theta = f_L(n)$$

transforms the differential form

$$\frac{1}{\sin \theta} \frac{d}{d\theta}\left[\sin \theta \frac{d}{d\theta} F(\cos \theta) \right] \qquad (0 < \theta < \pi)$$

into $-n(n + 1) f_L(n)$.

159   FURTHER PROPERTIES OF T_n

By replacing $n(n + 1)$ in property (7) above by $(n + \tfrac{1}{2})^2 - \tfrac{1}{4}$, we find that

$$(1) \qquad (n + \tfrac{1}{2})^2 f_L(n) = T_n\{\tfrac{1}{4}F(x) - [(1 - x^2)F'(x)]'\}.$$

Since $P_0(x) = 1$ and P_0 is orthogonal to P_n when $n = 1, 2, \ldots$, then, if C is a constant,

$$(2) \qquad T_n\{F(x) + C\} = \begin{cases} f_L(n) & \text{when } n \neq 0 \\ f_L(0) + 2C & \text{when } n = 0. \end{cases}$$

Suppose that a sectionally continuous function F has been adjusted by the addition of a constant so that $f_L(0) = 0$; that is,

$$\int_{-1}^1 F(x) \, dx = 0.$$

Then by solving the differential equation

$$[(1 - x^2)Y'(x)]' = -F(x)$$

for Y and requiring Y and Y' to be continuous on the interval $-1 \leqq x \leqq 1$, we obtain the property

$$(3) \qquad \frac{f_L(n)}{n(n + 1)} = T_n\left\{ \int_0^x \frac{1}{s^2 - 1} \int_{-1}^s F(t) \, dt \, ds + C \right\} \qquad (n = 1, 2, \ldots).$$

The value of the transform on the right when $n = 0$ depends on F and the choice of the constant C.

Legendre polynomials have the product property[1]

(4)
$$P_n(\cos \beta)P_n(\cos \theta) = \frac{1}{\pi} \int_0^\pi P_n(\cos v)\, d\alpha$$

where

(5)
$$\cos v = \cos \beta \cos \theta + \sin \beta \sin \theta \cos \alpha.$$

The following convolution property of T_n has been established with the aid of that property.[2]

Let $F(x)$ and $G(x)$ be sectionally continuous functions with Legendre transforms $f_L(n)$ and $g_L(n)$. Then

(6)
$$f_L(n)g_L(n) = T_n\{X_{FG}(x)\} = \int_0^\pi X_{FG}(\cos \theta)P_n(\cos \theta) \sin \theta\, d\theta$$

where X_{FG} is this *convolution* of F and G:

(7)
$$X_{FG}(\cos \theta) = \frac{1}{\pi} \int_0^\pi F(\cos \beta) \sin \beta \int_0^\pi G(\cos v)\, d\alpha\, d\beta,$$

$\cos v$ being given by formula (5). The convolution here is not a simple one to use.

Some useful transforms of particular functions can be found from the representation

(8)
$$(1 - 2tx + t^2)^{-\frac{1}{2}} = \sum_{n=0}^{\infty} t^n P_n(x) \qquad (-1 < t < 1)$$

of a generating function for P_n. When t is fixed, the series converges uniformly with respect to x, and it follows that

(9)
$$T_n\{(1 - 2tx + t^2)^{-\frac{1}{2}}\} = \frac{2t^n}{2n + 1}$$

when $-1 < t < 1$. Equation (9) is also valid when $t = \pm 1$. We can differentiate the members of the equation with respect to t to write

$$T_n\{(1 - 2tx + t^2)^{-\frac{3}{2}}(tx - t^2)\} = t^n - \frac{t^n}{2n + 1}.$$

[1] MacRobert, T. M., "Spherical Harmonics," p. 138, 1927.

[2] Churchill, R. V., and C. L. Dolph, Inverse Transforms of Products of Legendre Transforms, *Proc. Amer. Math. Soc.*, vol. 5, pp. 93–100, 1954. In formula (5), β, θ, and v can be interpreted as sides of a spherical triangle on the hemisphere $X^2 + Y^2 + Z^2 = 1$, $Z \geq 0$, and α as the angle between the first two sides.

Then with the aid of transform (9) it follows that

$$(10) \qquad T_n\{(1 - 2tx + t^2)^{-\frac{3}{2}}\} = \frac{2t^n}{1 - t^2} \qquad (-1 < t < 1).$$

Transforms of a few other functions and some applications of T_n are given in the author's paper, The Operational Calculus of Legendre Transforms, *Jour. Math. Physics*, vol. 33, pp. 165–178, 1954. Brief treatments of Legendre transforms are given in the books by Sneddon and Tranter listed in the Bibliography.

160   LEGENDRE TRANSFORMS ON THE INTERVAL $(0,1)$

Let $F(x)$ be defined over the interval $(0,1)$ and let F_1 be the odd extension of F to the interval $(-1,1)$. Then the product $F_1(x)P_n(x)$ is an even function when n is an odd integer; it is odd if n is zero or even. Thus the Legendre transform of F_1 on the interval $(-1,1)$ can be written

$$T_{2n+1}\{F_1(x)\} = 2\int_0^1 F(x)P_{2n+1}(x)\,dx, \qquad T_{2n}\{F_1(x)\} = 0,$$

where $n = 0, 1, 2, \ldots$. Let us use the symbol $\overline{T}_{2n+1}\{F\}$ to denote the integral transformation of F that appears here; that is,

$$(1) \qquad \overline{T}_{2n+1}\{F(x)\} = \int_0^1 F(x)P_{2n+1}(x)\,dx = \bar{f}_L(2n + 1)$$

$$(n = 0, 1, 2, \ldots).$$

We call $\bar{f}_L(2n + 1)$ the *Legendre transform of F on the interval $(0,1)$ with odd indices $2n + 1$.*

In terms of T_n on the interval $(-1,1)$

$$(2) \qquad \overline{T}_{2n+1}\{F(x)\} = \tfrac{1}{2}T_{2n+1}\{F_1(x)\}.$$

Since $T_{2n}\{F_1\} = 0$, the Legendre series representation (10), Sec. 158, of the function F_1 becomes

$$F_1(x) = \sum_{n=0}^{\infty} (2n + 1 + \tfrac{1}{2})T_{2n+1}\{F_1\}P_{2n+1}(x) \qquad (-1 < x < 1).$$

This becomes an *inversion formula for* $\overline{T}_{2n+1}$:

$$(3) \qquad F(x) = \sum_{n=0}^{\infty} (4n + 3)\bar{f}_L(2n + 1)P_{2n+1}(x) \qquad (0 < x < 1),$$

because $F_1(x) = F(x)$ when $0 < x < 1$. The formula is valid if F' is sectionally continuous over the interval $(0,1)$.

The orthogonality of the set of polynomials $P_{2n+1}(x)$ on the interval $(0,1)$ follows from Eq. (2) by writing $F(x) = P_{2m+1}(x) = F_1(x)$ and recalling that the set $P_n(x)$ is orthogonal on the interval $(-1,1)$.

In case F and F' are continuous on the interval $0 \leq x \leq 1$, then at the point $x = 0$ the function F_1 has the jump $2F(0)$, but the jump in F'_1 there is zero; elsewhere on the interval $-1 \leq x \leq 1$ the functions F_1 and F'_1 are continuous. Then if F'' is sectionally continuous, formula (8), Sec. 158, shows that

(4) $$T_n\{\mathscr{D}_2 F_1\} = -n(n+1)T_n\{F_1\} + 2P'_n(0)F(0),$$

where $\mathscr{D}_2 F_1 = [(1-x^2)F'_1]' = (1-x^2)F''_1 - 2xF'_1$. Since $\mathscr{D}_2 F_1$ is an odd function, then $T_{2n+1}\{\mathscr{D}_2 F_1\} = 2\overline{T}_{2n+1}\{\mathscr{D}_2 F\}$, while both members of Eq. (4) vanish when $n = 0, 2, 4, \ldots$. It follows that

$$\overline{T}_{2n+1}\{\mathscr{D}_2 F\} = -(2n+1)(2n+2)\overline{T}_{2n+1}\{F\} + P'_{2n+1}(0)F(0).$$

From the formula

(5) $$(1-x^2)P'_n(x) = n[P_{n-1}(x) - xP_n(x)] \qquad (n = 1, 2, \ldots)$$

we find that $P'_{2n+1}(0) = (2n+1)P_{2n}(0)$. Therefore

(6) $$\overline{T}_{2n+1}\{[(1-x^2)F'(x)]'\} = -(2n-1)(2n+2)\bar{f}_L(2n+1)$$
$$+ (2n+1)P_{2n}(0)F(0).$$

Formula (6), the *basic operational property of* $\overline{T}_{2n+1}$, shows how the form $\mathscr{D}_2 F$ on the interval $(0,1)$ is transformed in terms of $\overline{T}_{2n+1}\{F\}$ and the initial value $F(0)$.

In like manner we find that the *Legendre transformation with even indices, on the interval* $(0,1)$,

(7) $$\overline{T}_{2n}\{F(x)\} = \int_0^1 F(x)P_{2n}(x)\,dx = \bar{f}_L(2n) \qquad (n = 0, 1, 2, \ldots),$$

is related to the transformation of the even extension F_2 of F on the interval $(-1,1)$ by the formula

(8) $$\overline{T}_{2n}\{F(x)\} = \tfrac{1}{2}T_{2n}\{F_2(x)\} \qquad (n = 0, 1, 2, \ldots),$$

and $T_{2n+1}\{F_2\} = 0$. It follows that $\overline{T}_{2n}$ has the *inversion formula*

(9) $$F(x) = \sum_{n=0}^{\infty} (4n+1)\bar{f}_L(2n)P_{2n}(x) \qquad (0 < x < 1),$$

and the *basic operational property*

(10) $$\overline{T}_{2n}\{[(1-x^2)F'(x)]'\} = -2n(2n+1)\bar{f}_L(2n) - P_{2n}(0)F'(0)$$

that resolves the differential form $\mathscr{D}_2F$ on the interval $(0,1)$ in terms of $\overline{T}_{2n}\{F\}$ and the initial value $F'(0)$ of F'. Also, the set of functions P_{2n} is orthogonal on the interval $(0, 1)$.

161 DIRICHLET PROBLEMS FOR THE SPHERE

The problem of determining the harmonic function in a region which assumes prescribed values on the boundary is called a *Dirichlet problem*. If instead the normal derivative of the harmonic function is prescribed on the boundary, the problem is called a *Neumann problem*.

Consider the Dirichlet problem for the region interior to the unit sphere $r = 1$ such that the harmonic function $V(r,\cos\theta)$ is independent of the spherical coordinate ϕ. Then

$$\nabla^2 V = \frac{1}{r^2}\left[(r^2 V_r)_r + \frac{1}{\sin\theta}(\sin\theta V_\theta)_\theta \right] = 0 \qquad (r < 1, 0 < \theta < \pi),$$

and $V \rightarrow F(\cos\theta)$ as $r \rightarrow 1$; also V and its partial derivatives of first and second order are to be continuous interior to the sphere. We write $x = \cos\theta$. Then

$$(1) \qquad\qquad [r^2 V_r(r,x)]_r + [(1 - x^2)V_x(r,x)]_x = 0 \qquad\qquad (r < 1),$$

$$(2) \qquad\qquad V(1 - 0, x) = F(x) \qquad\qquad (-1 < x < 1).$$

If $v_L(r,n)$ is the Legendre transform of V with respect to x on the interval $(-1,1)$,

$$(3) \qquad\qquad v_L(r,n) = \int_{-1}^{1} V(r,x)P_n(x)\,dx = T_n\{V\} \qquad (n = 0, 1, 2, \ldots),$$

and $f_L(n) = T_n\{F(x)\}$, the above problem transforms into the problem

$$(4) \qquad\qquad r^2\frac{d^2 v_L}{dr^2} + 2r\frac{dv_L}{dr} - n(n + 1)v_L = 0, \qquad v_L(1,n) = f_L(n).$$

In problem (4) v_L is to be a continuous function of r when $0 \leqq r < 1$. The solution with that continuity is

$$(5) \qquad\qquad v_L(r,n) = f_L(n)r^n \qquad (0 \leqq r < 1, n = 0, 1, 2, \ldots),$$

and so the formal solution of the problem in V is

$$(6) \qquad\qquad V(r,x) = \sum_{n=0}^{\infty} (n + \tfrac{1}{2})f_L(n)r^n P_n(x) \qquad (r \leqq 1, -1 < x < 1).$$

This form of the solution is also obtained when the problem in V is solved by the method of separation of variables.

Another representation of the solution follows from our convolution property for T_n. According to transform (10), Sec. 159,

$$r^n = \frac{1 - r^2}{2} T_n\{(1 + r^2 - 2rx)^{-3/2}\} \qquad (0 \leqq r < 1).$$

The product $f_L(n)r^n$ in Eq. (5) is the transform of the convolution (7), Sec. 159, of the functions F and $T_n^{-1}\{r^n\}$; that is,

$$(7) \quad V(r,\cos\theta) = \frac{1 - r^2}{2\pi} \int_0^\pi F(\cos\beta)\sin\beta \int_0^\pi (1 + r^2 - 2r\cos v)^{-3/2}\, d\alpha\, d\beta$$

where $\cos v = \cos\beta\cos\theta + \sin\beta\sin\theta\cos\alpha$.

Formula (7) is known as the *Poisson integral formula* for the harmonic function interior to the sphere $r = 1$ in this special case where V is prescribed on the boundary as a function $F(\cos\theta)$ of θ only. It is of some interest to note, especially when V is interpreted as steady temperatures in a solid sphere $r \leqq 1$, that either representation (6) or (7) gives the value of V at the center $r = 0$ of the sphere as the mean value of F over the surface.

In the corresponding Dirichlet problem for the bounded harmonic function $W(r,\cos\theta)$ in the region $r > 1$ *exterior to the unit sphere*, where $W(1 + 0, \cos\theta) = F(\cos\theta)$, the solution of the transformed problem is

$$w_L(r,n) = \frac{1}{r}f(n)\left(\frac{1}{r}\right)^n \qquad (n = 0, 1, 2, \ldots).$$

Therefore, in terms of the function V found above,

$$(8) \qquad W(r,\cos\theta) = \frac{1}{r}V\left(\frac{1}{r},\cos\theta\right) \qquad (r > 1).$$

PROBLEMS

1. Using orthogonality properties of $P_n(x)$ on the interval $(-1,1)$, display the orthogonality and norms for the set of polynomials $P_{2n+1}(x)$, $n = 0, 1, 2, \ldots$, on the interval $(0,1)$ by showing that for each m $(m = 0, 1, 2, \ldots)$

$$\bar{T}_{2n+1}\{P_{2m+1}(x)\} = \begin{cases} 0 & \text{if } n \neq m \\ (4m + 3)^{-1} & \text{if } n = m \end{cases}$$

where (Sec. 160) $\bar{T}_{2n+1}\{F\} = \int_0^1 F(x)P_{2n+1}(x)\, dx$.

2. Display the orthogonality and norms of the set of polynomials $P_{2n}(x)$, $n = 0, 1, 2, \ldots$, on the interval $(0,1)$ by showing that for each m $(m = 0, 1, 2, \ldots)$

$$\bar{T}_{2n}\{P_{2m}(x)\} = \begin{cases} 0 & \text{if } n \neq m \\ (4m + 1)^{-1} & \text{if } n = m \end{cases}$$

where (Sec. 160) $\bar{T}_{2n}\{F\} = \int_0^1 F(x)P_{2n}(x)\, dx$.

3. Use properties (6) and (10), Sec. 160, to show that

(a) $\overline{T}_{2n+1}\{1\} = \dfrac{P_{2n}(0)}{2n + 2}$,

(b) $\overline{T}_{2n}\{x\} = -\dfrac{P_{2n}(0)}{(2n - 1)(2n + 2)}$.

4. The *Neumann problem* in a function $U(r,\cos\theta)$ independent of the spherical coordinate ϕ, for the interior of the unit sphere, is

$$\nabla^2 U(r,x) = 0 \quad \text{when } r < 1, \qquad U_r(1 - 0, x) = F(x) \qquad (-1 < x < 1),$$

where $x = \cos\theta$. Since U is to be continuous interior to the sphere, at the center $r = 0$ in particular, its Legendre transform $u_L(r,n)$, with respect to x, must be continuous when $0 \leq r < 1$. Show that, as a consequence, the prescribed values $F(x)$ of the normal derivative of U on the surface $r = 1$ must be such that $f_L(n) = 0$ when $n = 0$; that is, the mean value of $F(x)$ must be zero, or

(a) $\displaystyle\int_0^\pi F(\cos\theta)\sin\theta\, d\theta = 0.$

Also if U is interpreted as steady temperatures and $U_r(1,x)$ as flux of heat, note why condition (a) is to be expected in order that temperatures remain steady.

Under condition (a) derive the formulas

$$u_L(r,n) = \frac{1}{n} f_L(n) r^n \qquad \text{if } n = 1, 2, \ldots, u_L(r,0) = C,$$

$$U(r,\cos\theta) = \frac{1}{2}C + \frac{1}{2}\sum_{n=1}^{\infty} \frac{2n + 1}{n} f_L(n) r^n P_n(\cos\theta)$$

where $r < 1$, $0 \leq \theta \leq \pi$, and C is an arbitrary constant.

5. In the Dirichlet problem in $V(r,x)$ represented by Eqs. (1) and (2), Sec. 161, let $F(\cos\theta)$ satisfy condition (a) in Prob. 4 above; then $v_L(r,0) = 0$, and $v_L(r,n) = f_L(n) r^n$ when $n = 1, 2, \ldots$. From the transform of the solution $U(r,x)$ of the corresponding Neumann problem with the same function F, found in Prob. 4, show that

$$u_L(r,n) = \int_0^r \frac{1}{\rho} v_L(\rho,n)\, d\rho \qquad \text{if } n = 1, 2, \ldots, u_L(r,0) = C,$$

and deduce this relation between U and V:

$$U(r,\cos\theta) = \int_0^r \frac{1}{\rho} V(\rho,\cos\theta)\, d\rho + \frac{1}{2}C$$

$$= \int_0^1 \frac{1}{t} V(rt,\cos\theta)\, dt + \frac{1}{2}C \qquad\qquad (r < 1),$$

where C is an arbitrary constant. The relation also follows from the series representations of U and V. It can be verified even if F, U, and V are not independent of ϕ, provided that

$$\int_0^\pi F(\phi,\cos\theta)\sin\theta\, d\theta = 0.$$

6. Let $U(r,\cos\theta)$ be harmonic interior to the sphere $r = 1$ and satisfy this condition at the surface:

$$U_r(1,x) + (k + 1)U(1,x) = F(x)$$

where $x = \cos\theta$ and the constant $k + 1$ is positive. Show that the Legendre transform $u_L(r,n)$ on the interval $-1 < x < 1$ is related to the transform $v_L(r,n) = f_L(n)r^n$ in the corresponding Dirichlet problem (Sec. 161) in $V(r,\cos\theta)$ by the equation

$$r^{-k}\frac{d}{dr}[r^{k+1}u_L(r,n)] = v_L(r,n).$$

Thus deduce the relation

$$U(r,\cos\theta) = \int_0^1 V(rt,\cos\theta)t^k\, dt,$$

which can be verified even when F, U, and V are not independent of the spherical coordinate ϕ.

7. Heat is generated at a steady and uniform rate throughout a solid hemisphere $0 \le r < c, 0 \le \theta < \pi/2$. The entire boundary is kept at temperature zero. Thus the steady temperatures $V(r,x)$, where $x = \cos\theta$, satisfy the conditions

$$\nabla^2 V(r,x) + A = 0 \qquad\qquad (r < c, 0 < x < 1),$$

$$V(c,x) = V(r,0) = 0,$$

where A is a constant. With the aid of the transform $\bar{T}_{2n+1}\{1\}$ found in Prob. 3, derive the formula

$$V(r,\cos\theta) = \sum_{n=0}^{\infty} (4n + 3)\bar{v}_L(r, 2n + 1)P_{2n+1}(\cos\theta)$$

where $r \le c, 0 \le \theta \le \pi/2$ and

$$\bar{v}_L(r, 2n + 1) = \frac{Ac^2 P_{2n}(0)}{4(n + 1)(2n^2 + 3n - 1)}\left[\left(\frac{r}{c}\right)^2 - \left(\frac{r}{c}\right)^{2n+1}\right].$$

8. Let $U(r,\cos\theta)$ be the steady concentration of a substance diffusing through a porous hollow hemisphere $c < r < 1, 0 < \theta < \pi/2$ whose hemispherical boundaries are kept at zero concentration. If the flux of the substance through the base $\theta = \pi/2$, $c < r < 1$, is uniform and constant, then

$$\frac{\partial U}{r\,\partial\theta}\Bigg]_{\theta=\pi/2} = \frac{\sin\theta}{r}\frac{\partial U}{\sin\theta\,\partial\theta}\Bigg]_{\theta=\pi/2} = -\frac{1}{r}U_x(r,0) = B,$$

where $x = \cos\theta$ and B is a constant. Thus

$$\nabla^2 U(r,x) = 0 \qquad\qquad (c < r < 1, 0 < x \le 1),$$

$$U(1,x) = U(c,x) = 0, \qquad U_x(r,0) = -rB.$$

Note that $\bar{T}_{2n}\{x\}$ is given in Prob. 3 and show that

$$\bar{T}_{2n}\{U(r,x) + Brx\} = \frac{B\bar{T}_{2n}(x)}{1 - c^{4n+1}}\left[(1 - c^{2n+2})r^{2n} + (c - c^{2n})\left(\frac{c}{r}\right)^{2n+1}\right].$$

In the special case $c = 0$, show that

$$U(r, \cos \theta) = - Br \cos \theta + B \sum_{n=0}^{\infty} (4n + 1)\overline{T}_{2n}\{x\}r^{2n}P_{2n}(\cos \theta).$$

162 LAGUERRE TRANSFORMS

Operational properties have been developed for another integral transformation whose kernel is a family of orthogonal polynomials, the Laguerre
polynomials.[1] The Laguerre transformation is one over the half line $x > 0$,
with weight function e^{-x}. Its properties are presented only briefly here
because the differential form associated with this transformation is not one
that is frequently encountered in differential equations.

Laguerre polynomials of degree n can be written

$$(1) \qquad L_n(x) = \frac{e^x}{n!} \frac{d^n}{dx^n}(x^n e^{-x}) = \sum_{j=0}^{n} \frac{(-1)^j n! x^j}{(j!)^2(n-j)!},$$

where $n = 0, 1, 2, \ldots$. In particular

$$L_0(x) = 1, \qquad L_1(x) = 1 - x, \qquad L_2(x) = 1 - 2x + \tfrac{1}{2}x^2.$$

They satisfy the differential equation

$$(2) \qquad xL_n''(x) + (1 - x)L_n'(x) + nL_n(x) = 0$$

which has the self-adjoint form

$$\frac{d}{dx}[xe^{-x}L_n'(x)] = -ne^{-x}L_n(x).$$

Thus $L_n(x)$ are eigenfunctions and $\lambda = -n$ are eigenvalues of the singular
eigenvalue problem consisting of the Sturm-Liouville equation

$$[xe^{-x}y'(x)]' - \lambda e^{-x}y(x) = 0 \qquad\qquad (x > 0)$$

and the requirement that y be of class C'' on the half line $x \geqq 0$.

We find that the *Laguerre integral transformation*

$$(3) \qquad G_n\{F(x)\} = \int_0^{\infty} F(x)e^{-x}L_n(x)\, dx = f_G(n) \quad (n = 0, 1, 2, \ldots)$$

has the *operational property*

$$(4) \qquad G_n\{e^x(xe^{-x}F')'\} = G_n\{xF''(x) + (1 - x)F'(x)\} = -nf_G(n),$$

valid if F is of class C'' when $x \geqq 0$ and if F is $\mathcal{O}(e^{\alpha x})$ as $x \to \infty$, where $\alpha < 1$.

---

[1] McCully, Joseph, The Laguerre Transform, *SIAM Review*, vol. 2, pp. 185–191, 1960. The
author includes a convolution property for the transformation.

The set of polynomials $L_n(x)$, $n = 0, 1, 2, \ldots$, is orthonormal on the half line $x > 0$ with weight function e^{-x}; that is, when m has one of the values $0, 1, 2, \ldots$, then

$$G_n\{L_m(x)\} = 0 \quad \text{if } n \neq m, \qquad G_m\{L_m(x)\} = 1.$$

The corresponding generalized Fourier series representation of a function F can be written

$$(5) \qquad F(x) = \sum_{n=0}^{\infty} f_G(n)L_n(x) \qquad (x > 0).$$

That *inversion formula* for the transformation G_n is valid where F is continuous if F' is sectionally continuous on each bounded interval and if F is $\mathcal{O}(e^{\alpha x})$ as $x \to \infty$, where $\alpha < \frac{1}{2}$.

The convolution of two functions, whose transform is the product of the transforms, can be represented by an iterated integral; but the integral is a complicated one.

163 MELLIN TRANSFORMS

Our formal procedure of designing an integral transformation shows that if the differential form $xF'(x)$ on the half line $x > 0$ is to transform into $-sf_M(s)$, where $f_M(s)$ is the transform of F itself, then the kernel of the transformation can be taken as the function x^{s-1}. The transformation is the *Mellin transformation*

$$(1) \qquad M_s\{F(x)\} = \int_0^{\infty} F(x)x^{s-1}\, dx = f_M(s)$$

with the *operational property*

$$(2) \qquad M_s\{xF'(x)\} = -sM_s\{F\} = -sf_M(s).$$

Formally, from property (2) we find that

$$(3) \qquad M_s\{x(xF')'\} = M_s\{x^2F''(x) + xF'(x)\} = s^2 f_M(s)$$

and so

$$(4) \qquad M_s\{x^2F''(x)\} = (s^2 + s)f_M(s).$$

For further iterations of the operator $x\, d/dx$, property (2) leads to the formula

$$(5) \qquad M_s\left\{\left(x\frac{d}{dx}\right)^n F(x)\right\} = (-1)^n s^n f_M(s) \qquad (n = 1, 2, \ldots).$$

The differential form $\rho(\rho V_\rho)_\rho$ with respect to ρ can be isolated in Poisson's equation in $V(\rho,\phi)$ where ρ and ϕ are *polar coordinates in the plane*. The forms $r^2 V_{rr}$ and rV_r with respect to r can be isolated in Poisson's equation in $V(r,\phi,\theta)$ where r, ϕ, and θ are spherical coordinates. Thus the Mellin transformation with respect to the radius vector ρ in the first case, or which respect to r in the second, will remove those differential forms, provided that ρ and r *range through all positive values*. That range of values of ρ and r and conditions under which the Mellin transforms exist restrict the types of problems to which M_s applies.

Let us assume as usual that F is at least sectionally continuous over each bounded interval of the positive x axis. If $s = \sigma + i\alpha$ where σ and α are real, then

$$x^s = \exp(s\log x) = \exp(\sigma\log x)\exp(i\alpha\log x)$$

and thus $|x^s| = x^\sigma$ when $x > 0$. The integrand of the Mellin integral (1) is then $\mathcal{O}(x^{\sigma-1})$ as $x \to 0$ and we shall assume $\sigma > 0$ to ensure the convergence of the integral at its lower limit. Suppose that $F(x)$ is $\mathcal{O}(x^{-k})$ as $x \to \infty$, where $k > 0$. Then the integrand is $\mathcal{O}(1/x^{k-\sigma+1})$ as $x \to \infty$, so *the Mellin transform $f_s(\sigma + i\alpha)$ exists* when $k - \sigma > 0$ and $\sigma > 0$, that is, *when* $0 < \sigma < k$, provided that $F(x)$ is $\mathcal{O}(x^{-k})$ as $x \to \infty$, where $k > 0$.

When F is continuous $(x \geqq 0)$ and F' is sectionally continuous, an integration by parts shows that $M_s\{xF'\}$ exists and that property (2) is valid if $0 < \sigma < k$.

The Mellin transformation can be described *in terms of the exponential Fourier transformation E_α*. Writing $s = \sigma + i\alpha$ and substituting $x = e^{-t}$, we find that

$$M_s\{F(x)\} = \int_{-\infty}^{\infty} F(e^{-t})e^{-\sigma t}e^{-i\alpha t}\,dt\,;$$

that is,

$$(6) \qquad\qquad M_s\{F(x)\} = f_M(\sigma + i\alpha) = E_\alpha\{F(e^{-x})e^{-\sigma x}\}.$$

Thus, known exponential transforms furnish corresponding Mellin transforms, and some properties of M_s follow from properties of E_α. But the relationship also suggests the substitution $x = e^{-t}$ in order *to change the differential form $xF'(x)$ and its iterates into forms to which E_α applies*.

According to relation (6) and our inversion formula for E_α (Sec. 132), when F' is sectionally continuous, then

$$F(e^{-t})e^{-\sigma t} = \frac{1}{2\pi}\lim_{\beta\to\infty}\int_{-\beta}^{\beta} f_M(\sigma + i\alpha)e^{i\alpha t}\,d\alpha$$

if $|F(e^{-t})|e^{-\sigma t}$ is integrable from $t = -\infty$ to $t = \infty$, and we find this is true if $F(x)$ is $\mathcal{O}(x^{-k})$ as $x \to \infty$ and if $0 < \sigma < k$, where $x = e^{-t}$. Thus by writing $e^{-t} = x$, we have an *inversion formula* for M_s:

$$(7) \qquad F(x) = \frac{1}{2\pi} \lim_{\beta \to \infty} \int_{-\beta}^{\beta} f_M(\sigma + i\alpha)x^{-\sigma - i\alpha}\, d\alpha$$

$$= \frac{1}{2\pi i} \lim_{\beta \to \infty} \int_{\sigma - i\beta}^{\sigma + i\beta} f_M(s)x^{-s}\, ds.$$

The *kernel-product* property

$$(8) \qquad y^{s-1}f_M(s) = M_s\left\{\frac{1}{y}F\left(\frac{x}{y}\right)\right\} \qquad\qquad (y > 0)$$

is easily derived. The corresponding *convolution* of two functions $F(x)$ and $G(x)$ is (Chap. 10)

$$(9) \qquad X_{FG}(x) = \int_0^\infty G(y)F\left(\frac{x}{y}\right)\frac{dy}{y}$$

such that, if $M_s\{G\} = g_M(s)$, then

$$(10) \qquad M_s\{X_{FG}(x)\} = f_M(s)g_M(s).$$

Tables of Mellin transforms, including some of the properties of M_s, are given in A. Erdélyi *et al.*, "Tables of Integral Transforms," vol. 1. Other treatments of the theory of the transformation can be found in E. C. Titchmarsh, "Theory of Fourier Integrals" and G. Doetsch, "Handbuch der Laplace-Transformation," vol. 1. Other books listed in the Bibliography, under authors Sneddon and Tranter, include some applications of the transformation.

Bibliography

Bochner, S.: "Lectures on Fourier Integrals," Princeton University Press, Princeton, N.J., 1959.

Carslaw, H. S., and J. C. Jaeger: "Operational Methods in Applied Mathematics," Oxford University Press, London, 1941.

———: "Conduction of Heat in Solids," 2d ed., Oxford University Press, London, 1959.

Churchill, R. V.: "Fourier Series and Boundary Value Problems," 2d ed., McGraw-Hill Book Company, New York, 1963.

Doetsch, G.: "Theorie und Anwendung der Laplace-Transformation," Springer-Verlag, Berlin, 1937.

———: "Handbuch der Laplace-Transformation," vol. 1, Verlag Birkhäuser, Basel, 1950; vol. 2, 1955; vol. 3, 1956.

———: "Einführung in Theorie und Anwendung der Laplace-Transformation," 2d ed., Verlag Birkhäuser, Basel, 1970.

Kaplan, W.: "Operational Methods for Linear Systems," Addison-Wesley Publishing Company, Inc., Reading, Mass., 1962.

Lighthill, M. J.: "Fourier Analysis and Generalized Functions," Cambridge University Press, London, 1958.

Mikusiński, J.: "Operational Calculus," Pergamon Press, New York, 1959.

van der Pol, B., and H. Bremmer: "Operational Calculus Based on the Two-sided Laplace Integral," 2d ed., Cambridge University Press, London, 1955.

Sneddon, I. N.: "Fourier Transforms," McGraw-Hill Book Company, New York, 1951.

———: "Functional Analysis," Handbuch der Physik, vol. 2, pp. 198–348, Springer-Verlag, Berlin, 1955.

Titchmarsh, E. C.: "Theory of Fourier Integrals," Oxford University Press, London, 1937.

———: "Eigenfunction Expansions," Oxford University Press, London, 1946.

Tranter, C. J.: "Integral Transforms in Mathematical Physics," 2d ed., Methuen and Co., Ltd., London, 1956.

Voelker, D., and G. Doetsch: "Die Zweidimensionale Laplace-Transformation," Verlag Birkhäuser, Basel, 1950.

Widder, D. V.: "The Laplace Transform," Princeton University Press, Princeton, N.J., 1941.

Wiener, N.: "The Fourier Integral," Cambridge University Press, London, 1933.

Tables

Campbell, G. A., and R. M. Foster: "Fourier Integrals for Practical Applications,"
    D. Van Nostrand Company, Inc., New York, 1948.

Ditkin, V. A., and A. P. Prudnikov: "Integral Transforms and Operational Calculus,"
    Pergamon Press, New York, 1965.

Doetsch, G., H. Kniess, and D. Voelker: "Tabellen zur Laplace-Transformation und
    Anleitung zum Gebrauch," Springer-Verlag, Berlin, 1947.

Erdélyi, A., W. Magnus, F. Oberhettinger, and F. Tricomi: "Tables of Integral
    Transforms," vols. 1 and 2, McGraw-Hill Book Company, New York, 1954.

Oberhettinger, F.: "Tabellen zur Fourier Transformation," Springer-Verlag, Berlin,
    1957.

Appendix A
Tables of Laplace Transforms

Table A.1 Operations[1]

	$f(s)$ $\quad$ (Re $s > \alpha$)	$F(t)$ $\quad$ ($t > 0$)
1	$\displaystyle\int_0^\infty e^{-st}F(t)\,dt = L\{F\}$	$F(t)$
2	$f(s)$	$\dfrac{1}{2\pi i}\displaystyle\lim_{\beta \to \infty}\int_{\gamma - i\beta}^{\gamma + i\beta} e^{tz}f(z)\,dz$
3	$sf(s) - F(0)$	$F'(t)$
4	$s^n f(s) - s^{n-1}F(0)$ $\quad - s^{n-2}F'(0) - \cdots - F^{(n-1)}(0)$	$F^{(n)}(t)$
5	$\dfrac{1}{s}f(s)$	$\displaystyle\int_0^t F(\tau)\,d\tau$
6	$e^{-bs}f(s)$ $(b > 0)$	$\begin{cases} F(t - b) & \text{if } t > b \\ 0 & \text{if } t < b \end{cases}$
7	$f(s)g(s)$	$\displaystyle\int_0^t F(\tau)G(t - \tau)\,d\tau$
8	$f'(s)$	$-tF(t)$
9	$f^{(n)}(s)$	$(-1)^n t^n F(t)$
10	$\displaystyle\int_s^\infty f(x)\,dx$	$\dfrac{1}{t}F(t)$
11	$f(s - a)$	$e^{at}F(t)$
12	$f(cs)$ $(c > 0)$	$\dfrac{1}{c}F\left(\dfrac{t}{c}\right)$
13	$\dfrac{\int_0^a e^{-st}F(t)\,dt}{1 - e^{-as}}$ $(a > 0)$	$F(t)$ if $F(t + a) = F(t)$

[1] See Chaps. 1, 2, and 6 for details and other operations.

458

	$f(s)$ $(\mathrm{Re}\ s > \alpha)$	$F(t)$ $(t > 0)$
14	$\dfrac{\int_0^a e^{-st} F(t)\,dt}{1 + e^{-as}}$ $(a > 0)$	$F(t)$ if $F(t + a) = -F(t)$
15	$\displaystyle\sum_{n=0}^{\infty} \dfrac{a_n}{s^{n+k}}$ $(k > 0)$	$\displaystyle\sum_{n=0}^{\infty} \dfrac{a_n}{\Gamma(n + k)} t^{n+k-1}$

Table A.2   Laplace transforms[1]

	$f(s)$	$F(t)$ $(t > 0)$
1	$\dfrac{1}{s}$	1
2	$\dfrac{1}{s^2}$	t
3	$\dfrac{1}{s^n}$ $(n = 1, 2, \ldots)$	$\dfrac{t^{n-1}}{(n-1)!}$
4	$\dfrac{1}{\sqrt{s}}$	$\dfrac{1}{\sqrt{\pi t}}$
5	$\dfrac{1}{s\sqrt{s}}$	$2\sqrt{\dfrac{t}{\pi}}$
6	$s^{-(n+\frac{1}{2})}$ $(n = 1, 2, \ldots)$	$\dfrac{2^n t^{n-\frac{1}{2}}}{1 \times 3 \times 5 \cdots (2n-1)\sqrt{\pi}}$
7	$\dfrac{\Gamma(k)}{s^k}$ $(k > 0)$	t^{k-1}
8	$\dfrac{1}{s-a}$	e^{at}
9	$\dfrac{1}{(s-a)^2}$	te^{at}
10	$\dfrac{1}{(s-a)^n}$ $(n = 1, 2, \ldots)$	$\dfrac{1}{(n-1)!} t^{n-1} e^{at}$
11	$\dfrac{\Gamma(k)}{(s-a)^k}$ $(k > 0)$	$t^{k-1} e^{at}$
12[2]	$\dfrac{1}{(s-a)(s-b)}$	$\dfrac{1}{a-b}(e^{at} - e^{bt})$
13[2]	$\dfrac{s}{(s-a)(s-b)}$	$\dfrac{1}{a-b}(ae^{at} - be^{bt})$

[1] For more extensive tables of Laplace transforms see the books by A. Erdélyi *et al.*, vol. 1, G. Doetsch *et al.*, or V. A. Ditkin and A. P. Prudnikov, listed under Tables in the Bibliography.
[2] Here a, b, and (in entry 14) c represent distinct constants.

Table A.2   (continued)

	$f(s)$	$F(t)$ $(t > 0)$
14	$\dfrac{1}{(s-a)(s-b)(s-c)}$	$-\dfrac{(b-c)e^{at} + (c-a)e^{bt} + (a-b)e^{ct}}{(a-b)(b-c)(c-a)}$
15	$\dfrac{1}{s^2 + a^2}$	$\dfrac{1}{a}\sin at$
16	$\dfrac{s}{s^2 + a^2}$	$\cos at$
17	$\dfrac{1}{s^2 - a^2}$	$\dfrac{1}{a}\sinh at$
18	$\dfrac{s}{s^2 - a^2}$	$\cosh at$
19	$\dfrac{1}{s(s^2 + a^2)}$	$\dfrac{1}{a^2}(1 - \cos at)$
20	$\dfrac{1}{s^2(s^2 + a^2)}$	$\dfrac{1}{a^3}(at - \sin at)$
21	$\dfrac{1}{(s^2 + a^2)^2}$	$\dfrac{1}{2a^3}(\sin at - at \cos at)$
22	$\dfrac{s}{(s^2 + a^2)^2}$	$\dfrac{t}{2a}\sin at$
23	$\dfrac{s^2}{(s^2 + a^2)^2}$	$\dfrac{1}{2a}(\sin at + at \cos at)$
24	$\dfrac{s^2 - a^2}{(s^2 + a^2)^2}$	$t \cos at$
25	$\dfrac{s}{(s^2 + a^2)(s^2 + b^2)}$ $(a^2 \neq b^2)$	$\dfrac{\cos at - \cos bt}{b^2 - a^2}$
26	$\dfrac{1}{(s-a)^2 + b^2}$	$\dfrac{1}{b}e^{at}\sin bt$
27	$\dfrac{s-a}{(s-a)^2 + b^2}$	$e^{at}\cos bt$
28	$\dfrac{3a^2}{s^3 + a^3}$	$e^{-at} - e^{at/2}\left(\cos\dfrac{at\sqrt{3}}{2} - \sqrt{3}\sin\dfrac{at\sqrt{3}}{2}\right)$
29	$\dfrac{4a^3}{s^4 + 4a^4}$	$\sin at \cosh at - \cos at \sinh at$
30	$\dfrac{s}{s^4 + 4a^4}$	$\dfrac{1}{2a^2}\sin at \sinh at$
31	$\dfrac{1}{s^4 - a^4}$	$\dfrac{1}{2a^3}(\sinh at - \sin at)$

	$f(s)$	$F(t)$ $\quad(t > 0)$
32	$\dfrac{s}{s^4 - a^4}$	$\dfrac{1}{2a^2}(\cosh at - \cos at)$
33	$\dfrac{8a^3 s^2}{(s^2 + a^2)^3}$	$(1 + a^2 t^2)\sin at - at\cos at$
34[1]	$\dfrac{1}{s}\left(\dfrac{s-1}{s}\right)^n$	$L_n(t) = \dfrac{e^t}{n!}\dfrac{d^n}{dt^n}(t^n e^{-t})$
35	$\dfrac{s}{(s-a)\sqrt{s-a}}$	$\dfrac{1}{\sqrt{\pi t}}e^{at}(1 + 2at)$
36	$\sqrt{s-a} - \sqrt{s-b}$	$\dfrac{1}{2\sqrt{\pi t^3}}(e^{bt} - e^{at})$
37	$\dfrac{1}{\sqrt{s}+a}$	$\dfrac{1}{\sqrt{\pi t}} - ae^{a^2 t}\,\text{erfc}\,(a\sqrt{t})$
38	$\dfrac{\sqrt{s}}{s-a^2}$	$\dfrac{1}{\sqrt{\pi t}} + ae^{a^2 t}\,\text{erf}\,(a\sqrt{t})$
39	$\dfrac{\sqrt{s}}{s+a^2}$	$\dfrac{1}{\sqrt{\pi t}} - \dfrac{2a}{\sqrt{\pi}}e^{-a^2 t}\displaystyle\int_0^{a\sqrt{t}}\exp(\lambda^2)\,d\lambda$
40	$\dfrac{1}{\sqrt{s}(s-a^2)}$	$\dfrac{1}{a}e^{a^2 t}\,\text{erf}\,(a\sqrt{t})$
41	$\dfrac{1}{\sqrt{s}(s+a^2)}$	$\dfrac{2}{a\sqrt{\pi}}e^{-a^2 t}\displaystyle\int_0^{a\sqrt{t}}e^{\lambda^2}\,d\lambda$
42	$\dfrac{b^2 - a^2}{(s-a^2)(b+\sqrt{s})}$	$e^{a^2 t}[b - a\,\text{erf}\,(a\sqrt{t})] - be^{b^2 t}\,\text{erfc}\,(b\sqrt{t})$
43	$\dfrac{1}{\sqrt{s}(\sqrt{s}+a)}$	$e^{a^2 t}\,\text{erfc}\,(a\sqrt{t})$
44	$\dfrac{1}{(s+a)\sqrt{s+b}}$	$\dfrac{1}{\sqrt{b-a}}e^{-at}\,\text{erf}\,(\sqrt{b-a}\sqrt{t})$
45	$\dfrac{b^2 - a^2}{\sqrt{s}(s-a^2)(\sqrt{s}+b)}$	$e^{a^2 t}\left[\dfrac{b}{a}\,\text{erf}\,(a\sqrt{t}) - 1\right] + e^{b^2 t}\,\text{erfc}\,(b\sqrt{t})$
46[2]	$\dfrac{(1-s)^n}{s^{n+\frac12}}$	$\dfrac{n!}{(2n)!\sqrt{\pi t}}H_{2n}(\sqrt{t})$
47	$\dfrac{(1-s)}{s^{n+1}\sqrt{s}}$	$-\dfrac{n!}{\sqrt{\pi}(2n+1)!}H_{2n+1}(\sqrt{t})$

[1] $L_n(t)$ is the Laguerre polynomial of degree n (Sec. 162).

[2] $H_n(x)$ is the Hermite polynomial, $H_n(x) = e^{x^2}(d^n/dx^n)(e^{-x^2})$.

Table A.2 (continued)

	$f(s)$	$F(t)$ $(t > 0)$
48^1	$\dfrac{\sqrt{s + 2a}}{\sqrt{s}} - 1$	$ae^{-at}[I_1(at) + I_0(at)]$
49	$\dfrac{1}{\sqrt{s + a}\sqrt{s + b}}$	$e^{-\frac{1}{2}(a+b)t}I_0\left(\dfrac{a-b}{2}t\right)$
50	$\dfrac{\Gamma(k)}{(s + a)^k(s + b)^k}$ $(k > 0)$	$\sqrt{\pi}\left(\dfrac{t}{a-b}\right)^{k-\frac{1}{2}}e^{-\frac{1}{2}(a+b)t}\ I_{k-\frac{1}{2}}\left(\dfrac{a-b}{2}t\right)$
51	$\dfrac{1}{\sqrt{s + a}\sqrt{s + b}(s + b)}$	$te^{-\frac{1}{2}(a+b)t}\left[I_0\left(\dfrac{a-b}{2}t\right) + I_1\left(\dfrac{a-b}{2}t\right)\right]$
52	$\dfrac{\sqrt{s + 2a} - \sqrt{s}}{\sqrt{s + 2a} + \sqrt{s}}$	$\dfrac{1}{t}e^{-at}I_1(at)$
53	$\dfrac{(a - b)^k}{(\sqrt{s + a} + \sqrt{s + b})^{2k}}$ $(k > 0)$	$\dfrac{k}{t}e^{-\frac{1}{2}(a+b)t}I_k\left(\dfrac{a-b}{2}t\right)$
54	$\dfrac{(\sqrt{s + a} + \sqrt{s})^{-2v}}{\sqrt{s}\sqrt{s + a}}$ $(v > -1)$	$\dfrac{1}{a^v}e^{-\frac{1}{2}at}I_v\left(\dfrac{1}{2}at\right)$
55	$\dfrac{1}{\sqrt{s^2 + a^2}}$	$J_0(at)$
56	$\dfrac{(\sqrt{s^2 + a^2} - s)^v}{\sqrt{s^2 + a^2}}$ $(v > -1)$	$a^v J_v(at)$
57	$\dfrac{1}{(s^2 + a^2)^k}$ $(k > 0)$	$\dfrac{\sqrt{\pi}}{\Gamma(k)}\left(\dfrac{t}{2a}\right)^{k-\frac{1}{2}}J_{k-\frac{1}{2}}(at)$
58	$(\sqrt{s^2 + a^2} - s)^k$ $(k > 0)$	$\dfrac{ka^k}{t}J_k(at)$
59	$\dfrac{(s - \sqrt{s^2 - a^2})^v}{\sqrt{s^2 - a^2}}$ $(v > -1)$	$a^v I_v(at)$
60	$\dfrac{1}{(s^2 - a^2)^k}$ $(k > 0)$	$\dfrac{\sqrt{\pi}}{\Gamma(k)}\left(\dfrac{t}{2a}\right)^{k-\frac{1}{2}}I_{k-\frac{1}{2}}(at)$
61	$\dfrac{e^{-ks}}{s}$	$S_k(t) = \begin{cases} 0 \text{ when } 0 < t < k \\ 1 \text{ when } t > k \end{cases}$
62	$\dfrac{e^{-ks}}{s^2}$	$\begin{cases} 0 \quad \text{when } 0 < t < k \\ t - k \text{ when } t > k \end{cases}$
63	$\dfrac{e^{-ks}}{s^\mu}$ $(\mu > 0)$	$\begin{cases} 0 \qquad\quad \text{when } 0 < t < k \\ \dfrac{(t - k)^{\mu - 1}}{\Gamma(\mu)} \text{ when } t > k \end{cases}$

$^1 I_n(x) = i^{-n}J_n(ix)$, where J_n is Bessel's function of the first kind.

	$f(s)$	$F(t)$ $(t > 0)$
64	$\dfrac{1 - e^{-ks}}{s}$	$\begin{cases}1 \text{ when } 0 < t < k \\ 0 \text{ when } t > k\end{cases}$
65	$\dfrac{1}{s(1 - e^{-ks})} = \dfrac{1 + \coth \frac{1}{2}ks}{2s}$	$1 + [t/k] = n$ when $(n - 1)k < t < nk$ $(n = 1, 2, \ldots)$ (Fig. 5)
66	$\dfrac{1}{s(e^{ks} - a)}$	$\begin{cases}0 \text{ when } 0 < t < k \\ 1 + a + a^2 + \cdots + a^{n-1} \\ \quad \text{when } nk < t < (n + 1)k \\ \hfill (n = 1, 2, \ldots)\end{cases}$
67	$\dfrac{1}{s}\tanh ks$	$M(2k,t) = (-1)^{n-1}$ when $2k(n - 1) < t < 2kn$ $(n = 1, 2, \ldots)$ (Fig. 9)
68	$\dfrac{1}{s(1 + e^{-ks})}$	$\dfrac{1}{2}M(k,t) + \dfrac{1}{2} = \dfrac{1 - (-1)^n}{2}$ when $(n - 1)k < t < nk$
69	$\dfrac{1}{s^2}\tanh ks$	$H(2k,t)$ (Fig. 10)
70	$\dfrac{1}{s \sinh ks}$	$F(t) = 2(n - 1)$ when $(2n - 3)k < t < (2n - 1)k$ $(t > 0)$
71	$\dfrac{1}{s \cosh ks}$	$M(2k, t + 3k) + 1 = 1 + (-1)^n$ when $(2n - 3)k < t < (2n - 1)k$ $(t > 0)$
72	$\dfrac{1}{s}\coth ks$	$F(t) = 2n - 1$ when $2k(n - 1) < t < 2kn$
73	$\dfrac{k}{s^2 + k^2}\coth\dfrac{\pi s}{2k}$	$|\sin kt|$
74	$\dfrac{1}{(s^2 + 1)(1 - e^{-\pi s})}$	$\frac{1}{2}(\sin t + |\sin t|)$
75	$\dfrac{1}{s}e^{-(k/s)}$	$J_0(2\sqrt{kt})$
76	$\dfrac{1}{\sqrt{s}}e^{-(k/s)}$	$\dfrac{1}{\sqrt{\pi t}}\cos 2\sqrt{kt}$
77	$\dfrac{1}{\sqrt{s}}e^{k/s}$	$\dfrac{1}{\sqrt{\pi t}}\cosh 2\sqrt{kt}$
78	$\dfrac{1}{s\sqrt{s}}e^{-(k/s)}$	$\dfrac{1}{\sqrt{\pi k}}\sin 2\sqrt{kt}$
79	$\dfrac{1}{s\sqrt{s}}e^{k/s}$	$\dfrac{1}{\sqrt{\pi k}}\sinh 2\sqrt{kt}$
80	$\dfrac{1}{s^\mu}e^{-(k/s)}$ $(\mu > 0)$	$\left(\dfrac{t}{k}\right)^{(\mu-1)/2} J_{\mu-1}(2\sqrt{kt})$

Table A.2 (continued)

	$f(s)$	$F(t) \quad (t > 0)$
81	$\dfrac{1}{s^{\mu}}e^{k/s} \quad (\mu > 0)$	$\left(\dfrac{t}{k}\right)^{(\mu-1)/2} I_{\mu-1}(2\sqrt{kt})$
82	$e^{-k\sqrt{s}} \quad (k > 0)$	$\dfrac{k}{2\sqrt{\pi t^3}}\exp\left(-\dfrac{k^2}{4t}\right)$
83	$\dfrac{1}{s}e^{-k\sqrt{s}} \quad (k \geqq 0)$	$\operatorname{erfc}\left(\dfrac{k}{2\sqrt{t}}\right)$
84	$\dfrac{1}{\sqrt{s}}e^{-k\sqrt{s}} \quad (k \geqq 0)$	$\dfrac{1}{\sqrt{\pi t}}\exp\left(-\dfrac{k^2}{4t}\right)$
85	$\dfrac{1}{s\sqrt{s}}e^{-k\sqrt{s}} \quad (k \geqq 0)$	$2\sqrt{\dfrac{t}{\pi}}\exp\left(-\dfrac{k^2}{4t}\right) - k\operatorname{erfc}\left(\dfrac{k}{2\sqrt{t}}\right)$
86	$\dfrac{ae^{-k\sqrt{s}}}{s(a+\sqrt{s})} \quad (k \geqq 0)$	$-e^{ak}e^{a^2t}\operatorname{erfc}\left(a\sqrt{t}+\dfrac{k}{2\sqrt{t}}\right) + \operatorname{erfc}\left(\dfrac{k}{2\sqrt{t}}\right)$
87	$\dfrac{e^{-k\sqrt{s}}}{\sqrt{s}(a+\sqrt{s})} \quad (k \geqq 0)$	$e^{ak}e^{a^2t}\operatorname{erfc}\left(a\sqrt{t}+\dfrac{k}{2\sqrt{t}}\right)$
88	$\dfrac{e^{-k\sqrt{s(s+a)}}}{\sqrt{s(s+a)}}$	$\begin{cases} 0 & \text{when } 0 < t < k \\ e^{-\frac{1}{2}at}I_0(\frac{1}{2}a\sqrt{t^2-k^2}) & \text{when } t > k \end{cases}$
89	$\dfrac{e^{-k\sqrt{s^2+a^2}}}{\sqrt{s^2+a^2}}$	$\begin{cases} 0 & \text{when } 0 < t < k \\ J_0(a\sqrt{t^2-k^2}) & \text{when } t > k \end{cases}$
90	$\dfrac{e^{-k\sqrt{s^2-a^2}}}{\sqrt{s^2-a^2}}$	$\begin{cases} 0 & \text{when } 0 < t < k \\ I_0(a\sqrt{t^2-k^2}) & \text{when } t > k \end{cases}$
91	$\dfrac{e^{-k(\sqrt{s^2+a^2}-s)}}{\sqrt{s^2+a^2}} \quad (k \geqq 0)$	$J_0(a\sqrt{t^2+2kt})$
92	$e^{-ks} - e^{-k\sqrt{s^2+a^2}}$	$\begin{cases} 0 & \text{when } 0 < t < k \\ \dfrac{ak}{\sqrt{t^2-k^2}}J_1(a\sqrt{t^2-k^2}) & \text{when } t > k \end{cases}$
93	$e^{-k\sqrt{s^2-a^2}} - e^{-ks}$	$\begin{cases} 0 & \text{when } 0 < t < k \\ \dfrac{ak}{\sqrt{t^2-k^2}}I_1(a\sqrt{t^2-k^2}) & \text{when } t > k \end{cases}$
94	$\dfrac{a^{\nu}e^{-k\sqrt{s^2+a^2}}}{\sqrt{s^2+a^2}(\sqrt{s^2+a^2}+s)^{\nu}} \quad (\nu > -1)$	$\begin{cases} 0 & \text{when } 0 < t < k \\ \left(\dfrac{t-k}{t+k}\right)^{\frac{1}{2}\nu} J_{\nu}(a\sqrt{t^2-k^2}) & \text{when } t > k \end{cases}$
95	$\dfrac{1}{s}\log s$	$\Gamma'(1) - \log t \quad [\Gamma'(1) = -0.5772]$
96	$\dfrac{1}{s^k}\log s \quad (k > 0)$	$t^{k-1}\left\{\dfrac{\Gamma'(k)}{[\Gamma(k)]^2} - \dfrac{\log t}{\Gamma(k)}\right\}$

	$f(s)$	$F(t)$ $\quad (t > 0)$
97^1	$\dfrac{\log s}{s - a}$ $\quad (a > 0)$	$e^{at}[\log a + E_1(at)]$
98^2	$\dfrac{\log s}{s^2 + 1}$	$\cos t \, \mathrm{Si}\, t - \sin t \, \mathrm{Ci}\, t$
99^2	$\dfrac{s \log s}{s^2 + 1}$	$-\sin t \, \mathrm{Si}\, t - \cos t \, \mathrm{Ci}\, t$
100^1	$\dfrac{1}{s}\log(1 + ks)$ $\quad (k > 0)$	$E_1\!\left(\dfrac{t}{k}\right)$
101	$\log \dfrac{s - a}{s - b}$	$\dfrac{1}{t}(e^{bt} - e^{at})$
102	$\dfrac{1}{s}\log(1 + k^2 s^2)$	$-2\,\mathrm{Ci}\!\left(\dfrac{t}{k}\right)$
103	$\dfrac{1}{s}\log(s^2 + a^2)$ $\quad (a > 0)$	$2\log a - 2\,\mathrm{Ci}\,(at)$
104	$\dfrac{1}{s^2}\log(s^2 + a^2)$ $\quad (a > 0)$	$\dfrac{2}{a}[at \log a + \sin at - at\,\mathrm{Ci}\,(at)]$
105	$\log \dfrac{s^2 + a^2}{s^2}$	$\dfrac{2}{t}(1 - \cos at)$
106	$\log \dfrac{s^2 - a^2}{s^2}$	$\dfrac{2}{t}(1 - \cosh at)$
107	$\arctan \dfrac{k}{s}$	$\dfrac{1}{t}\sin kt$
108	$\dfrac{1}{s}\arctan \dfrac{k}{s}$	$\mathrm{Si}\,(kt)$
109	$e^{k^2 s^2}\,\mathrm{erfc}\,(ks)$ $\quad (k > 0)$	$\dfrac{1}{k\sqrt{\pi}}\exp\!\left(-\dfrac{t^2}{4k^2}\right)$
110	$\dfrac{1}{s}e^{k^2 s^2}\,\mathrm{erfc}\,(ks)$ $\quad (k > 0)$	$\mathrm{erf}\!\left(\dfrac{t}{2k}\right)$
111	$e^{ks}\,\mathrm{erfc}\,\sqrt{ks}$ $\quad (k > 0)$	$\dfrac{\sqrt{k}}{\pi\sqrt{t(t + k)}}$
112	$\dfrac{1}{\sqrt{s}}\mathrm{erfc}\,(\sqrt{ks})$	$\begin{cases} 0 & \text{when } 0 < t < k \\ (\pi t)^{-\frac{1}{2}} & \text{when } t > k \end{cases}$
113	$\dfrac{1}{\sqrt{s}}e^{ks}\,\mathrm{erfc}\,(\sqrt{ks})$ $\quad (k > 0)$	$\dfrac{1}{\sqrt{\pi(t + k)}}$

[1] The exponential-integral function $E_1(t)$ is defined in Sec. 35. For tables of this function and other integral functions, see, for instance, Jahnke and Emde, "Tables of Functions."
[2] The cosine-integral function is defined in Sec. 35. Si t is defined in Sec. 22.

Table A.2   (continued)

	$f(s)$	$F(t) \quad (t > 0)$
114	$\operatorname{erf}\left(\dfrac{k}{\sqrt{s}}\right)$	$\dfrac{1}{\pi t}\sin(2k\sqrt{t})$
115	$\dfrac{1}{\sqrt{s}}e^{k^2/s}\operatorname{erfc}\left(\dfrac{k}{\sqrt{s}}\right)$	$\dfrac{1}{\sqrt{\pi t}}e^{-2k\sqrt{t}}$
116	$\pi e^{-ks}I_0(ks)$	$\begin{cases}[t(2k-t)]^{-\frac{1}{2}} & \text{when } 0 < t < 2k \\ 0 & \text{when } t > 2k\end{cases}$
117	$e^{-ks}I_1(ks)$	$\begin{cases}\dfrac{k-t}{\pi k\sqrt{t(2k-t)}} & \text{when } 0 < t < 2k \\ 0 & \text{when } t > 2k\end{cases}$
118	$e^{as}E_1(as)$	$\dfrac{1}{t+a}\ (a>0)$
119	$\dfrac{1}{a} - se^{as}E_1(as)$	$\dfrac{1}{(t+a)^2}\ (a>0)$
120	$\left(\dfrac{\pi}{2} - \operatorname{Si} s\right)\cos s + \operatorname{Ci} s\sin s$	$\dfrac{1}{t^2+1}$
121	$\left(\dfrac{\pi}{2} - \operatorname{Si} s\right)\sin s - \operatorname{Ci} s\cos s$	$\dfrac{t}{t^2+1}$

Appendix B

Tables of Finite Fourier Transforms[1]

Table B.1   Finite sine transforms

	$f_s(n) \quad (n = 1, 2, \ldots)$	$F(x) \quad (0 < x < \pi)$
1	$\displaystyle\int_0^\pi F(x) \sin nx\, dx = S_n\{F\}$	$F(x)$
2	$f_s(n)$	$\dfrac{2}{\pi} \displaystyle\sum_{n=1}^\infty f_s(n) \sin nx$
3	$-n^2 f_s(n) + n[F(0) - (-1)^n F(\pi)]$	$F''(x)$
4	$\dfrac{1}{n^2} f_s(n)$	$\dfrac{x}{\pi} \displaystyle\int_0^\pi (\pi - r)F(r)\, dr - \int_0^x (x - r)F(r)\, dr$
5	$(-1)^{n+1} f_s(n)$	$F(\pi - x)$
6	$2 f_s(n) \cos nc$	$F_1(x + c) + F_1(x - c)$
7	$\dfrac{2}{n} f_s(n) h_s(n)$	$\displaystyle\int_0^\pi H(y) \int_{x-y}^{x+y} F_1(r)\, dr\, dy$
8	$\dfrac{\pi}{n}$	$\pi - x$
9	$\dfrac{1}{n}(-1)^{n+1}$	$\dfrac{x}{\pi}$
10	$\dfrac{1}{n}[1 - (-1)^n]$	1
11	$\dfrac{\pi}{n} \cos nc \quad (0 < c < \pi)$	$\begin{cases} -x & \text{if } x < c \\ \pi - x & \text{if } x > c \end{cases}$
12	$\dfrac{\pi}{n^2} \sin nc \quad (0 < c < \pi)$	$\begin{cases} (\pi - c)x & \text{if } x \leq c \\ (\pi - x)c & \text{if } x \geq c \end{cases}$

[1] See Chap. 11 for details and operations not tabulated here.

Table B.1   (continued)

	$f_s(n)$ $(n = 1, 2, \ldots)$	$F(x)$ $(0 < x < \pi)$
13	$\dfrac{6\pi}{n^3}$	$x(\pi - x)(2\pi - x)$
14	$\dfrac{2}{n^3}[1 - (-1)^n]$	$x(\pi - x)$
15	$\dfrac{\pi^2}{n}(-1)^{n+1} - \dfrac{2}{n^3}[1 - (-1)^n]$	x^2
16	$\pi(-1)^n\left(\dfrac{6}{n^3} - \dfrac{\pi^2}{n}\right)$	x^3
17	$\dfrac{n}{n^2 + c^2}[1 - (-1)^n e^{c\pi}]$	e^{cx}
18	$\dfrac{n}{n^2 + c^2}$ $(c \neq 0)$	$\dfrac{\sinh c(\pi - x)}{\sinh c\pi}$
19	$\dfrac{n}{n^2 + c^2}[1 - (-1)^n \cosh c\pi]$	$\cosh cx$
20	$\dfrac{n}{n^2 - k^2}$ $(k \neq 0, +1, +2, \ldots)$	$\dfrac{\sin k(\pi - x)}{\sin k\pi}$
21	0 if $n \neq m$; $f_s(m) = \dfrac{\pi}{2}$	$\sin mx$ $(m = 1, 2, \ldots)$
22	$\dfrac{n}{n^2 - k^2}[1 - (-1)^n \cos k\pi]$	$\cos kx$ $(k \neq \pm 1, \pm 2, \ldots)$
23	$n\dfrac{1 - (-1)^{m+n}}{n^2 - m^2}$ if $n \neq m$; $f_s(m) = 0$	$\cos mx$ $(m = 1, 2, \ldots)$
24	$\dfrac{2kn}{(n^2 - k^2)^2}$ $(k \neq 0, \pm 1, \pm 2, \ldots)$	$\dfrac{\partial}{\partial k}\left[\dfrac{\sin k(\pi - x)}{\sin k\pi}\right]$
25	$\dfrac{2cn}{(n^2 + c^2)^2}$ $(c \neq 0)$	$\dfrac{\partial}{\partial c}\left[\dfrac{\sinh c(x - \pi)}{\sinh c\pi}\right]$
26	$\dfrac{4c^2(\sin^2 c\pi + \sinh^2 c\pi)n}{n^4 + 4c^4}$ $(c \neq 0)$	$\sin cx \sinh c(2\pi - x)$ $\qquad\qquad - \sin c(2\pi - x) \sinh cx$
27	$\dfrac{2k^2 n(-1)^n}{n^4 - k^4}$ $(k \neq 0, \pm 1, \pm 2, \ldots)$	$\dfrac{\sinh kx}{\sinh k\pi} - \dfrac{\sin kx}{\sin k\pi}$
28	$\dfrac{c^2(-1)^n}{n(n^2 + c^2)}$ $(c \neq 0)$	$\dfrac{\sinh cx}{\sinh c\pi} - \dfrac{x}{\pi}$
29	$\dfrac{k^2(-1)^n}{n(n^2 - k^2)}$ $(k \neq 0, \pm 1, \pm 2, \ldots)$	$\dfrac{x}{\pi} - \dfrac{\sin kx}{\sin k\pi}$
30	b^n $(-1 < b < 1)$	$\dfrac{2}{\pi}\dfrac{b \sin x}{1 + b^2 - 2b \cos x}$

$f_s(n)$ $\quad$ $(n = 1, 2, \ldots)$	$F(x)$ $\quad$ $(0 < x < \pi)$
31 $\quad e^{-ny}$ $\quad (y > 0)$	$\dfrac{1}{\pi}\dfrac{\sin x}{\cosh y - \cos x}$
32 $\quad \dfrac{b^n}{n}$ $\quad (-1 < b < 1)$	$\dfrac{2}{\pi}\arctan\dfrac{b\sin x}{1 - b\cos x}$
33 $\quad \dfrac{1}{n}e^{-ny}$ $\quad (y > 0)$	$\dfrac{2}{\pi}\arctan\dfrac{\sin x}{e^y - \cos x}$
34 $\quad \dfrac{1 - (-1)^n}{n}b^n$ $\quad (-1 < b < 1)$	$\dfrac{2}{\pi}\arctan\dfrac{2b\sin x}{1 - b^2}$
35 $\quad \dfrac{1 - (-1)^n}{n}e^{-ny}$ $\quad (y > 0)$	$\dfrac{2}{\pi}\arctan\dfrac{\sin x}{\sinh y}$

Table B.2 Finite cosine transforms

$f_c(n)$ $\quad$ $(n = 0, 1, 2, \ldots)$	$F(x)$ $\quad$ $(0 < x < \pi)$
1 $\quad \displaystyle\int_0^\pi F(x)\cos nx\, dx = C_n\{F\}$	$F(x)$
2 $\quad f_c(n)$	$\dfrac{f_c(0)}{\pi} + \dfrac{2}{\pi}\displaystyle\sum_{n=1}^\infty f_c(n)\cos nx$
3 $\quad -n^2 f_c(n) - F'(0) + (-1)^n F'(\pi)$	$F''(x)$
4 $\quad \dfrac{1}{n^2} f_c(n)$ if $n = 1, 2, \ldots$	$\displaystyle\int_x^\pi (x - r)F(r)\, dr + \dfrac{f_c(0)}{2\pi}(x - \pi)^2 + A$
5 $\quad (-1)^n f_c(n)$	$F(\pi - x)$
6 $\quad 2f_c(n)\cos nc$	$F_2(x + c) + F_2(x - c)$
7 $\quad 2S_n\{F\}\sin nc$	$F_1(x + c) - F_1(x - c)$
8 $\quad \dfrac{1}{n}S_n\{F\}$ if $n = 1, 2, \ldots$	$-\displaystyle\int_0^x F(r)\, dr + A$
9 $\quad 2f_c(n)h_c(n)$	$\displaystyle\int_{-\pi}^\pi F_2(x - r)H_2(r)\, dr$
10 $\quad f_c(n)$ if $n \ne 0$; $f_c(0) + A\pi$ if $n = 0$	$F(x) + A$
11 $\quad 0$ if $n = 1, 2, \ldots$; $f_c(0) = \pi$	1
12 $\quad \dfrac{1}{n}$ if $n = 1, 2, \ldots$; $f_c(0) = 0$	$-\dfrac{2}{\pi}\log\left(2\sin\dfrac{x}{2}\right)$
13 $\quad \dfrac{2}{n}\sin nc$ if $n \ne 0$; $f_c(0) = 2c - \pi$	$\begin{cases} 1 & \text{if } 0 < x < c \\ -1 & \text{if } c < x < \pi \end{cases}$

	$f_s(n)$ $(n = 0, 1, 2, \ldots)$	$F(x)$ $(0 < x < \pi)$
14	$\dfrac{(-1)^n - 1}{n^2}$ if $n \neq 0$; $f_c(0) = \dfrac{\pi^2}{2}$	x
15	$\dfrac{2\pi}{n^2}(-1)^n$ if $n \neq 0$; $f_c(0) = \dfrac{\pi^3}{3}$	x^2
16	$\dfrac{1}{n^2}$ if $n \neq 0$; $f_c(0) = 0$	$\dfrac{(\pi - x)^2}{2\pi} - \dfrac{\pi}{6}$
17	$-\dfrac{24\pi}{n^4}(-1)^n$ if $n \neq 0$; $f_c(0) = 0$	$x^4 - 2\pi^2 x^2 + \tfrac{7}{15}\pi^4$
18	$\dfrac{1}{n^2 + c^2}$ $(c \neq 0)$	$\dfrac{\cosh c(\pi - x)}{c \sinh c\pi}$
19	$\dfrac{(-1)^n e^{c\pi} - 1}{n^2 + c^2}$ $(c \neq 0)$	$\dfrac{1}{c}e^{cx}$
20	$\dfrac{1}{n^2 - k^2}$ $(k \neq 0, \pm 1, \pm 2, \ldots)$	$-\dfrac{\cos k(\pi - x)}{k \sin k\pi}$
21	$\dfrac{(-1)^n \cos k\pi - 1}{n^2 - k^2}$ $(k \neq 0, \pm 1, \pm 2, \ldots)$	$\dfrac{\sin kx}{k}$
22	$\dfrac{(-1)^{m+n} - 1}{n^2 - m^2}$ if $n \neq m$; $f_c(m) = 0$	$\dfrac{\sin mx}{m}$ $(m = 1, 2, \ldots)$
23	$\dfrac{(-1)^{n+1}}{n^2 - 1}$ if $n \neq 1$, $f_c(1) = -\dfrac{1}{4}$	$\dfrac{1}{\pi}x \sin x$
24	0 if $n \neq m$; $f_c(m) = \dfrac{\pi}{2}$	$\cos mx$ $(m = 1, 2, \ldots)$
25	$\dfrac{2c(-1)^{n+1}}{(n^2 + c^2)^2}$ $(c \neq 0)$	$\dfrac{\partial}{\partial c}\left(\dfrac{\cosh cx}{c \sinh c\pi}\right)$
26	b^n $(-1 < b < 1, b \neq 0)$	$\dfrac{1}{\pi}\dfrac{1 - b^2}{1 + b^2 - 2b \cos x}$
27	e^{-ny} $(y > 0)$	$\dfrac{1}{\pi}\dfrac{\sinh y}{\cosh y - \cos x}$
28	$\dfrac{b^n}{n}$ if $n \neq 0$; $f_c(0) = 0$ $(-1 < b < 1)$	$-\dfrac{1}{\pi}\log(1 + b^2 - 2b \cos x)$
29	$\dfrac{\pi}{n}e^{-ny}$ if $n \neq 0$; $f_c(0) = 0$ $(y > 0)$	$y - \log(2\cosh y - 2\cos x)$
30	ne^{-ny} $(y > 0)$	$\dfrac{1}{\pi}\dfrac{\partial}{\partial y}\left(\dfrac{\sinh y}{\cos x - \cosh y}\right)$
31	$\dfrac{\pi}{2}\dfrac{b^n}{n!}$ if $n \neq 0$; $f_c(0) = \pi$	$\exp(b \cos x)\cos(b \sin x)$

Appendix C

Table of Exponential Fourier Transforms[1]

Table C.1 Experimental Fourier transforms

	$f_e(\alpha) \quad (-\infty < \alpha < \infty)$	$F(x) \quad (-\infty < x < \infty)$
1	$\displaystyle\int_{-\infty}^{\infty} F(x)e^{-i\alpha x}\,dx = E_\alpha\{F\}$	$F(x)$
2	$f_e(\alpha)$	$\displaystyle\lim_{\beta \to \infty} \frac{1}{2\pi}\int_{-\beta}^{\beta} f_e(\alpha)e^{i\alpha x}\,d\alpha$
3	$(i\alpha)^m f_e(\alpha) \quad (m = 1, 2, \ldots)$	$F^{(m)}(x)$
4	$i f'_e(\alpha)$	$xF(x)$
5	$e^{ic\alpha} f_e(\alpha) \quad (c \text{ real})$	$F(x + c)$
6	$f_e(\alpha - c) \quad (c \text{ real})$	$e^{icx} F(x)$
7	$f_e(c\alpha) \quad (c \text{ real}, c \neq 0)$	$\dfrac{1}{|c|}F\left(\dfrac{x}{c}\right)$
8	$f_e(-\alpha)$	$F(-x)$
9	$\overline{f_e(-\alpha)}$	$\overline{F(x)}$
10	$f_e(\alpha)g_e(\alpha)$	$\displaystyle\int_{-\infty}^{\infty} F(t)G(x - t)\,dt$
11[2]	$\dfrac{1}{c + i\alpha} \quad (c > 0)$	$e^{-cx}S_0(x)$
12	$\dfrac{1}{c - i\alpha} \quad (c > 0)$	$e^{cx}S_0(-x)$
13	$\dfrac{1}{(c + i\alpha)^2} \quad (c > 0)$	$xe^{-cx}S_0(x)$

[1] See Chap. 12 for details and Sec. 136 for references to other tables.
[2] $S_0(x) = 0$ if $x < 0$, $S_0(x) = 1$ if $x > 0$.

OPERATIONAL MATHEMATICS

Table C.1 (continued)

	$f_e(\alpha)$ $(-\infty < \alpha < \infty)$	$F(x)$ $(-\infty < x < \infty)$
14	$\dfrac{1}{(c - i\alpha)^2}$ $(c > 0)$	$-xe^{cx}S_0(-x)$
15	$\dfrac{2c}{c^2 + \alpha^2}$ $(c > 0)$	$\exp(-c|x|)$
16	$\dfrac{2}{1 + \alpha^2}e^{-ic\alpha}$ $(c\ \text{real})$	$\exp(-|x - c|)$
17	$\dfrac{4ic\alpha}{(c^2 + \alpha^2)^2}$ $(c > 0)$	$-x\exp(-c|x|)$
18	$\dfrac{n!}{(p + i\alpha)^{n+1}}$ $(\text{Re } p > 0)$	$x^n e^{-px}S_0(x)$ $(n = 1, 2, \ldots)$
19	$\pi\exp(-c|\alpha|)$ $(c > 0)$	$\dfrac{c}{c^2 + x^2}$
20	$2\sqrt{\pi c}\exp(-c\alpha^2)$ $(c > 0)$	$\exp\left(-\dfrac{x^2}{4c}\right)$
21	$i\pi\alpha\exp(-c|\alpha|)$ $(c > 0)$	$-\dfrac{2cx}{(c^2 + x^2)^2}$
22	$i\sqrt{\pi}\alpha\exp\left(-\dfrac{\alpha^2}{4c^2}\right)$ $(c > 0)$	$-2c^3 x\exp(-c^2 x^2)$
23	$\dfrac{2}{\alpha}\sin c\alpha$ $(c > 0)$	$S_0(x + c) - S_0(x - c)$
24	$2\dfrac{1 - \cos c\alpha}{i\alpha}$ $(c > 0)$	$\begin{cases} -1 \text{ if } -c < x < 0, \\ 1 \quad \text{if } 0 < x < c,\ 0 \text{ if } |x| > c \end{cases}$
25	$\dfrac{4c^2}{\alpha(4c^2 - \alpha^2)}\sin\dfrac{\pi\alpha}{2c}$ $(c > 0)$	$\begin{cases} \cos^2 cx \text{ if } |x| < \dfrac{\pi}{2c} \\ 0 \qquad \text{if } |x| > \dfrac{\pi}{2c} \end{cases}$

Appendix D

Tables of Fourier Sine and Cosine Transforms[1]

Table D.1 Sine transforms on the half line

	$f_s(\alpha) \quad (\alpha > 0)$	$F(x) \quad (x > 0)$
1	$\displaystyle\int_0^\infty F(x) \sin \alpha x \, dx = S_\alpha\{F(x)\}$	$F(x)$
2	$f_s(\alpha)$	$\displaystyle\frac{2}{\pi}\int_0^\infty f_s(\alpha) \sin \alpha x \, d\alpha = \frac{2}{\pi}S_x\{f_s(\alpha)\}$
3	$-\alpha^2 f_s(\alpha) + \alpha F(0)$	$F''(x)$
4	$\dfrac{i}{2}E_\alpha\{F_1(x)\} \qquad (E_\alpha \text{ in Appendix C})$	$F(x)$
5	$f_s(\alpha k) \qquad (k > 0)$	$\dfrac{1}{k}F\left(\dfrac{x}{k}\right)$
6	$f_s''(\alpha)$	$-x^2 F(x)$
7	$\dfrac{2}{\alpha}f_s(\alpha)g_s(\alpha)$	$\displaystyle\int_0^\infty F(r) \int_{|x-r|}^{x+r} G(t) \, dt \, dr$
8	$2f_s(\alpha) \cos \alpha k$	$F_1(x + k) + F_1(x - k)$
9	$f_s(\alpha + k) + f_s(\alpha - k)$	$2F(x) \cos kx$
10	$f_c'(\alpha) \qquad (f_c \text{ in Table D.2})$	$-xF(x)$
11	$\alpha f_c(\alpha)$	$-F'(x)$
12	$2f_c(\alpha) \sin \alpha k$	$F_2(x - k) - F_2(x + k)$
13	$f_c(\alpha - k) - f_c(\alpha + k)$	$2F(x) \sin kx$

[1] For details and properties not listed here see Chap. 13. References to other tables are given in Sec. 142.

Table D.1   (continued)

	$f_s(\alpha)$ $(\alpha > 0)$	$F(x)$ $(x > 0)$
14	$2f_s(\alpha)g_c(\alpha)$	$\displaystyle\int_0^\infty F(r)[G(|x - r|) - G(x + r)]\,dr$ $\displaystyle= \int_0^\infty G(t)[F(x + t) + F_1(x - t)]\,dt$
15	$\dfrac{\pi}{2}$	$\dfrac{1}{x}$
16	$\dfrac{\alpha}{c^2 + \alpha^2}$ $(c > 0)$	e^{-cx}
17	$\dfrac{\alpha}{1 + \alpha^4}$	$\exp\left(-\dfrac{x}{\sqrt{2}}\right)\sin\dfrac{x}{\sqrt{2}}$
18	$\dfrac{2c\alpha}{(c^2 + \alpha^2)^2}$ $(c > 0)$	xe^{-cx}
19	$\dfrac{8c^3\alpha}{(c^2 + \alpha^2)^3}$ $(c > 0)$	$x(1 + cx)e^{-cx}$
20	$\dfrac{1}{\sqrt{\alpha}}$	$\sqrt{\dfrac{2}{\pi}}\dfrac{1}{\sqrt{x}}$
21	$e^{-c\alpha}$ $(c > 0)$	$\dfrac{2}{\pi}\dfrac{x}{c^2 + x^2}$
22	$\alpha e^{-c\alpha}$ $(c > 0)$	$\dfrac{4c}{\pi}\dfrac{x}{(c^2 + x^2)^2}$
23	$\alpha^2 e^{-c\alpha}$ $(c > 0)$	$\dfrac{4x}{\pi}\dfrac{3c^2 - x^2}{(c^2 + x^2)^3}$
24	$1 - e^{-c\alpha}$ $(c > 0)$	$\dfrac{2}{\pi x}\dfrac{c^2}{c^2 + x^2}$
25	$\dfrac{1 - e^{-c\alpha}}{\alpha}$ $(c > 0)$	$\dfrac{2}{\pi}\arctan\dfrac{c}{x}$
26	$\alpha\exp(-c\alpha^2)$ $(c > 0)$	$\dfrac{x}{2c\sqrt{\pi c}}\exp\left(-\dfrac{x^2}{4c}\right)$
27	$\dfrac{1 - \exp(-c\alpha^2)}{\alpha}$ $(c > 0)$	$\operatorname{erfc}\dfrac{x}{2\sqrt{c}}$
28	$\dfrac{1 - \cos c\alpha}{\alpha}$ $(c > 0)$	$\begin{cases} 1 \text{ if } 0 < x < c \\ 0 \text{ if } x > c \end{cases}$
29	$e^{-\alpha}\sin\alpha$	$\dfrac{4}{\pi}\dfrac{x}{4 + x^4}$
30	$\dfrac{\sin c\alpha}{\alpha}$ $(c > 0)$	$\dfrac{1}{\pi}\log\dfrac{x + c}{|x - c|}$

Table D.2 Cosine transforms on the half line

	$f_c(\alpha)$ $(\alpha > 0)$	$F(x)$ $(x > 0)$
1	$\displaystyle\int_0^\infty F(x)\cos\alpha x\,dx = C_\alpha\{F(x)\}$	$F(x)$
2	$f_c(\alpha)$	$\dfrac{2}{\pi}\displaystyle\int_0^\infty f_c(\alpha)\cos\alpha x\,d\alpha = \dfrac{2}{\pi}C_x\{f_c(\alpha)\}$
3	$-\alpha^2 f_c(\alpha) - F'(0)$	$F''(x)$
4	$\frac{1}{2}E_\alpha\{F(|x|)\}$	$F(x)$
5	$f_c(\alpha k)$ $(k > 0)$	$\dfrac{1}{k}F\!\left(\dfrac{x}{k}\right)$
6	$f_c''(\alpha)$	$-x^2 F(x)$
7	$2f_c(\alpha)g_c(\alpha)$	$\displaystyle\int_0^\infty F(r)[G(x+r) + G(|x-r|)]\,dr$
8	$2f_c(\alpha)\cos\alpha k$ $(k > 0)$	$F(|x-k|) + F(x+k)$
9	$f_c(\alpha+k) + f_c(\alpha-k)$	$2F(x)\cos kx$
10	$f_s'(\alpha)$ $(f_s$ in Table D.1$)$	$xF(x)$
11	$\dfrac{1}{\alpha}f_s(\alpha)$	$\displaystyle\int_x^\infty F(r)\,dr$
12	$2f_s(\alpha)\sin\alpha k$ $(k > 0)$	$F(x+k) - F_1(x-k)$
13	$f_s(\alpha+k) - f_s(\alpha-k)$	$2F(x)\sin kx$
14	$2f_s(\alpha)g_s(\alpha)$	$\displaystyle\int_0^\infty F(r)[G(x+r) - G_1(x-r)]\,dr$
15	$\dfrac{c}{c^2+\alpha^2}$ $(c > 0)$	e^{-cx}
16	$\dfrac{2}{(1+\alpha^2)^2}$	$(1+x)e^{-x}$
17	$\dfrac{1}{1+\alpha^4}$	$\exp\!\left(-\dfrac{x}{\sqrt{2}}\right)\sin\!\left(\dfrac{\pi}{4}+\dfrac{x}{\sqrt{2}}\right)$
18	$\dfrac{\alpha^2}{1+\alpha^4}$	$\exp\!\left(-\dfrac{x}{\sqrt{2}}\right)\cos\!\left(\dfrac{\pi}{4}+\dfrac{x}{\sqrt{2}}\right)$
19	$\dfrac{1}{\sqrt{\alpha}}$	$\sqrt{\dfrac{2}{\pi}}\dfrac{1}{\sqrt{x}}$
20	$e^{-c\alpha}$ $(c > 0)$	$\dfrac{2}{\pi}\dfrac{c}{c^2+x^2}$
21	$\alpha e^{-\alpha}$	$\dfrac{2}{\pi}\dfrac{1-x^2}{(1+x^2)^2}$

Table D.2 (continued)

	$f_c(\alpha)$ $(\alpha > 0)$	$F(x)$ $(x > 0)$
22	$\dfrac{e^{-\alpha} - e^{-c\alpha}}{\alpha}$ $(c > 0)$	$\dfrac{1}{\pi} \log \dfrac{c^2 + x^2}{1 + x^2}$
23	$\dfrac{1 - e^{-c\alpha}}{\alpha}$ $(c > 0)$	$\dfrac{1}{\pi} \log\left(1 + \dfrac{c^2}{x^2}\right)$
24	$\exp(-c\alpha^2)$ $(c > 0)$	$\dfrac{1}{\sqrt{\pi c}} \exp\left(-\dfrac{x^2}{4c}\right)$
25	$\dfrac{1}{\alpha} e^{-\alpha} \sin \alpha$	$\dfrac{1}{\pi} \arctan \dfrac{2}{x^2}$
26	$\dfrac{1}{\alpha} \sin c\alpha$ $(c > 0)$	$\begin{cases} 1 \text{ if } 0 < x < c \\ 0 \text{ if } x > c \end{cases}$
27	$\dfrac{1}{\alpha} \arctan \alpha$	$E_1(x)$ $(E_1 \text{ in Sec. } 35)$
28	$\arctan \dfrac{2}{\alpha^2}$	$\dfrac{2}{x} e^{-x} \sin x$

Index